Scanning Electron Microscopy of Vascular Casts:
Methods and Applications

ELECTRON MICROSCOPY IN BIOLOGY AND MEDICINE

Current Topics in Ultrastructural Research

SERIES EDITOR: P.M. MOTTA

Already published in this series

Vol. 1 Motta, P.M. (ed.): Ultrastructure of Endocrine Cells and Tissues.
  ISBN: 0-89838-568-7.

Vol. 2 Van Blerkom, J. and Motta, P.M. (eds.): Ultrastructure of Reproduction: Gametogenesis.
  Fertilization, and Embryogenesis. ISBN: 0-89838-572-5.

Vol. 3 Ruggeri, A. and Motta. P.M. (eds.): Ultrastructure of the Connective Tissue Matrix.
  ISBN: 0-89838-600-4

Vol. 4 Motta, P.M. and Fujita, H. (eds.): Ultrastructure of the Digestive Tract.
  ISBN: 0-89838-893-7.

Vol. 5 Van Blerkom, J. and Motta, P.M. (eds.): Ultrastructure of Human Gametogenesis
  and Early Embryogenesis. ISBN: 0-89838-383-8.

Vol. 6 Riva, A. and Motta, P.M. (eds.): Ultrastructure of the Expatraparietal Glands of the Digestive
  Tract. ISBN: 0-7923-0303-2.

Vol. 7 Bonucci, E. and Motta, P.M.(eds.): Ultrastructure of Skeletal Tissue.
  ISBN: 0-7923-0373-3.

Vol. 8 Motta, P.M. (ed.): Ultrastructure of Smooth Muscle. ISBN: 0-7923-0480-2.

Vol. 9 Familiari, G., Makabe, S., and Motta, P.M. (eds.): Ultrastructure of the Ovary.
  ISBN: 0-7923-1003-9.

Scanning Electron Microscopy of Vascular Casts: Methods and Applications

edited by

Pietro M. Motta, M.D., Ph.D.
Department of Human Anatomy, Faculty of Medicine, University of Rome "La Sapienza," Rome, Italy

Takuro Murakami, M.D., Ph.D.
Department of Anatomy, Okayama University School of Medicine, Okayama, Japan

and

Hisao Fujita, M.D., Ph.D.
Department of Anatomy Osaka University Medical School, Osaka, Japan

SPRINGER SCIENCE+BUSINESS MEDIA, LLC

Library of Congress Cataloging-in-Publication Data

Scanning electron microscopy of vascular casts: methods and
applications / edited by Pietro M. Motta, Takuro Murakami,
and Hisao Fujita.
    p. cm. — (Electron microscopy in biology and
medicine; 10)
    Includes bibliographical references and index.
   ISBN 978-0-7923-1297-0   ISBN 978-1-4615-3488-4 (eBook)
   DOI 10.1007/978-1-4615-3488-4
   1. Microcirculation.   2. Blood-vessels —
Imaging.   3. Scanning electron microscopes. I. Motta,
Pietro M. II. Murakami, Takuro. III. Fujita, Hisao,
1928–1942. IV. Series: Electron Microscopy in biology and
medicine; EMBM 10.
   [DNLM: 1. Corrosion Casting. 2. Microcirculation —
ultrastructure. 3. Microscopy, Electron, Scanning.
W1 EL33E v. 10 / WG 104 S283]
QP106.6.S33   1991
612.1'3 — dc20
DNLM/DLC
for Library of Congress

                          91-7096
                            CIP

Printed on acid-free paper.

Preface

In the last part of the "Ode to a Grecian Urn," written by the 19th-century English poet John Keats, there is a famous verse that says, "beauty is truth, truth beauty." About 25 years ago, some noted Japanese electron microscopists also had a saying aimed at giving natural scientists the basis for some philosophical considerations; the saying stated, "the form of Nature is beautiful and often, well prepared electron micrographs are nothing but an aspect of this beauty."

Recently, attention has been called to the role that microvascular organization plays in the functional morphology of all organs and tissues, both in normal and pathological conditions.

Since its development by Murakami, the corrosion cast method for scanning electron microscopy has come to be considered as one of the most efficient means of clarifying the three-dimensional features of the microcirculation of organs and tissues. The present volume was planned to supply fundamental and new information regarding microcirculation studies to general biologists, anatomists, pathologists, and clinicians.

After much consultation, we selected the authors for this volume from among numerous researchers who have published excellent papers in this field. In response to our invitation to contribute special articles, we were happy to receive many positive replies. Most of the researchers contacted agreed to cooperate with us and, as a result, many interesting and well-prepared manuscripts were sent to us. These contained original findings and excellent electron micrographs obtained by using recently improved corrosion cast methods. It was in collecting this material and looking at the illustrations of this book that Keats' famous verse came to mind; clearly, the photographs seemed to be saying that "beauty is truth, truth beauty."

We hope and believe that this book, consisting of a rich variety of papers, will be useful to many and will provide both the basic and clinically oriented readers with plenty of good ideas, suggestions, and some original and worthwhile information.

The fine cooperation of Mr. J.K. Smith, publisher, and Mrs. J. Bencivenga and Ms. J. Pereira of the production staff of Kluwer Academic Publishers during all stages of the preparation of this work is gratefully acknowledged.

P.M. Motta, T. Murakami, and H. Fujita

Contents

Contributing authors

Adachi, K., Department of Anatomy, Niigata University, School of Medicine, Asahi-Machi 1, Niigata, 951 Japan

Albrecht, R.M., Department of Veterinary Science, University of Wisconsin, 1655 Linden Drive, Madison, Wisconsin 53706, USA

Azzali, G., Department of Human Anatomy, Faculty of Medicine, University of Parma, Via Gramsci 14, 43100 Parma, Italy

Bugajski A., Department of Urology, N. Copernicus Academy of Medicine, Gtzegótzecka 18 , PL-31-531 Kraków, Poland

Burri, P.H., Institute of Anatomy, Department of Developmental Biology, University of Berne, Bühltrasse 26, Postfach 139 CH 3000 Bern 9, Switzerland

Busch, L.C., Institute für Anatomie, Medizinische Universität zu Lübeck, Ratzeburger Allee 160, D 2400 Lübeck 1, Germany

Caggiati, A., Department of Human Anatomy, Faculty of Medicine, University "La Sapienza," Via A. Borelli 50, 00161 Roma, Italy

Castenholz, A., Department of Human Biology (FB 19), University of Kassel, Heinrich-Plett-Strasse 40, D 3500 Kassel, Germany

Christofferson, R.H., Department of Human Anatomy, University of Uppsala, Biomedical Center, Box 571, S-751 23 Uppsala, Sweden

Fujita, H., Department of Anatomy, Osaka University Medical School, 2-2 Yamadaoka, Suita-city, Osaka-fu 565, Japan

Fujita, T., Department of Anatomy, Niigata University, School of Medicine, Asahi-Machi 1, Niigata, 951 Japan

Fryczkowski, A.W., The Ohio State University, Columbus, Department of Ophthalmology, 456 West Tenth Avenue, Columbus, Ohio 43210, USA

Gaudio, E., Department of Anatomy, Faculty of Medicine, University of L'Aquila, 67100 L'Aquila, Italy

Groom, A.C., University of Western Ontario, Department of Medical Biophysics, Health Sciences Centre, London, Ontario, N6A 5C1 Canada

Grunt, T.W., Laboratory for Applied and Experimental Tumor Cell Biology, Division of Oncology, Department of Internal Medicine, University of Vienna, Waehringer Guertel 18–20, A-1090 Vienna, Austria

Imada, M., Department of Anatomy, Osaka University Medical School, 2-2 Yamadaoka, Suita-city, Osaka-fu 565, Japan

Itoshima, T., Department of Internal Medicine, Okayama Saiseikdi General Hospital, Ifuku-cho 1-17-18, Okayama, 700 Japan

Iwaku, F., First Department of Oral Anatomy, Asahi University, School of Dentistry, 1851-1 Hozumi-cho Motosu-gun, Gifu, 501-02 Japan

Kashimura, M., First Department of Internal Medicine, Matsudo City Hospital, Kamihongo 4206, Matsudo, Chiba, 271, Japan

Kikuta, A., Department of Anatomy, Okayama University, Medical School, 2-5-1 Shikata-cho, Okayama, 700 Japan

Koob, B., Institut für Veterinar-Anatomie, Histologie und Embryologie der Justus-Liebig-Universität, Frankfürter Strasse 98, D 6300 Giessen, Germany

Kühnel, W., Institut für Anatomie, Medizinische Universität zu Lübeck, Ratzeburger Allee 160, D 2400 Lübeck 1, Germany

Kuś, J., Department of Otolaringology, N. Copernicus Academy of Medicine, Kopernika 23a, PL 31-501 Kraków, Poland

Lametschwandtner, A., Institite of Zoology, Department of Experimental Zoology, University of Salzburg, Hellbrunnerstrasse 34, A 5020 Salzburg, Austria

Lametschwandtner, U., Institute of Zoology, Department of Experimental Zoology, University of Salzburg, Hellbrunnerstrasse 34, A 5020 Salzburg, Austria

Leiser, R., Institut für Veterinar-Anatomie, Histologie und Embryologie der Justus-Liebig-Universität, Frankfürter Strasse 98, D 6300 Giessen, Germany

Macchiarelli, G., Department of Human Anatomy, Faculty of Medicine, University "La Sapienza," Via A. Borelli 50, 00161 Rome, Italy

Maggioni, A., Department of Human Anatomy, Faculty of Medicine, University "La Sapienza," Via A. Borelli 50, 00161 Rome, Italy

Marinozzi, G., Department of Human Anatomy, Faculty of Medicine, University "La Sapienza," Via A. Borelli 50, 00161 Rome, Italy

Miodoński, A.J., Department of Otolaringology, N. Copernicus Academy of Medicine, Kopernika 23a, PL 31-501 Kraków, Poland

Motta, P.M., Department of Human Anatomy, Faculty of Medicine, University "La Sapienza," Via A. Borelli 50, 00161 Rome, Italy

Murakami, T., Department of Anatomy, Okayama University, School of Medicine, 2-5-1 Shikata-cho, Okayama, 700 Japan.

Nilsson, B.O., Department of Human Anatomy, University of Uppsala, Biomedical Center, Box 571, S-751 23 Uppsala, Sweden

Nottola, S.A., Department of Human Anatomy, Faculty of Medicine, University "La Sapienza," Via A. Borelli 50, 00161 Rome, Italy

Nowogrodzka-Zagórska, M., Department of Otolaringology, N. Copernicus Academy of Medicine, Kopernika 23a, PL 31-501 Kraków, Poland

Ohtani, O., Department of Anatomy, Faculty of Medicine, Toyama Medical and Pharmaceutical University, 2630 Sugitani, Toyama, 930-01 Japan

Ohtsuka, A., Department of Anatomy, Okayama University, School of Medicine, 2-5-1 Shikata-cho, Okayama, 700 Japan

Owen, R.L., Cell Biology and Aging Section (151E) VA Medical Center, 4150 Clement Street, San Francisco, California 94121 USA

Olszewski, E., Department of Otolaringology, N. Copernicus Academy of Medicine, Kopernika 23a, PL 31-501 Kraków, Poland

Pannarale, L., Department of Human Anatomy, Faculty of Medicine, University "La Sapienza," Via A. Borelli 50, 00161 Rome, Italy

Potter, R.F., University of Western Ontario, Department of Medical Biophysics, Health Sciences Centre, London, Ontario, N6A 5C1 Canada

Schraufnagel, D.E., Section of Respiratory and Critical Care Medicine, Department of Medicine (M/C 787), University of Illinois at Chicago, P.O. Box 6998, Chicago, Illinois 60680, USA

Steeber, D.A., Department of Veterinary Science, University of Wisconsin, 1655 Linden Drive, Madison, Wisconsin 53706, USA

Taguchi, T., Department of Anatomy, Okayama University, School of Medicine, 2-5-1 Shikata-cho, Okayama, 700 Japan

Tsuji, T., First Department of Internal Medicine, Okayama University, Medical School, Shikata-cho, 2-5-1 Okayama, 700 Japan

[†]**Walmsley, J.G.**, University of Illinois, College of Medicine at Rockford, Department of Biomedical Sciences, 1601 Parkview Avenue, Rockford, Illinois 61107 USA

Yamamoto K., First Department of Internal Medicine, Okayama University, Medical School, Shikata-cho, 2-5-1 Okayama, 700 Japan

Scanning Electron Microscopy of Vascular Casts:
Methods and Applications

CHAPTER 1

Historical Review and Technical Survey of Vascular Casting and Scanning Electron Microscopy

ALOIS LAMETSCHWANDTNER & URSULA LAMETSCHWANDTNER

1. Introduction

A number of techniques (angiography, scintigraphy, computer tomography, ultrasonics, and nuclear magnetic resonance) are used to diagnose and localize pathological changes in the principal drainage and supply vessels. However, the power of resolution of these techniques is too low to study the microvasculatory bed. Thus, two techniques have usually been used to study the terminal vascular bed: light microscopy of India-ink injected and cleared tissue specimens [1] and scanning electron microscopy (SEM) of corrosion casts (corrosion cast/SEM method). Intravital microscopy of living tissues is limited solely to the observation of the microcirculation patterns in the superficial and translucent layers. Reconstruction from serial sections and analysis of microvascular beds, especially those in large organs and tissues, are usually troublesome. This chapter reviews the corrosion cast/SEM method, which facilitates the three-dimensional and wide-ranged visualization or analysis of the microvascular bed with good resolution. A brief history and other advantages of this method are also included.

2. History of Vascular Casting and SEM

In 1935, the polymerizing resin Plastoid was introduced to cast the vascular system [2]. Apart from Bidloo's (1649–1713) and Lieberkühn's (1711–1756) works, with primarily historical interest [3], Schummer's preparations [2] resulted in the first durable corrosion casts. The term *corrosion cast* indicates that the injection medium withstands the corrosive treatments necessary to remove the tissue elements. Because of their high viscosity, the plastoids could not be used sufficiently to replicate the microvascular beds [4].

At about the same time when plastoids were introduced, the basic principles of scanning electron microscopy (SEM) were developed [5]. However, it lasted until 1970, when the scanning electron microscope (SEM) was used to examine the airway system of the avian lung cast with a crude latex medium (Cementex N-1971-S) [6]. One year later, a fine methyl methacylate corrosion casting technique was developed for detailed SEM analysis of the microvascular beds [7]. Since this work, many authors have used the corrosion cast/SEM method and have studied the microangioarchitecture of various organs and tissues under different conditions and during development and aging [8,9]. These authors also have confirmed that SEM of corrosion casts is a useful tool to analyze the complicated angioarchitecture, especially its micromeshes, (Figs. 1-1 and 1-2), or to investigate the luminal and mural structures of the vessels (Figs. 1-3–1-5) [8,9]. Furthermore, it has been shown that the corrosion cast/SEM method is useful for quantitative analysis of the vascular beds or vessels [8,9]. Thus, almost a thousand vascular casting/SEM studies have been published [8–15].

The (vascular) corrosion cast/SEM method is sometimes referred to as *vascular cast/SEM*,

Motta, P.M., Murakami, T., and Fujita, H. (eds.), Scanning Electron Microscopy of Vascular Casts: Methods and Applications.

2

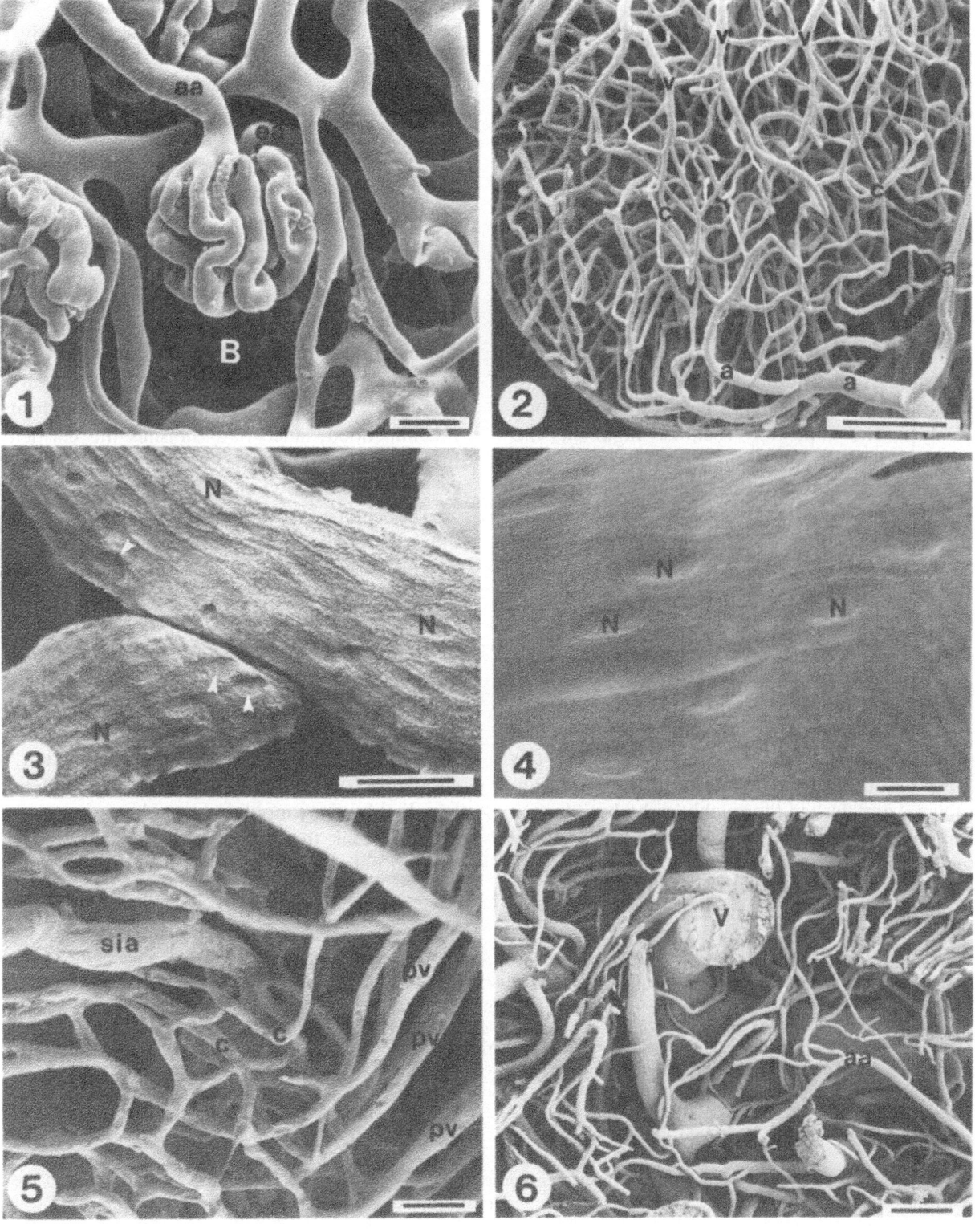

injection replica/SEM, or *microcorrosion cast/ SEM method* [7,12]. This method can be used in the study of other delicate tubular systems; recently refined injections with specially prepared low-viscosity casting media have prepared good casts of intestinal lymphatics and liver bile canaliculi [48].

3. Cast Preparation

The vascular corrosion cast/SEM method consists of precasting treatment (cleaning of vascular lumen or removal of blood), injection of casting medium, corrosive treatment (maceration/ digestion), dissection, mounting, conductive treatment, and scanning electron microscopy [7–15]. The following sections highlight the theories behind these important steps in producing good vascular casts or producing their beautiful images. These sections may also help the user to adapt simple steps in cast preparation to the particular SEM needs, including quantitative analysis.

3.1. Precasting Procedure

Complete removal of blood is of special importance and is necessary to inject the casting medium into the blood vascular system [7,15]. For this purpose, the target organs or tissues are cannulated and flushed thoroughly with physiological saline or other rinsing solutions [8].

It is usual that some anticoagulants (heparin and/or liquemin) are administered systemically and/or added to the rinsing solutions to prevent blood clotting. In some cases, either vasodilatory (acetylcholine, isoxsuprine) or spasmolytic agents (papaverine) are added to the solutions to prevent vasospasm and/or compression of blood vessels by contraction of the surrounding skeletal muscles. These careful precasting treatments are very important in the preparation of good casts of arteries, capillaries, and veins in normal, experimental, and pathological conditions [16].

Few data on anesthesia are available in casting. Perfusion fixation with low concentrated fixatives, such as 0.5–1.0% glutaraldehyde, is sometimes attempted after precasting flushing in order to replicate the endothelial surface structures (see below). However, such fixation is not always recommended, since it diminishes the elasticity of the vessels and hinders sufficient injection of casting media into vascular beds [48].

The choice of the cannulation site is also important. One may cannulate the vessels close to the target organ or tissue. These vessels, including minute vessels of surgically excised human materials, can be dissected free under a dissection light microscope (Fig. 1-6) [23,24]. This choice has an advantage in that it prepares satisfactory casts with minimal loss of injection pressure and also with a minimal amount of casting medium. Prior to the perfusion or injection, it should be confirmed that the target is completely supplied by the cannulated vessel.

In general, the casting media are injected via the ventricle, the aorta, or large arterial branches,

Figure 1-1. Renal glomerulus. *Bombina variegata* (Amphibia, Anura). This image demonstrates the high depth of focus, which makes the scanning electron microscope a useful tool in the study of vascular corrosion casts. Note different diameters of afferent (aa) and efferent arteriole (ea) and the space of Bowman's (B) capsule. Methyl methacrylate (MMA). Gold coating. Bar = 50 μm.
Figure 1-2. Midbrain. *Bufo bufo* L. (Amphibia, Anura). This specimen documents good filling and shows the transition from a feeding artery (a) via capillaries (c) to draining venules (v). MMA. Bar = 200 μm.
Figure 1-3. Brain artery with branch. *Eptatretus stouti* (Lock.) (Cyclostomes). "Arterial type" imprints of endothelial-cell nuclei (N). Note the deep circular constriction at the origin of the branch and the rather "venous-type" imprints of endothelial cell nuclei (arrowheads) at this site. Mercox-Cl-2B (M). Bar = 40 μm.
Figure 1-4. Cerebral vein. *Bufo bufo* L. (Amphibia, Anura). "Venous-type" imprints of endothelial cell nuclei (N). MMA. Bar = 25 μm.
Figure 1-5. Lateral region of the hypophysial median eminence. *Bufo bufo* (L) (Amphibia, Anura). Change of endothelial-cell nuclei imprint patterns from the superficial infundibular artery (sia) (with narrow, elongated nuclear imprints paralleling the long axis of the vessel) to capillaries (c) of the median eminence and to the portal veins (pv) (with ovoid to roundish nuclear imprints). MMA. Bar = 25 μm.
Figure 1-6. Vascular pattern of a keloid (human). Detail. Specimen excised during plastic surgery and cast 1 hour later. Note small arteriole (aa) and a vein (v) with endothelial-cell nuclei imprint patterns. M. Bar = 200 μm.

since the cannulation close to the target is rather difficult or troublesome. Such cannulation far from the target, however, sometimes results in incomplete injection of the casting medium. This is typical in casting the fish brain via the ventricle-ventral aorta; the intensive gill vasculature dramatically decreases the inflow of casting medium into the brain.

Various rinsing solutions with or without anti-coagulants and/or vasoactive substances have been used. Ionic strength, osmolarity (adjusted with high molecular dextrans) [11], pH, temperature, perfussion pressure, flow rate, rinsing volume, and perfusion time may be important variables in the rinsing or perfusion procedure. Unfortunately, however, data on these variables are scarce [15,17].

It is generally accepted that perfusion pressure should be equal to the mean arterial blood pressure (physiological pressure) at the site of injection. However, it was recently shown that in large arteries of the elastic type (descending aorta), pressure fixation prepares the casts with good geometric and dimensional information [18]. Casting or imprinting of mural structures, such as nuclear protrusions and cell boundaries of the vascular endothelium, is affected by shear stress forces acting on the vascular wall [19]. This imprinting allows a clear discrimination of arteries from veins [20] and also intracapillary blood flow assessment [15,20].

In casting the peripheral vascular territories, the pressure in the target organ is more important than that recorded centrally (at the injection site). The law of Hagen-Poiseuille states that a high-viscosity medium flows slowly in a long and wide pathway. Thus, the use of short, narrow, and thin-walled cannula, as well as a low-viscosity casting medium, is recommended in each casting, especially in casting the vascular system of larvae or embryos. Injection under different pressures may be necessary for good replication of the luminal surface details and of altered vascular beds [15,21,22].

Perfusion fixation prior to the casting (prefixation) induces vasospasms, with several oscillations in the intra-arterial pressure, or impairs sufficient injection of casting medium into the capillary beds [15]. However, prefixation (safeguarding the vascular walls) is useful to avoid the undesired leakages of the casting media, especially low-viscosity media, from the vascular lumen into the tissue spaces [25,26] and to improve the replication of endothelial surface structures [15]. Such a safeguard may be of special use to cast delicate vessels, such as the liver bile canaliculi [48]; injection studies done with the avian egg chorioallantoic membrane capillary plexus, however, neither rinse nor prefix this system [50]. The fixatives used in this safeguard are basically the same as those used for routine scanning and transmission electron microscopy [15,25,26,48].

3.2. Casting media and their Physico-Chemical Properties

Casting media are supplemented with catalysts to accelerate their polymerization in vessels prior to injection. The casting or injection media have to satisfy several criteria [8,11,12,15]. They should (1) be nontoxic to the investigator, (2) cause neither morphological nor physiological damages to the tissues or their vessels, (3) be of sufficient low viscosity or particle size to pass through capillaries, (4) show no marked leakage into the tissues or their spaces, (5) polymerize within 3–15 minutes, (6) replicate fine vascular connections as well as delicate luminal and endothelial structures, (7) show no marked shrinkage during curing (solidifying) or hardening, (8) permit microdissection with intact surrounding tissue, (9) be resistant to corrosion procedures, (10) be visible in the dissection light microscope, (11) retain structural configuration during drying, (12) be suitably for microdissection after drying, (13) be electron conductive or show no marked structural changes during conductive treatment, (14) be resistant to electron bombardment, (15) produce well-highlighted and well-contrasted SEM images, and (16) be useful for quantitative analysis.

From the broad spectrum of resins [12], laboratory-prepared methacrylate mixtures (Murakami's mixtures) [7,26,48] or commercially available resins, such as Mercox CL [14], Batson's No. 17 plastic [27,28], araldite CY 223 [15,29], and tardoplast [30], are usually used as the casting medium. Although physico-chemical properties of these casting media greatly define casting quality, few studies deal with the detailed impact of viscosity, shrinkage behavior, replication quality, electrical conductivity, and other items on the outcome of casting experiments [11,15,31,32].

A volume shrinkage of 6% in Mercox CL medium diluted with methyl methacrylate (4:1 v/v) [14], of 20% in Murakami's low-viscosity methyl methacrylate mixture [7], and of 1% in Batsons's no. 17 plastic [32] has been reported [18]. Because of different test conditions, these data cannot be compared directly. The viscosity is also different in each casting medium [15,33]. The degree of viscosity and volume shrinkage during polymerization are inversely related; the less viscous the injected resin the greater the final shrinkage will be. Ideally, the viscosity of casting medium should mimick blood viscosity (about 5 cS in humans and rat). Recent experiments have shown that sufficient vascular filling and good endothelial replication are possible with some low-viscosity media [15].

Mercox CL and methyl methacrylate casting media both sufficiently reproduce microridges of fish epidermal cells and the height of conductive layers in microchips down to 250 nm [31]. The protruding endothelial cell nuclei and deepened cell borders are also clearly imprinted, allowing the diagnosis of cast vessels as arteries or veins [20]. As far as we know, no authors replicated finer mural structures, such as endothelial fenestrations [14,15]. The insect tracheal system down to the tubulus, with a diameter of 70 nm, could be cast with Batson's plastic and could be clearly observed with the field emission high-resolution scanning electron microscope [34].

During tempering and maceration, corrosion casts are exposed to warm or hot solutions [7,8]. Thus, the casts must withstand this treatment without any changes in their structure and angio-architectural integrity [31].

3.3. Injection of Casting Medium

The working life of the casting medium is an additional variable. Depending on the medium, the working life ranges from 2 to 60 minutes [7,8,12,15]. In each casting, one must remember that the initial viscosity (grade of prepolymerization [33]) and the amount of hardener (catalyst or accelerator) define the life and volume available for injection. Adjustment of flow rate, injection pressure, and the dimension of the injection line and cannula are also important. The casting media are injected manually or by using a perfusion apparatus with a flow meter [15,21,22] or manometer [15]. The subjective control criteria of injection is sometimes referred to as the "dry nose criteria" [14].

Coloration — if colored injection media are used — indicates filling conditions in the target organs, although it does not guarantee complete filling [14]. However, the use of colored media is sometimes helpful as an injection index. For example, complete coloration of eye choroidea indicates good filling of the brain vasculature [14]. Additional or limited injection of colored media after the full injection may be useful for the discrimination of arteries and veins in the cast samples [10]. Sudan black B may be useful for such coloration of the casting media. Mercox CL-2R is colored red, and Mercox CL-2B is colored blue.

3.4. Polymerization of Casting Medium

Curing or polymerization of the casting medium starts right after supplementation with catalyst or accelerator; during injection, the medium continues to polymerize and increases its viscosity. Thus, the injection medium should be prepared after the precasting flushing or perfusion (see above), and promptly injected into the target blood vascular bed [7,31].

The injected specimens are immersed for 12 hours or longer in a hot water bath (60°C) to accelerate or complete the polymerization of the perfused casting media (tempering) before the tissues or their vessels begin to degenerate [7,31]. This immersion has another advantage in that it keeps the injected specimens in their natural form [7].

3.5. Dissection of Injected Specimen or Cast Sample

Dissection right after hardening (polymerization, curing, solidification) of the casting medium is done in whole-body preparations. Target organs are carefully excised, leaving sufficient tissues

6

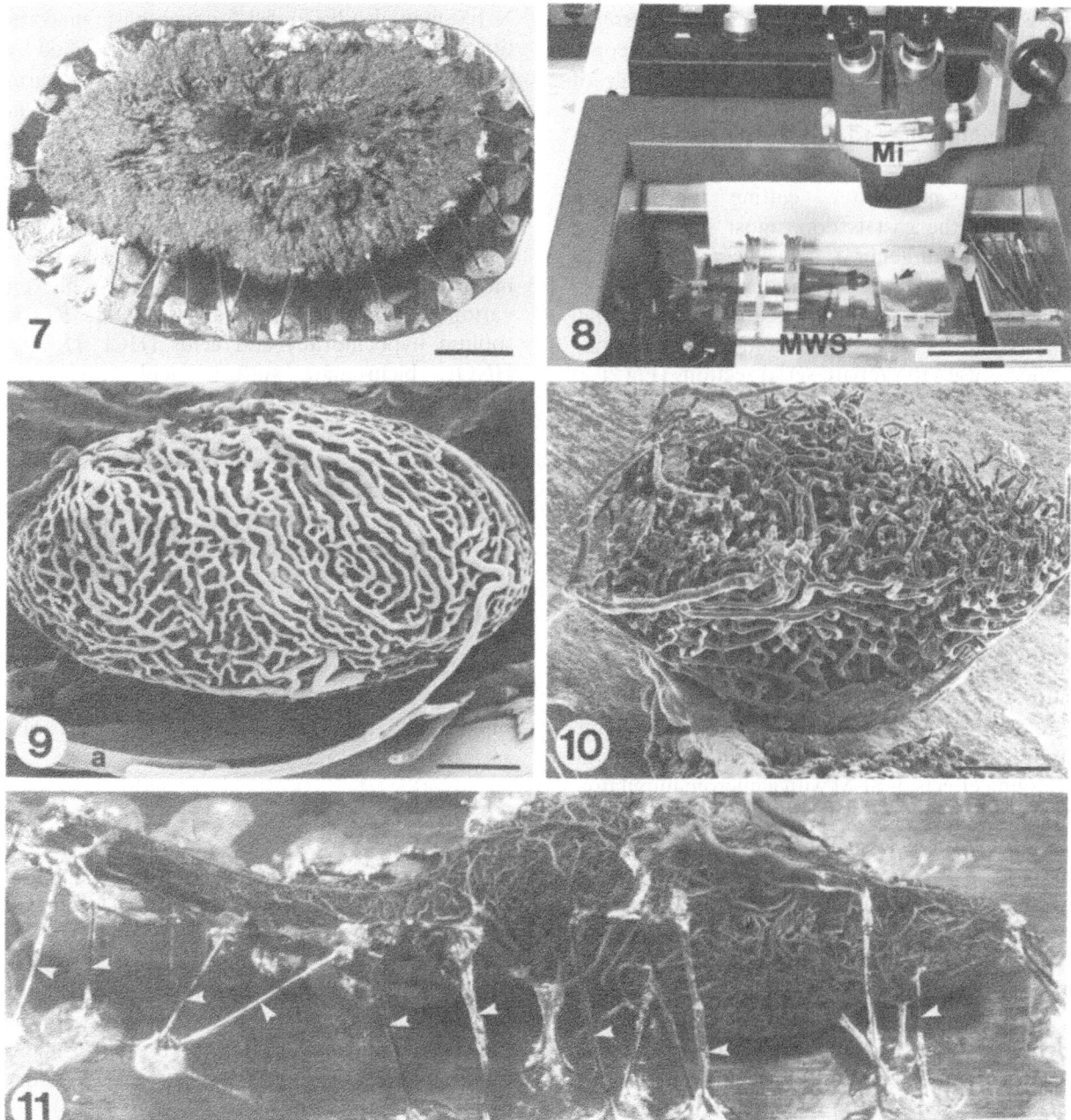

Figure 1-7. Kidney slice. *Sus scrofa* forma domestica (Mammalia). The kidney was cut in 5-mm thick serial sections by a small carpenter band saw. M. Bar = 15 mm.

Figure 1-8. Mini wheel saw (MWS) positioned in a cryomicrotome (Ultracut E). The device is cooled to −20°C, and then small ice blocks with embedded casts can be sectioned under microscopic control (Mi). The arrow points to the diamond wheel saw. Bar = 10 cm.

Figure 1-9. Corpuscle of Stannius of a teleost fish, *Blennius pavo* (L.). The corpuscle was isolated by manual dissection. (a) artery. M. Bar = 100 μm.

Figure 1-10. Corpuscle of Stannius. The specimen was sectioned tangentially with the mini wheel saw. Controlled cutting enables one to relate the surface to the parenchymal vascular patterns. M. Bar = 100 μm.

Figure 1-11. Brain. *Pseudemys scripta elegans* (Reptilia). Mounting was by means of the "conductive bridges" method [38]. Arrowheads mark "conductive bridges" (fine copper wires), which ground electrons and minimize specimen charging. Rostral to the right. M. Bar = 5 mm.

around them. Dissection after maceration or during washing (rinsing) in water enables controlled manipulation under the dissecting microscope, which avoids undesired damage or breakage of cast structures.

Embedding the casts in ice after washing offers several advantages in dissection. If the plane of cutting does not matter, "cutting" the ice block containing the cast is done most simply with a razor blade. In general, however, a specific area of interest has to be exposed under the dissection microscope. Controlled cutting with a small carpenter's bandsaw or a multi- bladed cutting device is used to prepare serial sections (Fig. 1-7) [8]. Very small specimens may be cut with a mini wheel saw placed in the chamber of a cryomicrotome (Figs. 1-8–1-10).

Dissection of dry corrosion casts is best done freehands with sharpened needles or forceps after casts are mounted or fixed on metal stubs or specimen holders (see below) [8]. This dissection is usually repeated under the dissection microscope for thorough exposure of the part of interest. Prior to each dissection, the part of interest has to be documented by high-quality light or SEM micrographs [8]. The osmium-impregnated casts can be dissected without the charging effects of SEM with a micromanipulator.

3.6. Removal of Tissue Elements

Tissue elements must be removed partially [35] or totally [7] to observe the cast blood vascular system with the SEM. Total maceration completely destroys the topographical relations between the cast vessels and tissue elements. This deficiency may be supplied by other methods, such as light microscopy, intravital microscopy, SEM, or transmission electron microscopy of tissue specimens [47].

Different protocols of maceration have been given to obtain clear casts that are free of tissue remnants. We attribute these differences to (1) specimen size, (2) tissue composition, (3) vascular density, and (4) precasting procedures. A recent controlled study or Latin-square test of the injected rat lung tissue has confirmed that highly concentrated alkalis (10–20% NaOH or KOH) sufficiently corrode the tissues and that their maceration potentials are markedly enhanced at high temperatures (60°C or higher) [36]. A hot

mixture of 5–10% NaOH and 5–20% Triton X-100 may also be useful for such basic maceration. Fatty or lipoid tissue elements saponified by alkali treatment may be removed in a hot or warm neutral detergent bath. The tissue maceration can be optimized by a gradient of corrosive agent around the specimen. This is best achieved by suspending the injected specimen in the corrosive solution and stirring with a magnetic stirrer [16].

The casts must be stable in the corrosive treatments. However, some authors who checked variously concentrated alkalis (NaOH, KOH, sodium hypochlorite) and acids (HCl, H_2SO_4, HNO_3, bichromate sulfuric acid, chromium trioxide, hydrogen peroxide, and HCOOH) have shown that some highly concentrated agents, such as HNO_3 cause severe damage to the casts, especially their surface structures [15,20]. It is well known that methacrylate casts are melted away by HNO_3.

Bony material is removed, before or after the corrosive treatment, by decalcification with hydrochloric acid [15].

After the maceration or decalcification, the casts are cleaned with formic acid to remove the small tissue remnants that may mask the cast surface details [15,37].

3.7. Drying of Cast

Air drying is recommended only in strong casts that can resist the high surface tensions produced by water evaporation. Short-term immersion in ethanol may be useful to decrease the danger of structural alterations during air drying. Prolonged ethanol immersion softens and deforms the casts, though it is sometimes useful for manipulation and results in little breakage of vascular connections.

Freeze drying with minimized surface tensions is the best method. In this drying, the casts are immersed in distilled water and frozen very slowly to produce homogenous ice or to avoid undesired breakages of the embedded casts due to ice cleavage [8,48]. Critical-point drying is rarely used [35].

3.8. Mounting of Cast

The cast samples for SEM are mounted on metal stubs under the dissecting microscope. If the commercially available stubs are too small, they are enlarged by soldering a piece of copper foil on them. Many conductive or mounting media are commercially available. These media should be used carefully; otherwise the details of vascular meshworks or surface structures may be masked or hidden.

To avoid such troublesome mounting with conductive paint, we devised a "conductive bridging" method, which consists of fixation of the cast to the metal stub with many pieces of thin copper wires (Figs. 1-7 and 1-11) [38]. Under binocular control, the pieces (conductive bridges) can be stuck to the individual vessels. If desired, the casts can be removed and remounted, with neither breakage nor contamination.

3.9. Conductive Treatment of Cast

Plastics used in the corrosion cast/SEM method are all insulators. Such electron-conductive plastics as iodine-doped resins have not yet been tested for casting.

Chemical metal staining and physical metal coating render the resin casts sufficiently conductive. The former staining, vapor-phase osmium-hydrazine hydrate, or wet-phase osmium-thiocarbohydrazide-osmium staining [39], is rarely used.

Physical metal (gold or carbon/gold) coating in a vacuum chamber is routinely used because it is easy to perform. A 1–2 nm thick gold or chromium film can be formed by high-energy sputtering. Chromium grains, as well as gold grains, sputtered onto the casts are clearly visible with a high-resolution or field-emission SEM, especially with an in-lens apparatus (S-900, Hitachi, Japan). Such high-energy sputtering and high-resolution SEM [34] are not necessary in casts, unless their replication quality surpasses 250 nm [15,31].

3.10. Storage of Cast

The mounted and coated specimens can be stored for a long time in a dry, dust-protected environment (dessicator, plastic box). The nonmounted and uncoated specimens are placed in the similar environment or frozen in ice. Storage in wet conditions should be for only a short time; in this storage, acidified solutions containing antibiotics are used to prevent fungal and bacterial growth [8,15].

4. SEM Inspection of Cast

After examination of the main drainage and supply vessels in the dissecting microscope, the mounted and coated casts are observed with the SEM. The use of a low accelerating voltage may produce good SEM images without charging effects. The casts can be moved along the x-, y-, and z-axes, tilted along the x- or y- axis, and rotated around the vertical (z-) axis. A rotary specimen holder is occasionally inserted into the SEM to inspect the whole superficial structures of round specimens, such as the kidney glomeruli and Stannius corpuscles. The specimens are sometimes tilted 45° to get optimal signals from the specimens. Use of an eucentric stage is useful to avoid specimen adjustment after tilting. If measurements for quantification (see below) are necessary, the tilting has to be corrected. Rotation becomes important in consecutive SEM sessions and rotation stereomicroscopy.

When an overview of the casts is needed, a photomontage is produced. To get a high-quality SEM photomontage, the micrographs (1) must have identical brightness, contrast, and magnification; (2) overlap each other up to one third in length and height; and (3) be gradually thinned out at their margins before they are mounted together. Stereomicrographs can be recorded by tilting or rotating the specimens (see above). Usually, three recordings are made with tilting angles (rotation) of 4–5°.

5. Diagnosis and Interpretation of Cast Structure

The most valuable and reliable findings obtained by SEM of corrosion casts may be the details of vascular connections, arrangements, and distributions [7,15]. In the beginning of this microscopy, the arteries and veins were discriminated mainly by tracing them to their origins or parent vessels, or by additional injection of colored casting medium into the arteries (see above).

The arterial and venous casts have some characteristic endothelial imprint patterns (imprints of endothelial cell nuclei and cell borders) on their surfaces, respectively (Figs. 1-3 and 1-4) [20]. The arterial surface is characterized by ovoid endothelial nuclear imprints oriented parallel to the long axis of the vessel (Fig. 1-3), while the venous surface reveals shallower, roundish endothelial nuclear imprints with no particular orientation (Fig. 1-4) [15,40]. The capillaries gradually change their surface patterns from the "arterial type" to the "venous one" (Fig. 1-5) [15]. These arterial and venous imprint patterns are constantly and obviously found throughout the vertebrates.

The mean ratio (D/d) of nuclear imprints estimated by measuring their large diameter (D) and small diameter (d) also clearly differentiates the arterial (D/d > 2.9) and venous (D/d < 2.0) vessels [15]. True capillaries (diameter <8 μm) have the mean ratios changing from the arterial side to the venous side [15]. It is difficult to distinguish between arterioles and venules in the casts, since mural tissue structures are removed by maceration. Further studies on the D/d-ratio or other imprint patterns may be needed for such vascular denomination in the casts.

Venous valves, endothelial cushions, sphincter structures, and other histological or mural structures of vessels are replicated in the casts [14, 16,41]. Muscle tones and other mural resistances against the injection of casting medium are also replicated as a broad scale of imprint patterns. Arterial contractions during the injection is imprinted as mixed longitudinal and circular microwaves on the cast surface.

In some organs, the cast vessels sometimes show blind endings, extravasations, or strong dilations [15,21,22,49]. It is questioned, whether these structures are real or artificial [15]. This matter is quite important in casting the embryonic blood vascular system and also in studying tumor angiogenesis [15,21,22,24,49]. Several criteria for discriminating artificial products have been proposed [15,21,22,25] that may deserve further discussion. Recently vascular neogenesis or sprouting in the mesentery of normal young rats was studied by intravital microscopy and a fine structural analysis. This study may be useful as a model for casting the developing vessels.

6. Quantification of Cast

Estimations of specific density (d) and weight (w) allow calculation of the volume (v) of the whole cast (v = w/d) [31,47]. Measurements of diameter (d) and length (2) also allow a similar calculation (v) of the cast vessels (v = 1/4 d l) [33,41–43]. Both approaches rely on optimal casting. Furthermore, SEM measurements of length, diameter, angle, and distance may be useful for the rheology of the cast vessels.

Other quantifications, such as planimetry, point counting, sterephotogrammetry, and image analysis, can be done on optimal casts. Planimetry is the simplest method in which micrographs or projected negative films are used [44]. Stereophotogrammetry needs a specific stereoscope with a parallax measuring system [45]. Point counting uses the principles of stereology [46]. Modern systems of image analysis can quantify all sorts of replicated structures [17,43]. The respiratory surfaces of teleost gills have been estimated by this analysis [43].

7. Concluding Remarks

In 20 years of application, the corrosion cast/SEM method has become a standard technique to study the fine angioarchitecture of tissues and organs in normal, experimental, and pathological states, as well as in development and aging. Provided careful precasting, casting, and postcasting procedures are performed, reliable structural and functional information is received. Since the physico-chemical properties of casting media are still not yet known in full detail, their impact on casting outcome may be tremendous; future work must concentrate on this topic. In the meantime we still have to interpret our results very carefully, especially with regard to quantitative cast analysis.

Acknowledgments

This work was supported by the Stiftungs- und Förderungsgesellschaft der Paris-Lodron- Universität Salzburg, who financed the personal computer used for word processing, and by

10

the Jubiläumsfonds der Oesterreichischen Nationalbank (Project No. 1868). The authors are also grateful to Karin Bernatzky and Rudolf Hametner for their photographic assistance.

References

1. Spalteholz W. *Über das Durchsichtigmachen von menschlichen und tierischen Präparaten.* 2. Auflage, Leipzig, 1914.

2. Schummer A. Ein neues Mittel, (Plastoid) und Verfahren zur Herstellung korrosionsanatomischer Präparate. *Anat Anz Jena* 81:177–201, 1935.

3. Cole FJ. The history of anatomical injections. In: Singer C (ed.), *Studies in the History and Method of Science*, Vol. II, Clarendon Press, Oxford, pp 285–343, 1921.

4. Tompsett DH. *Anatomical Techniques.* E & S Livingstone, Edinburgh, 1970.

5. Von Ardenne M. *Das Elektronen-Rastermikroskop. Z Phys* 109:553–572, 1938.

6. Nowell JA, Pangborn J, Tyler WS. Scanning electron microscopy of the avian lung. *Scann Electron Microsc* 249–256, 1970.

7. Murakami T. Application of the scanning electron microscope to the study of the fine distribution of blood vessels. *Arch Histol. Jpn* 32:445–454, 1971.

8. Lametschwandtner A, Lametschwandtner U, Weiger T. Scanning electron microscopy of vascular corrosion casts — technique and applications. *Updated Scann Microsc Review*, 4:889–941, 1990.

9. Schraufnagel DE. Microvascular corrosion casting of the lung. A state-of-the-art review. *Scann Microsc* 1: 1733–1747, 1987.

10. Murakami T. Methyl-methacrylate injection replica method. In: Hayat M (ed.), *Principles and Techniques of Scanning electron microscopy. Biological applications*, Vol. 2, Van Nostrand-Reinhold, New York, pp 159–169, 1978.

11. Gannon BJ. Vascular casting. In: Hayat M (ed.), *Principles and Techniques of Scanning Electron Microscopy*, Vol. 6, Van Nostrand-Reinhold, New York, pp 170–193, 1987.

12. Hodde KC, Nowell JA. SEM of micro-corrosion casts. *Scann Electron Microsc*, II:88–106, 1980.

13. Ohtani O. Microcirculation studies by the injection replica method with special reference to the portal circulations. In: Allen DJ, Motta PM, and Didio LJA (eds.), *Three Dimensional Microanatomy of Cells and Tissue Surfaces*, Elsevier North Holland, pp 51–70, 1981.

14. Hodde KC. *Cephalic Vascular Patterns in the Rat.* Akademisch Proefschrift, Rodopi, Amsterdam, 1981.

15. Christofferson RH. Angiogenesis as Induced by Trophoblast and Cancer Cells. PhD Thesis, University of Uppsala, 1988.

16. Motti EDF, Imhof HG, Garza JM, Yasargil GM. Vasospastic phenomena of the luminal replica of rat brain vessels. *Scann Microsc* 1:207–222, 1987.

17. Schraufnagel DE, Schmid A. Microvascular casting of the lung: Vascular lavage. *Scann Microsc* 2:1017–1020, 1988.

18. Kratky RG, Zeindler CM, Dorian KCL, Roach MR. Quantititive measurement from vascular casts. *Scann Microsc* 3:937–943, 1989.

19. Olson KR. Effects of perfusion pressure on the morphology of the central sinus in the trout gill filament. *Cell Tissue Res* 232:319–325, 1983.

20. Miodonski AJ, Hodde KC, Bakker C. Rasterelektronmikroskopie von Plastik-Korrosions-Präparaten; morphologische Unterschiede zwischen Arterien und Venen. *BEDO* 9:435–422, 1976.

21. Grunt TW, Lametschwandtner A, Karrer K. The characteristic structural features of the blood vessels of the Lewis lung carcinoma. *Scann Electron Microsc*, II:575–589, 1986.

22. Grunt TW, Lametschwandtner A, Karrer K, Staindl O. The angioarchitecture of the Lewis lung carcinoma in laboratory mice. *Scann Electron Microsc*, II:557–573, 1986.

23. Grunt T, Lametschwandtner A, Staindl O. Die Angioarchitektur der Haut des retroaurikulären Bereiches des Menschen. Eine rasterelektronmikroskopische Untersuchung an Korrosionspräparaten. *Hals- Nasen-Ohrenheilk* 30:420–425, 1982.

24. Miodonski AJ, Kus J, Nowogrodzka-Zagorska M, Olszewski J, Bugajski A. Angiomorphology of the human larynx and renal carcinoma: A comparative study. *Scanning Electron Microscopy of Vascular Casts: Methods and Applications* X: 1991.

25. Schraufnagel DE, Schmid A. Microvascular casting of the lung: Effects of various fixation protocols. *J Electron Microsc Techn* 8:185–191, 1988.

26. Murakami T. Pliable methacrylate casts of blood vessels. Use in scanning electron microscope study of the microcirculation in rat hypophysis. *Arch Histol Jpn* 38: 151–168, 1975.

27. Batson OV. Corrosion specimens prepared with a new material (abstract). *Anat Rec* 121:425, 1955.

28. Nopanitaya W, Aghajani JG, Gray LD. An improved plastic mixture for corrosion casting of the gastro-intestinal microvascular system. *Scann Electron Microsc* 751–755, 1979.

29. Hanstede JG, Gerrits PO. A new plastic for morphometric investigation of blood vessels, especially in large organs such as the human liver. *Anat Rec* 203: 307–315, 1982.

30. Amselgruber W, Sinowatz F, König HE. Ein neuer Kunststoff zur Anfertigung korrosionsanatomischer Präparate für die Rasterelektronenmikroskopie. Poster. 82. Versamml. Anat. Gesellschaft in Leipzig, 1987.

31. Weiger T, Lametschwandtner A, Stockmayer P. Technical parameters of plastics (Mercox CL-2B and various methylmethacrylates) used in scanning electron microscopy of vascular corrosion casts. *Scann Electron Microsc*, I:243–252, 1986.

32. Kratky RG, Roach MR. Shrinkage of Batson's and its relevance to vascular casting. *Atherosclerosis* 51: 339–341, 1984.

33. Gannon BJ. Preparation of microvascular corrosion casting media: Procedure for partial polymerization of methylmethacrylate using ultra violet light. *Biomed Res* 2:227–233, 1981.

34. Meyer EP. Corrosion casts as a method for investigation of the insect tracheal system. *Cell Tiss Res* 256:1–6, 1989.

35. Castenholz A. The outer surface morphology of blood vessels as revealed in scanning electron microscopy in resin cast non-corroded tissue specimens. *Scann Electron Microsc*, IV:1955–1962, 1983.

36. Schraufnagel DE. Ranking corrosion efficiency: A Latin square study of rat lung microvascular corrosion casts. *Scann Microsc* 3:299–304, 1989.

37. Miodonski A, Kus J, Tyrankiewicz R. SEM blood vessel casts analysis. In: Didio JA, Motta PM, Allen DJ (eds.), *Three-Dimensional Microanatomy of Cells and Tissue Surfaces*. Elsevier North Holland, pp 71–87, 1981.

38. Lametschwandtner A, Miodonski A, Simonsberger P. On the prevention of specimen charging in scanning electron microscopy of vascular corrsion casts by attaching conductive bridges. *Mikroskopie* 36:270–273, 1980.

39. Murphy JA. Non-coating techniques to render biological specimens conductive. *Scann Electron Microsc*, I:175–193, 1979.

40. Castenholz A. Interpretation of structural patterns appearing on corrosion casts of small blood and initial lymphatic vessels. *Scann Microsc* 3:315–325, 1989.

41. Hossler Fe, West RF. Venous valve anatomy and morphometry: Studies on the duckling using vascular corrosion casting. *Am J Anat* 181:425–432, 1988.

42. Anderhuber F, Weiglein, A, Pucher RK. Ein Beitrag zur Hirndurchblutung des Menschen anhand von Gefässausgüssen des Karotissystems. *Acta Anat* 136: 42–48, 1989.

43. Pohla H, Bernroider G, Lametschwandtner A, Goldschmid A. Computerized measurements of gill respiratory surface area with corrosion casts of gill vasculature (abstract). Proc. 5th Congr. Europ. Ichtyol., Stockholm, p 126, 1985.

44. Lametschwandtner A, Weiger T, Bernroider G. Morphometry of corrosion casts. In: Motta PM (ed.), *Cells and Tissues. A Three-Dimensional Approach by Modern Techniques in Microscopy*. Alan R. Liss, New York, pp 427–433, 1989.

45. Cavanagh B, Gannon BJ, Randall DJ. Stereophotogrammetry of scanning electron micrographs of vascular casts for three-dimensional measurement of microvascular lumina (abstract). *J Anat* 124:537, 1977.

46. Weibel ER. *Stereological Methods*, Vol I. Academic Press, London, 1979.

47. Murakami T, Ohtani Ohtsuka A, Kikuta A. Injection replication and scanning electron microscopy. In: Hodges GH, Carr KE (eds.), *Biomedical Research Applications of Scanning Electron Microscopy*, Vol. 3, Academic Press, London, pp 1–30, 1983.

48. Murakami T, Itoshima T, Hitomi K, Ohtsuka A, Jones AL. A monomeric methyl and hydroxypropyl injection medium and its utility in casting blood capillaries and liver bile canaliculi for scanning electron microscopy. *Arch Histol Jpn* 47:223–237, 1984.

49. Murakami T, Kikuta A, Kaneshige T, Naito I. Minute structure of human kidney glomerulus, its embryonic development and age-related changes: A scanning electron microscope study of vascular casts. In: Seno S, Copley AL, Venkatachalam MA, Hamashima Y, Tsujii T (eds.), *Glomerular Dysfunction and Biopathology of Vascular Wall*. Academic Press, Tokyo, pp 103–117, 1985.

50. Burton GJ, Palmer ME. The chorioallantoic capillary plexus of the chicken egg: A microvascular corrosion casting study. *Scann Microsc* 3:549–558, 1989.

Authors' address:
Prof. Alois Lametschwandter
  and Dr. Ursula Lametschwandter
University of Salzburg
Institute of Zoology
Department of Experimental Zoology
Hellbrunnerstrasse 34
A-5020, Salzburg, Austria

CHAPTER 2

Routine Methods for Vascular Casting and SEM

OSAMU OHTANI & TAKURO MURAKAMI

1. Introduction

Most morphological investigations of tubular, alveolar, and cavity systems have been made by examining tissue slices under the light and transmission electron microscope, by reconstruction of serial sections, and by image-analyzing computers. However, during the past two decades our comprehension of these systems has been greatly facilitated by scanning electron microscopic (SEM) examination of corrosion casts.

Corrosion casting is by no means a new technique, but has been used by anatomists to reveal the complicated internal morphology since the 16th century, when Leonardo da Vinci (1452–1519) made a wax replica of the cerebral ventricles and the chambers of the heart [1]. During the intervening centuries, many improvements have been made in replicating media, and in methods of injection and removing surrounding tissues. However, it was as recent as the 1970s that the methods for SEM observation of corrosion casts were introduced [2,3], and thus our knowledge on the tubular, alveolar, and cavity systems, and the blood vascular system, in particular, has begun to increase dramatically.

This chapter deals with (1) the vascular corrosion casting/SEM methods that are currently used in the routine examination of the vascular system and (2) the morphological characteristics of corrosion casts that facilitate the identification of the vessel types.

2. Microvascular Corrosion Casting/SEM Method

The microvascular corrosion casting/SEM method consists of the injection of casting medium into the vascular system, corrosion of tissue elements, and observation of the resulted casts under SEM (Table 2-1). This method allows the examination of the three-dimensional organization of microvessels, including the blood capillaries. During the past two decades, this method has been extensively employed to reveal the blood microvascular architecture of a variety of organs and tissues in humans and in a large number of experimental animals. This method is also applicable for the demonstration of the three-dimensional organization of the lymphatic system [4]. Even a fine vascular system such as the bile canaliculi in the liver can be reproduced in corrosion casts that can be observed under SEM [5] (Fig. 2-1).

2.1. Casting Media

Currently used casting media are rubber compounds and polyester resins that are either laboratory prepared or commercially produced. The rubber compounds have some advantages and disadvantages, as reported by Hodde and Nowell [6]. Because the vulcanization of the compounds takes place with a change in pH, there is a minimal increase in viscosity during injection. The

Motta, P.M., Murakami, T., and Fujita, H. (eds.), Scanning Electron Microscopy of Vascular Casts: Methods and Applications.

14

Table 2-1. Steps in the microvascular corrosion casting/SEM method

---

1. Perfuse animals (or organs) with saline or Ringers solution and inject the casting medium.
2. Place the specimens in a hot water bath (60°C) to restore them to their original form and to accelerate polymeriation (for several hours).
3. Macerate in a concentrated (15–20%) NaOH or KOH solution (60°C, overnight or longer).
4. Wash in running warm water.
5. Cut into blocks, air dry or freeze dry, and mount on specimen holder with conductive paste.
6. Coat with heavy metal and observe in a SEM with an accelerating voltage of 5–10 kV.
7. Microdissect under a binocular light microscope to expose deeper structures of interest.
8. Coat with heavy metal and observe in a SEM.
9. Repeat steps 7–8, until interesting structures are satisfactorily observed.

---

opacity of the media allows the orientation of the specimen during dissection under the light microscope. The vulcanized media are elastic enough to allow virtually distortion-free gross dissection. However, the rubber compounds do not consistently replicate luminal surface microstructures. The corrosion casts of the rubber compounds, such as Cementex [2], Vultex [7], and Geon latex [8], need freeze drying or critical-point drying. For these reasons, the rubber compounds have only been applied for casting the airways [2,7] and pulmonary microvessels [7,8]. Because microcorrosion casts of silicone rubber (Microfil) are extremely fragile and tend to disintegrate during corrosion, the use of this medium has been confined to light microscopy of methyl salicylate-cleared tissue sections.

The methyl methacrylate mixture or commercially produced polyester resin are currently most widely used for replicating microvasculatures. The methyl methacrylate mixture for SEM corrosion casting was initially introduced by Murakami (1971) [3]. Later he made several modifications to his original formula for making the brittle casts [9], rendering the casts conductive with OsO_4-hydrazine hydrate vapor [10], and lowering the viscosity by supplementing 30–50% monomeric methacrylate to the base resin. A further development was made by Gannon [11]

in which the methyl methacrylate mixture was partially polymerized with ultraviolet light, rather than being applied heat (Table 2-2). This partial polymerization procedure yields very consistent and low viscosity, which can be adjusted by the time of irradiation of ultraviolet light.

The commercially produced partially polymerized methacrylates that can be used for corrosion casting are Batson's plastic no. 17, Technovit 8001, Mercox, Trylon, and Araldite CY223. It is convenient that these casting media can be used without the rather laborious partial polymerization procedure. However, the viscosity of these media is, in general, higher than the desired viscosity and is frequently different from batch to batch. The viscosity of Mercox has been lowered by the addition of 20–30% methyl- or hydroxy-ethyl methacrylate monomer [12] and that of Batson's with monomeric methacrylate, Sevitron [13].

2.2. Preparation of Animals or Human Materials

Prior to injecting the casting medium into the blood vasculature, it is essential to remove all blood cells in order to fill the entire vascular bed, as well as to obtain endothelial cell impressions on the surface of the cast. The blood of animals or organs is removed prior to replication with saline,

Table 2-2. Preparation procedure of a methyl methacrylate partial polymer

---

1. In a suitable quantity (e.g., 150 ml) of methyl methacrylate monomer is dissolved 1% w/v of 2–4 dichloro-benzoyl peroxide paste initiator.
2. Six (25 ml size) screw-capped glass scintillation vials, each with 20 ml of monomer plus initiator, are lined up against an erythremal fluorescent tube.
3. Two "blank" vials, each containing 20 ml of water, are placed at each end of the line of vials containing methacrylate against the fluorescent tube.
4. The lamp is turned on; each vials is shaken briefly for about 10 minutes.
5. The lamp is turned off after the time chosen to produce the desired viscosity. (A viscosity of 3–5 cS gives the best microvascular casts). The vials containing partially polymerized methacrylate are stored at −10°C for later use (for up to several weeks).
6. The actual viscosity achieved is monitored at 20°C with a modified Ostward viscometer.

From Gannon [11] with permission.

Ringer's solution, or Krebs solution containing heparin (10^4 U/l) at 37°C via an afferent cannulated artery at a physiological pressure. When the afferent artery is too thin to cannulate, as in small organs obtained by surgical operation, perfusion with saline followed by injection of casting medium can be performed through a cannula inserted into the main efferent vein of the organ. The use of heparin can be omitted without any significant differences in the resulting casts. Adding vasodilators, such as papaverin (10^{-7} g/ml), 0.5% procaine, $NaNO_2$ (10^{-3} mol/l), or sodium bicarbonate (0.5 mol), to the perfusate dilates the vasculature and produces more complete casts. Perfusion is continued until all the blood has been washed from the animal or organ (usually 5–8 minutes).

Perfusion fixation of the animals or organs prior to the injection of casting medium is usually omitted, because fixation causes vascular constriction, which hinders filling of the medium. However, in order to obtain casts with distinct imprints of endothelial cells, it is better to perfuse with 10% formalin or 1% glutaraldehyde in phosphate-buffered saline prior to the injection of casting medium.

For replication of the lymphatic vessels, it is not necessary to remove the blood [4]. However, when replicating the lymphatic and blood vessels simultaneously, the blood should be removed, as in replicating the blood vascular system [4].

For replication of tubular and alveolar organs, such as airways [7], glands, and the biliary tree [5], the duct should be exposed and a cannula should be inserted into it.

2.3. Injecting the Casting Medium into the Blood Vascular System

Immediately after perfusing the animal or organ with physiological perfusate, the injection medium (resin) is prepared. The resin consists of 14 ml methyl methacrylate partial polymer (Table 2-2), 6 ml hydroxypropyl methacrylate monomer, 0.2 g benzoyl peroxide (or 2-4 dichloro-benzoyl peroxide), and 0.3 ml N-n dimethyl aniline (Table 2-3) [11].

The resin is delivered to the vasculature by a hand-controlled syringe or a syringe connected to a perfusion apparatus [14] until the resin leaking

Table 2-3. Composition of the injection medium

1. 14 ml methyl methacrylate partial polymer of viscosity of 3–5 cS at 20°C
2. 6 ml hydroxypropyl methacrylate monomer
3. 0.2 g of benzoyl peroxide (or 2–4 dichloro- benzoyl peroxide)
4. 0.3 ml of N, N dimethyl aniline (accelerator)

From Gannon [11] with permission.

out from the effluent vein(s) becomes virtually water-free. The use of the injection apparatus allows monitoring of the pressure of compressed air that drives the resin into the vasculature. The intraarterial pressure in the animal can be recorded in the artery with a strain-gauge transducer connected to a recorder [14]. According to Christofferson and Nilsson [14], in the rat a resin infusion pressure ranging from 90 to 120 mmHg gives the best filling and endothelial cell replication.

Partial casting of blood vessels is sometimes useful in studying the distribution pattern of the arterial and/or venous systems. A cannula is inserted into either the main afferent artery or efferent vein, and is perfused with physiological perfusate through the cannula. Then the organ is perfused with 10% formalin in 0.1 mol/l phosphate buffered saline, after which small amount of the resin (e.g., Mercox) is infused with a hand-controlled syringe until the arterial or venous system is filled.

2.4. Injecting the Casting Medium into the Lymphatics

Methods have recently developed for making lymphatic corrosion casts. Retrograde injection of partially polymerized methyl methacrylate (e.g., Mercox) into the bile duct may fill the lymphatic system located in the portal canal of the liver [15]. The lymphatic system of the lymph node can be replicated by injection of resin into the afferent lymphatic vessels of the lymph node [16]. Intraparenchymal puncture-injection of low-viscosity resin (e.g., Mercox supplemented with methyl methacrylate monomer to give a viscosity of 1.5–2.0 cS) fills the lymphatic vessels in some organs. In the case of the gastrointestinal tract, a

16

small volume of low-viscosity resin (see above) is injected intraparenchymally into the submucosa and deep mucosa by needle puncture [4].

Scanning electron microscopic observation of the simultaneously cast blood and lymphatic vessels clearly show the spatial relationship between the two vascular systems [4]. Under anesthesia, a cannula is inserted into the main afferent artery of the organ, and the blood vascular system is perfused through the cannula with physiological perfusate. A small volume of low-viscosity resin (Mercox diluted with monomeric methyl methacrylate, see above) is injected intraparenchymally by needle puncture, followed by the infusion of Mercox into the blood vascular system.

In newborn animals, the intraarterial injection of low-viscosity resin (see above) fills the lymphatic vessels, as well as the blood vessels and interstitial tissue spaces [17].

2.5. Injection of the Casting Medium into the Biliary System

The method of corrosion casting of the biliary system, including bile canaliculi, was developed by Murakami and his colleagues [5]. The casting medium consists of methyl methacrylate monomer and 2-hydroxypropyl methacrylate monomer, which is mixed in a volume ratio of 5:4 to 6:5. This mixture is supplemented with N,N-dimethyl aniline to give a concentration of 1.5% shortly before injection.

Under anesthesia, the thoracic aorta (in the case of small animals such as rats) and the common bile duct are cannulated. The animal or liver is perfused with physiological saline through the cannulated blood vessel, which is followed by perfusion with 2% glutaraldehyde in 0.1 mol/l phosphate buffer. Immediately after this procedure, the resin is injected retrogradely into the cannulated common bile duct by a syringe with applied force at a pressure of about 70–80 mmHg. The injection is continued until the casting medium in the syringe hardens. Injected organs are treated in the same way as in blood vascular corrosion casting (see below), except that the corrosion casts of the bile canaliculi are freeze dried.

2.6. Tissue Corrosion

Maceration of the tissue from the replica can be accomplished by immersing the resin-injected tissues into 15–20% NaOH or KOH (60°C). Christofferson and Nilsson [14] have reported that 20% KOH, 40% KOH, or 18% HCl macerate tissues most effectively, followed by 20% NaOH. Usually the resin-injected tissues are placed in warm water (60°C) for several hours to 1–2 days before maceration, which increases the efficacy of corrosive agents, thus reducing the corrosion time considerably. The time required to macerate the tissues varies from several hours to several days, depending on the size of the resin-injected tissues and the corrosive agent used.

When obtaining microvascular corrosion casts of bones, decalcification should be accomplished with either 2% HCl, 6% trichloro-acetic acid, or Plank-Rychlo solution, which consists of 850 ml distilled water, 70 g aluminum chloride, 50 ml formic acid (90%), and 80–90 ml HCl (37%).

Following maceration, the casts is washed in gently running warm water. Maceration of tissues rich in lipid with NaOH frequently causes depositions of white saponified materials, which are barely washed away by running tap water. Immersing the casts in warm water (40–60°C) for several hours or a day or two usually dissolves such saponified materials. Washing in neutral detergent is also useful to clean the casts. Ultrasonic cleaning is not recommended, since microvascular corrosion casts may be destroyed by ultrasonic vibration. Thorough washing with running warm water followed by distilled water is recommended to remove all salts, which could crystallize on the dry casts.

2.7. Dissection of the Corrosion Casts

Initial dissection can be done on the wet specimen, before, during, or after corrosion. Before corrosion but after immersion in warm water (60°C) for several hours to overnight is the best time for gross dissection, because the supportive tissues surrounding the casts not only serve as anatomical landmarks, but are soft enough to be dissected with tweezers and scissors without breaking or dislocating the microvascular casts.

A large mass of corrosion casts can be frozen

in water and cut into smaller blocks with a saw. Final trimming can be done by freezing the cleaned casts in water and cutting with a razor blade. Alternately, the cleaned cast can be embedded in water-soluble wax (Aquax, Gurr 80900), from which desired areas are cut without the risk of cracking [6]. The dried and mounted specimens can be cut with a laser beam [6].

In order to satisfactorily observe the microvascular organization, microdissection should be performed to expose the structures of interest. Microdissection is usually performed on mounted specimens with fine tweezers and needles under a binocular microscope. Alternately, the mounted specimens are embedded in ice and the superficial part of the casts are cut away with a razor blade. Microdissection of conglomerated microvascular casts such as renal glomeruli can be performed with fine forceps in a warm alcohol bath [18].

Usually after initial observations in a SEM, microdissection, coating, and observation in a SEM are repeated until the microvascular organization of interest is fully disclosed.

Microdissection can be performed in the SEM using a micromanipulator, as described by Pawley and Nowell [19], on casts that have been rendered conductive by OsO_4 and hydrazine hydrate vapor [10] (see below).

2.8. Drying the Casts

The blood and lymphatic microvascular corrosion casts can usually be dried in air without any detectable distortion or dislocation. However, air drying is not recommended if fine networks consist mostly of capillaries, because surface tension dislocates these most fragile parts, thus destroying the spatial relationships. Critical-point drying omitting amylacetate, which softens the cast, can be used to drying such fragile casts, but the casts may be destroyed or dislocated during the drying procedure. It is recommended that fragile casts such as those of capillaries and of bile canaliculi be freeze dried [5].

2.9. Rendering the Cast Conductive

In order to observe the casts in a SEM, the casts must be electron conductive. A sputtered 40-nm gold coat is suitable for microvascular corrosion

Table 2-4. Steps for chemically rendering the cast conductive

1. Place the dried casts in OsO_4 vapor for 24–48 hours.
2. Place in air for 15 minutes to sublimate superfluous osmium vapor.
3. Place in vaporized hydrazine hydrate for 12–24 hours.

From Murakmi et al. [10].

casts for observations at accelerating voltages between 4 and 7 kV [14]. The recommended alternative is rendering the bulk specimen chemically conductive [10] (Table 2-4). Osmium impregnation from OsO_4 crystals gives acceptable secondary electron images at a TV scan rate without bulk charging or thermal movements, regardless of the magnification and accelerating voltages (4–20 kV) [14]. However, as vessel edges are frequently charged at a low scan rate, it is recommended that the osmium-impregnated casts also be lightly sputter coated with gold in order to produce virtually charging-free micrographs.

2.10. Observation in the SEM

Full advantage should be taken of the depth of field by using long working distances and the smallest final aperture for documentation [6]. Observations of stereoscopic pairs of micrographs frequently offer much additional information about the spatial relationships between vessels, as well as the surface morphology of casts. Stereoscopic pairs of micrographs can be taken with a tilt separation of 4–7°. When initial observations do not fully reveal the microvascular organization of the area of interest, microdissection followed by SEM observations should be repeated until the desired area is satisfactorily disclosed (see above).

2.11. Artifacts

The artifacts sometimes encountered are incomplete corrosion, incomplete filling, and extravasation of injected resin.

Incomplete corrosion. The most common results of incomplete corrosion are tissue remnants on vessel casts. The tissue remnants are arterial endothelial nuclei [14] and "plastic strips" that

18

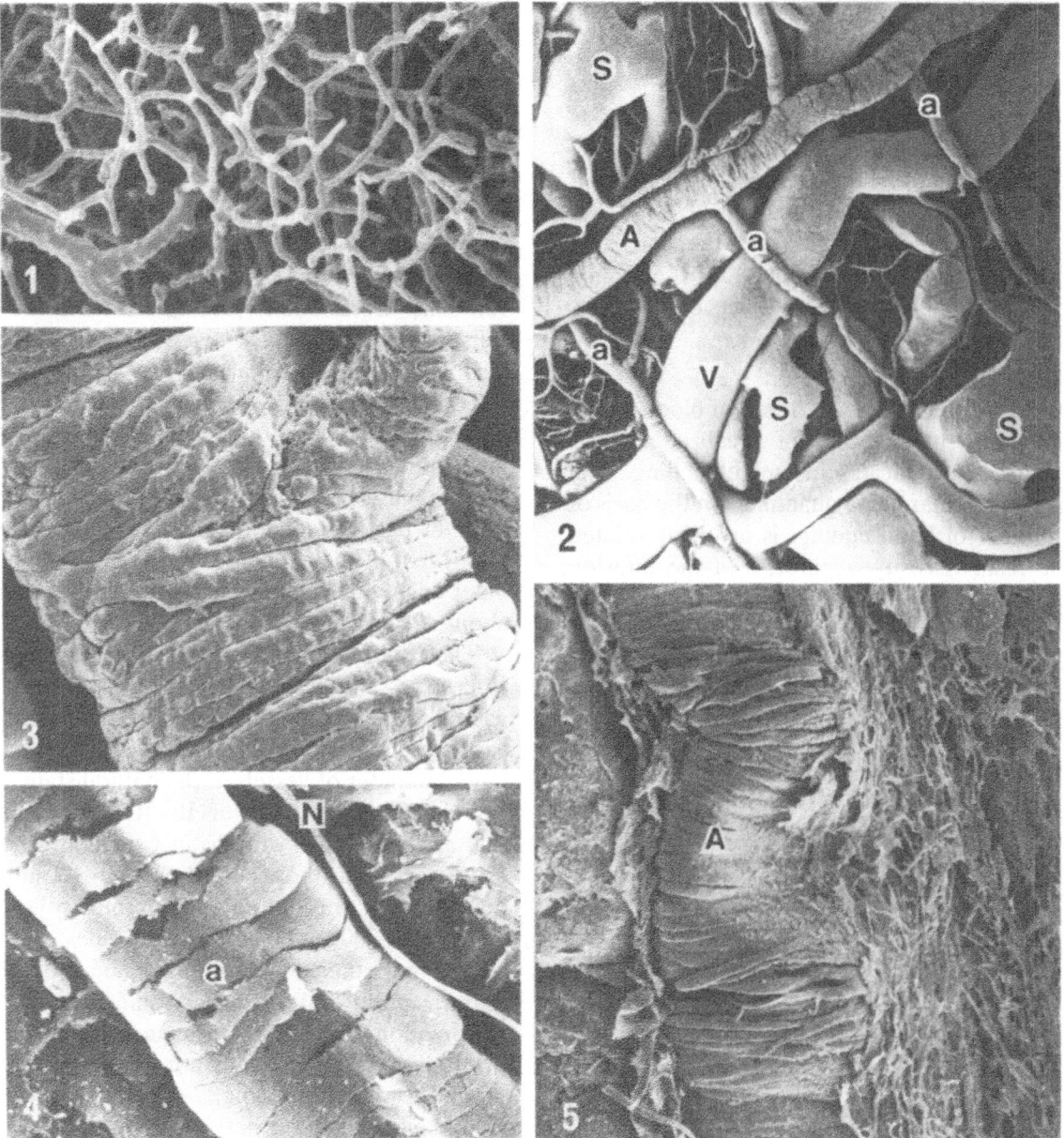

Figure 2-1. SEM micrograph of corrosion cast of the bile canalicular system in the rat liver. ×870.

Figure 2-2. SEM micrograph of lymphatic and blood vascular corrosion cast of the rabbit appendix viewed from the serosal side. Note that so-called plastic strips surround arterial (A) and arteriolar casts (a). Lymphatic sinuses (S) surrounding the lymphoid follicles are partially filled with resin. V = vein. ×30.

Figure 2-3. Closer SEM view of the so-called plastic strips surrounding the arterial cast. The morphological properties of the strips are those of the smooth muscle cells of artery (compare with Fig. 2-5). ×640.

Figures 2-4 and 2-5. SEM micrographs of alkali-collagenase-treated rabbit appendix showing circumferentially arranged smooth muscles of an arteriole (a in Fig. 2-4) and of an artery (A in Fig. 2-5). Note that the shape of the smooth muscles resembles that of the plastic strips seen in Fig. 2-3. N = Nerve fiber. Fig. 2-4: ×1600; Fig. 2-5: ×300.

have a spiral-like appearance with a similar periodicity. However, "plastic strips" are no more than uncorroded smooth muscles. This is supported by two reasons: "plastic strips" are only seen around arterial and arteriolar casts (Figs. 2-2 and 2-3), and the strips correlate well with those of smooth muscles revealed by SEM of alkali-collagenase treated tissues [20] (Figs. 2-4 and 2-5). Etching damage from the corrosion procedures may result in small (10–500 nm), round resin defects on the surface of the casts [14].

Incomplete filling. The occurrence of blood-vessel casts ending blindly and of localized defects in cast vessels indicates incomplete filling. The degree of vascular filling quantified on the basis of the number of blind endings is inversely related to the resin viscosity, regardless of the infusion pressure [5,11,14].

Extraversation. Injection of casting medium at a pressure much higher than the physiological level (about 200 mmHg in the rat) may result in randomly distributed extraversations of varying size. The appearance of the extraversated resin is "scrambled egg"-like, with globular masses lacking tubular structures and without impressions of endothelial cell nuclei.

Extraversations of injected resin are rather frequently encountered in such areas as the lymphoid tissues, gastrointestinal mucosa, and the implantation-site endometrium during early pregnancy [14]. The capillaries of these areas have fenestrations and a high permeability. Thus, the occurrence of the extraversation of injected resin may, to some extent, reflect the vascular permeability, given that the low-viscosity resin is injected at a pressure within the physiological range.

3. Morphology of Corrosion Casts

3.1. Identification of Blood Vessel Types

According to the extensive light and transmission-electron microscopy performed by Rhodin [21, 22], various sections of the microvascular bed can be defined as follows. *Arteriole*: Small arteries ranging between 50 and 100 μm in inner diameter that have more than one smooth muscle layer. *Terminal arteriole*: A vessel with an inner diameter of less than 50 μm that has only a single muscle layer. *Postcapillary venule*: Microvessels with a diameter of about 8–30 μm that are continuous with venous capillaries (up to 8 μm). *Collecting venule* (30–50 μm): Vessels with one complete layer of pericytes and a complete layer of veil cells. Occasionally primitive smooth muscle cells occur. *Muscular venule* (50–100 μm): Nearly all periendothelial cells are smooth muscle cells that sometimes form two layers. *Small collecting veins* (100–300 μm): Vessels with a prominent media consisting of a continuous layer of smooth muscle cells. *Capillaries* consist of a single layer of flattened endothelial cells and pericytes [23]. Most of the capillaries are 4–10 μm in their inner diameter, but some are much thicker (up to about 40 μm).

These definitions cannot be directly applied to cast vessels, because corrosion casts have no vascular wall. However, given the fact that the cast vessels are very close in size to the actual inner diameter of vessels in the natural state, one can define vessel casts of 10–100 μm diameter that are continuous with distinct small arteries as arterioles, and those of about 10–100 μm of diameter that are continuous with small collecting veins (100–300 μm diameter) as venules.

The endothelial cell imprint and branching patterns are useful for distinguishing vessel types of corrosion casts [14,24]. The nuclear imprints of arteriolar endothelial cells are usually elongated along the vessel axis (Fig. 2-6). Nuclear polarity [25] is occasionally observed: the pointed end is downstream and the round end is upstream. The arteriolar branching pattern is either symmetrical (daughter vessels of approximately equal diameter) or asymmetrical [14]. Capillary casts also show ovoid nuclear imprints and shallow linear imprints of endothelial boundaries (Fig. 2-7). In the arterial end of the capillary (arterial capillary), nuclear imprints of endothelial cells are elongated, while in the venous end (venous capillary) they are rounded. Endothelial nuclear imprints of venular and venous casts are shallow and round, or ovoid (Fig. 2-6). The branching pattern of venules is in many cases asymmetrical,

20

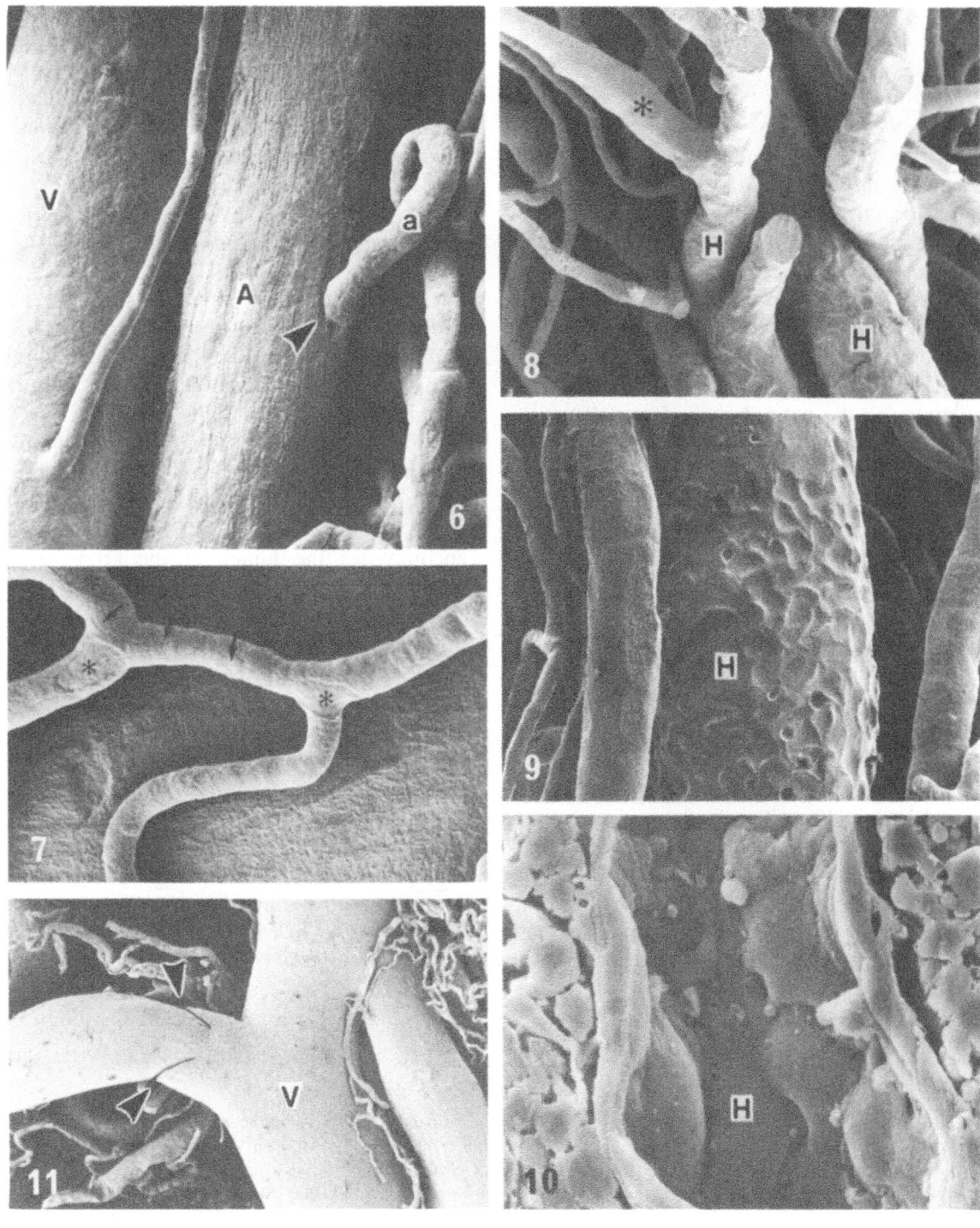

with venules receiving tributaries as large as themselves or as small as capillaries [14]. There are frequently arterio-arteriolar and venulo-venular anastomoses, as seen in the submucosal layer of the gastrointestinal tract.

High endothelial venules. In lymphoid organs there are specially differentiated venules, termed *high endothelial venules* (HEVs), through which lymphocytes migrate from the blood stream into the lymphoid tissues [26]. Scanning electron microscopy of tissues has revealed that the endothelial cells of HEVs protrude into the lumen with their rounded cell bodies, showing up as a cobblestone pattern [27,28]. The surface of cast HEVs (Figs. 2-8 and 2-9) reflects their luminal morphology in the natural state (Fig. 2-10) and shows deep imprints of endothelial cells and irregular ridges or furrows between the imprints (Figs. 2-7 and 2-8). Round deep depressions are frequently observed on cast HEVs, which probably reflect the lymphocytes within the lumen or the wall of the HEVs (Fig. 2-9).

3.2. Venous Valves

Venous valves usually consist of a pair of intimal folds or leaflets projecting into the vessel lumen from opposite sides of the vessel wall. Most valves are bicuspid, but they can be unicuspid, tricuspid, or quadricuspid [29]. The leaflets are covered on both sides by endothelium. Valves are most abundant in such regions as the extremities, which are subject to gravitational backflow. It is widely believed that the hepatic portal vein or its tributaries has no valves. However, we have noticed that valves do exist in the tributaries of the hepatic portal vein in the dog (Fig. 2-11). SEM of vascular corrosion casts has revealed that the casts typically exhibit slight expansions at valve sinuses and deep slits at the site of the valve (Fig. 2-11). Imprints of endothelial-cell nuclei on the medial surface of the cast valve leaflet are oriented parallel to the long axis of the vessels, whereas those on the lateral surface are oriented perpendicular to the axis [30].

3.3. Intraarterial Cushions

Intrarterial cushions, variously termed *internal cushions*, *subendothelial cushions*, or *arterial branch pads* (*polster*), are the longitudinal thickenings of arterial intima that occur at branching sites and constitute the so-called valves or cushions. Intraarterial cushions have been found in a variety of organs [31]. Three-dimensional reconstruction [32] has shown that the cushion is an elevated, asymmetric ring of tissues that encompasses the lumen of the artery at sites where the vessel gives rise to collateral branches. Since the intraarterial cushion contains bundles of longitudinally arranged smooth muscle, the cushion seems to provide sphincterlike control over blood flow to the collateral branch [33]. SEM of vascular corrosion casts of the rat uterus [34] and rat kidney [35] have shown two kinds of configurations of cushion impressions on the casts: One is pyriform in shape, with the tapering point directed retrograde to the blood flow, and the other is a sphincter ring surrounding the orifice of the arterial branch and lacking a tapering point. Our SEM of corrosion casts has confirmed intraarterial cushions in a variety of organs (Fig. 2-6).

←

Figure 2-6. SEM micrograph of blood vascular corrosion cast of the rat pancreas. The cast of the small artery (A) shows elongated imprints of endothelial-cell nuclei, while that of vein (V) exhibits shallow round or oval imprints. There is a constriction (arrowhead) at the branching point of an arteriole (a), which is suggestive of the location of an intraarterial cushion. ×500.

Figure 2-7. SEM micrograph of corrosion cast of the capillary showing imprints of nuclei (*) and cell boundaries (arrows) of the endothelium. ×500.

Figures 2-8 and 2-9. SEM micrographs of blood vascular corrosion casts of rabbit appendix. The postcapillary venules (*) abruptly increase their diameter and lead into HEVs (H) with irregular impressions of endothelial cells. There are also round, deep depressions indicative of the locations of lymphocytes. Fig. 2-8: ×80; Fig. 2-9: ×400.

Figure 2-10. SEM micrograph of the luminal view of the HEV (H) in the rabbit appendix. ×2000.

Figure 2-11. SEM micrograph of efferent veins (V) of the dog stomach. Note a pair of slitslike depressions (arrowheads) indicative of the valve location. ×65.

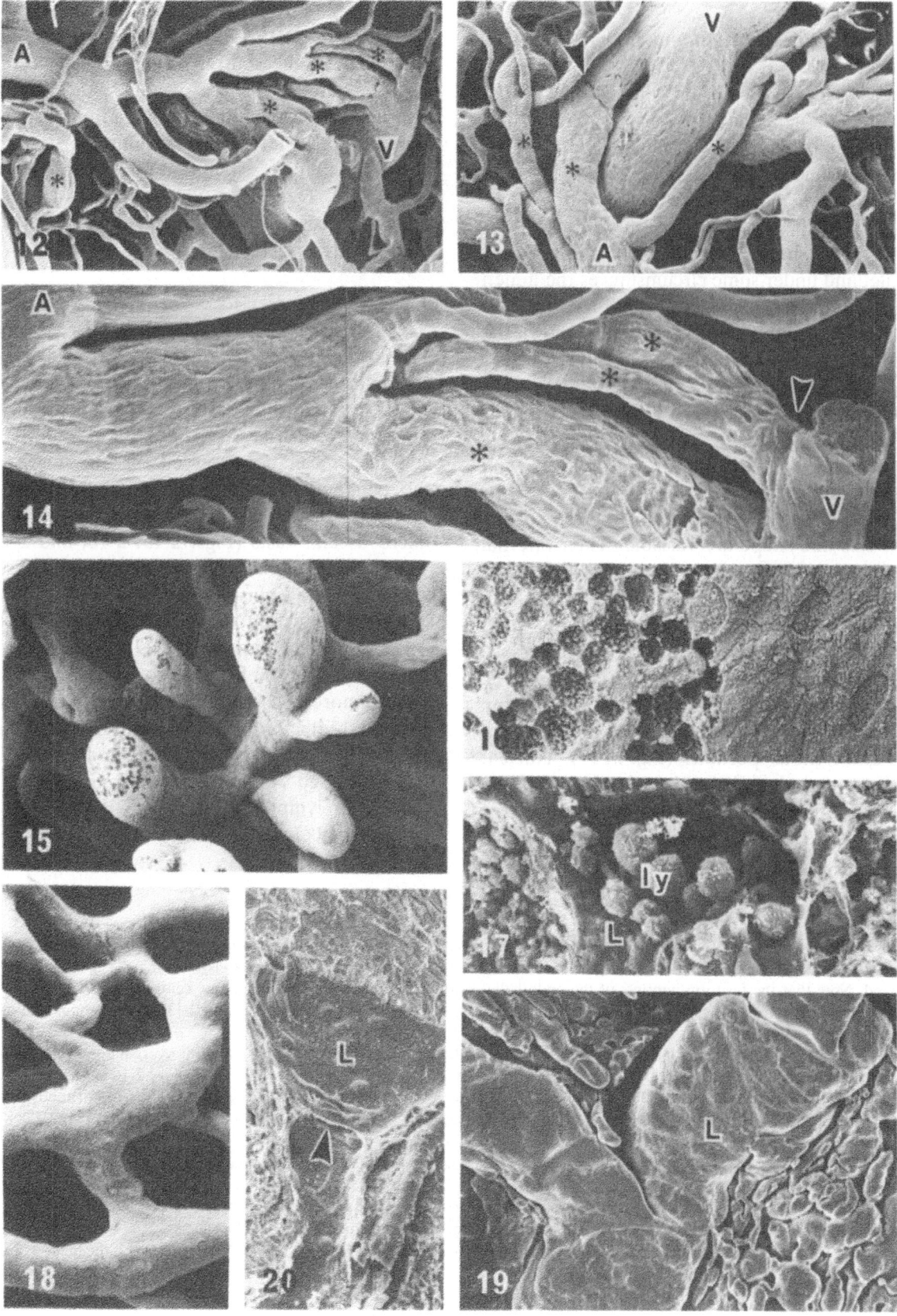

3.4. Arterio-Venous Anastomoses (AVAs)

AVAs are described light microscopically in cutaneous vascular beds of various regions that the hindlimb, bird foot, human ear, rabbit ear, nasal mucosa, and tongue. AVAs are presumed to play a major role in temperature regulation [36,37,39]. In vivo observations of rabbit-ear AVAs have shown that they are indeed contractile, and under some circumstances the lumen may be completely occluded [37,38]. Recent SEM studies of vascular corrosion casts have shown the detailed organization of AVAs in the rabbit ear [40] and dog tongue [41]. AVAs in other regions have not been demonstrated by SEM of vascular corrosion casts.

According to Morris and Bevan [40], and our own studies (Figs. 2-12–2-14), AVAs occur singly, in pairs, or in clusters of three to six. Their size and shape are variable. Their internal diameter varies from 20 to 100 µm. They follow straight courses (Figs. 2-12 and 2-13), make a U-turn at the venous portion to lead into veins (Fig. 2-13), or are coiled around adjacent vessels. Based on the surface morphology of cast AVAs, AVAs can be divided into three portions from the arterial to the venous side: the arterial, central, and venous portions (Fig. 2-14). Nuclear imprints of endothelial cells of AVAs are, in general fusiform, oriented along the vascular axis, but are more randomly arranged than those of veins and arteries. Nuclear imprints are most prominent in the central portion. They become slightly rounder in the venous portion. The arterial portion sometimes shows a funnel-shaped appearance, whereas the venous portion frequently exhibits circumferential constrictions (Fig. 2-14). Deep slits suggestive of valve locations are occasionally observed at the venous ends of AVAs (Fig. 2-13).

3.5. Lymphatic Vessels

The corrosion casts of lymphatic vessels also have a tubular appearance. Their diameters vary considerably, and they are usually much thicker than blood microvessels. Imprints of endothelial cell nuclei can be observed on the surface of the lymphatic vessel casts, but they are more randomly arranged than those seen on blood microvascular casts (Figs. 2-15, 2-16, and 2-18). The casts of lymphatic vessels in lymphoid tissues frequently show clusters of round impressions of lymphocytes within the vessels (Figs. 2-15 and 2-16). SEM of tissues shows aggregations of lymphocytes in the lymphatic vessels (Fig. 2-17). Furthermore, the lymphatic vessels around the lymphoid follicles form wide sinuses. Undulating cell boundaries of the endothelium are occasionally impressed on the cast surface. Some casts of the initial lymphatics start with blind ends as central lacteals in the small intestine [4] (Fig.

Figures 2-12–2-14. SEM micrographs of blood vascular corrosion casts of the rabbit ear showing AVAs (*) that intervene between the artery (A) and vein (V). The casts have been microdissected, and most of the microvessels covering the AVAs have been removed. Note that the fusiform imprints of the endothelial cell nuclei are most prominent in the central portion of the AVAs. There are constrictions (arrowhead in Fig. 2-14) and deep slitlike depressions (arrowhead in Fig. 2-13) at the venous end of the AVAs. Fig. 2-12: ×65; Fig. 2-13: ×120; Fig. 2-14: ×330.

Figures 2-15 and 2-16. SEM micrographs of corrosion cast of the initial lymphatics in the samll villi overlying the interfollicular regions of the rabbit sacculus rotundus (aggregation of lymphoid follicles at the terminal ileum). Figure 2-16 is a closer view of the top of the cast of the blind-ended lymphatic vessel that is seen in the top center of Figure 2-15. On the top of the lymphatic cast, there are clusters of round depressions of lymphocytes, as well as ovoid imprints of endothelial cell nuclei. Fig. 2-15: ×120, Fig. 2-16: ×800.

Figure 2-17. SEM micrograph of the interfollicular region of the rabbit appendix. There are many lymphocytes (ly) in the lymphatic vessel (L). ×1600.

Figure 2-18. SEM micrograph of lymphatic corrosion cast in the deep mucosa of the dog stomach. These initial lymphatics with various diameters form a meshwork. There are some imprints of endothelial-cell nuclei. ×320.

Figure 2-19. SEM micrograph of the corrosion cast of the collecting lymphatic vessel (L) that is located between the inner and outer muscular coat of the rat stomach. The lymphatic cast shows pairs of slitlike impressions indicative of valve locations. Scrambled-egg-like masses represent the interstitial tissue spaces replicated. ×640.

Figure 2-20. SEM micrograph showing a luminal view of the collecting lymphatic vessel (L) with valves (arrowhead). This lymphatic vessel is located between the inner and outer muscular coats of the rat stomach. There are randomly distributed swellings of endothelial-cell nuclei. ×400.

2-15), while others form meshworks, as in the deep mucosa of the stomach [42] (Fig. 2-18). The corrosion casts of the collecting lymphatics show slitlike impressions or notches (Fig. 2-19). Since the slitlike impressions correspond well with valves observed under SEM (Fig. 2-20), the impressions indicate the locations of valves. Most of the valves are bicuspids. The intervals between the valves vary considerably depending on the tissues and organs.

4. Conclusions

The corrosion casting/SEM method [3] has now been widely used to routinely investigate the three-dimensional organization of the blood and lymphatic microvascular bed. In addition, this method has been applied to a variety of studies: endothelial cell morphology and distribution [43], intraarterial cushions [34,35], angiogenesis [44], alterations in perfusion distribution [45], blood-vessel diameter [46], casts as tissue skeleton [47], the bile canalicular system [5], the internal structure of hard tissues [48], the insect tracheal system [49], etc. The corrosion casting/SEM method can perhaps be applied to many other investigations. Recently, morphometry has been applied to corrosion casts [50]. However, in order to obtain accurate quantitative data, we have to know more about chemical and physical properties of the casting media and the effects the casting media exert upon vessels during the casting procedure. We also need better casting media that can replicate the luminal surface ultrastructures of vessels more precisely than those media currently in use.

References

1. Tompsett DH. *Anatomical Techniques*, 2nd ed, E & S Livingstone, London, 1970.
2. Nowell JA, Pangborn J, Tyler WS. Scanning electron microscopy of the avian lung. *Scann Electron Microsc* 249–256, 1970.
3. Murakami T. Application of the scanning electron microscope to the study of the fine distribution of the blood vessels. *Arch Histol Jpn* 32:445–454, 1971.
4. Ohtani O, Ohtsuka A. Three-dimensional organization of lymphatics and their relationship to blood vessels in rabbit small intestine. A scanning electron microscopic study of corrosion casts. *Arch Histol Jpn* 48:255–268, 1985.
5. Muramaki T, Itoshima T, Hitomi K, Ohtsuka A, Jones AL. A monomeric methyl and hydroxypropyl methacrylate injection medium and its utility in casting blood capillaries and liver bile canaliculi for scanning electron microscopy. *Arch Histol Jpn* 47:223–237, 1984.
6. Hodde KC, Nowell JA. SEM of micro-corrosion casts. *Scann Electron Microsc* II:89–106, 1980.
7. Nowell JA, Pangborn J, Tyler WS. Stabilization and replication of soft tubular and alveolar systems. A SEM study of the lung. *Scann Electron Microsc* 305–312, 1972.
8. Frasca JM, Carter HW, Schaffer WA. An improved latex injection method for replicating the pulmonary microvasculature. *Scann Electron Microscopy* II:485–489, 1978.
9. Murakami T. Pliable methacrylate casts of blood vessels: Use in a SEM study of the microcirculation in rat hypophysis. *Arch Histol Jpn* 38:151–168, 1975.
10. Murakami T, Unehira M, Kawakami H, Kubotsu K. Osmium impregnation of methyl-methacrylate vascular casts for SEM. *Arch Histol Jpn* 36:119–124, 1973.
11. Gannon BJ. Preparation of microvascular corrosion casting media: Procedure for partial polymerization using ultraviolet light. *Biomed Res* 2 (Suppl.):227–233, 1981.
12. Ohtani O, Murakami T. Peribiliary portal system in the rat liver as studied by the injection replica scanning electron microscope method. *Scann Electron Microsc* II:241–244, 1978.
13. Nopanitaya W, Aghajanian JG, Gray LD. An improved plastic mixture for corrosion casting of the gastro-intestinal microvascular system. *Scann Electron Microsc* III:751–755, 1979.
14. Christofferson RH, Nilsson BO. Microvascular corrosion casting with analysis in the scanning electron microscope. *Scanning* 10:43–63, 1988.
15. Yamamoto K, Phillips MJ. Three-dimensional observation of the intrahepatic lymphatics by scanning electron microscopy of corrosion casts. *Anat Rec* 214:67–70, 1986.
16. Kurokawa T, Ogata T. A scanning electron microscopic study on the lymphatic microcirculation of the rabbit mesenteric lymph node. A corrosion casts study. *Acta Anat* 107:439–466, 1980.
17. Ohtani O, Murakami T. Lymphatics and myenteric plexus in the muscular coat in the rat stomach: A scanning electron microscopic study of corrosion casts made by intra-arterial injection. *Arch Histol Jpn* 50:87–93, 1987.
18. Murakami T. Vascular arrangement of the rat renal glomerulus. A scanning electron microscope study of corrosion casts. *Arch Histol Jpn* 34:87–107, 1972.
19. Pawley JB, Nowell JA. Microdissection of biological SEM samples for further study in the TEM. *Scann Electron Microsc* 333–340, 1973.
20. Miller BG, Wood RI, Bohlen HG, Evan AP. A new morphological procedure for viewing microvessels: A scanning electron microscopic study of the vasculature of small intestine. *Anat Rec* 203:493–503, 1982.
21. Rhodin JA. The ultrastructure of mammalian arterioles and precapillary sphincters. *J Ultrastruc Res* 18:181–223, 1967.

22. Rhodin JA. Ultrastructure of mammalian venous capillaries, venules and small collecting veins. *J Ultrastruc Res* 25:452–500, 1968.
23. Wolff JR. Ultrastructure of the terminal vascular bed as related to function. In: Kaley G, Altura BM (eds.), *Microcirculation*, Vol. 1, University Park Press, Baltimore, pp 95–130, 1977.
24. Hodde KC, Miodonski A, Bakker C, Veltman WAM. SEM of microcorrosion casts with special attention on arteriovenous differences and application to the rat's cochlea. *Scan Electron Microsc* II:477–484, 1977.
25. Burger PC, Chandler DB, Klintworth GK. Scanning electron microscopy of vascular casts. *J Electron Microsc Technol* 1:341–348, 1984.
26. Gowans JL, Knight EJ. The route of recirculation of lymphocytes in the rat. *Proc R Soc Lond B* 159:257–282, 1964.
27. Umetani Y. Postcapillary venule in rabbit tonsil and entry of lymphocytes into its endothelium: A scanning electron microscope study. *Arch Histol Jpn* 40:77–94, 1977.
28. Cho Y, De Bruyn PPH. The endothelial structure of the postcapillary venules of the lymph node and the passage of lymphocytes across the venule wall. *J Ultrastruc Res* 69:13–21, 1979.
29. Maros T. Data regarding the typology and functional significance of the venous valves. *Morphol Embryol* 27:195–214, 1981.
30. Hossler FE, West RF. Venous valve anatomy and morphometry: Studies on the duckling using vascular corrosion casting. *Am J Anat* 181:425–432, 1988.
31. Moffat DB. A regulatory mechanism in the posterior ciliary arteries of the dog. *Nature* 169:1015–1016, 1952.
32. Velican C, Velican D. Studies of human coronary arteries. *Acta Anat* 99:337–385, 1977.
33. Moffat DB. An intra-arterial regulating mechanism in the uterine artery of the rat. *Anat Rec* 134:107–123, 1959.
34. Kardon RH, Farley DB, Heidger PM Jr., Van Orden DE. Intraarterial cushion of the rat uterine artery: A scanning electron microscope evaluation utilizing vascular casts. *Anat Rec* 203:19–29, 1982.
35. Casella D, Dupont M, Jover B, Mimran A. Scanning electron microscopic study of arterial cushions in rats: A novel application of the corrosion-replication technique. *Anat Rec* 203:419–428, 1982.
36. Grant RT, Bland EF. Observations on arteriovenous anastomoses in human skin and in the bird's foot with special reference to cold. *Heart* 15:385–411, 1930.
37. Grant RT. Observations on direct communications between arteries and veins in the rabbit's ear. *Heart* 15:281–303, 1930.
38. Clark ER, Clark EL. Observations on living arteriovenous anastomoses as seen in transparent chambers introduced into the rabbit's ear. *Am J Anat* 54:229–286, 1934.
39. Hales JRS, Fawcett AA, Bennett JW, Needham AD. Thermal control of bloodflow through capillaries and arteriovenous anastomoses in skin of sheep. *Pflügers Arch* 378:55–63, 1978.
40. Morris JL, Bevan RD. Development of the vascular bed in the rabbit ear: Scanning electron microscopy of vascular corrosion casts. *Am J Anat* 171:75–89, 1984.
41. Kishi Y, So S, Harada Y, Takahashi K. Three-dimensional SEM study of arteriovenous anastomoses in the dog's tongue using corrosive resin casts. *Acta Anat* 132:17–27, 1988.
42. Ohtani O. Corrosion casts in liver and stomach microcirculation. In: Motta PM (ed.), *Cells and Tissues: A Three-Dimensional Approach by Modern Techniques in Microscopy*. Alan R. Liss, New York, pp 317–326, 1989.
43. Levesque MJ, Cornhill JF, Nerem RM. Vascular casting — a new method for the study of the arterial endothelium. *Atherosclerosis* 34:457–467, 1979.
44. Burger PC, Chandler DB, Klintworth GK. Corneal neovascularization as studied by scanning electron microscopy of vascular casts. *Lab Invest* 48:169–180, 1983.
45. Olsen KR. Application of corrosion casting procedures in identification of perfusion distribution in a complex microvasculature. *Scann Electron Microsc* III:357–364, 1980.
46. Rieke GK. Thalamic arterial pattern: An endocast and scanning electron microscopic study in normotensive rat. *Am J Anat* 178:45–54, 1987.
47. Rogers PAW, Gannon BJ. The microvascular cast of a three-dimensional tissue skeleton: Visualization of rapid morphological changes in tissues of the rat uterus. *J Microsc* 131:241–247, 1983.
48. Weber DF. An improved technique for producing casts of the internal structure of hard tissues, including some observations on human dentis. *Arch Oral Biol* 28:885–891, 1983.
49. Meyer EP. Corrosion casts as a method for investigation of the insect tracheal system. *Cell Tissue Res* 256:1–6, 1989.
50. Haefeli-Bleuer B, Weibel ER. Morphometry of the human pulmonary acinus. *Anat Rec* 220:401–414, 1988.

Authors' addresses:
Prof. Osamu Ohtani
Department of Anatomy
Faculty of Medicine
Toyama Medical and Pharmaceutical University
2630 Sugitani, Toyama 930-01 Japan
and
Prof. Takuro Murakami
Department of Anatomy
Okayama University
School of Medicine
Shikata-Cho 2-5-1
Okayama 700, Japan

CHAPTER 3

Microvascular Corrosion Casting in Angiogenesis Research

ROLF H. CHRISTOFFERSON & B. OVE NILSSON

1. Introduction

Angiogenesis, i.e., the formation of new blood vessels, is a crucial phenomenon in human embryogenesis and placentation, and in the development of certain diseases. The process of angiogenesis is often investigated in essentially two-dimensional tissues (e.g., tadpole tails, chick chorioallantoic membranes, rat corneas) in experimental animals. By infusion of a liquid resin that solidifies in the microcirculation, three-dimensional vascular casts of parenchymatous organs during angiogenesis can be produced. Microvascular casts can also be obtained from surgically resected human organs. When observed in a scanning electron microscope, microvascular casts permit the identification of individual vessels, such as arterioles, capillaries, or venules; vascular sprouts can be identified with certainty, and in series of casts the establishment and even regression of new blood vessels can be observed. Microvascular casting in angiogenesis research is particularly suitable for studying vascular growth in organs whose interior is difficult to expose in vivo. The major disadvantages of the casting technique are (1) that it can only give information on the microvascular anatomy at a given instant, i.e., at the moment of casting and (2) that it does not yet permit quantification of the angiogenesis observed, for reasons of incompleteness of preparations, irreproducibility of cast sectioning, complexity of the specimen surface topography, and unevenness of specimen conductivity. It seems hopeful that recent developments in stereology and computer-aided image analysis will lead to algorithms for the quantification of vascular density and distribution, and of angiogenesis.

2. Definition of Angiogenesis

Angiogenesis means the formation of new blood vessels. The term was coined by Hertig in 1935 during studies of the development of fetal blood vessels in the human chorion and of the placenta in the macaque monkey [1]. Two types of angiogenesis have been described, namely, angiogenesis *in situ* [2–4], when new capillaries are formed by the coalescence of blood islands, i.e., angioblastic mesenchyma, and angiogenesis by *sprouting*, when new capillaries are formed as initially blindly ending sprouts from preexisting capillaries and small venules, sprouts that later connect [5–8]. This presentation will focus upon the sprouting type of angiogenesis, which seems to be the predominant type in mature tissues. For an extensive description of the sprouting process, the reader is referred to the review by Rhodin and Fujita [8].

3. The Role of Angiogenesis in Human Embryogenesis, Placentation, and Disease

Under physiological conditions, angiogenesis is strictly limited in time and extent to embryonic

---

This work is dedicated to the memory of Professor Andrew W. Rogers, University of Sheffield, UK, who died in July 1989.

Motta, P.M., Murakami, T., and Fujita, H. (eds.), Scanning Electron Microscopy of Vascular Casts: Methods and Applications.

28

morphogenesis and growth, rebuilding of the endometrium during the proliferative phase of the ovarian cycle in menstruating species, formation of the corpus luteum in mammals, formation of the placenta in eutherian species, wound healing, and adaptation to physical exercise [9,10]. The capacity for physiological angiogenesis (successful placentation, embryological development and growth, wound healing, and adaptation to stress and a high-altitude habitat) has probably been a Darwinian advantage for the individual and its species during evolution. Under this assumption, it is tempting to propose four hypotheses concerning the process of new blood-vessel formation, hypotheses that are partly interdependent: (1) Angiogenesis has a high level of biochemical redundancy [11], i.e., several mediators of different cellular origin are involved; (2) since angiogenesis is strictly limited in time and extent but highly efficient when triggered, it is a "cascade reaction" involving several activating and inhibiting steps [12], analogous to blood clotting; (3) the mediators of angiogenesis are similar or related in the animal series, which means that the use of animal experimental models is valid; and (4) angiogenesis is simpler to inhibit specifically than to induce specifically, i.e., in order to mimic physiological angiogenesis several mediators have to be administered in a proper sequence and dosage at the right time in the target compartment, while specific inhibition can be induced by blocking one limiting step.

On the other hand, both overshooting angiogenesis and insufficient angiogenesis are often seen under pathological conditions. It is not known whether excessive or a lack of angiogenesis constitute the pathogenesis in such diseases, but it is likely that this is partly responsible for the clinical picture, either primarily or secondarily [13]. Some disease entities in which excessive angiogenesis have been reported are atherosclerosis, solid cancers, diabetic retinopathy, psoriasis, and rheumatoid arthritis, while insufficient angiogenesis is a characteristic feature in chronic leg ulcers, nonhealing fractures, and developmental anomalies, such as hemifacial microsomia [13]. Several more diseases can probably be added to these groups. There is clearly a therapeutic potential in drugs that are able to modify angiogenesis, either by inhibition or by stimulation.

4. Techniques Employed in Investigations of Angiogenesis

In attempts to identify mediators involved in angiogenesis, animal experimental models, and tissue culture and cell culture techniques have been developed, and various methods of analysis have been tried. Among the different experimental models developed, the chick chorioallantoic membrane assay, the rat corneal pocket assay, the rat mesentery assay, and the culture of endothelial cells may be mentioned; and methods of analysis have included intravital microscopy, quantitative light microscopy, scanning and transmission electron microscopy, and autoradiography (for references, see Hudlicka and Tyler [9]). The three major problems encountered in this development have concerned (1) the specificity and sensitivity of the experimental models, (2) the specificity and sensitivity of the methods of analysis, and (3) the quantification of the angiogenesis observed. At present it is doubtful whether there is any one combination of model and method that permits reproducible quantification of angiogenesis alone. However, comparisons between different models and methods have increased our knowledge of the mediators involved in the formation of new blood vessels. Several factors have been shown to be angiogenic in at least two different models, namely, angiogenin, tumor angiogenesis factor, acidic and basic fibroblast growth factor, transforming growth factors α and β, tumor necrosis factor α, and seleniomethionine [9–11,13,14]. Specific and powerful inhibition of angiogenesis has been observed in two different angiogenic models following treatment with a combination of heparin and a steroid, U42129, lacking mineralocorticoid and glucocorticoid activities [10,15].

The chick chorioallantoic membrane assay is probably the most widely used model for demonstrating angiogenic or angiostatic activity. The method of analysis consists in subjective appreciation in a stereomicroscope (magnification × 7–10) of the vascular density of the fetal membrane in the area of deposition of the substance tested. Since this is a reference method, its specificity and sensitivity cannot be determined [16]. There are at least five theoretical arguments against the chorioallantoic membrane assay as a reference method in angiogenesis research. First, the membrane itself is vascularized with an extensive capillary bed [17], which undergoes spontaneous angiogenesis during incubation [18], thereby reducing the specificity and sensitivity of the assay. Secondly, the analysis is performed at a magnification at which individual capillaries cannot be discriminated, reducing the sensitivity of the method to congestion or hyperemia of preexisting arterioles and venules. The fact that it is not the sprouting process per se that is observed may explain the resigned view that "... the observed subjective vascular responses do correlate to some extent with new vessel formation" [16]. Thirdly, the vascular pattern that is considered to be pathognomic for angiogenesis can be elicited not only by angiogenic substances, but also by unspecific inflammation and by thermal, chemical, and mechanical trauma [16], thereby reducing the specificity of the assay. Fourthly, the membrane is essentially two-dimensional, which means that little information is obtained on how new vessels extend, merge, differentiate, and regress in three dimensions.

Regarding the methods of analysis of angiogenesis, some sort of microscopical analysis is necessary for the identification of vessels that are budding sprouts (parent vessels) and of elongating sprouts, since the microcirculation consists of vessels less than $100\,\mu m$ in diameter. Stereomicroscopes or intravital microscopes offer a possibility of examining the dynamic process of sprouting in vivo, but are restricted to flat preparations or the surfaces of parenchymatous organs. Light microscopy of sectioned material permits the identification of sprouts in serial sections only (what may appear as a sprout in one section often proves to be an obliquely sectioned capillary in the next section). Transmission electron microscopy (TEM) is excellent for demonstrating vascular sprouts, but this requires serial sections [8]. One way to increase the specificity of TEM for visualizing vascular sprouts is to perform initial light microscopy on thick ($50\,\mu m$) vibratome sections stained for horseradish peroxidase, and to cut ultrathin sections of selected sprouts from these and process them for TEM [19]. Despite its high specificity and sensitivity, TEM is probably too cumbersome a technique for screening the activity of the proposed angiogenic factors. Light microscopy of organs injected intravascularly with India ink, with subsequent immersion of the organs in oil of wintergreen, for example, to make them transparent, permits three-dimensional analysis of the vascular tree; but what may be interpreted as sprouts may in fact be incompletely injected capillaries. Also, continuous adjustments of the focal plane are required to establish the origin and direction of probable sprouts.

5. Microvascular Corrosion Casting in Angiogenesis Research

The above objections imply a need for more specific, sensitive, and quantitative methods for detecting new blood-vessel formation. One way to increase the specificity and sensitivity in angiogenesis assays is to cast the microcirculation of an organ subjected to a physiological or pathological angiogenic stimulus and to examine the cast in an electron microscope. By this means budding of new vascular sprouts, elongation and merging of sprouts, and the organization and regression of newly formed vessels can be observed at high magnification. This method is called *microvascular corrosion casting*, and the morphological analysis is performed in a scanning electron microscope (SEM) [20,21].

Our experience with this method of morphological analysis is based on angiogenesis in four different situations: (1) during normal placentation in the rat uterus [22,23], (2) during growth of a human cancer in the rat uterus [24], (3) during subcutaneous growth of a human cancer in the rat hindleg [25], and (4) following cerebral infarction and transplantation of cortical neurones from rat embryos in the adult rat brain [Grabowski et al., in preparation]. Also, we have analyzed the use of different preparational techniques with this method [21]. It is beyond the scope of this review article to describe the designs of the above animal experimental models, but the casting procedure we have used throughout is as follows:

The rats are anesthetized with 6 mg sodium pentothal (Mebumal®, ACO, Sweden) per 100 mg body weight given intraperitoneally. An arterial cannula is inserted either in the femoral artery (when casting the rat uterus or hindleg) or the brachial artery (when casting the rat brain) and connected to a pressure transducer and a recorder, calibrated to record intraarterial pressures of between -15 and $200-300$ mmHg. Heparin, 300 IU in 1 ml saline, is given through the arterial cannula prior to blood wash-out to avoid intravascular coagulation. This precaution makes it possible to produce perfect casts, even if the animal dies before injection of the resin, or if the resin cannula is displaced during casting.

The thoracic cavity is opened and a cannula is inserted into the thoracic aorta (when casting the rat uterus or hindleg) or the ascending aorta (when casting the rat brain). The right atrium of the heart is incised to permit free escape of the venous blood. Blood wash-out is then performed with a wash-out medium consisting of $200-400$ ml Dulbecco's phosphate-buffered saline with 10,000 IU heparin/l, 4 mg papaverine (ACO)/l, and 58.7 g polyvinyl pyrrolidone (molecular weight approx. 40,000; Fluka AG, Buchs, Switzerland), pH $7.3-7.4$, 300 mOsm/l and heated to 39°C to ensure maximal vasodilation. Blood wash-out is performed at an intra-arterial pressure of $80-120$ mmHg.

The base resin consists of 60 ml of methyl methacrylate (Fluka), stabilized by the manu-facturer with 25 ppm hydroquinone. It is import-ant to note that the following procedure is only suitable for methacrylate preparations with 25 ppm hydroquinone; more hydroquinone inter-feres with the polymerization of the methacrylate. Then, 1.2 g dibenzoyl peroxide (Fluka) is dis-solved in the methacrylate, and the solution is prepolymerized with UV light to a viscosity of $2-4$ cS [26] (water has a viscosity of 1 cS and blood $2-3$ cS). During blood washout, 1.2 ml n, n-dimethyl aniline (Fluka) is added to the base resin to accelerate polymerization. The casting medium is infused at an intraarterial pressure of $70-120$ mmHg.

After gelling of the resin, the organs of interest are removed and sectioned with razor blades. Apart from vascular casts, conventional light microscopy of histological sections is performed

Table 3-1. Criteria for the ideal casting medium

The ideal casting medium should:
1. Be nontoxic both for the investigator and for the system cast
2. Be of sufficiently low viscosity or particle size to pass through tubules as narrow as 5 μm
3. Be physiologically inert in the system cast
4. Polymerize within $3-15$ minutes after activation
5. Not shrink during polymerization
6. Permit microdissection with the surrounding tissue intact
7. Be resistant to corrosion procedures
8. Be visible in the dissection microscope after corrosion
9. Retain its original configuration during drying
10. Be electron conductive
11. Be resistive to electron bombardment
12. Replicate all topographical detail on the luminal surface of the endothelium
13. Indicate the direction of flow in the system cast

Adapted from refs. 29 and $40-42$.

on cast, uncorroded specimens from all groups. The casts are corroded by "differential cor-rosion," [21] involving acid and basic hydrolysis and also mechanical removal of tissue. The remaining casts are then either sectioned in small blocks of ice with a razor blade or dried directly in air from distilled water, mounted on SEM stubs with silver paint, sputter coated with a $400-1000$ Å gold layer, and observed in a Philips 525 scanning electron microscope connected to a Link AN 10,000 computer and a Crystal image pro-cessor, at acceleration voltages of $5-15$ kV.

For the purpose of comparison, Table 3-1 lists the criteria for an ideal medium for vascular casting, and Table 3-2 shows the advantages and disadvantages of the casting procedure described above over light microscopy of histological sec-tions and casting by manual injection of com-mercially available resins.

In our experience, the advantages of micro-vascular corrosion casting with SEM analysis over intravital microscopy and light microscopy of histological sections are as follows:
- Vascular sprouts can be readily identified and discriminated from preexisting vessels [21,23, 24,27,28]. Sprouts appear as elongated, sharp-pointed, blindly ending extensions from the parent vessel. If the sprout has imprints of endothelial cell nuclei, it is unlikely to rep-resent an incompletely cast capillary [21] (Figs.

Table 3-2. Advantages and disadvantages of the applied casting technique

1. Advantages over light microscopy:
   - Three-dimensional representation
   - Specimens can be moved in all degress of freedom, including tilt and rotation
   - Rather large, intact, or transected specimens can be observed
   - High resolution power
   - Large depth of focus
   - Sprouts can be identified with certainty
2. Advantages over other casting techniques:
   - Better degree of filling of the microcirculation
   - Higher quality of endothelial cell replication
   - Tips of sprouts are replicated
   - Artifacts such as extravasations and incomplete corrosion have been minimized
3. Disadvantages compared with light microscopy:
   - Minimal information from the tissue surrounding the vessels
   - Only vessels less than 200–300 μm deep to the surface of the cast can be observed, while serial sections can easily be made for light microscopy
   - Cumbersome technique with several preparational steps
4. Disadvantages compared with other casting techniques:
   - Large (>100 μm) vessels often incompletely cast and distorted
   - Difficult to dissect because of lack of plasticizer
   - Transparent casting medium makes ocular appreciation of the degree of filling impossible during casting
   - Casting medium is not commercially available

3-1a, 3-1b, and 3-2). Preexisting vessels are often easily separated from newly formed ones, since each tissue exhibits its own, stereotypic angioarchitecture (see Lametschwandtner et al. [29] for a review). Detailed examinations of blindly ending vascular casts from non-angiogenic areas have shown either rounded tips (indicating incomplete vascular filling) or broken tips (indicating that they have been cut during sectioning or broken during specimen handling) [21].

- Vessels giving off sprouts (parent vessel) can be identified in terms of type of vessel (arteriolar, capillary, or venular), diameter, and location. Analysis of the branching pattern and imprints of the endothelial cell nuclei has revealed that all sprouts observed emanate from capillaries or small venules [21,24,27,28].
- The direction of budding and elongating sprouts in relation to the angiogenic stimulus can immediately be assessed [23,24,27,28].

- The three-dimensional arrangement of the newly forming capillary network can be observed [23,24,27,28] (Figs. 3-3a and 3-3b).
- The differentiation of newly formed vessels into arterioles and venules, and regression of newly formed vessels, can be observed [23,24, 27,28] (Fig. 3-4).
- Vascular sprouting can be studied in three dimensions within parenchymatous organs, and not only in flat preparations of organ surfaces [23,24].
- The probable flow within cast vessels, including capillaries, can be assessed by the shape of the imprints of endothelial-cell nuclei [21,30] (Fig. 3-5).

There are basically four disadvantages of the casting technique: (1) only information from the luminal surface of the vascular endothelium can be obtained, so that the cast findings must be correlated with the histological appearance of the tissue; (2) the preparation technique is cumbersome and is not a routine procedure, as, for example, in the preparation of histological section; (3) one cast only gives an instant representation of the microcirculation at the moment of gelling of the resin, necessitating the use of several preparations for following dynamic processes, such as angiogenesis; and (4) the angiogenesis observed cannot be quantified, but can only be semiquantified into absent, present, and redundant.

6. The Morphology of Angiogenesis

For reference, microvascular corrosion casting has been applied in several angiogenetic studies [e.g., 23,24,27,28,31–35]. Microvascular casting of the chick chorioallantoic membrane has been performed by Fuchs and Lindenbaum [17], and by Burton and Palmer [36].

Five steps in angiogenesis by sprouting have been observed in microvascular casts in the SEM, which are in general accordance with light microscopical and TEM observations [5–9]:

1. Sprouting from a parent capillary or small venule, the tips of the sprouts facing the angiogenic stimulus. Sprouts most often arise from the convexity of a parent vessel curvature [23,24,27,28].

32

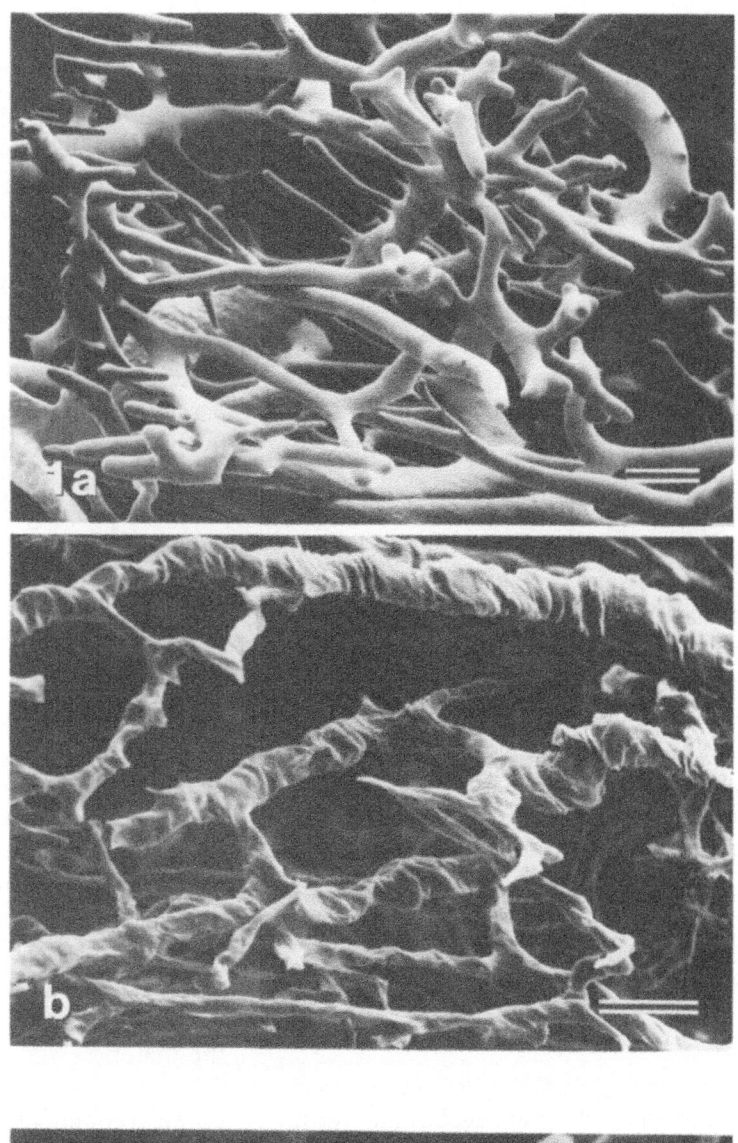

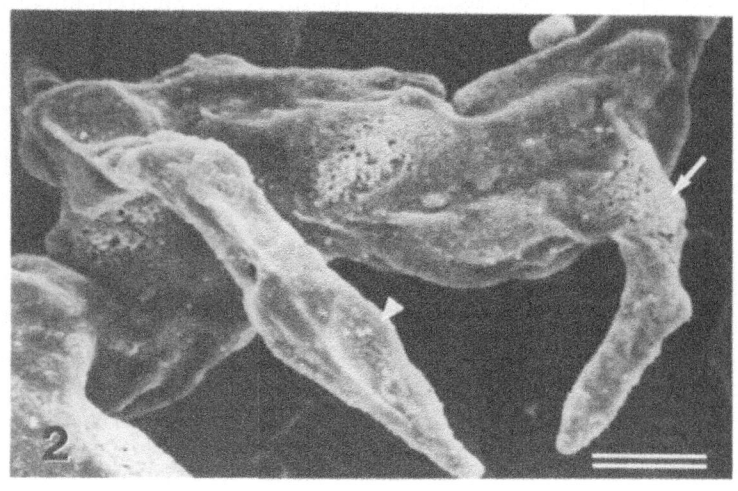

2. Elongation of vascular sprouts [23,24,27,28], most frequently towards each other, but also towards the initial stimulus [24].
3. Fusion of two or more elongated sprouts [23,24,27,28].
4. Differentiation of some newly formed vessels into afferent arterioles and efferent venules with an interposed capillary network [23,24, 27,28], giving rise not only to an anastomosing network of capillaries, but also to a functional microcirculation [23,24].
5. Regression of newly formed vessels, mainly capillaries and venules, if the angiogenic stimulus has ceased [27,28].

On comparing a physiological example of angiogenesis [23] with a pathological one [24], it was noted that the latter was continuous, i.e., not limited in time and extent, and less well organized, giving rise to extensive anastomoses between vessels of similar caliber with "nonsense loops" [25]. Also, the angioarchitecture of the newly formed vascular bed seems to correlate with the normal angioarchitecture of the tissue involved; i.e., endometrial angiogenesis gave rise to placental or endometrial vascular beds [23], while adenocarcinomatous angiogenesis partly mimicked the vascular bed of the original tissue [24,25].

7. Vascular Casting of Resected Human Specimens

Angiogenic research is a young discipline. Physiological angiogenesis in different tissues has yet to be established, and numerous putatively stimulating and inhibiting factors need to be assayed in different experimental models. The relevance of such experimental models must, in turn, be determined by correlative studies of human material — normal, pathological, and, if possible, experimental. In view of the additional information that can be obtained by microvascular corrosion casting compared with conventional methods of analysis, it is important that casts be made from human, resected specimens. Technically excellent human microvascular casts have been made by some authors [e.g., 37–39], but in our experience (based on casting of human uteri after hysterectomy), it is more difficult to obtain technically satisfactory casts from resected human specimens than from laboratory animals. The reason for this is not clearly understood, but a few possible complicating factors are (1) the comparatively larger size of human organs, demanding larger volumes of washout medium and casting medium, prolonging and impairing the corrosion, and involving more traumatization during specimen handling; (2) the difficulty in obtaining one afferent and one efferent vessel for infusion and draining of the casting medium, respectively — casting medium may escape from cut surfaces, in which case no proper infusion pressure is built up; and (3) the risk of vasospasm and intravascular coagulation during the interval between resection of the specimen and casting.

To facilitate the correlation between laboratory specimens and human specimens, we suggest that scientists reporting on casting of human specimens (1) cast as many specimens as possible in order to build up experience and to test alternative preparations; (2) pay special attention to the materials and methods section in the report, so that the procedure is easy to repeat; (3) incorporate a comment on the "success rate" of casting in the results section; and (4) present low-power micrographs in the results section, so that the degree of vascular filling in terms of the retainment of the gross anatomical outline of the organ can be appreciated.

←

Figure 3-1. SEM of endometrial microvascular casts. (a) Blind endings due to incomplete filling. Note the blunt blind endings and the absence of endothelial-cell nuclear imprints. Human uterus, late proliferative phase. Bar = 50 μm. (b) Blind endings due to angiogenesis. Note the sharp-pointed blind endings and the presence of endothelial-cell nuclear imprints. Nude rat uterus, 7 days after deposition of tumor cells. Animal treated with progesterone and estrogen. Bar = 50 μm.
Figure 3-2. SEM of nude rat uterine cast, 14 days after tumor deposition. The animal was treated with progesterone and estrogen. Two vascular sprouts are facing each other. This pattern was often observed and is interpreted as a sign of future merging between the sprouts. The center of the tumor is located in the bottom of the micrograph. Some endothelial-cell nuclei have not been removed from the parent venule during corrosion and remain on the surface of the cast (arrow); others have given rise to prominent imprints (arrowhead). Bar = 10 μm.

34

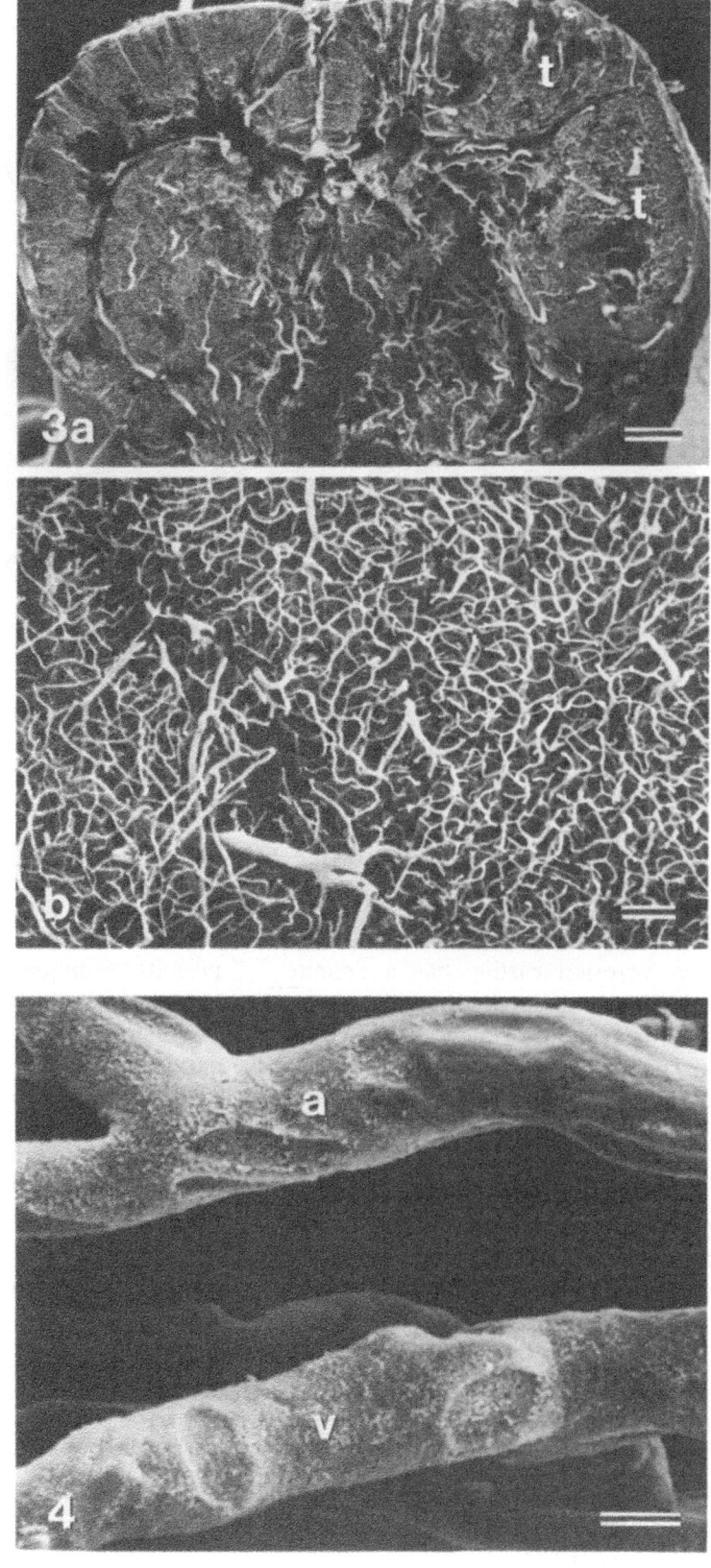

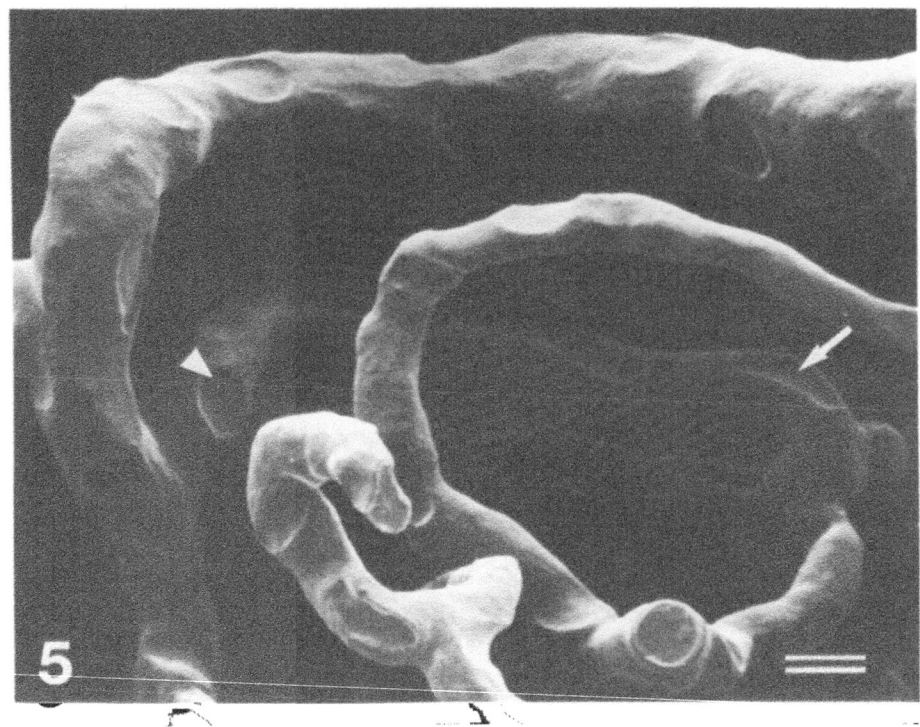

Figure 3-5. SEM of rat uterine cast. The origin and termination of a short longitudinal muscle capillary can be seen in the background. The arterial end of the capillary has the elongated, arteriolar type of imprint from an endothelial cell nucleus (arrow), while its venous end exhibits the rounded, venular type (arrowhead), suggesting that the flow within the capillary is from right to left. Bar = 10 μm.

8. Future Outlook

Microvascular corrosion casting has a definite place in angiogenesis research when the three-dimensional organization of new blood-vessel formation in parenchymatous organs is to be investigated. When applied with proper care, the technique is likely to have high sensitivity and specificity for angiogenesis. Because of the complexity of angiogenesis, the casting technique must be used in combination with other methods, since casting (1) only gives information on the microvascular morphology at the moment of cast-ing, (2) only gives information on the luminal surface of the vascular endothelium, and (3) only permits semiquantitative appreciation of the extent of new blood-vessel formation.

Quantification of angiogenesis would facilitate characterization of physiological angiogenesis, identification of inducers and inhibitors of the process, and communication between scientists in different fields of angiogenesis research. Use of the casting technique for such quantification offers several challenges, however, and a number of sources of error have to be taken into account. These include incompletely cast specimens,

Figure 3-3. SEM of brain cast of a spontaneously hypertensive rat, 3 months after ligation of the right medial cerebral artery followed by transplantation of isogenic, fetal cortical neurones 7 days later. (a) Overview, indicating the presence of vascularized transplants (t) in the right hemisphere. Bar = 1 mm. (b) Angioarchitecture of the transplant, exhibiting a complex microcirculation connected to arterioles and venules limiting the infarction cavity. Bar = 100 μm.
Figure 3-4. SEM of nude rat uterine cast, 14 days after tumor deposition. Animal treated with progesterone. In this area, corresponding to a capillary plexus in the intact endometrium, one arteriole (a) and one venule (v) are seen within the tumor margin. These vessels represent newly formed, differentiated vessels, as shown by the typical imprints of endothelial-cell nuclei (arrows). Bar = 10 μm.

damage to specimens during preparation for SEM, irreproducible specimen sectioning, complicated specimen surface topography, and uneven specimen conductivity; angular errors may also occur in the measurement of areas and volumes in the SEM.

It is our belief that such random and systematic errors can be overcome by focusing upon one experimental model and systematically documentating an optimal casting procedure and the most efficient preparation for SEM, and recording the normal vascular densities, lengths, and diameters. An angiogenic situation can then be introduced into the system, and the extent of new blood-vessel formation can be calculated in the light of the known parameters. From such a meticulously documented model, knowledge can probably be extrapolated to other models and organs. The recent developments in stereology and computer-aided image analysis offer powerful tools that can facilitate the cumbersome collection of data from specimens in the SEM [E. Weibel, personal communication, 1989]. For a given experimental model with strictly controlled procedures, it is likely that a system of algorithms for quantification of vascular density, vascular dimensions, and angiogenesis can be produced.

Acknowledgments

Excellent technical assistance was provided by Barbro Einarsson, Leif Ljung, Marianne Ljungkvist, and Lena Kårud. This work was supported by grants from the Swedish Medical Research Council, the Faculty of Medicine of Uppsala University, and the Swedish Society of Medicine.

References

1. Hertig AT. Angiogenesis in the early human chorion and in the primary placenta of the macaque monkey. *Carnegie Contri Embryol* 25:37–81, 1935.
2. Reagan FP. Vascularization phenomena in fragments of embryonic bodies completely isolated from yolk-sac blastoderm. *Anat Rec* 9:329–341, 1915.
3. Poole TJ, Coffin JD. Developmental angiogenesis: Quail embryonic vasculature. *Scanning Microsc* 2:443–448, 1988.
4. Risau W, Lemmon V. Changes in the vascular extracellular matrix during embryonic vasculogenesis and angiogenesis. *Develop Biol* 125:441–450, 1988.
5. Clark ER. Studies on the growth of blood-vessels in the tail of the frog larva — by observation and experiment on the living animal. *Am J Anat* 23:37–88, 1918.
6. Shoefl GI. Studies on inflammation. III. Growing capillaries: Their structure and permeability. *Virchows Archiv Pathol Anat* 337:97–141, 1963.
7. Ausprunk DH. Tumor angiogenesis. In: Houck JC (ed.), *Chemical Messengers of the Inflammatory Process*, Elsevier/North-Holland, Amsterdam, pp 317–351, 1979.
8. Rhodin JAG, Fujita H. Capillary growth in the mesentery of normal young rats. Intravital video and electron microscope analyses. *J Submicrosc Cytol Pathol* 21:1–34, 1989.
9. Hudlickà O, Tyler KR. *Angiogenesis: The Growth of the Vascular System*. Academic Press, London, 1986.
10. Folkman J. How is blood vessel growth regulated in normal and neoplastic tissue? G.H.A. Clowes Memorial Award Lecture. *Cancer Res* 46:467–473, 1986.
11. Folkman J, Klagsbrun M. A family of angiogenic peptides. *Nature* 239:671–672, 1987.
12. Millaway DS, Redmer DA, Kirsch JD, Anthony RV, and Reynolds LP. Angiogenic activity of maternal and fetal placental tissues of ewes throughout gestation. *J Reprod Fertil* 86:689–696, 1989.
13. Folkman J, Klagsbrun M. Angiogenic factors, *Science* 235:442–447, 1987.
14. McAuslan BR, Reilley W. Selenium-induced cell migration and proliferation: Relevance to angiogenesis and microangiopathy. *Microvasc Res* 32:112–120, 1986.
15. Ingber DE, Madri JA, Folkman J. A possible mechanism for inhibition of angiogenesis by angiostatic steroids: Induction of capillary basement membrane dissolution. *Endocrinology* 119:1768–1775, 1986.
16. Vu MT, Smith CF, Burger PC, Klintworth GK. Methods in laboratory investigation: An evaluation of methods to quantitate the chick chorioallantoic membrane assay in angiogenesis. *Labo Invest* 53:499–508, 1985.
17. Fuchs A, Lindenbaum ES. The two- and three-dimensional structure of the microcirculation of the chick chorioallantoic membrane. *Acta Anatom* 131:271–275, 1988.
18. DeFouw DO, Rizzo VJ, Steinfeld R, Feinberg RN. Mapping of the microcirculation in the chick chorioallantoic membrane during normal angiogenesis. *Microvasc Res* 38:136–147, 1989.
19. Mato M, Ookawara S, Namiki T. Studies on the vasculogenesis in rat cerebral cortex. *Anatom Rec* 224:355–364, 1989.
20. Murakami T. Application of the scanning electron microscope to the study of the fine distribution of the blood vessels. *Archiv Histol Jpn* 32:445–454, 1971.
21. Christofferson RH, Nilsson BO. Microvascular corrosion casting with analysis in the scanning electron microscope. *Scanning* 10:43–63, 1988.
22. Christofferson RH, Nilsson BO. Morphology of the endometrial microvasculature during early placentation in the rat. *Cell Tissue Res* 253:209–220, 1988.

23. Christofferson RH, Nilsson BO. Placentation in the rat: A SEM study of microvascular casts. In: Motta PM (ed.), *Development in Ultrastructure of Reproduction*, Prog in Clini and Biolo Res 296:435–442, 1989.

24. Christofferson RH, Nilsson BO. Angiogenesis in a human cancer growing in the nude rat endometrium. In: Christofferson RH (ed.), Angiogenesis as Induced by Trophoblast and Cancer Cells. PhD thesis, University of Uppsala, Uppsala, pp 73–81, 1988.

25. Ahlström H, Christofferson RH, Lörelius L-E. Vascularization of the continuous human colonic cancer cell line LS 174 T subcutaneously deposited in nude rats. *Acta Pathol Microbiolog Immunolog Scandin* 96:701–710, 1988.

26. Gannon BJ. Preparation of microvascular corrosion casting media: Procedure for partial polymerization using ultraviolet light. *Biomed Res* 2:227–233, 1981.

27. Burger PC, Chandler DB, Klintworth GK. Corneal neovascularization as studied by scanning electron microscopy of vascular casts. *Lab Invest* 48:169–180, 1983.

28. Burger PC, Chandler DB, Drysdale DB, Tano Y, Crapo JD, Freeman BA, Klintworth GK. Scanning electron microscopy of vascular casts in experimental ocular vasoproliferation. *Scanning Electron Microscopy/1984/IV*:1893–1898, 1984.

29. Lametschwandtner A, Lametschwandtner U, Weiger T. Scanning electron microscopy of vascular corrosion casts — technique and applications. *Scanning Electron Microscopy 1984/II*:663–695, 1984.

30. Nerem RM, Levesque MJ, Cornhill JF. Vascular endothelial morphology as an indicator of the pattern of blood flow. *J Biomech Eng* 103:172–176, 1981.

31. Tatematsu M, Cohen SM, Fukushima S, Shirai T, Shinohara Y, Ito N. Neovascularization in benign and malignant urinary bladder epithelial proliferative lesions of the rat observed in situ by scanning electron microscopy and autoradiography. *Cancer Res* 38:1792–1800, 1978.

32. Tano Y, Chandler DB, Machemer R. Vascular casts of experimental retinal neovascularization. *Am J Ophthalmol* 92:110–120, 1981.

33. Ohkuma H, Ryan SJ. Vascular casts of experimental subretinal neovascularization in monkeys. *Invest Ophthalmol Vis Sci* 24:481–490, 1983.

34. Grunt TW, Lametschwandtner A, Karrer K, Staindl O. The characteristic structural features of the blood vessels of the Lewis lung carcinoma. *Scann Electron Microsc 1986/II*:575–589, 1986.

35. Spanel-Borowski K. Amselgruber W, Sinowatz F. Capillary sprouts in ovaries of immature superstimulated golden hamsters: A SEM study of microcorrosion casts. *Anat Embryol* 176:387–391, 1987.

36. Burton GJ, Palmer ME. The chorioallantoic capillary plexus of the chicken egg: A microvascular corrosion casting study. *Scanning Microscc* 3:549–558, 1989.

37. Miodónski A, Kús J, Olszewski E, Tyrankiewicz R. Scanning electron microscopic studies on blood vessels in cancer of the larynx. *Arch Otolaryngol* 106:321–332, 1980.

38. Duvernoy HM, Delon S, Vannson JL. Cortical blood vessels of the human brain. *Brain Res Bull* 7:519–579, 1981.

39. Bugajski A, Nowogrodzka-Zagórska M, Lenko J, Miodónski AJ. Angiomorphology of the human renal clear cell carcinoma. *Virchows Archiv A Pathol Anat* 415:103–113, 1989.

40. Nowell JA, Lohse CL. Injection replication of the microvasculature for SEM. Scann Electron Microsc 1974/I:267–274, 1974.

41. Gannon BJ. Vascular casting. In: Hayat MA, (ed.), *Principles and Techniques of Scanning Electron Microscopy, Vol. 6: Biological Applications*. Van Nostrand Reinhold Co, New York, pp 170–193, 1978.

42. Hodde KC, Nowell JA. SEM of micro-corrosion casts. *Scann Electron Microsc 1980/II*:88–106, 1980.

Author's address:
Dr. Rolf H. Christofferson, MD, PhD
Department of Human Anatomy
Biomedical Center
Box 571
S-751 23 Uppsala
Sweden

CHAPTER 4

Corrosion Cast Technique Applied to Lymphatic Pathways

ANTON CASTENHOLZ

1. Introduction

Although vascular casting proves to be a well-established method in SEM and has been applied in the past to nearly all blood vessel systems of many animals and humans, the number of studies devoting to the lymphatic system is still rather scarce. This mainly depends upon the fact that the lymphatics in general have not been given as much attention to the morphology and related disciplines as have the blood vessels. Moreover, some sections of the lymphatic system, such as the initial part or its intricate intranodal pathways, need specific techniques for their representation in cast specimens. This is due to the fact that the lymphatic vessels do not form a closed circulatory system, like that of the blood vessels, but begin "blindly" with small rootlets everywhere in the tissue and are proximally connected with the venous side of the blood circulation. Due to several lymph-node stations interposed into the lymphatic pathways, the luminal width along a main drainage system repeatedly changes from a large caliber to a very minute one. All these facts make corrosion casting of lymphatic structures as a measure more than a routine procedure. Undoubtedly, whatever technique is employed in this connection, it demands in every case careful attention to the specific properties of the vascular area to be studied.

Nevertheless, filling the lymphatic pathways in various organs with different kinds of liquids, masses, dyes, and resins was attempted for quite a long time before the era of SEM began, and can be dated back to the time of the discovery of the lymphatics three centuries ago. The techniques applied were based either on direct injection of the medium into the vessel's lumen or, in the case of indirect filling, on interstitial injections of the medium into the connective tissue. These two possibilites can also be used in modern-day studies of SEM, although most investigators prefer to use indirect casting of the lymphatic structures of various organs.

Kobayashi et al. [1] were the first to report on corrosion casts of lymphatics of the thyroid gland. Lymphatic casts of the stomach were carried out by Tamura et al. [2] and of the small intestine by Miller et al. [3]. The relation between the prelymphatic spaces and the initial lymphatics was demonstrated in casts by Rodbard and Taller [4] and Casley-Smith and Vincent [5]. Collecting lymph vessels and their valves were examined by Gnepp and Green [6]. Corrosion casts of mesenteric lymph nodes were shown by Kurokawa and Ogata [7]. In 1981, relief formations on casts of lymphatic vessels of the heart were demonstrated by Karaganov et al. [8]. Bhalla et al. [9] studied the intestinal lymphoid follicles in Peyer's patches, and Rowing [10] studied the lymphatics of the head region in the eel using corrosion casting. Kardon and Kessel [11] described cast lymphoid tissue of the lymph node, thymus, and the perirectal region. The endothelium of small lymphatics of various organs of reptiles, birds, and mammals were studied on casts by Konitz et al. [12]. Lymphatics were studied using the corrosion cast technique in the small intestine [13],

Motta, P.M., Murakami, T., and Fujita, H. (eds.), Scanning Electron Microscopy of Vascular Casts: Methods and Applications.

in the stomach [14, 15].

Own investigations in rat, rabbit, and tupaia based on corrosion casts of lymphatic pathways refer to the lymphatic plexus of the mucous membrane and skin, as well as to the lymph node and collector vessels [16–23]. Moreover, the lymphatic system in prenatal and postnatal development was examined in casting studies. [24,25].

Today, casting of fine and large lymphatics utilizing favorable properties of modern plastics and the power of the SEM has led to new statements about both the overall organization of the lymphatic vessels and their fine structural features. Hence, SEM studies of lymphatic casts carried out in the last two decades have broadly enriched our knowlege on the morphology of that special vascular system and have also provided a base for a better understanding of its particular functions as the process of lymph formation and immunological defense.

2. General Organization of the Lymphatic Vascular System

The lymphatic system can be regarded a one-way drainage system that conveys fluid, small particles, and cells from the tissue back to the blood circulation. As shown in the schematic diagram (Fig. 4-1), the lymphatic system begins with fine rootlets, the *initial lymphatics* (also called *lymphatic capillaries*), which in many tissues coalesce to form plexuslike structures. The content of the initial lymphatic plexus, the *lymph*, is drained by small collecting ducts, the *precollectors*, which are the joining elements to the first lymph nodes. In the proper *collecting vessels*, several lymph nodes are inserted, which the lymph has to pass before it eventually reaches the *lymph stems*. The largest representative of this kind of vessel is the *thoracic duct*. At the angle, where the subclavian and jugular veins join, the thoracic duct connects with the blood system.

The thin wall of the initial lymphatics consists of only the endothelial layer, which is covered on the outside by a fine net of reticular fibers that contribute to form a small basement membrane. The diameter of initial lymphatics varies greatly in caliber. Its average width exceeds that of the blood capillary and reaches values of 50 μm and more. Nuclear portions and small or broad processes of the endothelium bulge into the vascular lumen and give the luminal surface an irregular profile. Flaplike valves occur in the lumen of the initial lymphatics as well. There are numerous open junctions along the endothelial boundaries. Most of these are covered by cytoplasmic structures of neighboring cells, thus forming pocketlike structures, which function as tiny inlet valves. The question of a direct open communication between the initial lymphatics and the tissue spaces of the connective tissue is still being debated. Recent in vivo investigations, however, have confirmed the existence of these openings, which are important in controlling a state of high lymphatic fluid load [26–28].

While the precollectors exhibit fine structural features scarcely deviating from that of initial lymphatics, the collecting lymphatics proper possess a structure similar to that of veins. Three layers form a constant element of the vascular wall: *intima* with the endothelium and a small connective-tissue sheet; *media* with a well-developed smooth muscle system; and *adventitia* with the connective tissue containing supporting vessels and nerve fibers. Very characteristic of this type of vessel is segmentation caused by well-established bicuspid valves. The segment between two valves is called *lymphangion*. The lymphangions exhibit a spontaneous rhythmic activity and are able to propel the lymph from the initial area of the lymphatic system towards the lymph stems. The latter bear the same morphological and functional characteristics reported for the collecting lymphatics. Special morphological conditions exist for the transport of the lymph through the intranodal pathways. The main lymph flow occurs through the intranodal lymph sinuses — the marginal, intermediar, and medullary sinus — which are lined by a single layer of regularly formed endothelial cells (littoral cells). Some lymph can enter into the spaces of the reticular tissue through the wall of the sinuses, forming the framework of the nodal parenchyma. As a rule, several afferent lymphatic vessels empty into the marginal sinus over the entire periphery of the node, whereas only one or two efferent vessels drain the lymph at the hilus.

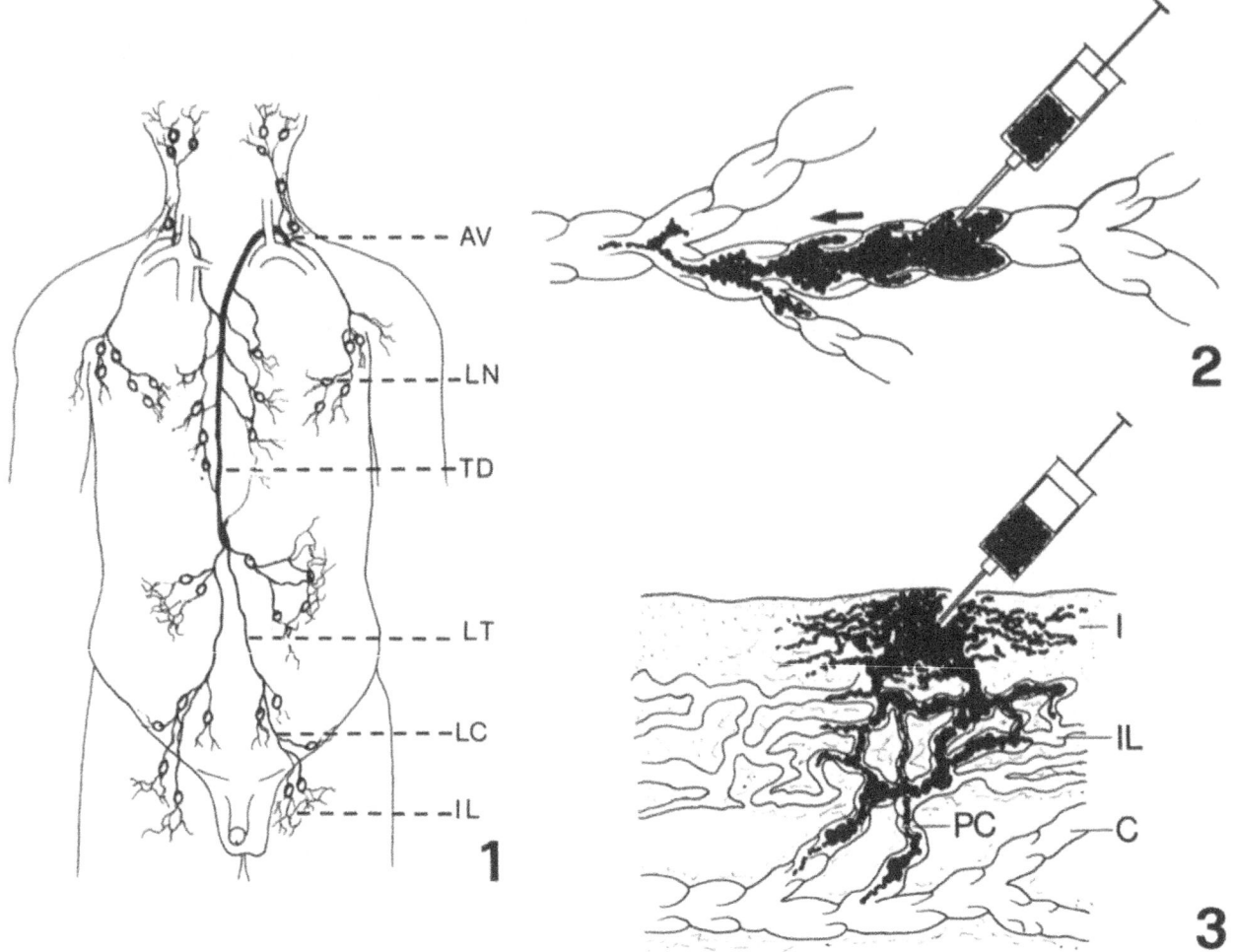

Figure 4-1. Diagram of the general organization of the lymphatic vascular system in man. IL = initial lymphatics; LC = lymph collector, LT = lumbal lymph trunc; TD = thoracic duct; LN = group of lymph nodes; AV = angulus venosus (main site of confluence of the lymphatic system into the blood circulation).

Figure 4-2. Direct casting of a lymph collector by intraluminal injection of resin. Note that, due to the well-developed valve system, casting is only possible in the proximal direction (arrow).

Figure 4-3. Indirect casting of initial lymphatics by interstitial injection of resin. From the deposit set into the interstitial spaces, the resin enters the initial lymphatic plexus via a few broad communications between the tissue and the vessel. I = interstice; IL = initial lymphatic plexus; PC = precollector; C = collector.

3. Methods

3.1. Preparation of the Tissue

For the casting of lymphatic vessels preparation procedures, that are commonly used in the casting of blood vessels, are not necessary. Hence, measures to make the sacrificed animal quickly bloodless and rinsing of the blood circulation with warm physiological solutions prior to resin injection must not be carried out. Prefixation of

the animal, or at least of the organ to be chosen for casting, can be useful. This technique is even essential, when the resin cannot be injected immediately into the sacrificed animal or into freshly dissected tissue. Prefixation also seems to be a method of choice in all cases in which vascular casts with imprint patterns on their surfaces that exactly reflect the fine structural details of the sample are desired.

It must be taken into account, however, that prefixation, either by immersion or via the blood circulation, can hinder an effective casting, i.e., a complete filling state, by creating hardening and shrinking of the tissue. In our studies we prefer a 1% glutaraldehyde solution for prefixation, a fixative that does not have the disadvantages mentioned.

After casting immersion, postfixation for at least 24 hours using 2.5% glutaraldehyde should be done in all instances to avoid postmortal changes of the tissue during long lasting polymerization of the resin.

3.2. The Resin

For corrosion casting of the lymphatics, the medium has to fulfill two criteria: It should have a low viscosity and a good replication quality. Thus, above all, a very liquid resin is necessary for the casting of prelymphatic tissue spaces and the initial lymphatic structures when using interstitial injection (see below). When using this technique, the resin has to be able to penetrate the very intricate spatial system and the tender structures of the initial lymphatic plexus. Thus, a sufficiently complete filling state of such a broadly scattered system can only be expected if the resin spreads within a short time. A similar situation occurs in the lymph node due to its special morphology. The broadly ramified system of sinuses and spaces in the lymph nodal parenchyma can be reached and filled only by a low-viscous resin.

Resins such as Batsons, Mercox, and methyl methacrylate possess such properties. We have obtained good results with Mercox (Vilene, Tokyo), which, similar to Batsons, combines both good replication quality and a low shrinkage rate (6%) [29,30]. As a rule, we do not use this medium in the original concentration delivered by the manufacturer, but rather in a concentration of 5:1 with methyl methacrylate. The diluted Mercox offers the advantage of prolonging the polymerization time, which otherwise lasts only 5–8 minutes. Ohtani and Ohtsuka [13] used Mercox diluted to 40–50% with methyl methacrylate monomer to get excellent casts of the small intestinal lymphatics, including the lacteals.

In casting larger collecting lymphatics using the intraluminal injection method (see below), we applied Plastoid, instead of Mercox, with good

success. Plastoid (Röhm., FRG) has a higher viscosity and a long polimerization time and also has good replication properties. The reduced warming of the tissue, due to the prolongation of the hardening process of Plastoid, should be mentioned as well. Another liquid resin is Tardoplast (Schumm, FRG), which exhibits elastic properties after polymerization. This may be an advantage when cutting off casts after corrosion [31].

3.3. Processing for SEM

Further procedures for the corrosion of tissue — cleaning, drying, and mounting — are done after the practice proved in corrosion casts of blood vessels: During postfixation the wholly injected animal or organ is kept at room temperature or at 40°C in an incubator. Then parts of the preparations are dissected and put into 10% KOH for maceration at 60°C. During the corrosion process, the KOH is exchanged several times. Being free of tissue, the casts are rinsed in tap water and then in distilled water. Drying of larger specimens is done as air drying in an incubator at 40°C. Small specimens can be dried in critical-point drying apparatus. Significant differences between using these two drying methods were not found in our studies. The vascular area of interest is then excised from the dried specimens under a binocular dissecting microscope and is mounted on stubs with conducting carbon, colloidal silver, or via adhesive metal tape, if there are very small samples, such as single vessels. After sputtering with gold, the cast specimens can be examined in the SEM at accelerating voltages of 5–15 kv.

3.4. Intraluminal Injection

This mode of injecting the resin into the lymphatics is reserved for bigger collectors and the lymph stems. The site of injection must be chosen, if possible, in a distal part of the vessel so that the resin can fill the vessel's lumen in the course of the lymph stream (Fig. 4-2). The propagation of the resin in the opposite direction, i.e., from the proximal side into the periphery, is hindered by the numerous valves of collector vessels or leads only to the incomplete filling of a short vascular portion.

Moreover, the vessel to be cast must be carefully prepared and laid open for intraluminal casting. Due to their colorless lymph content, even in living animals, collector vessels are not as easily identified as blood vessels. A bigger problem is the recognition of the lymphatics in a dead body, when the vascular lumina are free from lymph and vessels are in a collapsed state. Thus, in studies of small animals marking the lymphatic pathways seems to be necessary prior to casting. Marking can easily be carried out using dyes such as patent blue. After setting a small deposit of the dye into the tissue, the draining pathways soon become stained and can be easily recognized. The staining effect is very distinct in the living animal. However, it rapidly disappears or becomes indistinct due to the spreading of the dye into the paravasal zone of the vessels, when the blood circulation comes to a standstill. In such a case it is necessary to begin casting immediately. Casts of collector vessels produced by direct intraluminal injection of Mercox are shown in Figure 4-5.

3.5. Interstitial Injection

It is impossible to cast the initial part of the lymphatic system, such as the lymphatic plexus of the skin and mucous membranes, by intraluminal injection. This is due to the fact that with retrograde casting of a collecting vessel the resin never reaches the most distal parts of the initial lymphatic system, which remain either uncast or, at least, incompletely filled. The explanation for this phenomenon is that the resin advancing into the rootlets of the lymphatic system is never able to overcome the resistance obligatorily existing at the barrier between the lymphatic system and the tissue. As mentioned above, the initial lymphatics represent, with regard to their systematic and histotopographic position, a vessel type that is quite different from that of blood capillaries. The latter can be cast without any problems by intraluminal injection from either the arterial or venous side, because the resin can freely advance to the other side.

The only technique for casting initial lymphatics and precollectors with adequate results is *interstitial injection*. This indirect method is accomplished by positioning the cannula into the connective tissue, which permits the liquid resin to spread into the tissue spaces and then into the initial lymphatics and the draining precollectors. While casting, the resin takes the same pathway as do the interstitial fluid during the process of lymph formation (Fig. 4-3).

When applying the interstitial injection method, both systems are always cast, namely, that of the interstitial spaces of the connective tissue and that of the lymphatics in the related area. The achievement of the method, such as a complete filling state, avoidance of artifacts, and good reproducibility, depends decisively on the practice and experience of the investigator. The draining lymphatic vessels of a certain tissue area are distributed and arranged according to a principle that is specific for each species, organ, and tissue. Therefore, the knowledge of the location of the exact site from which the lymphatics can be well filled is very important. In the rat tongue, for example, good filling of the mucous initial lymphatic plexus is possible when the cannula is positioned in the loose fibrous connective tissue just beneath the endothelium; while good casts of the precollector system running near the base of the tongue are obtained when the cannula is inserted deeply into the body of the tongue (Figs 4-7 and 4-8c).

Casting by interstitial injection of the resin can be carried out with a high reproducibility. For good results, the amount of the injected resin and the pressure applied during casting are further parameters to be taken into account. In our experience no more than 0.5–1.0 ml of Mercox should be injected at once, and it should be applied with moderate pressure by hand or by means of a micrometer drive. The full "open time" of the resin has to be utilized for the injection, which for Mercox lasts about 5 minutes. Extreme tissue swelling, may damage vascular structures, so this should be avoided. Filling of the lymphatics can be improved by soft massage of the injected deposit when the resin is still liquid.

Another question is whether or not the tissue should be perfused with physiological solution prior to interstitial resin application. We were able to observe the single steps of casting initial lymphatics during interstitial injection of Mercox, either with or without preceding saline perfusion, under microscopical control in the living animal

44

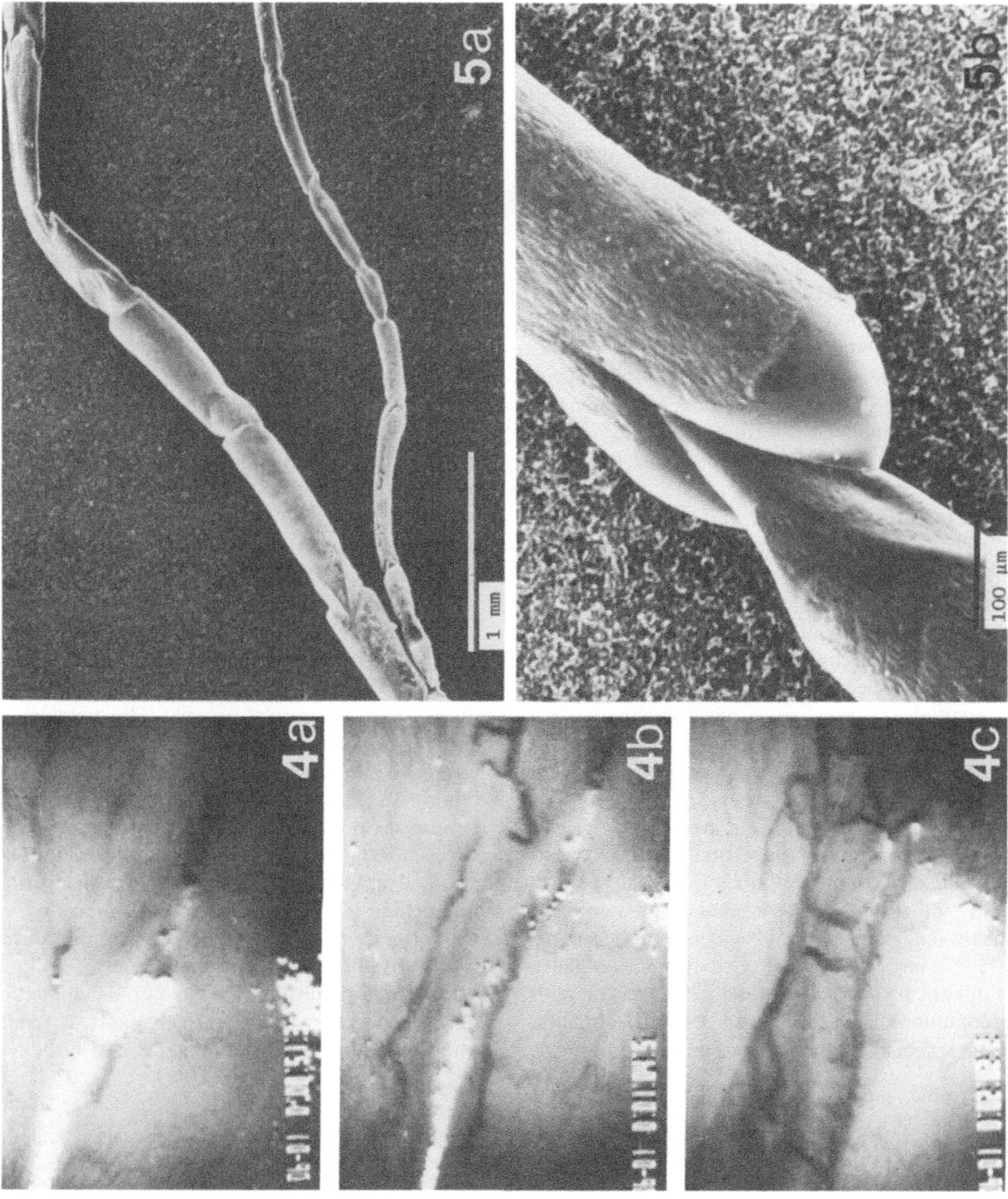

[32]. With this approach it became obvious that the cast pattern of the initial lymphatic plexus is established by several small portions of the resin fusing gradually to form a comprehensive system (Fig. 4-4). Integration of a single portion of resin was disturbed in those experiments in which a saline solution was injected before the resin was applied.

Casting of the intranodal lymphatic pathways is possible using intraluminal injection if filling is tried via the afferent prenodal lymphatics. However, the node is not usually completely filled. Better results are achieved by intranodal injection of the resin. Similar to the interstitial injection of connective tissue, in the case of intranodal injection the tip of the cannula is positioned in the nodal parenchyma, from which the resin also enters the sinus system. Direct intra-sinusoidal injections are also possible when the injecting cannula is inserted into the spatial marginal sinus, which is easy to reach and is located just under the node capsule.

Different minute ducts and spaces have recently been cast, even in such small animals as insects [33], using the vacuum-injection technique. This method seems to be a useful tool to cast the lymphatics and blood vessels in biopsies in experimental clinical research as well. Today, casts of such specimens are difficult to make using interstitial injection, since most lymphatics are not filled or are incompletely filled by the resin, due to their being in a collapsed state. Tissue pieces of biopsies have been successfully fixed by glutaraldehyde with the help of a special apparatus that makes interstitial injection possible under vacuum conditions [34]. The fixation easily spreads into the tissue and, as the lumina of the initial lymphatics are widened under these circumstances, the glutaraldehyde reaches these lymphatic pathways more easily than do common preparations. Thus, attempts to cast the lymphatic using a similar method should be very fruitful.

4. Interpretation of Lymphatic Casts

Casts of lymphatic vessels, such as those of blood vessels, are informative with regard to both the general vasoarchitecture of a special vascular area and the fine morphological details of single vascular structures [35]. Moreover, casts based on interstitial injection of the resin also reveal the organization of the tissue space system around the lymphatics and the histotopographic relationships between both systems. It has been already pointed out that, after intraluminal injection of the resin, only that part of a lymphatic vessel that is located proximal to the side of injection will be filled sufficiently. Therefore, only casts of that particular vascular area can reflect the true morphological conditions, whereas little information is obtained on the lymphatic pathways located distally from the side of the injection.

Some general criteria need to be considered in order to indentify lymphatic vessels in casts after interstitial injection. Special imprint patterns are exhibited by casts of *bigger collecting vessels* depending on the particular morphological features. A segmental unit of the lymphangions appears to be sharply marked between deep notches caused by the bicuspid valves (Fig. 4-5). The various patterns of nuclear imprints referring either to a valvular segment or an intervalvular portion of these vessels are very noticeable. The diameter of *initial lymphatic* casts varies considerably. The luminal width always exceeds that of the capillary blood vessels. This becomes very obvious in specimens simultaneously injected from the interstice and via the arterial system (Fig. 4-6). The initial lymphatics form bizarre plexuslike structures (Fig. 4-7) that are drained by *precollectors* with irregular lumina, and these features differ from those of venules and small veins.

In order to distinguish the initial lymphatics and precollectors from tissue channels, it seems important to consider the character of both sys-

←

Figure 4-4. Series of screen-off photographs taken from a living rat tongue during interstitial injection of Mercox. With this methodical approach, it is possible to control the mode and completeness of casting of mucous lymphatic structures. Three stages (a–c) of filling of the lymphatic plexus during the first 2 minutes of injection are shown. The cannula was positioned at the right side on the bottom. Note the stepwise fusion of single portions of resin to form a cast of the whole system.

Figure 4-5. a: Lymph collectors cast by intraluminal injection of Mercox. The segmentation of the vessels caused by valves of the bicuspid type is clearly visible. Each segment corresponds to a vascular unit, the lymphangion. b: Valve region of a collector at higher magnification showing more shallow indentations as replicas of the endothelium.

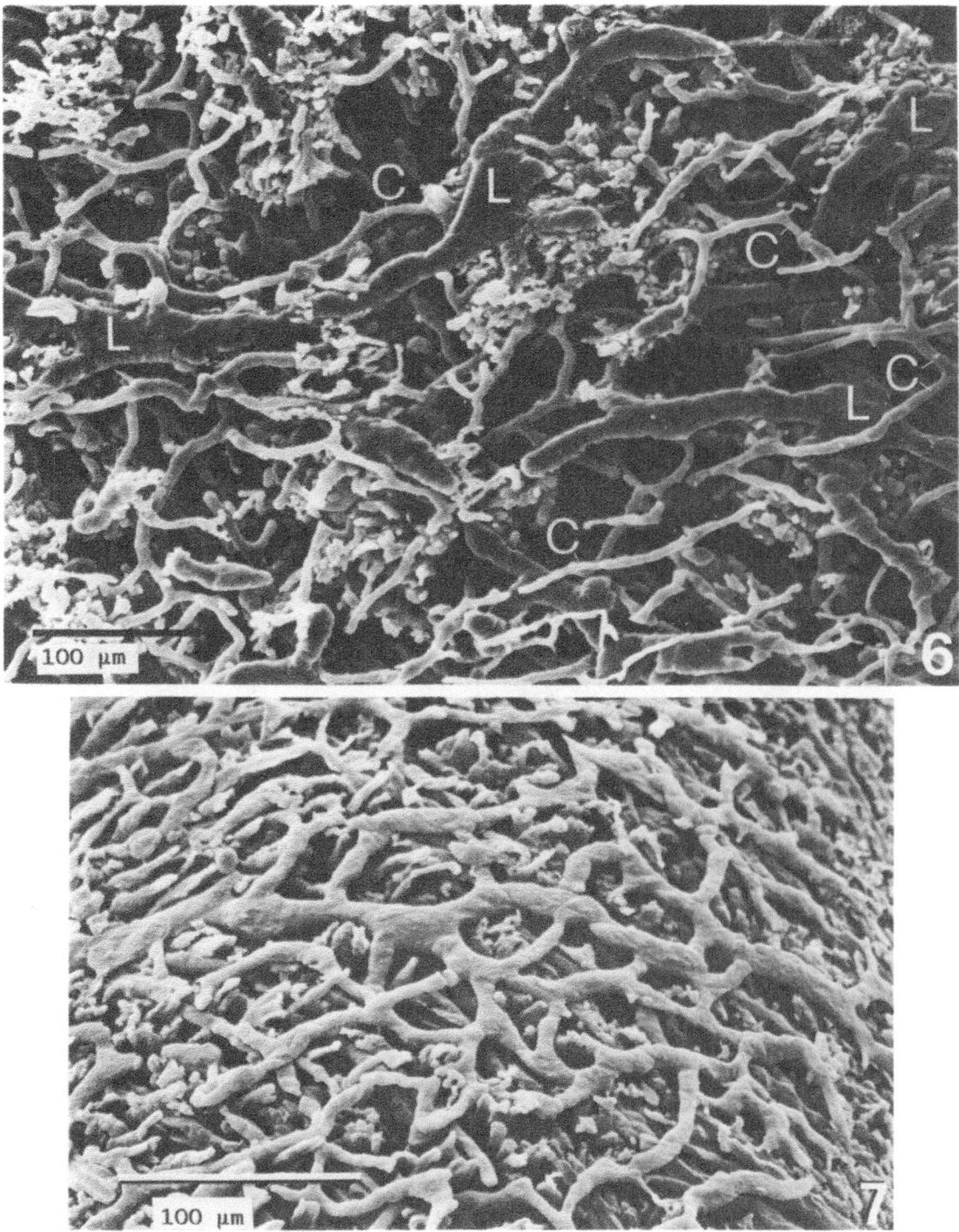

Figure 4-6. As a result of double filling of both the lymphatic and blood vascular system, this corrosion cast of the rat tongue exhibits initial lymphatic structures (L) and capillary blood vessels (C). The spaces between these structures are partially filled with the irregularly shaped pattern of the interstitial spaces. Note the wide luminal casts and the strongly varying diameter of the lymphatics, in contrast to the casts of the terminal blood vessels.

Figure 4-7. Corrosion cast of mucous lymphatics of the tongue of a young rat. The lymphatic structures form a plexuslike system overlying the irregular cast structures of the tissue spaces.

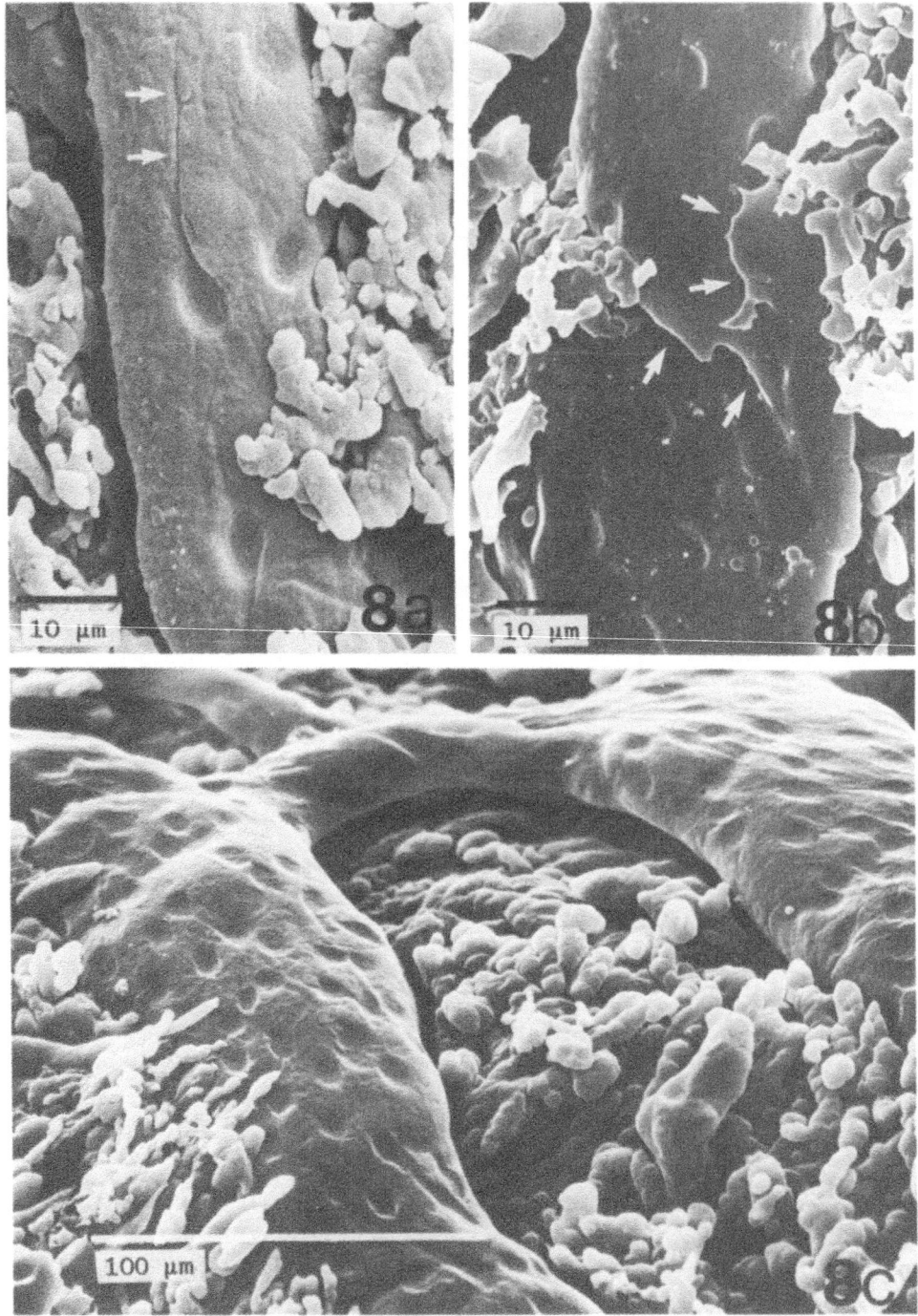

Figure 4-8. a: Cast of an initial lymphatic with imprints of the nuclear portions and cellular processes (arrows) of endothelial cells. b: Deep fissures and crestlike protrusions (arrows) in the cast of an initial lymphatic are created by primitive valve structures, trabeculae, and overlapping structures of the endothelium. c: A more regular pattern of nuclear imprints appears on the cast surfaces of precollectors, which can be used to identify this type of vessel.

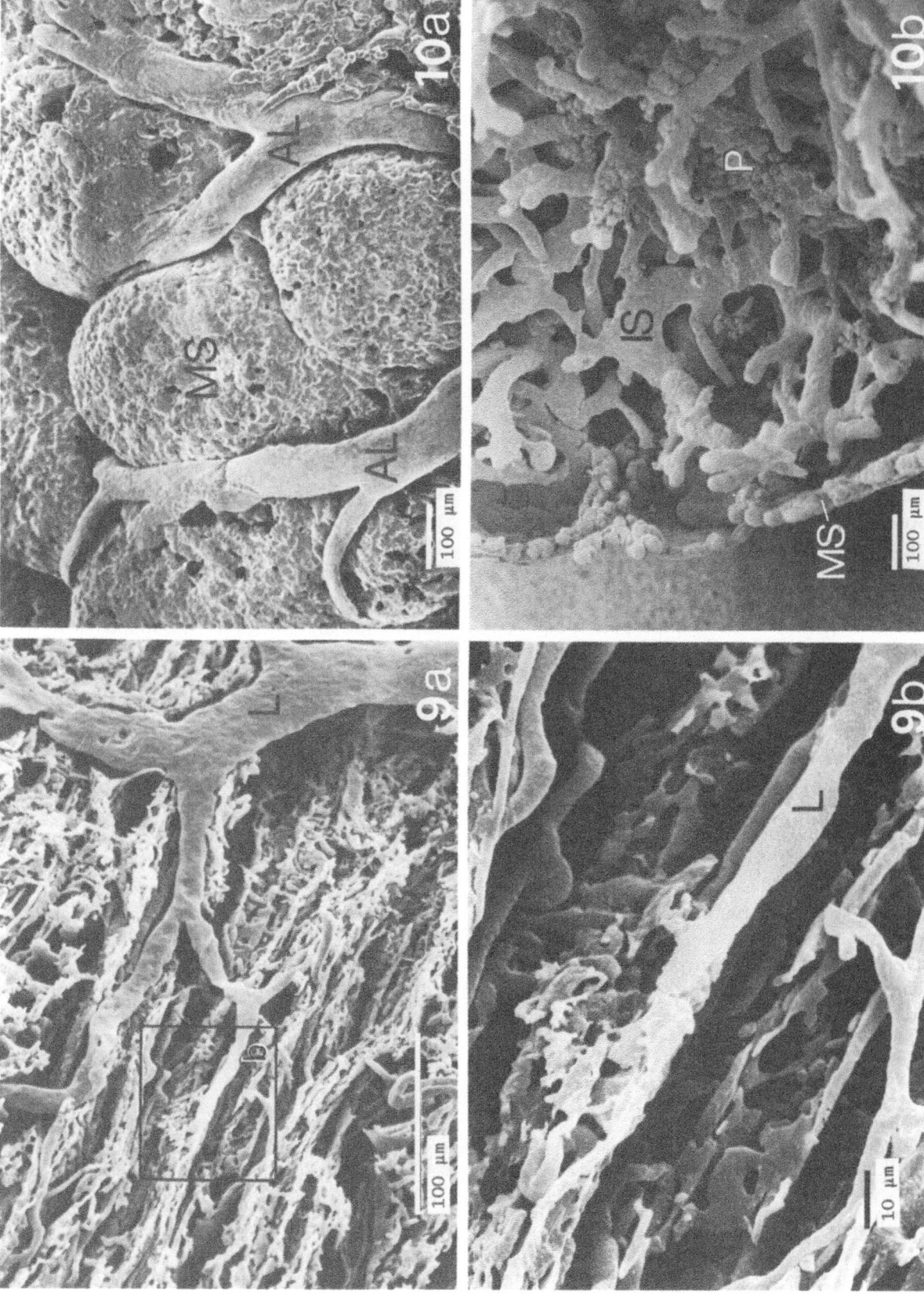

tems. The *tissue spaces* exhibit special patterns, corresponding to the specific fiber organization of the connective tissue of different organs. Cast patterns composed of squamous or roundish irregular space structures, are typical of loose connective tissue of the skin and mucous membranes, for example, while that of the stromal connective tissue of striated muscles is longish in shape. A very dense network of crisscrossed structures appears in the dense fiber connective tissue. To determine the morphological relationships between cast structures of corroded specimens and the very tissue elements surrounding and printing them, a method has been established in our laboratory [36,37] that enables the examination of casts in noncorroded tissue using the light microscope. In some cases fluorochrome was added to Mercox in order to make all vascular structures and tissue spaces easily visible under fluorescent light conditions. Thus, the exact location of cast formations within the tissue, and even their morphological differentiation against cellular elements was possible.

In good specimens the imprint pattern appearing on the cast surface of small lymphatics even reflects the endothelial fine morphology. Thus, oval and irregularly arranged elongated notches are created by the protruding nuclear portions and processes of endothelial cells of the initial lymphatics (Fig. 4-8a). Deep, sharply marked fissures are also frequently found on casts of both the initial lymphatics and the precollectors, which are replicated structures of primitive or already well-established valves. A more discrete pattern of softly indented structures is produced by the overlapping endothelial boundaries of the initial lymphatics (Fig. 4-8b). In precollectors the nuclear imprints assume a more regular distribution, and this corresponds to the well-arrayed pattern of their rhomboid-shaped cellular boundaries (Fig. 4-8c).

Realizing all these points for the recognition of true lymphatic vessels and prelymphatic tissue spaces, it is also possible to identify both structures in corrosion casts and to recognize those sites where the two systems interweave (Fig. 4-9). Direct prelymphatic-lymphatic transitions can be demonstrated in most casts after interstitial injection of the resin if they are carefully viewed with this in mind. Thus, we suggest that the lymphatic endothelium does not form a continuous sheet with the tissue, but that there exist broad communications between the tissue and the vascular lumen, which become very effective whenever the tissue is flooded with a large lymphatic load [21–23,28,35]. Other investigators have stated that the resin should invade the vessels lumen during interstitial injection through the system of open junctions of the lymphatic endothelium [15]. Small resin bridges between casts of the vessel's lumina and those of the surrounding tissue spaces could support this consideration. However, we found such structures in our preparations in only exceptional cases.

In casts of *lymph nodes*, the afferent and efferent lymphatic vessels are easy to identify as true vascular elements of the capsular zone. The appearance of sinus casts vary from broadly extended sheetlike casts of the marginal sinuses to the small tubelike casts of the intermediary sinuses in terms of their morphological features (Fig. 4-10). A plexuslike arrangement is typical for the medullary sinus system. The wall profile of all sinuses exhibits a very irregular appearance due to the numerous reticulum cells protruding or crossing the lumen of the sinus, a situation that is more or less replicated in casts of these spaces.

←

Figure 4-9. a: Corrosion cast exhibiting initial lymphatic pathways (L) at one site in direct connection with the interstice (rectangular area b). b: Transitorial area marked in Fig. 4-9a at a higher magnification. As seen, there is a continuous cast bridge between the scattered structures of the interstitial spaces and the beginning of the lymphatics (L). This indicates the site of resin entering the vessel.

Figure 4-10. a: View of the surface of a corrosion cast of a rabbit lymph node that was filled via the prenodal lymphatics. The marginal sinus (MS) and the afferent lymphatics (AL) fusing with the MS can be recognized. b: Cast of a mesenteric lymph node in rabbit produced by intranodal injection of the resin. In the cut plane, the casts of the marginal sinus (MS), of the tubular intermediar sinuses (IS), and the intricate spatial system of the nodal parenchyma (P) can be easily distinguished.

The bizarre pattern that is referred to the small spaces scattered over the nodal parenchyma can be well distinguished from the casts of sinuses.

5. Concluding Remarks

Casting lymphatic structures in combination with modern scanning electron microscopy is a very useful method in basic morphology and histophysiology. Taking into account the special structure of the prelymphatic tissue and the true lymphatic vessels, similar results, in terms of both the quality and information content of the corroded preparations, can be obtained using this casting technique as have been achieved in other fields of microvascular research. Furthermore, the approach indirectly representing the lymphatics, i.e., as replicated structures, depends on the particular subject of investigation chosen by the investigator. Therefore, interest in the method reported in this chapter should increase in the future.

New injection techniques are necessary to refine established methods and to widen their range of application. It seems important to apply corrosion casting to isolated tissue, such as biopsies in humans. The method can thus also be used in human pathology and clinical research. Other approaches that combine casting of tissue with other methods, such as fluorescent light microscopy [36], could be improved. In this manner, additional new methods may be opened to bridge the gap between the indirect casting techniques and the microscopy of fixed tissue commonly used in histology.

Although early researchers of the lymphatics used the injection technique with primitive equipment and described the gross anatomy of this system, our lack of knowledge of this area is still very great. The possibilities that corrosion casting offer today can help us to overcome this unsatisfactory situation and to shed new light on the differentiated structures of the lymphatic pathways in humans and animals.

Reference

1. Kobayashi S, Osatake H, Kashima Y. Corrosion casts of lymphatics. *Arch Histol Jpn* 39:177–181, 1976.

2. Tamura A, Scanning electron microscopic studies on the lymphatic vessels of the rat stomach. *J. Clin Electron Microsc* 11:426–427, 1978.

3. Miller BG, Woods RJ, Bohlen HG, Evan AP. A new morphological procedure for viewing microvessels: A scanning electron microscopic study of the vasculature of small intestine. *Anat Rec* 203:498, 1982.

4. Rodbard S, Taller S. Plastic casts of extracapsular spaces and lymphatic vessels. *Med Exp* 19:65–70, 1969.

5. Casley-Smith JR, Vincent AH. The quantitative morphology of interstitial channels in some tissues of the rat and rabbit. *Tissue Cell* 10:571–584, 1978.

6. Gnepp DR, Green FHY. Scanning electron microscopic study of canine lymphatic vessels and their valves. *Lymphology* 13:91–99, 1980.

7. Kurokawa F, Ogata T. A scanning electron microscopic study on the lymphatic microcirculation of the rabbit mesenteric lymph node. A corrosion cast study. *Acta Anat* 107:439–466, 1980.

8. Karaganov YL, Mironov WA. SEM studies of injection replicas of sequential segments of the rat omentum's microvasculature at normal and increased venous pressures. *Verh Anat Ges* 75:709–711, 1981.

9. Bhalla DK, Murakami T, Owen RD. Microcirculation of intestinal lymphoid follicles in rat Peyer's patches. *Gastroenterology* 81:481–491, 1981.

10. Rowing CGM. Interrelationships between arteries, veins, and lymphatics in head region of the eel, *Anguilla anguilla* L. *Acta Zool* (Stockholm) 62:159–170, 1981.

11. Kardon RH, Kessel RG. The microcirculation of lymphoid tissue in three dimensions. Scanning electron microscopy of corrosion casts of the lymph node thymus and perirectal lymphoid tissue. *Biomed Res* 2:173–180, 1981.

12. Konitz H, Berens v. Rautenfeld D, Klanke J. REM Darstellung der initialen Lymphstrombahn durch indirekte Applikation von Mercox oder Glutaraldehyd. *Beitr Electronenmikroskop Direktabb Oberfl* 18:249–256, 1985.

13. Ohtani O, Ohtsuka A. Three-dimensional organization of lymphatics and their relationship to blood vessels in rabbit small intestine. A scanning electron microscopic study of corrosion casts. *Arch Histol Jpn* 48:255–268, 1985.

14. Ohtani O, Murakami T. Lymphatics and myenteric plexus in the muscular coat in the rat stomach: A scanning electron microscpic study of corrosion casts made by intra-arterial injection. *Arch Histol Jpn* 50:87–93, 1987.

15. Wenzel-Hora BI, Berens v. Rautenfeld D, Partsch H. The effects on the opening apparatus of the initial lymphatics. *Lymphology* 20:134–144, 1987.

16. Castenholz A. Morphological characteristics of initial lymphatics in the tongue as shown by scanning electron microscopy. *Scann Electron Microsc*, II:1343–1352, 1984.

17. Castenholz A. The demonstration of lymphatics in casts and fixed tissue with the scanning electron microscope. In: Bollinger A, Partsch H, Wolfe JHW (eds.), *The Initial Lymphatics. New Methods and Findings*, Thieme/Thieme Stratton, Stuttgart/New York, pp 75–83, 1985.

18. Castenholz A. Observations on the structural and functional properties of initial lymphatics — light and electron

microscopic studies of the subepithelial lymphatic plexus in the rat tongue. In: Casley-Smith JR, Piller NB (eds.), *Progress in Lymphology*, University of Adelaide Press, pp 20–23, 1985.

19. Castenholz A. Corrosion cast technique applied in lymphatic pathways. *Scann Electron Microsc* II:599–605, 1986.

20. Castenholz A. Structural and functional properties of initial lymphatics in the rat tongue: Scanning electron microscopic findings. *Lymphology* 20:112–125, 1987.

21. Castenholz A. Structure of initial and collecting lymphatic vessels. In: Olszewski WL (ed.), *Lymph Stasis — Pathophysiology, Diagnosis and Treatment*. CRC Press, Boca Raton FL, 15–42, 1991.

22. Castenholz A. Architecture of the lymph node with regard to its function. In: Grundmann E, Vollmer E (eds.), *Current Topics in Pathology, Vol. 84, Reaction Patterns of the Lymph Node*. Springer Verlag Berlin/Heidelberg, 1–31, 1990.

23. Castenholz A. Correlative vital microscopic and scanning microscopic study of initial lymphatics. Excerpta Medica, Amsterdam/NY/Oxford, 311–312, 1990.

24. Castenholz A. SEM representation of initial lymphatic structures in early stages of ontogenetic development in rat. In: Casley-Smith JR, Piller NB (eds.), *Progress in Lymphology*. University of Adelaide Press, pp 24–26, 1985.

25. Castenholz A. The endothelium of initial lymphatics during postnatal development of the rat. *Scann Electron Microsc* III:1201–1208, 1985.

26. Castenholz A. Vitalmikroskopische Beobachtungen an den Lymphbahnen der Zungenschleimhaut der Ratte. In: *Lymphologica Jahresband 1989*. Medikon Verlag, München, pp 12–25, 1989.

27. Castenholz A. Fluoreszenzmikroskopische Unterorsuchungen zur Stoffpassage über die Blut-Lymphschranke. In: *Lymphologica 1989*, Jahresband, Medikon Verlag, München, 73–80, 1990.

28. Castenholz A, Hauck G. Zur strukturellen Grundlage des prälymphatisch-lymphatischen Übergangsbereiches. Manuskriptenband der 13. Jahrestagung der Gesellschaft für Mikrozirkulation, München, 216–220, 1989.

29. Weiger T. Die Verwendbarkeit der polymerisierenden Kunststoffe Methylmethacrylat und Mercox CL zur Herstellung von Korrosionspräparaten zur rasterelektronenmikroskopischen Untersuchung der Gefäßarchitektur tierischer Gewebe und Organe. Hausarbeit, Zoological Institute of the University, Salzburg, 1981.

30. Weiger T, Lametschwandtner A, Adam H. Methylmethacrylat und Mercox CL in der Rasterelektronenmikroskopie von Korrosionspräparaten. *Mikroskopie* 39:187–197, 1982.

31. Hunneshagen C. Das Lymphgefäßsystem der Rotwangen-Schmuckschildkröte unter Berücksichtigung des phylogenetischen Aspektes. Inaugural Dissertation. Tierärztliche Hochschule Hannover, 1988.

32. Castenholz A. The mode of casting lymphatic structures after interstitial injection. Correlative vital- and scanning electron microscopic studies. Abstralchenband 1st European Workshop on SEM of corrosion casting, Salzburg/Austria, 16, 1989.

33. Meyer E. Corrosion cast of the insect's tracheal system. Paper presented at the 1st European Workshop on SEM of corrosion casting, Sept. 5–7, 1989, Salzburg/Austria, 1989.

34. Berens v. Rautenfeld D, Lubach D, Wenzel-Hora BI, Hunneshagen C, Deutsch A. Vaccum injection — a new indirect fixing method for the demonstration of initial lymphatics in biopsy specimens of human organs. *Acta Anat* 136:172–176, 1989.

35. Castenholz A. Interpretation of structural patterns appearing on corrosion casts of small blood and initial lymphatics vessels. *Scan Microsc* III:315–325, 1989.

36. Castenholz A. REM-Darstellung von interstitiellen Raum systemen und Lymphbahnen mit der Kunststoffinjektion. *GIT* Suppl. 2:16–21, 1985.

37. Castenholz A. Zöltzer H, Erhardt H. Structures imitating myocytes and pericytes in corrosion casts of terminal blood vessels. A methodical approach to the phenomenon of "plastic strips" in SEM. *Mikroskopie* 39:95–106, 1982.

Author's address:
Prof. Anton Castenholz
Department of Human Biology (FB 19)
University of Kassel
Heinrich-Plett-Strasse 40
D-3500 Kassel
Germany

CHAPTER 5

Blood and Lymphatic Microvascular Organization of the Digestive Tract

OSAMU OHTANI & TAKURO MURAKAMI

1. Introduction

The blood vascular bed not only supplies the digestive tract, but is the ultimate source of digestive juices secreted into the alimentary tract and is also the recipient of absorbed digesta. On the other hand, the lymphatic system transports nutrient fluid absorbed from the gut lumen, excess tissue fluid, and cellular elements involved in the immune system to the lymph nodes and ultimately to the blood vascular system.

During the past two decades, scanning electron microscopy (SEM) of corrosion casts [1] has provided us with much knowledge on the three-dimensional organization of the blood and lymphatic microvasculatures of the digestive tract [2–17], as well as many other organs and tissues.

This article reviews the three-dimensional organization of the blood and lymphatic microvascular beds of the digestive tract as revealed by SEM of corrosion casts. Attention is largely placed on the mucosa of the stomach, small intestine, and colon, and on the gut-associated lymphoid tissues, including the palatine tonsils, Peyer's patches, and appendix.

2. Materials and Methods

The animals used were healthy rats (Wistar, adults and newborns), rabbits, and young mongrel dogs. Human palatine tonsils were obtained from patients operated for tonsillar hypertrophy. The blood vascular corrosion casts were made accord-

ing to the method of Murakami [1]. The lymphatic corrosion casts and the blood/lymphatic corrosion casts were reproduced by a method previously described [10,13–15]. The casting medium used was methacrylate resin (Mercox, Oken Shoji, Tokyo). For the preparation of lymphatic corrosion casts, the casting medium was diluted with monomeric methyl methacrylate to give a viscosity of 1.5–2.0 cS [13]. Observations were made under the JSM-U3 (JEOL) or S-2300 (Hitachi) scanning electron microscope with an accelerating voltage of 5–15 kV.

3. Blood and Lymphatic Architecture

3.1. Stomach

In the stomach, the arteries in the subserosa penetrate the muscular coat and reach the submucosa, where they repeat branching and anastomosing to form the submucosal arterial plexus (Figs. 5-1 and 5-2). The arterial branches or arterioles that arise from this plexus largely supply the mucosa. In the submucosa, there is also a venous plexus (Figs. 5-1 and 5-2), which receives venules from the mucosa and from the muscular coat, and is drained into the subserosal veins that accompany their counterpart arteries. Only large submucosal arteries that run transversely to the long axis of the stomach accompany their counterpart veins (Fig. 5-1).

The muscular coat is supplied by arterioles that arise from the subserosal arterial branches as they

Motta, P.M., Murakami, T., and Fujita, H. (eds.), Scanning Electron Microscopy of Vascular Casts: Methods and Applications.

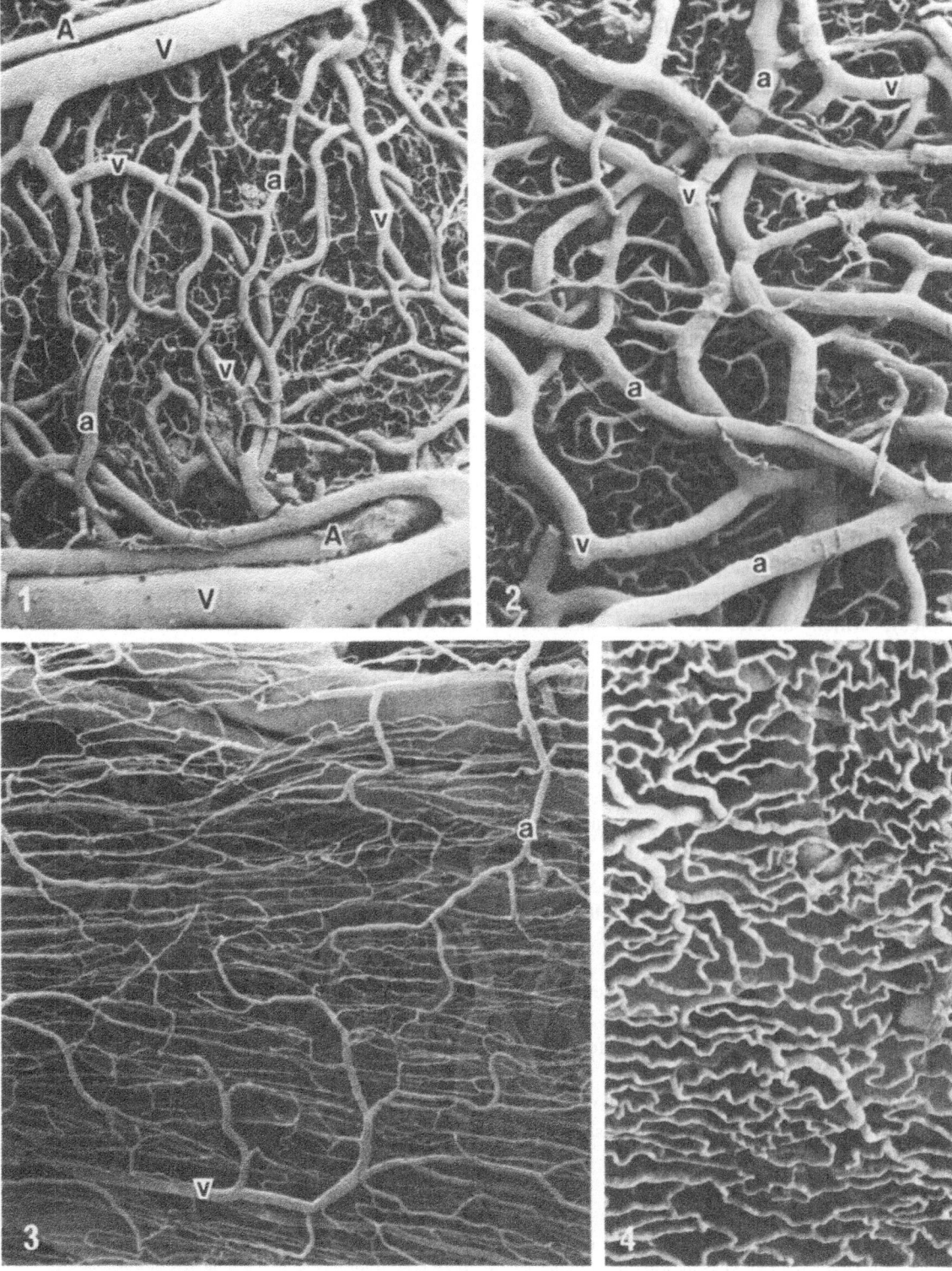

pierce the muscle before entering the submucosa or from the submucosal arterial plexus. In the rat and rabbit, two sets of capillaries are distinguished in the muscular coat, one deep and the other superficial, both running parallel to one another at 40–80 μm intervals [18]. Capillaries run along the muscle cells and frequently communicate with other capillaries on the same plane, as well as with others on different planes (Fig. 5-3). The arterioles in the muscular coat run independently of the venules (Fig. 5-3).

The microvascular pattern of the gastric mucosa has some regional and interspecies differences. In the rat forestomach (Fig. 5-4), there is a monolayered capillary network immediately below the squamous epithelial cells. This capillary network is supplied by arterioles that arise from the submucosal arterial plexus and gather into venules that lead into the submucosal venous plexus. In the rat gastric corpus (Figs. 5-5 and 5-6), the submucosal arterial plexus gives off numerous terminal arterioles (i.e., the mucosal arterioles), which ascend for short distances toward the glandular mucosa. Near the base of the fundic glands, the mucosal arterioles break up into capillaries, which in turn ascend toward the luminal surface and form the glandular capillary network around the fundic glands. The luminal surface of the corrosion casts of the rat stomach have a honeycomb appearance; each area corresponds to one foveola. The mucosal capillary network is collected into venules that originate immediately below the surface mucous cells. These venules gather into the mucosal venules, which descend perpendicularly to lead into the submucosal venous plexus (Fig. 5-5). The mucosal venules collect capillaries around the superficial part of the fundic glands, but receive few capillaries or venules in their courses.

In the rabbit stomach, the submucosal arterioles give off two types of arterioles, short and long, to the mucosa [16]. The short arterioles break up into capillaries around the base of the gastric glands. These capillaries ascend toward the luminal surface and form the glandular capillary network. The long arterioles ascend, without breaking up into capillaries en route, and reach the superficial mucosa, where they branch out into capillaries immediately below the surface mucous cells (Fig. 5-7). This superficial capillary network is denser than the deep network around the fundic glands (Fig. 5-8). In the glandular neck region, the superficial and deep capillary networks connect with each other and converge into common venules (i.e., the mucosal venules), which descend perpendicularly along the glands without receiving any capillaries or venules en route and lead into the submucosal venous plexus (Fig. 5-9). The microvascular architecture of the human stomach is essentially equivalent to that in the rabbit [17].

No general agreement has been reached as to whether or not arteriovenous anastomoses (AVAs) exist in the gastric wall. Since Barclay and Bently [19] proposed the existence of AVAs in the gastric wall, some authors described AVAs in various regions of the gastric wall, although these were denied by many others [18,20]. Careful and extensive SEM studies of the vascular plexus of completely filled specimens have failed to detect casts of any short-pathway, wide-caliber vessels passing between arterial and venous vessels in the stomach [6,7,11,16]. At least there is no AVA, as seen in the rabbit ear and dog tongue (see Chapter 2).

The SEM findings of blood vascular corrosion casts indicate that the microvasculature of the stomach is organized such that the blood in the glandular mucosal capillaries flows upward from the glandular bottom to the surface epithelium. Indeed, intravital microscopy of living rat stomach has confirmed this upward flow [6,

←

Figure 5-1. Blood vascular corrosion casts of the rat stomach.
Figure 5-2. The large arteries (A) along with their counterpart veins (V) run transversely to the long axis of the stomach (Fig. 5-1). Between these vessels, smaller arteries (a) and veins (v) form arterial and venous plexuses respectively (Fig. 5-1 and 5-2). There are many arterioles breaking up into capillaries at the base of the gastric glands (Fig. 5-2). Fig. 5-1. ×40; Fig. 5-2. ×50.
Figure 5-3. The muscular coat is supplied by capillaries that run parallel to one another and have frequent communications. The arterioles (a) travel independent of the venules (v). ×50.
Figure 5-4. The subepithelial capillary network of the forestomach. ×150.

56

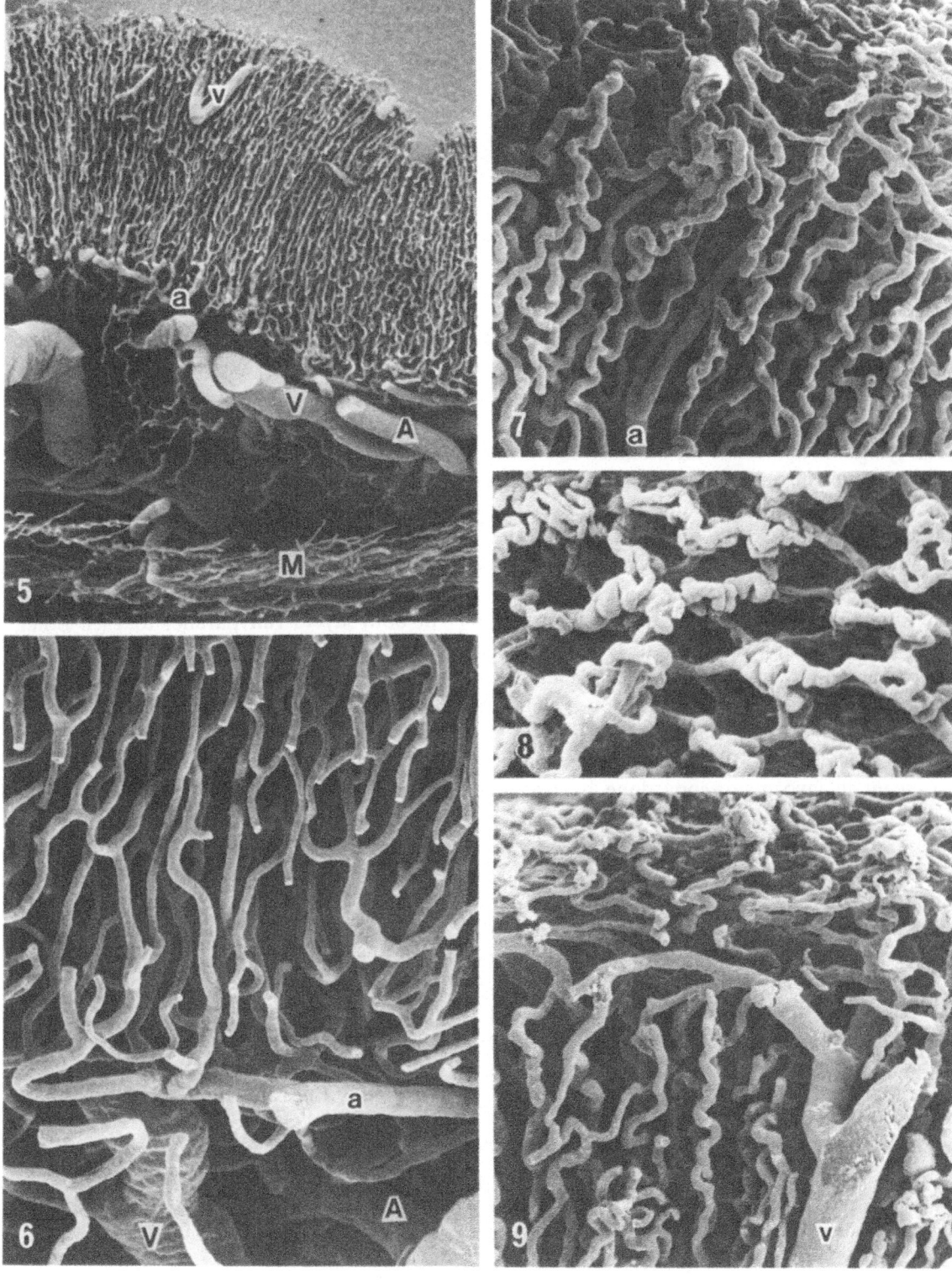

18]. This gastric microcirculation may have two physiological reasons for protecting the gastric mucosa from acid injury: (1) secretion of H^+ by the parietal cells is accompanied by the release of HCO_3^- into the adjacent interstitial space where deep segments of the mucosal capillaries exist. Released HCO_3^- may be transported to the surface epithelium by the mucosal blood flow. Thus, the capacity of the surface mucous cells to secrete HCO_3^--rich mucus or to neutralize backdiffusion of H^+ would be increased [7]. (2) The upward mucosal blood flow may also contribute to transportation of substances activating surface mucous cells, such as serotonin [21] or somatostatin [22], secreted by endocrine cells that are located in close proximity to capillaries in the deep segment of the gastric glands.

The lymphatic system of the stomach has been studied by injecting mercury, India ink, or Gerota mass; clearing the tissue; and then observing the lymphatics filled with these materials. The lymphatic corrosion casting/SEM method [11, 13–15] is more useful in demonstrating the three-dimensional architecture of the lymphatic system than previously employed techniques.

Between the base of the gastric glands and the muscularis mucosa in the rat and dog stomach, we found a single-layered network of lymphatic capillaries [11] (Fig. 5-10), which agrees with the fluorescent in vivo microscopy done by Nagata and Guth [23] and other injection studies. There is no distinct cast of lymphatics in the interglandular region. This suggests that the lymphatic capillaries exist only in deep mucosa, largely below the bottom of the gastric glands, but not in the superficial mucosa. The mucosal lymphatic capillaries gather into thicker lymphatic vessels that penetrate through the muscularis mucosa and lead into the thicker submucosal lymphatic vessels [11]. The submucosal lymphatic plexus drains into the lymphatic plexus located between the inner and outer layers of the muscular coat [11,15] (Fig. 5-11). The corrosion casts of the lymphatic vessels in the submucosa and muscular coat show notches indicative of valve locations (Fig. 5-11). SEM of tissues shows lymphatic-vessel valves positioned in these regions [11] (see Chapter 2).

3.2. Small Intestine

The subserosal arteries penetrate the muscular coat and reach the submucosa, where they run transversely to the long axis of the intestine, giving off branches that, in turn, repeat branching to form the submucosal arterial plexus. This plexus gives off arterioles that supply the mucosa and the muscular coat. In the submucosa, there is also a venous plexus that collects venules from the mucosa and the muscular coat, and leads into the subserosal veins.

The capillary network of the muscular coat consists of two layers: one deep and the other superficial. As in the stomach, the capillaries in both layers run parallel to each other along the muscle cells and have frequent connections between each other [9,16].

In the rat small intestine (Figs. 5-12 and 5-13), the mucosal arterioles, which arise from the submucosal arterial plexus, can be classified into two categories: (1) long or villous arterioles and (2) short or cryptal ones. The villous arterioles ascend in the villous connective tissue core and break up into capillaries, which form a basketlike network immediately below the villous epithelial basal lamina. The villous capillary network drains into the villous venule, which descends in the villous connective tissue core and leads into the submucosal venous plexus. The cryptal arterioles reach the bottom of the intestinal crypt or intestinal glands of Lieberkuhn and break up into capillaries that ascend along the crypt, making

←

Figures 5-5 and 5-6. Blood vascular corrosion casts of the rat stomach (tangential sections). At the level of the glandular base, the arterioles (a) from the submucosal arteries (A) break up into capillaries that ascend along the gastric glands toward the luminal surface. The mucosal venules (v) originate near the luminal surface, descend perpendicularly through the glandular layer, and lead into the submucosal veins (V). M = capillary network of the muscular coat. Fig. 5-5. ×45; Fig. 5-6. ×210.

Figures 5-7–5-9. Blood vascular corrosion casts of the rabbit stomach/Figs. 5-7 and 5-9: tangential sections; Fig. 5-8: luminal view). Two long mucosal arterioles (a) ascend through the glandular layer and reach the superficial mucosa (Fig. 5-7), where the arterioles form a network of convoluted capillaries (Figs. 5-7 and 5-8). The capillary drainage into the mucosal venules (v) initiates between the gastric pits and the gastric gland neck region (Fig. 5-9). Fig. 5-7. ×190; Fig. 5-8. ×200; Fig. 5-9. ×160.

58

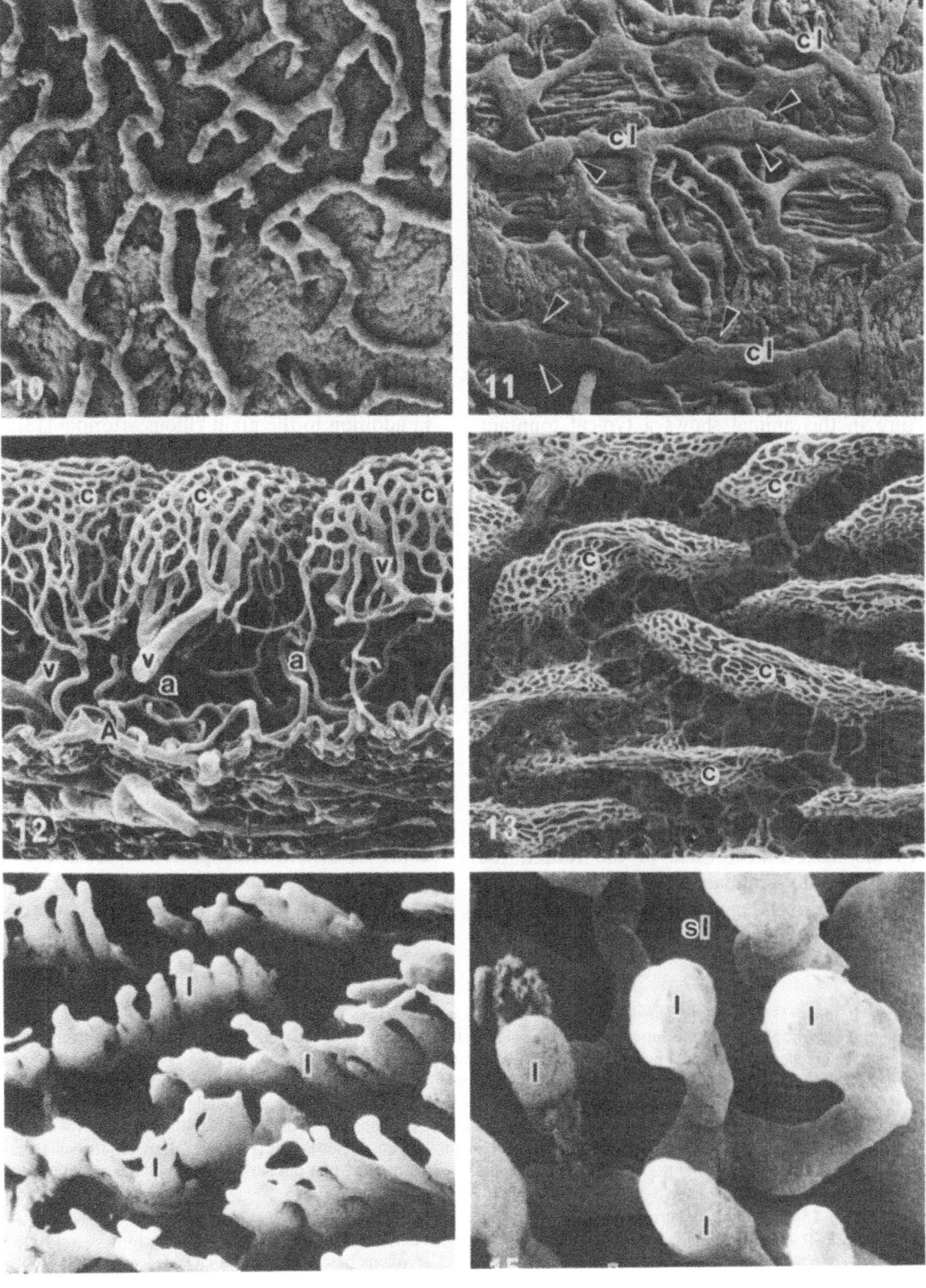

connections with each other in their courses, and thus form the pericryptal capillary plexus. Although there are some venules that drain the cryptal capillary network [8] into the submucosal venous plexus, most of the capillaries around the crypt connect with the villous capillary network [4,5,9,13].

The villous microvascular architecture shows some interspecies differences, which seem to reflect differences in the shapes of the intestinal villi [9]. In the cylindrical or conical villus of the cat, the arteriole ascends in the center of the villous connective tissue core and breaks up into capillaries at the villous tip to supply the villous capillary network. The villous capillaries converge, at or near the villous base, into the villous venule. Thus, the villous microvascular architecture in this animal shows a typical fountain pattern [5]. In the flat or tongue-shaped villus of the rat (Figs. 5-12 and 5-13), the villous arterioles ascend centrally through the villous connective tissue core and connect at the villous tip with the apically situated thicker capillaries, which in turn break up into the villous subepithelial capillary network [8,9]. In the jejunum of the rat, each villus has usually two villous venules that are symmetrical in location with respect to the villous arteriole. These venules converge into a single venule that descends through the cryptal layer to lead into the submucosal venous plexus. However, in rat ileum, in particular in aged rats, the villi are wider than those in the upper intestine and have three or more villous venules.

In the flat, conical villus of the rabbit, the villous arteriole terminates in a T-junction below the villous tip and gives rise to two marginal vessels, one of which passes up and around the villous tip, while the other travels back down the apical margin of the villus. Both of the marginal vessels, giving off branches to the villous sub-epithelial capillary network en route, ultimately connect with the villous venules [4,13]. The former marginal vessel usually connects with the villous venule that is situated asymmetrically with respect to the main villous arteriole in the villous connective tissue core, while the latter connects with the villous venules that are located half-way down the villus and collect capillaries of the lower half of the villous, as well as those of the upper half [13]. These marginal vessels seem to represent preferential pathways from the arteriolar to the venular side. Each villus capillary network is frequently supplied with several thin arterioles, in addition to the main villous arteriole [13]. The villous microvascular architecture in the human has been reported to be similar to that of the rabbit [4].

The level of connection of the villous subepithelial capillary network with the villous venules is 30–70% (rat), 20–40% (rabbit), and 15–25% (human) of the villous height below the tip [5,9]. In vivo microscopy of the villi in living animals shows that blood flows from the capillary plexus of the crypt to that of the villus [4,5]. Thus, the vascular connections between these two capillary plexuses may be regarded as portal vessels [8,9]. It has been postulated that hormones released from the endocrine cells (which are condensed at the deeper portion of the crypt) can be efficiently transported to act at the lower portion of the villi [8]. According to Gannon et al. [24], this microcirculatory pattern may

←

Figures 5-10 and 5-11. Lymphatic corrosion casts of the rat stomach. There is a monolayered lymphatic network between the bottom of the gastric glands and the muscularis mucosae (Fig. 5-10). In the muscular coat is a network of collecting lymphatic vessels (cl) that have notches (arrowheads) indicative of valve locations (Fig. 5-11). Fig. 5-10. ×75; Fig. 5-11. ×80.

Figures 5-12 and 5-13. Blood vascular corrosion casts of the rat small intestine (Fig 5-12: tangential section; Fig. 5-13: luminal view). The submucosal arteries (A) give off two types of arterioles, one breaking up into capillaries around the base of the intestinal glands and the other (a) ascending through the villous connective tissue core to connect with the villous capillary network (c) at the villous tip (Fig. 5-12). The villous venules (v) originate at around the middle of the villous height (Fig. 5-12). Between the villous capillary network are capillary rings around the openings of the crypts (Fig. 5-13). Note that the pericryptal capillaries are continuous with the villous capillaries (Fig. 5-12 and 5-13). Fig. 5-12. ×110; Fig. 5-13. ×100.

Figures 5-14 and 5-15. Lymphatic corrosion casts of the lower segment of the small intestine. Each villus in the rat has 5–10 or more central lacteals (l) aligned in a palisade fashion (Fig. 5-14), while that in the rabbit possesses one rodlike central lacteals (Fig. 5-15). The central lacteals in the rat fuse and form sinuses at the level of villous base. The central lacteals lead into the submucosal lymphatics (sl). Fig. 5-14. ×80; Fig. 5-15. ×220.

60

facilitate the process of washing intestinal contents and may serve as a solvent drag mechanism for carrying digested products towards the villi for absorption.

The duodenum is characterized by large, ridge-like villi, and particularly by Brunner's glands, which lie adjacent to the gastro-duodenal junction and tail off in the proximal duodenal segment [2]. The arteries that enter the submucosa give off arterioles to Brunner's glands, as well as to the crypts and villi. The former arterioles break up into capillaries that surround the Brunner's glands. The capillaries of the Brunner's gland mass drain into venules, which tend to be centrally placed in the glandular tissue and thus lie within the serosa [2]. These venules ultimately drain into the submucosal venous plexus. The microvascular architecture of the duodenal villi and crypts is, in general, similar to that of the jejunum and ileum [2] (see above). These, therefore, indicate that the microvascular bed of Brunner's glands is clearly a separate vascular bed from that of the rest of the duodenal mucosa and is arranged essentially parallel with it [2]. This separation of venous outflow from the glands may be important in maintaining the HCO_3^- protection of the duodenal surface from acid erosion [2]. Florey and Harding [25] found that the first part of the duodenum, into which the Brunner's glands secrete, is more resistant to acid erosion than the more distal small intestine. They also found that, following acid stimulation, intestinal luminal HCO_3^- concentration increased more in that portion of the duodenum containing Brunner's glands than it did in the adjacent duodenal region.

The lymphatics in the intestinal villi begin with blind-ended vessels, i.e., the central lacteals, which are located in the center of the connective tissue core of the villi [26] (Figs. 5-14 and 5-15). The central lacteals are surrounded externally by the villous capillary network [10,13]. At the base of the villus, the central lacteals in the rat fuse and form a wide sinus, from which two or three lymphatics originate and descend, leading into the submucosal lymphatics (Fig. 5-14). The sinus corresponds to the ampulla described by Ranvier [26]. The submucosal plexus consists of thin and thick lymphatics; the thin lymphatics form a network that fills in the coarse mesh of the thick collecting lymphatic vessels that have valves

[10,13]. The thick lymphatic vessels give off the efferent lymphatic vessels, which collect the lymphatics in the muscular coat en route and drain into the lymphatics in the mesentery.

The number and shape of the lacteals varies in each villi [10,13], which perhaps reflects the shapes of the villi. The cylindrical villus in the rabbit ileum usually contains one rod-like lacteals (Fig. 5-15), while the flat conical villus in the upper part of the rabbit small intestine contains two to five central lacteals with interconnections [13]. The flat and wide tongue-shaped villus of the rat small intestine possesses 3–10 or more central lacteals, depending upon the villous width (Fig. 5-14).

Since the central lacteals and the sinuses occupy a large portion of the villous connective tissue core, they seem to have great potential as a reservoir for fluid propelled into the lacteals [10,13]. During its contraction and elongation, the intestinal villus will cause shrinkage and expansion of the central lacteals, in particular the sinus, with the aid of lymphatic anchoring filaments [27] and the perpendicularly oriented smooth muscle cells in the villus [28]. Expansion of the central lacteals will create a pressure gradient across the endothelium of the lacteals, which facilitates lymph propulsion into the lacteals. On the other hand, the shrinkage propels lymph out of the central lacteals into the submucosal lymphatics. In the rabbit, the thick submucosal arteries and veins are intimately associated with the lymphatics. This arrangement suggests that blood vascular vasomotion may represent an important energy source, aiding lymph formation and transport [13], as in hamster skin [29] and rat skeletal muscle [30]. As the lymphatics, especially the collecting lymphatics with valves, are located in the muscular layers of the intestine, contraction of the layers during peristaltic movement of the organ would also press the lymphatics, and thus push the lymph towards the mesenteric lymphatics [10].

3.3. Colon

The colon secretes mucus and absorbs water, the amount of which is comparatively small under normal conditions. Normally about 90% of water secreted proximally into the gut is reabsorbed by the small intestine. Nevertheless, the colon removes 80–90% of the remaining water and is capable of absorbing far greater quantities of water when the amount passing through it is higher than normal [31].

The microvascular architecture in the colon has been investigated by light microscopy and SEM of corrosion casts [3]. Since the microvascular organization of the muscular coat and the submucosa is quite similar to that of the small intestine, we will describe only the mucosal microvascular architecture of the colon.

At the mucosal surface of the rat colon, the honeycomblike plexus of capillaries outlines the stromal confines of the colonic mucosal glands (Figs. 5-16 and 5-17). The arterioles originating from the submucosal arterial plexus break up into capillaries around the bottom of the colonic glands. These capillaries ascend along the glands, with frequent communication with each other (Fig. 5-17). The capillary drainage into venules occurs most frequently at the luminal aspect of the mucosa, but there are some venule collecting capillaries at the deeper portion of the mucosa (Fig. 5-17). These venules are collected into thicker venules, which in turn drain into the submucosal venous plexus.

It is widely believed that colonic lymphatics are not prominent in the mucosa but can be identified lying in proximity to the muscularis mucosae [32]. SEM of corrosion casts made by intraparenchymal puncture injection has revealed that blind-ended rodlike lymphatics originate in the superficial mucosa (Fig. 5-18). These lymphatics descend perpendicularly through the glandular layer and lead into the lymphatic network made up of incomplete rings of lymphatics in the deep mucosa close to the muscularis mucosae (Fig. 5-18). Immediately below the muscularis mucosae in the submucosa, there is a polygonal network of thick collecting lymphatic vessels with valves (Fig. 5-19). The submucosal collecting lymphatic vessels drain into thicker vessels, which, gathering those in the muscular coat en route, ultimately lead into mesenteric lymphatic vessels.

3.4. Gut-Associated Lymphoid Tissues

Lymphoid cells not only occur singly in the lamina propria and the intraepithelial spaces of the gut, but are also organized into distinct aggregates that are distributed all along the gastrointestinal tract [33]. The lymphoid aggregates in the gastrointestinal tract, which are collectively referred to as the gut-associated lymphoid tissue (GALT), are separated from the gut lumen by a single or multilayered squamous or cuboidal epithelium. The gut-associated lymphoid organs sample orally acquired antigens through specialized cells, termed *M cells* [34], and initiate immune responses [35–37].

The lymphocytes that proliferate and differentiate in the gut-associated lymphoid tissues are transported by the lymphatics into the bloodstream [38]. The lymphocytes recirculate from the bloodstream into the lymphoid tissues through high endothelial venules (HEVs) [39,40].

3.4.1. Palatine tonsils. The blood microvascular organization of the palatine tonsils in humans and rabbits has been demonstrated by SEM of vascular corrosion casts [12], but the lymphatic arrangement has only been studied by light microscopy [41].

The arteries of the human palatine tonsil travel in the connective tissue capsule and give off many branches. They further branch into arterioles, which, giving off capillaries in the connective tissue septa en route, enter either the follicle or the interfollicular region. These arterioles, giving off capillaries in the follicle and the interfollicular region en route, reach the subepithelial region, where they form the dense subepithelial capillary network (Fig. 5-20). The capillary plexus in the follicle is coarser than that in the subepithelial region. The subepithelial capillary network overlying the follicle protrudes hemispherically towards the crypt (Fig. 5-20). The network that overlies the interfollicular region possesses many switchback capillary loops, which project into the epithelium (Fig. 5-20). However, immediately above the follicle are few switchback capillary loops. The subepithelial capillaries are sinusoids (20–40 μm in diameter) that have numerous fenestrae with diaphragms.

The subepithelial capillaries gather into venules that run in the subepithelial region. These venules, along with the venules of the follicle and interfollicular region, rather abruptly connect with HEVs in the interfollicular region. The HEVs course down in the interfollicular region and ultimately lead into the ordinary veins that run in the septa or in the capsule. The corrosion

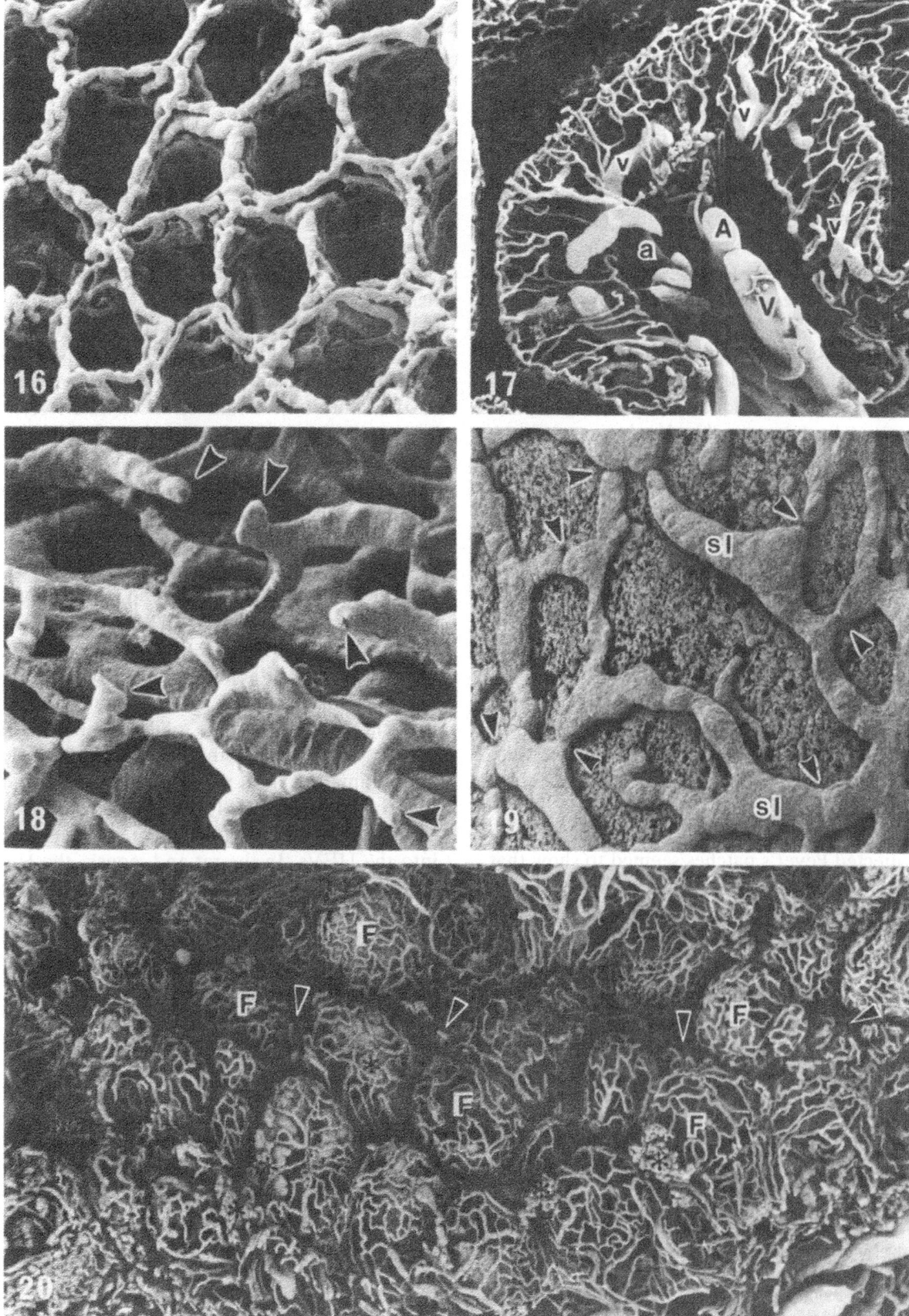

casts of the HEVs show irregular surfaces with numerous depressions and protrusions, reflecting the surface morphology of the HEVs.

It is noteworthy that the HEVs are downstream of the sinusoidal capillaries, which are located immediately below the epithelium infiltrated by numerous free cells, such as lymphocytes and macrophages. Such a microvascular organization of the tonsil seems to facilitate the transportation of certain substances (e.g., cytokins) taken up into the capillaries to the postcapillary venules, where their condensation can reach a certain level so that the substances can induce the differentiation of high endothelium [12]. Evidence has accumulated that endothelial specialization of HEVs may be determined by local microenvironmental factors, rather than by a selective population of particular vessels of distinct endothelial cell lineages [42–45].

The lymphatic vessels in the palatine tonsils exist in the interfollicular region and are well developed, as in other gut-associated lymphoid organs. The lymphatic vessels and sinuses form a plexus that surrounds the lateral and bottom surfaces of the follicle. The HEVs are located close to the lymphatic vessels. This intimate association suggests that fluids leaked out of HEVs during extravasation of lymphocytes can easily flow into the lymphatics.

3.4.2. Peyer's patches. Light microscopy of India ink-injected Peyer's patches seemed to show that germinal centers are relatively devoid of blood vessels. However, SEM of corrosion casts has revealed an elaborate system of blood vessels in the subepithelial regions, as well as in the follicle proper, of Peyer's patches in the rat [46] and mouse [47].

Peyer's patches are supplied by branches of the superior mesenteric artery that enter the small intestine along its mesenteric margin. Upon reaching the Peyer's patches, each artery divides into several branches, which enter the interfollicular region and run close to the serosa. These arteries repeat branching as they progress and give off two sets of arterioles: One set of arterioles run horizontally or parallel to the serosal surface and enter the follicle from the serosal surface, and the other set ascends vertically in the interfollicular region and enters the follicle from its lateral surface. Both sets of arterioles break up into capillaries in the follicle and the dome of the Peyer's patch. In addition to these arterioles, one or more relatively unbranched, prominent arterioles arise from the horizontal or the interfollicular arterioles, and ascend through the follicle and the dome to reach the subepithelial region of the dome, where they form an elaborate subepithelial capillary network. Some of the prominent arterioles in the interfollicular region ascend, supplying the pericryptal capillary network en route, and reach the tip of the villi, where the arterioles break up into villous subepithelial capillary networks.

The capillaries in the subepithelial region and those in the dome and the follicle drain into small postcapillary venules that lead into the HEVs and ultimately into the ordinary veins in the interfollicular region. The venules from the villi located over the interfollicular region also drain into the HEVs. In the rat and mouse, the HEVs are located around the serosal surface of the follicle, as well as in the interfollicular region. In the rabbit the HEVs occur invariably between adjacent follicles and a few appear around the bottom of the follicle. As in the palatine tonsils, the HEVs are downstream of the subepithelial capillary network of the dome (see above).

←

Figures 5-16–5-19. Blood (Figs. 5-16 and 5-17) and lymphatic (Figs. 5-18 and 5-19) corrosion casts of the rat colon. The luminal view of the blood microvasculature shows a honeycomb pattern (Fig. 5-16). The arterioles (a) from the submucosal artery (A) break up into capillaries at various level of the mucosa (Fig. 5-17). The mucosal venules (v) originate mostly in the superficial mucosa, and descend through the glandular layer to lead into the submucosal vein (V) (Fig. 5-17). Blind-ended lymphatics (large arrowheads) begin in the superficial mucosa and lead into the lymphatic network in the deep mucosa (Fig. 5-18). The mucosal lymphatics drain into the submucosal network (sl) made up of polygonal meshes of lymphatic vessels with notches (small arrowheads), indicative of valve locations (Fig. 5-19). Fig. 5-16. ×330; Fig. 5-17. ×80; Fig. 5-18. ×120; Fig. 5-19. ×40.

Figure 5-20. Blood vascular corrosion cast of the human palatine tonsils. The subepithelial capillary network overlying the follicle (F) protrudes hemispherically towards the lumen. Between these protrusions are switchback loops of capillaries (arrowheads). * = leaked resin. ×40.

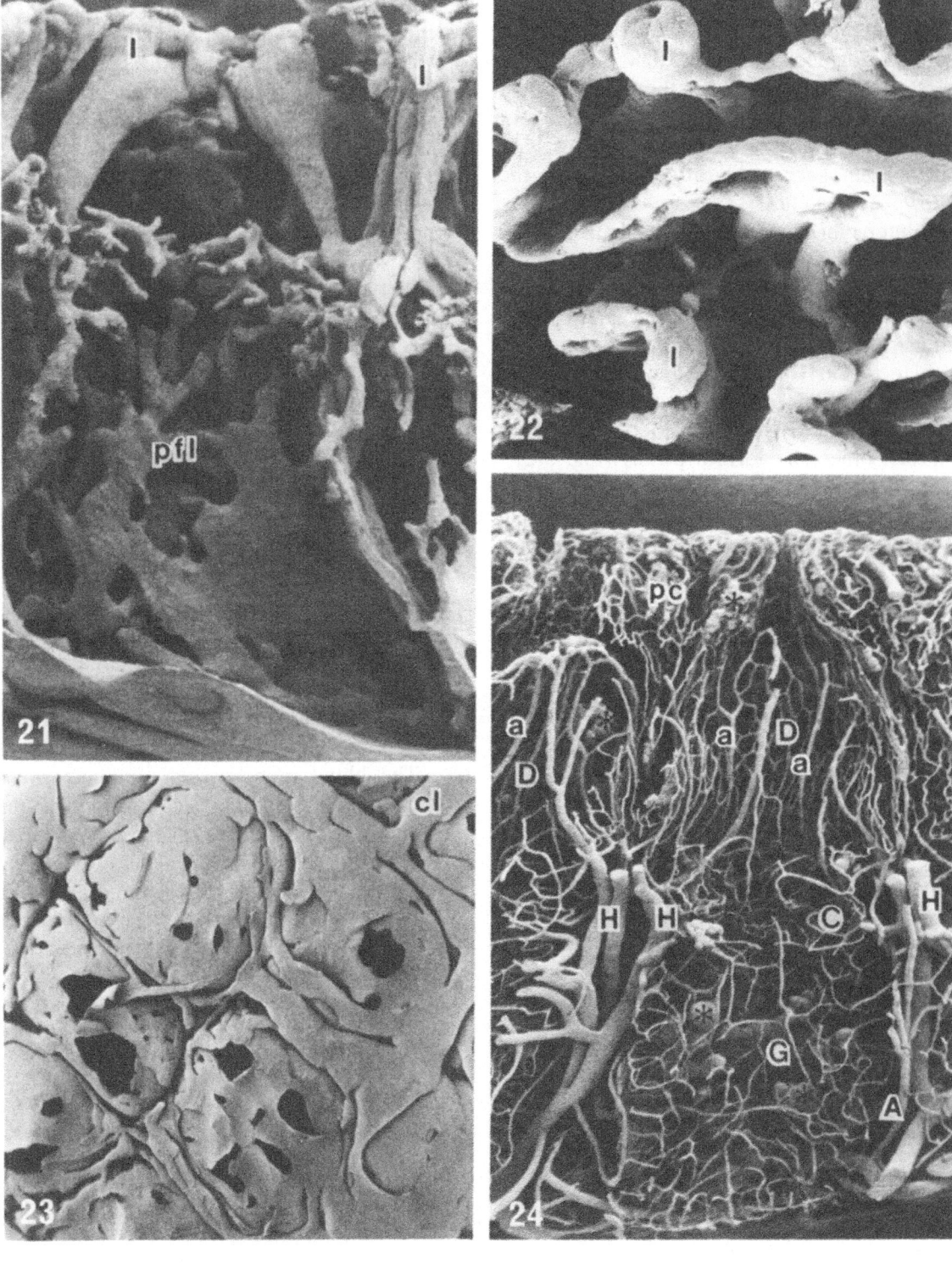

Some researchers in the 19th century described dense, irregular networks of the lymphatic vessels surrounding the follicles of the Peyer's patches in the cat and dog. However, these light microscopic studies did not reveal the three-dimensional architecture of the lymphatic system. SEM of corrosion casts has revealed a voluminous lymphatic system surrounding the follicle of the rabbit Peyer's patches.

The interconnected central lacteals in the villi overlying the interfollicular region of the rabbit Peyer's patches are connected across the glandular layer by straight lymphatic vessels, which continue into the lymphatic plexus in the interfollicular region (Figs. 5-21 and 5-22). In the upper part of the interfollicular region, there are many blind-ended lymphatics that interconnect with each other and form a plexus (Fig. 5-21). The plexus gradually fuses and becomes flat, and forms the perifollicular lymphatic network that surrounds the lateral surface of the individual follicles (Fig. 5-21). The perifollicular lymphatic vessels surrounding the lower half of the follicle become particularly wide and give the appearance of sinuses (Fig. 5-21). The sinuses extend and surround the large area of the follicle bottoms (Fig. 5-23). There are no lymphatic vessels or sinuses within the follicle and the dome. The perifollicular lymphatic network drains into short, relatively small lymphatic vessels with valves (Fig. 5-23). These vessels lead into larger collecting lymphatic vessels that run between adjacent bottoms of the follicles. The collecting lymphatic vessels in the submucosa gather the lymphatic in the muscular coat en route and ultimately lead into the mesenteric lymphatic vessels.

The perifollicular lymphatic networks are located between the blood capillary network of the follicle and the HEVs in the interfollicular region. The position of the HEVs is frequently extremely close to the lymphatic vessels. Such an intimate association of the lymphatic vessels with the HEVs suggests that fluids and macromolecules that leak out of the HEVs in association with lymphocyte migration drain promptly into the lymphatic vessels, thus maintaining tissue fluid homeostasis. It has been reported that HEVs are more leaky than other kinds of blood vessels [48]. Blood vessels run through the interruption of the perifollicular lymphatic network. The submucosal arteries and veins frequently penetrate through the lymphatic sinuses covering the bottom of the follicles.

3.4.3. Appendix. The blood microvascular system of the appendix has yet to be studied. The existence of a large lymphatic sinus around the follicle in the human appendix was described by Teichmann and later confirmed by Lockwood. India-ink injection studies by Frey and Crabb and Kelsall demonstrated the general architecture of the lymphatic system in the rabbit appendix. The perifollicular lymphatic sinus was also demonstrated by the intraparenchymal air-injection technique. Recently SEM of lymphatic corrosion casts has demonstrated the three-dimensional organization of the lymphatic system of the rabbit appendix in general [14].

The blood microvascular architecture of the rabbit appendix (Fig. 5-24) resembles that of rabbit Peyer's patches, except for the area surrounding the intestinal glands overlying the interfollicular region. The capillaries surrounding the intestinal glands drain into venules that descend through the isthmus and continue to the HEVs in the interfollicular region.

The organization of the lymphatics of the rabbit appendix (Figs. 5-25 and 5-26) also resembles that of the Peyer's patches, although there are some differences. Between the intestinal glands of the

←

Figures 5-21–5-23. Lymphatic corrosion casts of rabbit Peyer's patch (Fig. 5-21: tangential section; Fig. 5-22: luminal view; Fig. 5-23: serosal view). Interconnected central lacteals (l) in the villi overlying the interfollicular region (Figs. 5-21 and 5-22) drain into the perifollicular lymphatic network (pfl) (Fig. 5-21). The bottom of the follicle is largely covered by a perifollicular lymphatic sinus, which leads into the collecting lymphatics (cl) in the submucosa (Fig. 5-23). Fig. 5-21. ×80; Fig. 5-22; ×110, Fig. 5-23. ×30.
Figure 5-24. Blood vascular corrosion cast of the rabbit appendix (tangential section). Many arterioles enter the follicle from its lateral and basal surfaces, and form the capillary network of the follicle, that of the germinal center (G) being looser than that of the corona (C). Some prominent, relatively unbranched arterioles (a) arising from the interfollicular arteries (A) reach the subepithelial region of the dome (D), where the arterioles connect with the elaborate subepithelial capillary network. The venules from the dome and the follicle lead into HEVs (H) in the interfollicular region. pc = pericryptal blood vessels, * = leaked resin. ×60.

66

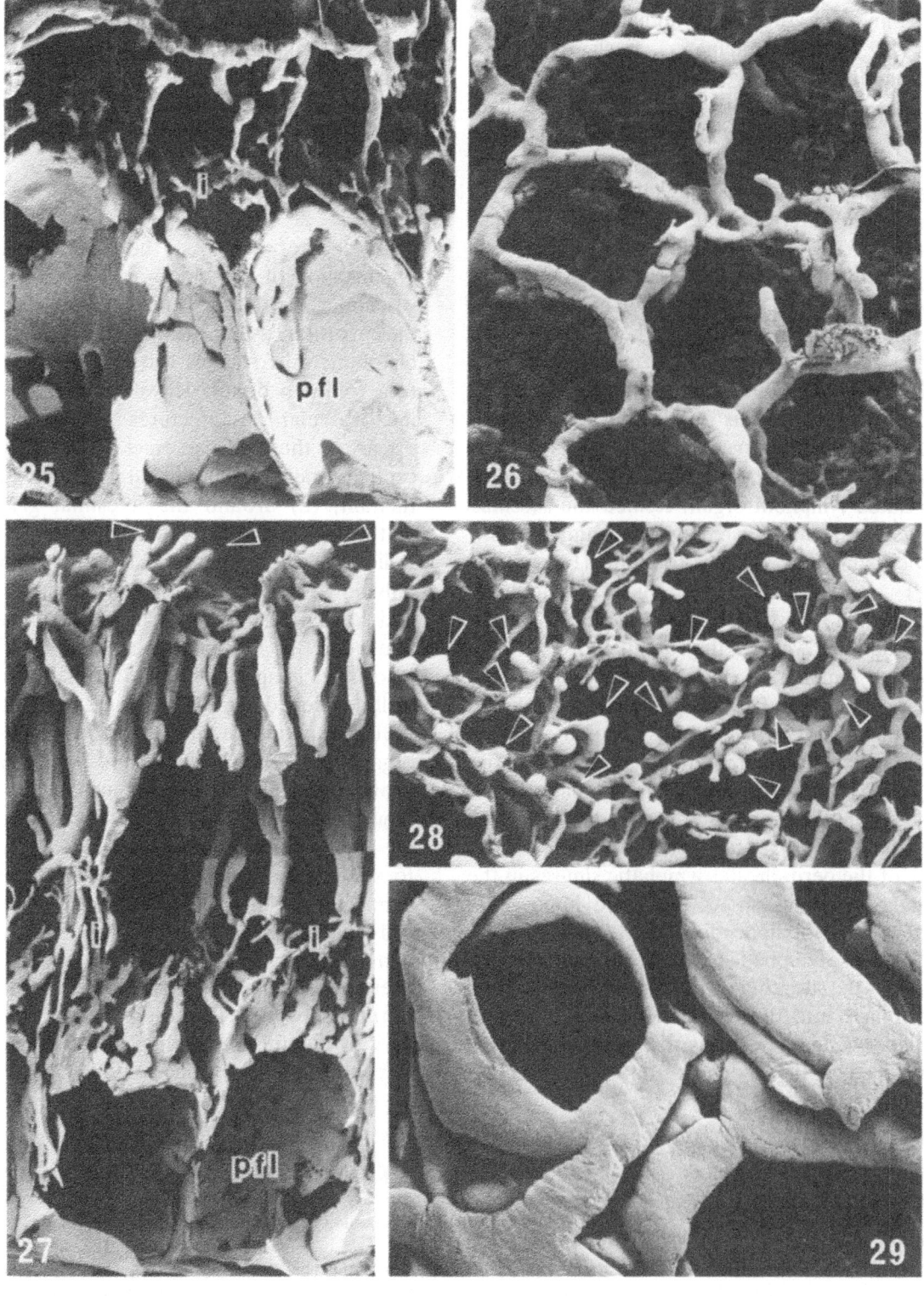

appendix mucosa, there is a network made up of interconnecting rings of lymphatics of various sizes (20–140 μm in diameter). Each side of the net frequently has one or two short side branches that project toward the lumen. From the network in the mucosa, straight lymphatics descend perpendicularly through the isthmus. At the base of the isthmus the lymphatics connect with those in the interfollicular region [i.e., thymus-dependent areas (TDAs)]. In the TDA there is a dense lymphatic plexus that is continuous with the upper part of the perifollicular lymphatic sinus surrounding each follicle. The perifollicular sinuses in the appendix are more developed than those in the Peyer's patches. However, the sinuses of the appendix have some interruptions through which blood vessels pass.

In the rabbit there is another prominent gut-associated lymphoid organ, termed the *sacculus rotundus*, which is located at the terminal portion of the ileum. The sacculus rotundus (Figs. 5-27–5-29) also possesses perifollicular lymphatic sinuses that are quite similar to those of the appendix. However, the architecture of the mucosal lymphatics differs from that of the Peyer's patches and of the appendix. Between the intestinal glands in the sacculus rotundus, there is a network made up of interconnecting lymphatics of various sizes, from which many short blind-ended lymphatics project perpendicularly into the small villi (Figs. 5-27 and 5-28). The serosal view of the perifollicular lymphatic sinuses in the sacculus rotundus shows the some honeycomb pattern as in the appendix (Fig. 5-29).

The dense lymphatic plexus in the interfollicular region is the site where lymphocytes migrate from the follicle into the lymphatics. As the perifollicular lymphatic sinuses are extremely well developed in the appendix and in the sacculus rotundus, as well as in the Peyer's patches, it appears that the lymphoid follicles are soaked in a vast lymphatic pool. Thus, the perifollicular lymphatic sinuses seem to have a great capacity as a reservoir and as a drainage route for fluids and lymphocytes [14].

4. Conclusion

This chapter reviews the three-dimensional organization of the blood and lymphatic microvasculature of the digestive tract, as revealed by the corrosion casting/SEM method [1]. Blood and lymphatic microvessels show a specific architecture for each organ. However, there are some microvascular architectures common to the organs of the digestive tract. These are (1) the capillary network immediately below the epithelium, (2) around the intestinal glands, (3) in the muscular coat, and (4) in the lymphoid follicles. In the stomach mucosa the lymphatics lie only in the deep mucosa, while in both small and large intestine the lymphatics initiate with blind ends in the superficial mucosa. The gut-associated lymphoid follicles are surrounded by well-developed lymphatic sinuses. In the interfollicular region of the gut-associated lymphoid organs are HEVs, which are located in close proximity to the lymphatics. Since the tissue fluids flow in the connective tissue spaces, our studies on the microvasculature should be extended to studies dealing with the organization of the connective tissue spaces. The recently established cell-maceration/SEM method [49,50], as well as conventional SEM methods, have proven to be useful to disclose the architecture of the collagen fibrillar networks. Thus these methods promise to be very helpful in unveiling the fine organization of connective tissue spaces.

←

Figures 5-25 and 5-26. Lymphatic corrosion casts of the rabbit appendix (Fig. 5-25: tangential section; 5-26: luminal view). The ringlike lymphatic network between the intestinal glands is connected across the isthmus with the lymphatic plexus (i) in the interfollicular region (Fig. 5-25). The plexus drains into well-developed perifollicular lymphatic sinuses (pfl) (Fig. 5-25). Fig. 5-25. ×30; Fig. 5-26. ×30.

Figures 5-27–5-29. Lymphatic corrosion casts of the rabbit sacculus rotundus (Fig. 5-27: tangential section; 5-28: luminal view; 5-29: serosal view). Many rodlike lymphatics (arrowheads) protrude towards the intestinal lumen from the lymphatic network in the glandular layer (Figs. 5-27 and 5-28). The network drains into perpendicularly oriented thick lymphatics, which in turn lead into the lymphatic plexus (i) in the interfollicular region (Fig. 5-27). The plexus is continuous with the well-developed perifollicular lymphatic sinus (pfl). The serosal view of the perifollicular sinus shows a honeycomb pattern (Fig. 5-29). Fig. 5-27. ×30; Fig. 5-28. ×30; Fig. 5-29. ×30.

68

References

1. Murakami T. Application of the scanning electron microscope to the study of fine distribution of the blood vessels. *Arch Histol Jpn* 32:445–454, 1971.
2. Browning J, Gannon B. The microvascular architecture of rat proximal duodenum, with particular reference to Brunner's glands. *Biomed Res* 5:245–258, 1984.
3. Browning J, Gannon B. Mucosal microvascular organization of the rat colon. *Acta Anat* 126:73–77, 1986.
4. Gannon BJ. The co-existence of fountain and tuft patterns of blood supply in individual intestinal villi of rabbit and man: Resolution of an old controversy. *Bibl Anat* 20:130–133, 1981.
5. Gannon BJ, Gore RW, Rogers PAW. Is there an anatomical basis for a vascular counter-current mechanism in rabbit and human intestinal villi? *Biomed Res* 2 (Suppl.): 235–241, 1981.
6. Gannon BJ, Browning J, O'Brien P. The microvascular architecture of the glandular mucosa of rat stomach. *J Anat* 133:667–183, 1982.
7. Gannon B, Browning J, O'Brien P, Rogers P. Mucosal microvascular architecture of the fundus and body of human stomach. *Gastroenterology* 86:866–875, 1984.
8. Ohashi Y, Kita S, Murakami T. Microcirculation of the rat small intestine as studied by the injection replica scanning electron microscope method. *Arch Histol Jpn* 39:271–282, 1976.
9. Ohtani O, Kikuta A, Ohtsuka A, Taguchi T, Murakami T. Microvasculature as studied by the microvascular corrosion casting/scanning electron microscope method. I. Endocrine and digestive system. *Arch Histol Jpn* 46:1–42, 1983.
10. Ohtani O. Three-dimensional organization of lymphatics and its relationship to blood vessels in rat small intestine. *Cell Tissue Res* 248:365–374, 1987.
11. Ohtani O. Corrosion casts in liver and stomach microcirculation. In: Motta PM (ed.), *Cell and Tissue: A Three-Dimensional Approach by Modern Techniques in Microscopy*, Alan R. Liss, New York, pp 317–326, 1989.
12. Ohtani O, Kikuta A, Terasawa K, Higashikawa T, Yamane T, Taguchi T, Masuda Y, Murakami T. Microvascular organization of human palatine tonsils. *Arch Histol Cytol* 52, in press.
13. Ohtani O, Ohtsuka A. Three-dimensional organization of lymphatics and their relationship to blood vessels in rabbit small intestine. A scanning electron microscopic study of corrosion casts. *Arch Histol Jpn* 48:255–268, 1985.
14. Ohtani O, Ohtsuka A, Owen RL, Three-dimensional organization of the lymphatics in the rabbit appendix. A scanning electron and light microscopic study. *Gastroenterology* 91:947–955, 1986.
15. Ohtani O, Murakami T. Lymphatics and myenteric plexus in the muscular coat in the rat stomach: A scanning electron microscopic study of corrosion casts made by intra-arterial injection. *Arch Histol Jpn* 50:87–93, 1987.
16. Ohtsuka A, Ohtani O. The microvascular architecture of the rabbit stomach corpus in vascular corrosion casts. *Scann Electron Microsc* IV:1951–1956, 1984.

17. Raschke M, Lierse W, van Ackeren H. Microvascular architecture of the mucosa of the gastric corpus in man. *Acta Anat* 130:185–190, 1987.
18. Guth PH, Robsenberg A. In vivo microscopy of the gastric microcirculation. *J Digest Dis* 17:391–398, 1972.
19. Barclay AE, Bentley FH. The microvascularization of the human stomach. *Br J Radiol* 22:62–69, 1949.
20. Piasecki C. Blood supply to the human gastroduodenal mucosa with special reference to the ulcer-bearing areas. *J Anat* 118:295–335, 1974.
21. Fujita H, Fujita T. *Textbook of Histology*, part 2, 3rd ed., Tokyo, Igaku-Shoin (in Japanese).
22. Johansson C, Aly A. Stimulation of gastric mucus output by somatostatin in man. *Eur J Clin Invest* 12:37–39, 1982.
23. Nagata H, Guth PH. In vivo observation of the lymphatic system in the rat stomach. *Gastroenterology* 86: 1443–1450, 1984.
24. Gannon B, Brownning J, Rogers P, Harper B. Microvascular organization in the intestine. In: Koo A, Lam SK, Smaje LH (eds.), *Microcirculation of the Alimentary Tract*, World Scientific, Singapore, pp 39–55, 1983.
25. Florey HW, Harding HE. Further observations on the secretion of Brunner's glands. *J Pathol Bacteriol* 39: 255–276, 1934.
26. Ranvier L. Des lymphatiques de la villosite intestinale ches le rat et le lapin. *CR Acad Sci Paris* 123:923–925, 1896.
27. Leak LV, Burke JF. Ultrastractural studies on the lymphatic anchoring filaments. *J Cell Biol* 36:129–149, 1968.
28. Papp M, Rohlich P, Rusznyák I, Toro I. An electron microscopic study of the central lacteals in the intestinal villus of the cat. *Z Zellforsch* 57:475–486, 1962.
29. Intaglietta M, Gross JF. Vasomotion, tissue fluid flow and the formation of lymph. *Int J Microcirc Clin Exp* 1:55–65, 1982.
30. Skalak TC, Schmid-Schönbein GW, Zweifach BW. New morphological evidence for a mechanism of lymph formation in skeletal muscle. *Microvasc Res* 28:95–112, 1984.
31. Schults SG. Ion transport by mammalian large intestine. In: Johnson (ed.), *Physiology of the Gastrointestinal Tract*, Vol. 2, Raven Press, New York, pp 991–1002, 1981.
32. Kveitys PR, Wilborn WH, Granger DN. Effects of net transmucosal volume flux on lymph flow in the canine colon. Structural-functional relationship. *Gastroenterology* 81:1080–1090, 1981.
33. Owen RL, Bhalla DK. Lympho-epithelial organs and lymph nodes. In: Hodges GM, Carr KE (eds.), *Biomedical Research Applications of Scanning Electron Microscopy*, Vol. 3, Academic Press, London, pp 79–169, 1983.
34. Owen RL, Jones AL. Epithelial cell specialization within human Peyer's patches: An ultrastructural study of intestinal lymphoid follicles. *Gastroenterology* 66:189–203, 1974.
35. Owen RL. Sequential uptake of horseradish peroxidase by lymphoid follicle epithelium of Peyer's patches in the

normal unobstructed mouse intestine: An ultrastractural study. *Gastroenterology* 72:440–451, 1977.

36. Wolf JL, Rubin DH, Finberg R, Kauffman RS, Sharpe AH, Trier JS, Fields BN. Intestinal M cells: A pathway for entry of reovirus into the host. *Science* 212:471–472, 1981.

37. Owen RL, Pierce NF, Apple RT, Cray WC Jr. M cell transport of vibrio cholera from the intestinal lumen into Peyer's patches: A mechanism for antigen sampling and for microbial transepithelial migration. *J Infectious Dis* 153:1108–1118, 1986.

38. Ford WL. Lymphocyte migration and immune responses. *Prog Allergy* 19:1–59, 1975.

39. Gowans JL, Knight EL. The role of re-circulation of lymphocytes in the rat. *Proc R Soc B* 159:257–282, 1964.

40. Stamper HB Jr., Woodruff JJ. Lymphocyte homing into lymph node: In vivo demonstration for high-endothelial venules. *J Exp Med* 144:828–833, 1976.

41. Hellman T. Der lymphatische Rachenring. In: Möllendorff WV (ed.), *Handbuch der Mikroskopischen Anatomie des Menschen*, V/1, Julias Springer, Berlin, pp 245–289, 1927.

42. Freemont AJ, Ford WL. Functional and morphological changes in post-capillary venules in relation to lymphocytic infiltration into BCG-induced granulomata in rat skin. *J Pathol* 147:1–12, 1985.

43. Bevilacqus MP, Pober JS, Wheeler ME, Cotran RS, Gimbrone MA Jr. Interleukin 1 acts on cultured human vascular endothelium to increase the adhesion of polymorphonuclear leukocytes, monocytes, and related leukocyte cell lines. *J Clin Invest* 76:2003–2011, 1985.

44. Cavender DE, Haskard DO, Joseph B, Ziff M. Interleukin 1 increases the binding of human B and T lymphocytes to endothelial cell monolayers. *J Immunol* 136:203–207, 1986.

45. Duijvestijn AM, Schreiber AB, Butcher EC. Interferon-γ regulates an antigen specific for endothelial cells involved in lymphocyte traffic. *Proc Natl Acad Sci USA* 83: 9114–9118, 1986.

46. Bhalla DK, Murakami T, Owen RL, Microcirculation of intestinal lymphoid follicles in rat Peyer's patches. *Gastroenterology* 81:481–491, 1981.

47. Yamaguchi K, Schoefl GI. Blood vessels of the Peyer's patch in the mouse: II. High endothelial venules. *Anat Rec* 206:419–438, 1983.

48. Yamaguchi K, Schoefl GI. Blood vessels of the Peyer's patch in the mouse: II In vivo observations. *Anat Rec* 206:403–417, 1983.

49. Ohtani O, Three-dimensional organization of the connective tissue fibers of the human pancreas: A scanning electron microscopic study of NaOH treated tissues. *Arch Histol Jpn* 50:557–566, 1987.

50. Ohtani O, Ushiki T, Taguchi T, Kikuta A. Collagen fibrillar networks as skeletal frameworks: A demonstration by cell-maceration/scanning electron microscope method. *Arch Histol Cytol* 51:249–261, 1988.

Author's address:
Prof. Osamu Ohtani
Department of Anatomy
Faculty of Medicine
Toyama Medical and
  Pharmaceutical University
2630 Sugitani, Toyama, 930-01
Japan

Three-Dimensional Architecture of Blood, Lymphatic, and Biliary Pathways in the Liver by SEM of Corrosion Casts

KAZUHIDE YAMAMOTO, TATSUYA ITOSHIMA, TAKAO TSUJI, & TAKURO MURAKAMI

1. Introduction

Blood supply and drainage are essential to maintain the normal function of organs. The liver is situated between the gastrointestinal tract and the heart, and receives both portal venous and hepatic arterial blood flow, which constitutes up to one fifth of the cardiac output [1]. Each portal venous and hepatic arterial blood supply plays a distinct role [1,2]. The portal venous blood contains nutrients and toxic substances absorbed in the intestine, which are metabolized and detoxified during its course in the hepatic microcirculation. The hepatic arterial blood flow, with its high pressure, brings oxygen to the liver. Both portal and hepatic arterial blood mix in the hepatic microcirculation, delivering oxygen and nutrients throughout the acinus. An exact understanding of the hepatic microvascular architecture is, therefore, fundamental to analyze the normal function, and also the pathological condition, of the liver. The hepatic lymphatic channels and the biliary tract are other important structures for maintaining the normal function of the liver and participate in draining tissue fluid and bile, respectively.

The three-dimensional arrangement of microvasculature has been studied with various methods. These include light microscopy of tissues injected with various dyes or resin into blood vessels, binocular microscopic observation of vascular casts, and light microscopic observation of serial sections. However, these methods have still had difficulty in revealing three-dimensional architecture with adequate resolution. Scanning electron microscopic (SEM) observation of corrosion casts has solved this problem [3]. This method provided (1) easy surveillance of the microvasculature under a wide range of magnification and (2) three-dimensional observation with sufficient depth of field. Using this method, great progress has been made in the anatomical and functional study of the microcirculation, and many excellent reviews are already in the literature [4–7]. This technique has been applied to the study of the hepatic microvasculature [8–15], the hepatic lymphatics [16], and the biliary tract [17–23].

In this paper, we will discuss the method for preparing corrosion casts, not only for the hepatic microvasculature, but also for the lymphatic and biliary channels. We will also review the three-dimensional architecture of the hepatic microvasculature, the lymphatics, and the biliary tract studied by the present method with reference to their functional implication and some pathological changes.

2. Methods for Preparing Corrosion Casts and SEM Observation

2.1. Preparation of Hepatic Microvascular Casts

Various types of casting media are available commercially to make suitable casts for scanning electron microscopy and have been reviewed in

Motta, P.M., Murakami, T., and Fujita, H. (eds.), Scanning Electron Microscopy of Vascular Casts: Methods and Applications.

72

the literature [4–7]. Mercox (Dainippon Ink Co., Ltd.) and Batson's plastic (Polyscience Inc.) are easy to use when the resin is mixed with catalyst just prior to injection. Laboratory–prepared resin is also available [3], i.e., semipolymerized resin is mixed with benzoyl peroxide just prior to injection. Mercox can be made less viscous by mixing with monomeric methylmethacrylate in the proper ratio [9]. The more dilute the resin is, the larger the amount of catalyst to be added.

Prior to injection of resin, the liver is perfused with physiological saline or a Ringer's solution in order to wash out the blood. In animals, heparin can be injected intravenously before perfusion. The animal liver is perfused through a catheter inserted retrogradely into the abdominal aorta [13]. The abdominal aorta is clamped above the celiac artery just after starting perfusion, and the inferior vena cava is cut above the liver. This perfusion enables a complete washout of blood, because the liver is perfused without interruption of the blood flow. In autopsy or surgically removed samples, the liver is perfused with the above solution from both the hepatic artery and the portal vein. Thorough perfusion is important for complete filling of resin into minute vessels. The perfusion pressure is maintained at each physiological pressure of the hepatic artery or the portal vein by monitoring the pressure with a manometer.

Injection of resin is performed through catheters used for liver perfusion. In animals, resin is injected through a catheter inserted into the abdominal aorta, and complete vascular casts are easily prepared because resin is injected from both the hepatic artery and the portal vein at the same time. The injecting pressure is maintained at a physiological pressure, as in liver perfusion. In autopsy or surgically removed liver, resin is injected from both the hepatic artery and the portal vein. The amount of resin can be controlled to fit with the purpose of the study. Partial injection is recommended when the relationship between portal blood vessels and sinusoids is to be studied. Injection of resin is continued until the resin polymerizes and becomes hard. The liver is bathed in water at 60°C for polymerization of resin overnight and is then macerated in a 30% NaOH solution at 60°C. Vascular casts are washed in running water and dried in air.

2.2. Preparation of Replica for the Hepatic Lymphatics

Resin or dyes can be injected into the hepatic lymphatics through the bile ducts [16]. An excess amount of resin or dyes injected into the biliary tract leaks at the terminal bile ducts into the surrounding interstitial space and drains into the portal lymphatics. The regurgitation of biliary materials into the portal lymphatics was first noted a century ago [24]. Using this route, a replica of the hepatic lymphatics can be prepared. After perfusion of the liver, resin diluted with monomeric methacrylate is injected into the bile duct. Injection into the biliary tract is continued at a pressure that is less than 16 mmHg until the resin polymerizes in the syringe. The lymphatic casts are treated as in vascular casts.

2.3. Preparation of Replica for the Biliary Tract and Bile Canaliculi

Less viscous resin can be injected retrogradely into the common bile duct [17,19]. Resin (Mercox) is diluted with monomeric methyl methacrylate at the ratio of 1:1 to 1:3 and is then injected into the common bile duct while monitoring to maintain the pressure below 16 mmHg. Various amounts of catalyst are added to the mixture to control the polymerization time. A small amount of catalyst is recommended because a slower polymerization time is necessary for complete filling of the biliary tree.

For injection into the bile canaliculi, a mixture of 50–60% (V/V) methyl methacrylate monomer and 40–50% (V/V) 2-hydroxy methacrylate monomer is prepared and supplemented with 1.5% (W/V) benzoyl peroxide and 1.5% N,N-dimethylaniline just prior to injection [18]. After perfusion with Ringer's solution or physiological saline, the liver is perfusion fixed with 2% glutaraldehyde. Immediately after glutaraldehyde fixation, the resin is infused into the common bile duct at a pressure of 70–80 mmHg. The injection is continued without a decrease in pressure until the medium is hardened in the syringe. As in vascular casts, the resin is polymerized in a water bath and tissue is macerated in a 30% NaOH solution. The delicate replica of the bile canaliculi is rinsed in water, with repeated changes, and is freeze dried to avoid surface tension.

2.4. Preparation of Casts for SEM

Corrosion casts are observed under a binocular microscope and are microdissected using a fine needle and forceps. Casts are sometimes frozen and cut with a razor blade or a saw, and the cut surface is, if necessary, further microdissected under a binocular microscope. Microdissection is the most important process for exposing an interesting area to be observed. Sometimes casts are microdissected after SEM observation. Repeated microdissection and SEM observations provide further details. Microdissected specimens are mounted on metal stubs with conductive paste, sputter-coated with gold or platinum, and observed in a scanning electron microscope with an accelerating voltage of 10–15 kV. Vapored osmium tetroxide can be used with or instead of a metal coating. Such specimens can be microdissected under a scanning electron microscope with a manipulator. Stereopairs at an angle of 5–10° must be taken for analysis of the three-dimensional interrelationship among branching casts.

3. Hepatic Microvascular Architecture

3.1. Normal Hepatic Microvasculature

Identification of each hepatic microvasculature can be made either by tracing back to the major branches or by the surface morphology with imprints of endothelial cells [6]. Arteries and arterioles have longitudinal indentations from endothelial-cell bodies and nuclei, whereas the portal or hepatic veins have scattered oval or ellipsoid indentations from nuclei.

The hepatic artery and portal vein are the two afferent blood vessels in the liver. In the portal canal, the hepatic artery branches off an arterial network around the bile duct, the peribiliary plexus, or the peribiliary portal system [8], and branches into the portal connective tissue [Fig. 6-1]. In the medium-sized and smaller portal tracts, these arterial blood vessels in the portal

tract branch into capillaries and terminate directly in the periportal sinusoids.

The peribiliary plexus in the large- and medium-sized portal canals is double layered, i.e., the inner arterial network and outer portal venous network, which finally drain into the periportal sinusoids [8,9]. The portal tract is larger, and the peribiliary plexus is thicker. In the small portal tract, the plexus is formed by a layer of blood vessels. The three-dimensional architecture of the plexus was clearly demonstrated for first time by SEM of microvascular casts [8,9].

The functional significance of the peribiliary arterial plexus is still unknown. The most probable function is that it may be involved in the modification of bile. The plexus is found in most mammalian livers. In the lamprey, the bile ducts are not surrounded by the peribiliary plexus. Instead, the bile ducts themselves are convoluted, like tubules in the kidney [20]. Double injection of resin into vascular and biliary channels reveals that the bile ducts have an intimate relationship with vascular networks. The presence of an intimate biliovascular relationship, even in the lamprey liver, suggests that the bile ducts may exhibit a dynamic exchange of constituents between the blood and bile. Another possibility is feedback control of biliary secretion [25]. Some materials secreted into the bile are reabsorbed at the peribiliary plexus and drain back into the liver, and may contribute to control the secretory process of hepatocytes.

The manner of arterial termination has long been a matter of debate. SEM of microvascular casts settled the argument [9,10,13]. The manner of arterial termination differs among species, and such differences might be one of the reason for a long debate. In the rat liver, hepatic arterioles often terminate in the terminal portal vein or inlet venules, and these arterio-portal anastomoses are numerous (Figs. 6-2, and 6-3). However, in the human liver arterio-portal anastomosis is rare and arterial capillaries terminate [8] directly in periportal sinusoids (Fig. 6-4) [13]. In any species, intralobular arterioles [26] are rarely observed [9,13]. Arterial termination in the periportal sinusoids may provide a structural basis for the oxygen gradient in the acinus, i.e., high oxygen content in zone 1 and low oxygen in zone 3. The zonal oxygen gradient is reflected in the

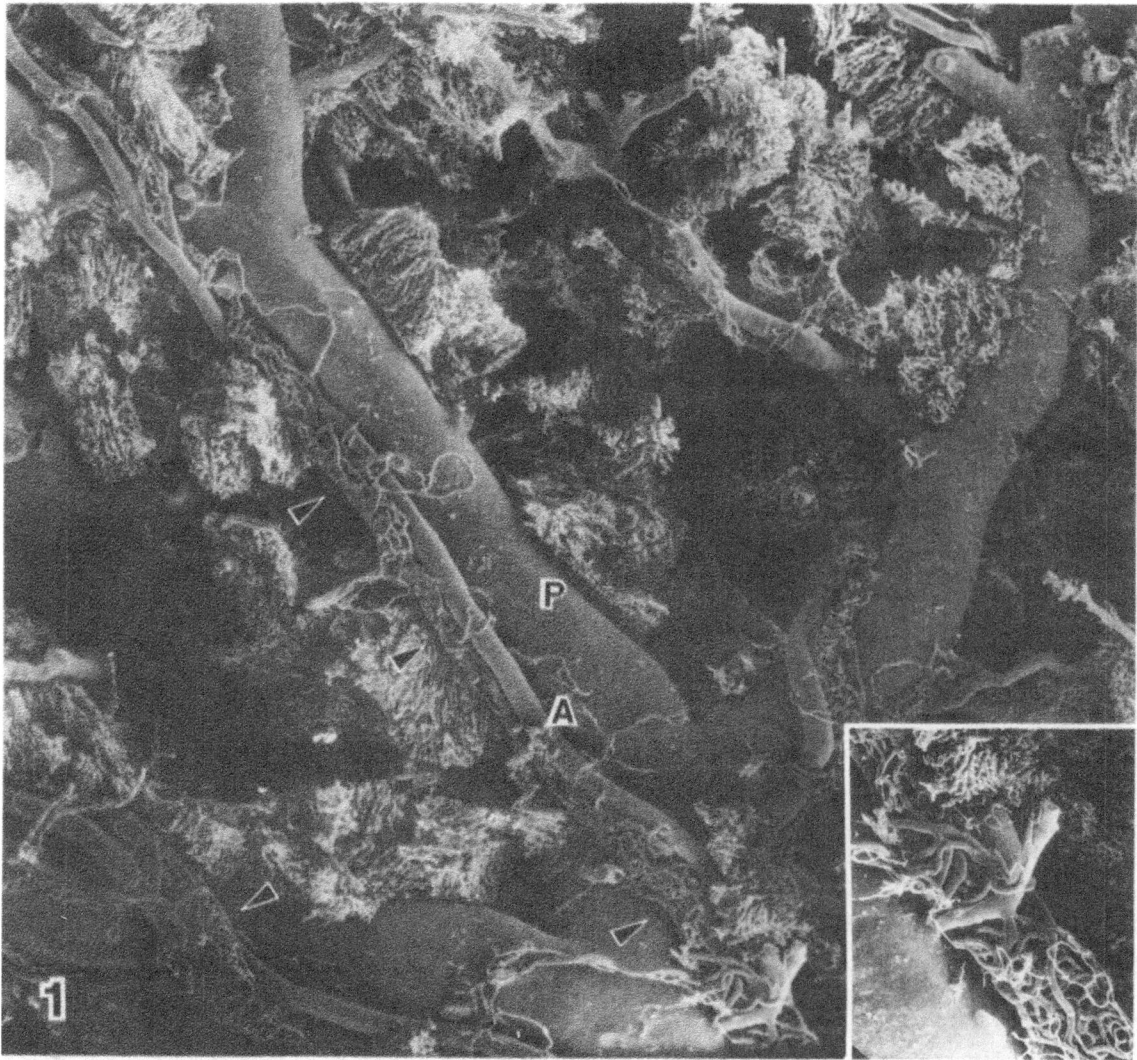

Figure 6-1. Scanning electron micrograph of vascular casts of the rat liver. The branching portal vein (P) is accompanied by the hepatic artery (A) and peribiliary plexus (arrowheads). ×40. Inset: Higher magnification of peribiliary plexus. Note rich network surrounding the bile duct. ×60.

compartment with metabolic enzymes in each zone [1].

The portal vein divides into terminal branches. The portal plexus is coarse and drains into the sinusoids, peribiliary efferent vessels, or portal terminal branches.

3.2. *Microvascular Alteration in the Pathological Liver*

This technique has been applied to elucidating microvascular changes in liver diseases, such as human liver cirrhosis (Fig. 6-5) [11,14] and experimental [12] and human hepatocellular carcinoma [15]. In general, arterioles are shown to proliferate in various pathological liver conditions. In liver cirrhosis, a rich arteriolar plexus surrounds cirrhotic nodules (Fig. 6-5) and arterioles connect with sinusoids, while portal venules are tortuous and show narrowing and dilatation. The arteriolar plexus is present, even in thin septa subdividing nodules. Arterio-portal anastomosis is clearly demonstrated in liver cirrhosis [14].

Arterial blood vessels have been suggested to

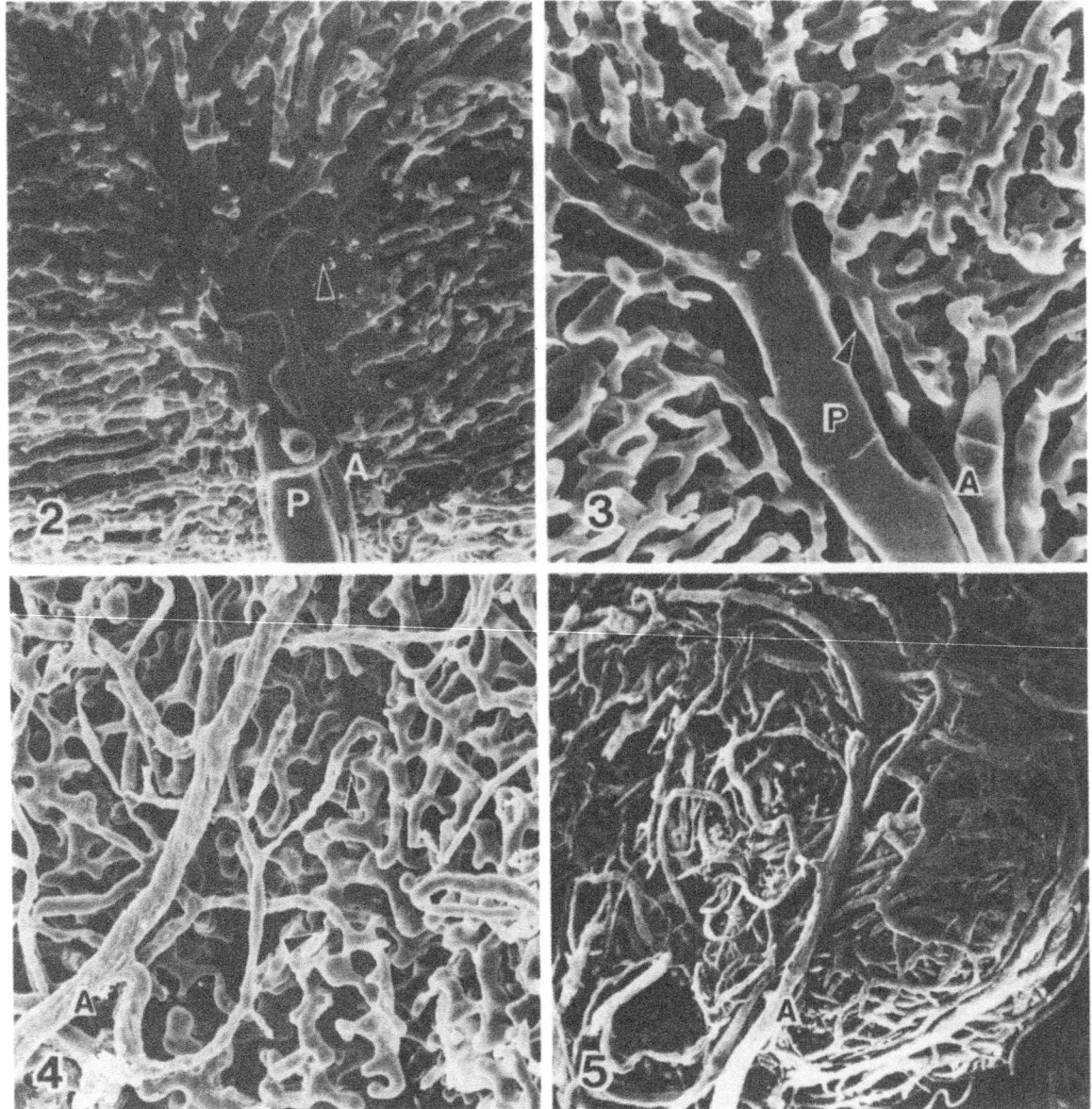

Figure 6-2. Terminal portal tract in the rat liver. An arteriole (A) anastomoses with a branch of the portal vein (P) (arrowhead). ×170. (Reproduced from Ref. 13 with permission.)

Figure 6-3. An arteriole (A) directly connects with sinusoids (S) and shares a common entrance with an inlet venule. P = portal vein. ×270.

Figure 6-4. Normal human liver. Arterioles (A) connect directly with periportal sinusoids (arrowheads). ×220.

Figure 6-5. Microvascular casts in human liver cirrhosis. Note rich arteriolar plexus surrounding a cirrhotic nodule. A = arteriole. ×30. (Reproduced from Ref. 11 with permission.)

play an important role as supplying vessels in some cancers, including hepatocellular carcinoma [12,15]. Proliferation of arterial blood vessels is demonstrated in experimental [12] and human hepatocellular carcinoma [15]. In experimental hepatocellular carcinoma, the hepatic arterial blood supply is shown to increase beginning with the stage of precancerous lesion [12]. SEM observation of corrosion casts will contribute to elucidating the role of angiogenesis in cancer.

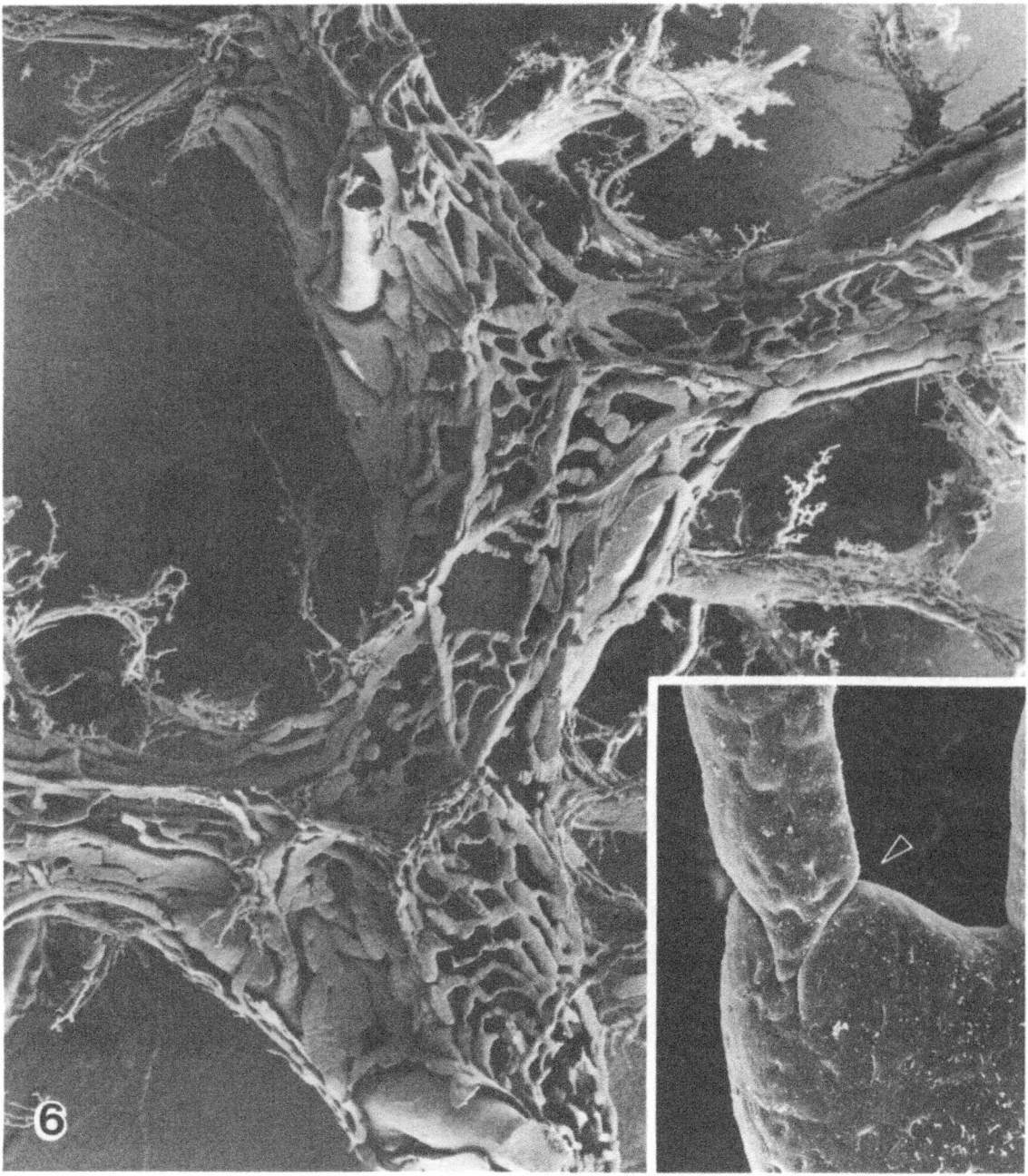

Figure 6-6. Lymphatics in the rabbit liver. Lymphatics form a network around the portal tract. ×80. Inset: Valvelike constriction (arrowhead) and numerous impressions formed by endothelial cell nuclei. ×350. (Reproduced from Ref. 16 with permission.)

4. Hepatic Lymphatics

Resin (Mercox diluted with monomeric methyl methacrylate) injected retrogradely into the bile duct does not get into the lumen of bile canaliculi.

Excess resin leaks into the interstitial space after filling the biliary tract (Fig. 6-6). The resin then drains into lymphatic channels in the portal tract. It has been noticed for over a century that dyes injected into the bile duct leak into the interstitial space of the portal tract and drain into lymphatic channels [16]. Lymphatic casts have demonstrated that the terminal portal lymphatics end blindly. The lymphatics found in the portal tract are composed of long, straight channels and interconnecting short branches (Fig. 6-6). Valvelike constrictions are sometimes observed (Fig. 6-6). Lymphatic networks are especially rich around the bile ducts and the bifurcation of the portal tract, and these ultimately drain into the lymphatics in the hilum of the liver.

Anastomoses between the portal lymphatics and the capsular lymphatics, or those accompanying the hepatic vein, have been reported in the case of ligation of the thoracic duct [27]. Capsular lymphatics become engorged in patients with hepatic cirrhosis because the portal lymph flow is blocked by fibrosis and nodule formation [28]. Such anastomoses are not elucidated with the present injection method, because the direction of draining is physiological. Retrograde injection from the hilar lymphatics may reveal the relationship.

The route through which resin leaks and drains into the lymphatics may have some clinical implication. The thoracic duct lymph in biliary obstruction is known to contain biliary materials much earlier than bilirubin becomes elevated in the blood. Biliary material might leak from the bile duct into the interstitial space, then get into the portal lymphatics, and finally drain into the thoracic duct. Cholangiovenous reflux is also noted by scanning electron microscopy of casts injected retrogradely into the bile duct. In this case, resin leaks into the spaces of Mall and Disse, and drains into the sinusoids [21].

5. Bile Canaliculi and Biliary Tract Casts

5.1. Bile Canaliculi

Bile canaliculi are the channels formed by two or three adjacent hepatocytes, and sealed with tight junctions from the intercellular space. Their detailed two-dimensional structure has been extensively studied by transmission electron microscopy [29,30]. Their three-dimensional structure was demonstrated for the first time by SEM of corrosion casts [18]. Resin prepared

by mixing methyl methacrylate with 2-hydroxy methacrylate monomer can be injected into the terminal bile ductules and also deep into bile canaliculi (Fig. 6-7). The three-dimensional arrangement of bile canaliculi looks like a chicken-wire meshwork composed of interconnecting hexagonal or pentagonal frameworks. Each hexagonal or pentagonal framework is on a plane, but adjoining frameworks are on different planes, sharing only one side in common. This spatial arrangement can be easily discerned using stereo-micrographs. Each side of the bile canalicular framework consists of straight channels, which show some narrowing and dilatation during their course.

The bile canalicular networks are intercalated with sinusoidal networks. The two channels are countercurrent, and some ions and water may diffuse following concentration or pressure gradients between them [31]. The interdigitating bile canalicular and sinusoidal networks might form a structural basis for the lobular gradient of materials secreted into bile canaliculi. Periportal hepatocytes are bathed in plasma containing higher concentrations of constituents than the cells in the perivenular zone. This concentration gradient in the blood is reflected in the solute contents of bile.

5.2. Canaliculoductular Junction

Bile canaliculi connect with bile ductules at the periphery of the portal tract. The transition from bile canaliculi to bile ductules is not uniform. Some canalicular networks drain into a wider channel, i.e., the canal of Hering, which ultimately connects with bile ductules. Ampullary dilatation at the junction is sometimes observed, whereas others abruptly connect with bile ductules. The nature of canaliculo-ductular junctions has been a matter of debate. SEM observation of both fractured tissue [32] and bile canalicular casts [18] demonstrate two types of junction. In the first type, several bile canaliculi coalesce into a thick channels corresponding to the canal of Hering, in which the wall is formed partly by hepatocytes and partly by ductular cells. In the rat liver, ampullary dilatation is observed at this type of termination. The other type consists of bile canaliculi terminating directly into portal bile

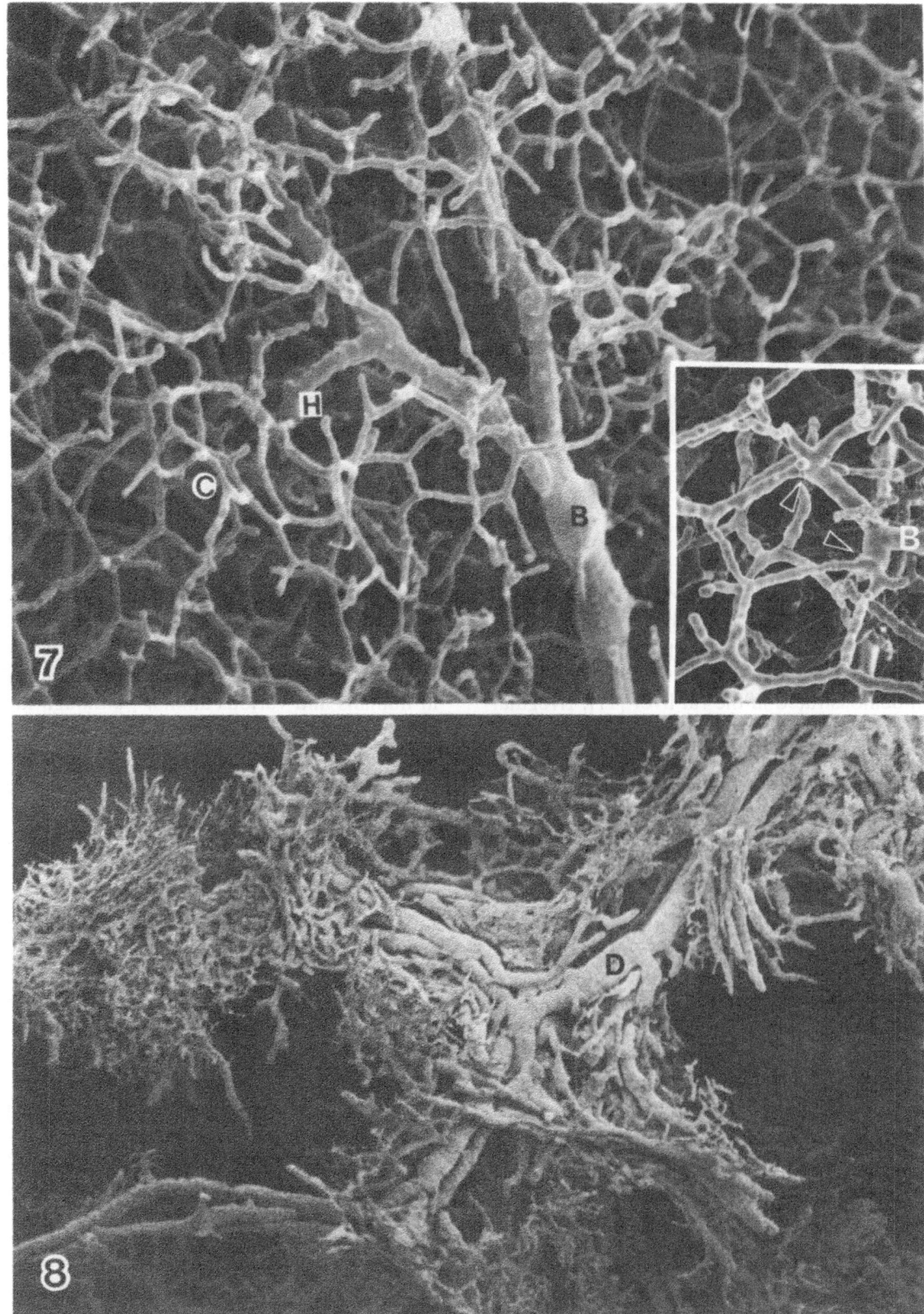

ductules without ampullary dilatation.

During chemical hepatocarcinogenesis in rodents, ductular cells called *oval cells* are known to proliferate. Since oval cells have properties that belong to both ductular cells and hepatocytes [33], it has been suggested that oval cells might be derived from cells that form the canaliculo-ductular junction. They form ductular structures with a basement membrane and produce α-fetoprotein [33]. The three-dimensional arrangement of oval cells is demonstrated with corrosion casting and the SEM method (Fig. 6-8) [23]. Ductular structures formed by oval cells are connected with the portal bile ducts, suggesting that oval cells have characteristics that are more similar to those of biliary epithelia than hepatocytes.

5.3. Intrahepatic Bile Duct

Casts of larger bile ducts are also prepared with less viscous resin by diluting Mercox with methyl methacrylate. Human bile ducts exhibit many irregular side branches and pouches (Fig. 6-9) [19], while a peculiar bile ductular plexus has been demonstrated in the rat liver [17]. Irregular side branches and pouches correspond to parietal sacculi or periductal glands [19,24]. Periductal glands are also present around the common bile duct [36]. Intrahepatic bile ducts are suggested to arise as ductal plates and to transform into bile ducts [34]. The bile ductular plexus in the rat liver might be a remnant of the ductal plate during the development of the bile ducts [35].

The functional implication of such side branches and biliary pouches is not known. Plexiform side branches may function as an anastomosis between the large bile ducts [19]. Storage and modification of bile are also suggested [36,37]. Recent immunohistochemical studies demonstrate both immunoglobulin A and a secretory component in the glandular structures around the large bile duct [38] and suggest the involvement of such structures in the protective function of the biliary tract. This possibility is supported by the fact that such glandular structures are known to increase in patients with long-standing biliary infection or cholestasis [39].

Bile ducts and ductules are known to proliferate in chronic liver diseases, such as chronic hepatitis and liver cirrhosis [40–42]. Anastomosing bile ducts and ductules have been demonstrated by light microscopic observation of serial sections [41] or corrosion casts prepared by injecting latex into the common bile duct [42]. SEM of corrosion casts of the biliary tract easily reveals the three-dimensional arrangement of proliferated bile ducts and ductules. Anastomosing bile ducts and ductules are tortuous and show narrowing and dilatation (Fig. 6-10).

6. Conclusion

This paper reviewed the method for preparing corrosion casts for the hepatic microvascular, lymphatic, and biliary channels, and also discussed the results obtained with this technique. The technique has been intensively applied to the anatomical study of the hepatic microvasculature in humans and various animals, although human studies are still scarce because of limited availability of materials.

It was clearly demonstrated that the vascular system has an intimate relationship with the biliary system through the peribiliary plexus [8,9]. Such a relationship is not clarified between the vascular system and the lymphatic system or the biliary and lymphatic systems. Furthermore, few data are available to give a structural basis for studying the hepatic microcirculation in the acinus. Another important area is its application to pathological conditions [11,12,14,15]. For example, a corrosion cast study demonstrated a rich arterial plexus around cirrhotic nodules that had not been recognized with other methods [11].

Limited studies have been reported on the lymphatic and biliary channels. Both lymphatic and biliary channels are difficult to fill with resin. Lymphatic casts could be prepared by injecting

←

Figure 6-7. Replica of bile canaliculi of the rat liver. Bile canaliculi (C) drain into a canal of Hering (H) and then into bile ductules. (B). ×960. Inset: Higher magnification of canaliculoductular junction (arrowheads). ×1400. (Reproduced from Ref. 18 with permission.)

Figure 6-8. Anastomosing ductular structure formed by proliferated oval cells in the rat liver. D = bile duct. ×80. (Reproduced from Ref. 23 with permission.)

80

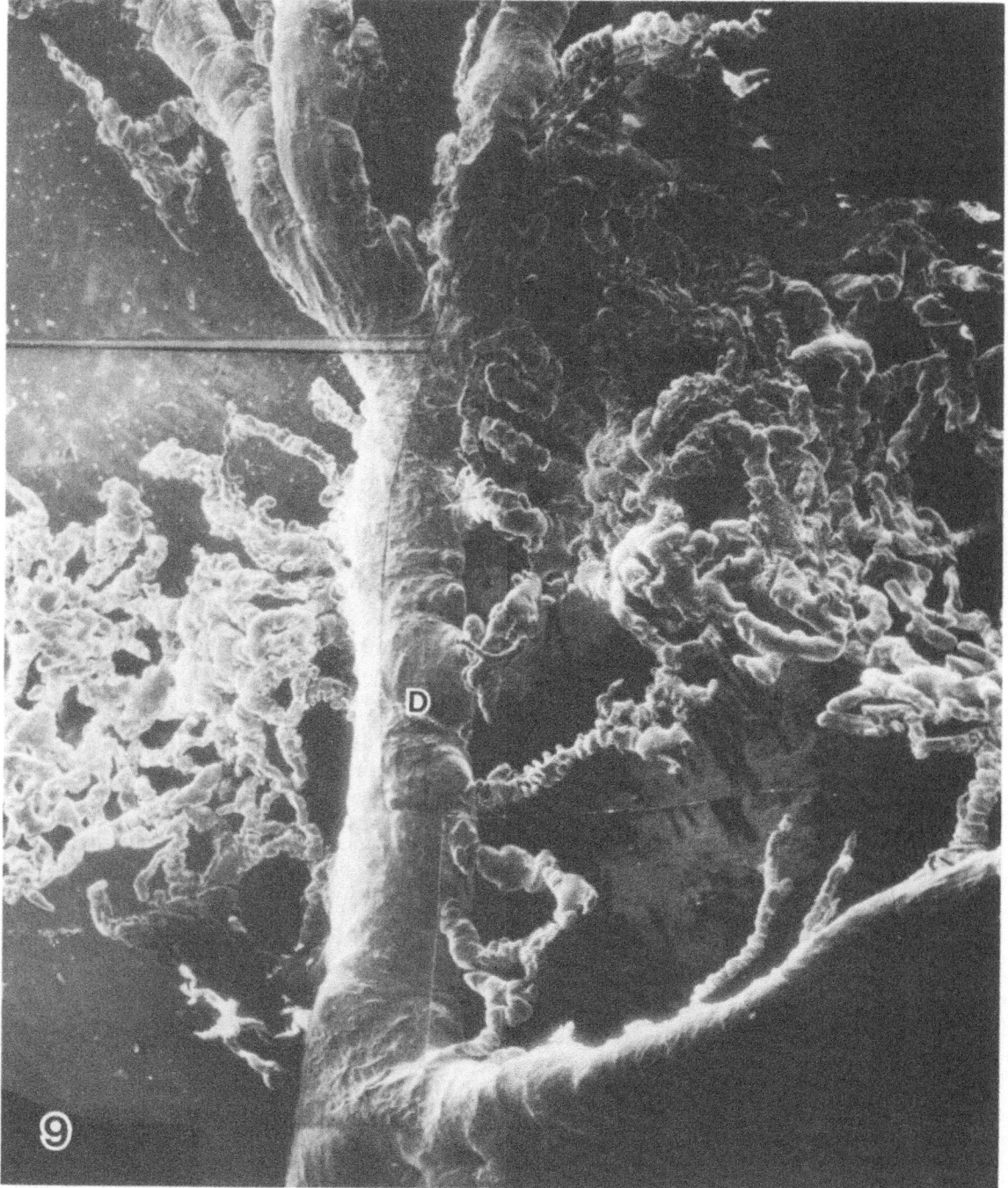

Figure 6-9. Bile duct casts in normal human liver. Note the rich plexiform side branches and sacculi on the opposite sides of the bile duct (D). ×15. (Reproduced from Ref. 19 with permission.)

resin through the bile duct [16]. Retrograde injection of resin into the hilar lymphatics will reveal different features of the portal lymphatics. Biliary tract casts could be prepared by injecting resin retrogradely into the bile duct [17,19], and less viscous resin has been injected up to minute bile canaliculi [18]. Little is known about the three-dimensional structure of lymphatic and biliary

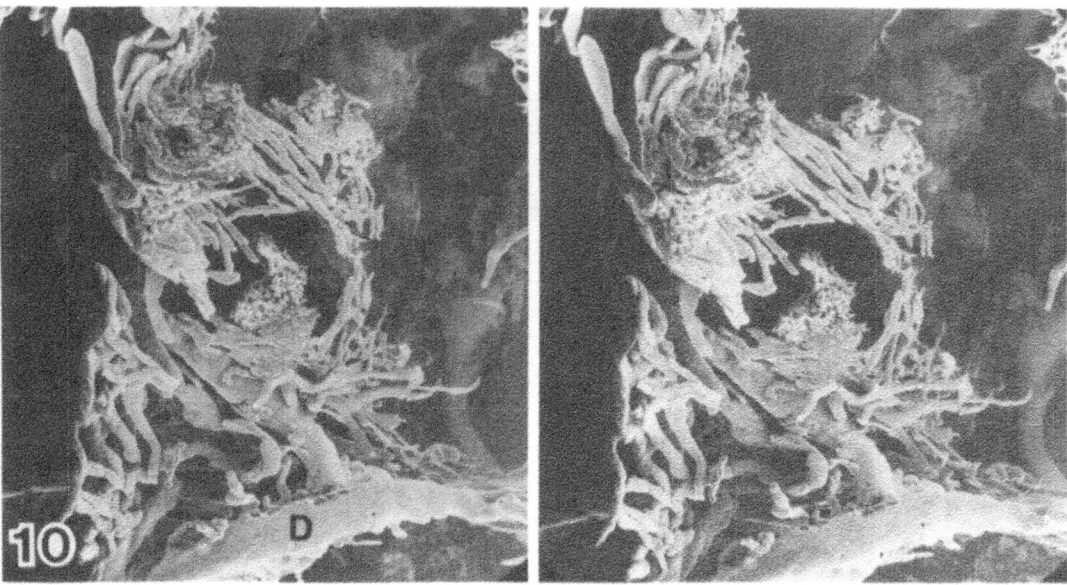

Figure 6-10. Stereomicrographs of bile duct casts in human liver cirrhosis. Note anastomosing bile ducts around cirrhotic nodules. D = bile duct. ×50.

channels in pathological livers and much work remains to be performed in this area.

References

1. Rappaport AM. Physioanatomic consideration. In: Schiff L, Schiff E (eds.), *Diseases of the Liver*, Lippincott, London, pp 1–46, 1987.
2. Jones AL. Anatomy of the normal liver. In: Zakim D, Boyer TD (eds.), *Hepatology. A Textbook of Liver Disease*, Saunders, Philadelphia, pp 3–31, 1982.
3. Murakami T. Application of the scanning electron microscope to the study of the fine distribution of the blood vessels. *Arch Histol Jpn* 32:445–454, 1971.
4. Gannon BJ. Vascular casting. In: Hayat MA, (ed.), *Principles and Techniques of Scanning Electron Microscopy*, Vol. 6, Van Nostrand Reinhold, New York, pp 170–190, 1978.
5. Hodde K, Nowell JA. SEM of micro-corrosion casts. *Scann Electron Microsc II*:88–106, 1980.
6. Lametschwandtner A, Lametschwandtner U, Weiger T. Scanning electron microscopy of vascular corrosion casts — technique and applications. *Scann Electron Microsc II*:663–695, 1984.
7. Ohtani O. Review of scanning electron and light microscopic methods in microcirculation research and their application in pancreatic studies. In: *Scann Electron Microsc* II:653–661, 1984.
8. Murakami T, Itoshima T, Shimada Y. Peribiliary portal system in the monkey liver as evidenced by the injection replica scanning electron microscope method. *Arch Histol Jpn* 37:245–260, 1974.
9. Ohtani O, Murakami T. Peribiliary portal system in the rat liver as studied by the injection replica scanning electron microscope method. *Scann Electron Microsc* II: 241–244, 1978.
10. Kardon RH, Kessel RG. Three-dimensional organization of the hepatic microcirculation in the rodent as observed by scanning electron microscopy of corrosion casts. *Gastroenterology* 79:72–81, 1980.
11. Yamamoto K, Kobayashi T, Phillips MJ. Perinodular arteriolar plexus in liver cirrhosis. Scanning electron microscopy of microvascular casts. *Liver* 4:50–54, 1984.
12. Tsuda H, Tamano S, Imaida K, Ohshima M, Kitahori Y, Ito N. Three-dimensional observations by scanning electron microscopy on the blood supply and organization of vasculature during hepatocarcinogenesis in rats. *Acta Pathol Jpn* 34:957–970, 1984.
13. Yamamoto K, Sherman I, Phillips MJ, Fisher MM. Three-dimensional observations of the hepatic arterial terminations in rat, hamster and human liver by scanning electron microscopy of microvascular casts. *Hepatology* 5:452–456, 1985.
14. Hirooka N, Iwasaki I, Horie H, Ide G. Hepatic microcirculation of liver cirrhosis studied by corrosion cast/scanning electron microscope examination. *Acta Pathol Jpn* 36:375–387, 1986.
15. Kita K, Itoshima T, Yamamoto K, Makino K, Tsuji T. Three dimensional observation of microvascular casts of human hepatocellular carcinoma. In: Tsuchiya M, Mishima Y, Asano M, Oda M (eds.), *Microcirculation, an Update*, Excerpta Medica, Amsterdam, pp 391–392, 1987.
16. Yamamoto K, Phillips MJ. Three-dimensional observation of the intrahepatic lymphatics by scanning electron microscopy of corrosion casts. *Anat Rec* 214:67–70, 1986.

82

17. Yamamoto K, Phillips MJ. A hitherto unrecognized bile ductular plexus in normal rat liver. *Hepatology* 4: 381–385, 1984.
18. Murakami T, Itoshima T, Hitomi K, Ohtsuka A, Jones AL. A monomeric methyl and hydroxypropyl methacrylate injection medium and its utility in casting blood capillaries and liver bile canaliculi for scanning electron microscopy. *Arch Histol Jpn* 47:223–237, 1984.
19. Yamamoto K, Fisher MM, Phillips MJ. Hilar biliary plexus in human liver. A comparative study of the intra-heptatic bile ducts in man and animals. *Lab Invest* 52:103–106, 1985.
20. Yamamoto K, Sargent PA, Fisher MM, Youson JH. Convoluted bile ducts in the liver of the larval lamprey, *Petromyzon marinus* L. *Anat Embryol* 173:355–359, 1986.
21. Stewart L, Pellegrini CA, Way LW. Cholangiovenous reflux pathways as defined by corrosion casting and scanning electron microscopy. *Am J Surg* 155:23–28, 1988.
22. Gaudio E, Pannarale L, Carpino F, Marinozzi G. Micro-corrosion casting in normal and pathological biliary tree morphology. *Scann Microsc* 2:471–475, 1988.
23. Makino Y, Yamamoto K, Tsuji, T. Three-dimensional arrangement of ductular structures formed by oval cells during hepato-carcinogenesis. *Acta Med Okayama* 42:143–150, 1988.
24. Beale LS. *The Liver. Lecture on the Principles and Practice of Medicine.* Churchill, London, pp 47–52, 1989.
25. Jones AL, Schmucker DL, Renton RH, Murakami T. The architecture of bile secretion. A morphological perspective of physiology. *Dig Dis Sci* 25:609–629, 1980.
26. Elias H. A re-examination of the structure of the mammalian liver. II. The hepatic lobule and its relation to the vascular and biliary system. *Am J Anat* 85:379–456, 1949.
27. Lee FC. On the lymph vessels of the liver. *Contrib Embryol Carneg Inst* 15:63–71, 1923.
28. Yoffey JM, Courtice FC. Lymphatics. In: *Lymph and lymphomyeloid Complex*, Academic Press, London, pp 229–236, 1970.
29. Steiner JW. Carruthers JS. Studies on the fine structure of the terminal branches of the biliary tree. I. The morphology of normal bile canaliculi, bile pre-ductule (ducts of Hering) and bile ductules. *Am J Pathol* 38: 639–661, 1961.
30. Biava CG. Studies on cholestasis. A re-evaluation of the fine structure of normal human bile canaliculi. *Lab Invest* 13:840–864, 1964.
31. Erlinger S. Bile secretion. In: Schiff L, Schiff ER (eds.), *Diseases of the Liver*, J.B. It Lippincott, Philadelphia, pp 77–101, 1987.
32. Itoshima T, Kiyotoshi S, Kawaguchi K, Yoshino K, Munetomo F, Ohta W, Shimada Y, Nagashima H. Scanning electron microscopy of the rat bile caniculo-ductular junction. *Scann Electron Microsc* III:373–378, 1980.
33. Sell S. Distribution of α-fetoprotein and albumin-containing cells in the livers of Fischer rats fed four cycles of N-2-fluorenylacetamide. *Cancer Res* 38:3107–3113, 1978.
34. Jorgensen MJ. The ductal plate malformation. *Acta Path Microbiol Scand* 257 (Suppl.):1–88, 1977.
35. Desmet VJ. Intrahepatic bile ducts under the lens. *J Hepatol* 1:545–559, 1985.
36. Bouchier IAD, Cooperband SR, El Kodsi BM. Mucous substances and viscosity of normal and pathological human bile. *Gastroenterology* 49:343–353, 1965.
37. Spitz L, Petropoulos A. The development of the glands of the common bile duct. *J Pathol* 128:213–220, 1979.
38. Terada T, Nakanuma Y, Ohta G. Glandular elements around the intrahepatic bile ducts in man; their morphology and distribution in normal livers. *Liver* 7:1–8, 1987.
39. Chou ST, Gibson JB. The histochemistry of biliary mucins and the changes caused by infection with *Clonorchis sinensis*. *J Pathol* 101:185–197, 1970.
40. International group. Histopathology of the intrahepatic biliary tree. *Liver* 3:161–175, 1983.
41. Yamada S. A study in morphological changes of inter-lobular bile duct in viral hepatitis. *Jpn J Med* 19:9–18, 1980.
42. Masuko K, Popper H. Proliferation of bile ducts in cirrhosis. *Arch Pathol* 78:421–431, 1964.

Author's address:
Prof. Kazuhide Yamamoto
First Department of Internal Medicine
Okayama University School of Medicine
2-5-1 Shikata-Cho
Okayama 700, Japan

Morphological Features of Absorbing Peripheral Lymphatic Vessels Studied by TEM, SEM, and Three-Dimensional Models

GIACOMO AZZALI

1. Introduction

The lymphatic vascular system is made up of two fundamental parts: (1) the absorbing peripheral sector (vessels with high or low absorptive capacity); (2) the sector of conduction and outflow of the lymph towards lympho-venous openings.

The *absorbing peripheral sector* is represented by those vessels (called by Ottaviani [1] the absorbing lymphatic peripheral apparatus, ALPA) assigned to keep the interstitial balance. This kind of vessel, also called the *lymphatic capillary*, *initial lymphatic*, or *terminal lymphatic* [2–6], includes, in our opinion, the part of the lymphatic system that extends from the initial segment up to but not including the precollecting lymphatic vessel. The ALPA vessel has a strong absorptive power, and removes fluids and plasma proteins from the interstitium. Wherein the interstitial fluids and the particles undergo absorption or release from cells, otherwise may enter the venous or lymphatic stream, by means of migratory transendothelial processes. An important role, in keeping this balance, must also be ascribed to the interstitial fluid streams, which are ruled by both active and passive processes. This condition comprises the "avascular" circulatory system, the importance of which [7,8] becomes evident when it is impaired by pathological conditions.

The *conduction and outflow sector* consists of the precollecting (with low absorptive power) and prenodal collecting vessels. The latter, together with the postnodal collecting vessels, drain the lymph into the lymphatic trunks that join the thoracic duct. The absorptive power of prenodal and postnodal collecting vessels, and also of lymphatic trunks, is restricted to membrane diffusion and micropinocytosis.

2. Precollecting Vessel, Prenodal and Postnodal Collecting Vessels, and Lymphatic Trunks

2.1. The Precollecting Lymphatic Vessel

This vessel connects the absorbing peripheral and lymphatic collecting vessels. It can also be functionally regarded, as an absorbing and conducting vessel [9–12]. The precollecting lymphatic vessel is located within organs (e.g., the interlobar perivascular connective tissue of the kidney [13]), but it is mostly found on their outer surfaces, such as in the connective tissue of the mesothelia (diaphragm, intestine, lung).

The structure of this vessel is simple and consists of a continuous endothelium lying on a clear, uninterrupted basal lamina, which, in the first tract of the vessel, is surrounded by a row of smooth muscle cells [13,14]. This small-caliber vessel is provided with bicuspid valves. The thin leaflets consist of an axis of collagen and reticular fibers coated by a monolayer of endothelial cells, whose peripheral edges are mostly joined by end-to-end or overlapping junctions. Berens v. Rautenfeld and Wenzel-Hora [15] assert that precollecting vessels are "vasa fibrotypica" and are seldom provided with valves (not pocket-

Motta, P.M., Murakami, T., and Fujita, H. (eds.), Scanning Electron Microscopy of Vascular Casts: Methods and Applications.

shaped as the valves in collecting ones), inter-endothelial openings, and anchoring fibers.

2.2. Prenodal and Postnodal Collecting Vessels

These two groups of vessels, together, comprise the *lymphatic trunks*, the outflow route for the lymph. They are propelling valved vessels [11, 14,16]. Their endothelium is similar to that lining the ALPA vessels but has fewer free, uncoated vesicles (see Section 4); they are provided with a thick, continuous basal lamina coated by four to six rows of smooth muscle cells [5,17]. These vessels are permeable to small molecules, but not to serum albumin and larger molecules, in both directions [7,9,11]. Transendothelial transport occurs by membrane diffusion and micropinocytosis, as O'Morchoe et al. [18] and Jones et al. [19] demonstrated by means of horseradish peroxidase particles in the collecting vessels of the renal hilum. Wenzel-Hora et al. [20] reported on open interendothelial junctions in the precollecting and prenodal collecting vessels of human skin. These "open junctions" are usually rare in postnodal collecting vessels. The valves are mostly bicuspid and are anchored to the vessel wall by folds, like mesenteric collecting vessels [21]. Collecting vessels have a segmental organization: segments of widely muscular wall alternate regularly with segments provided with valves and a few muscle cells. A segment bounded by two valves identifies the morphofunctional unit [22] that Mislin [23, 24] called the *lymphangion*. Lymphangions have intrinsic contractile properties and are coordinated both by myogenic and nervous factors. Lymphangions propel lymph, and each works as a systo-diastolic pump. The valves ensure a unidirectional flow, which is determined by intrinsic and extrinsic factors [14, 25]. The intrinsic factors include valves and vas-cular contractility, elasticity, and permeability. The extrinsic factors include all the activities outside the vessel, such as the contraction of skeletal muscles, respiratory movements, and intestinal peristalsis, which contribute to increase the luminal lymphatic pressure and thus to enhance lymph drainage toward the venous blood stream.

3. Paralymphatic Channels

It is deemed that ALPA vessels are preceded by *lacunar paravascular channels*, also called *prelymphatic channels* [4,6,26–29] or *para-lymphatic channels* [9,30]. These elements play an important role in filling the ALPA vessel. Their relation with lymphatic vessels is supported by both in vivo and electron microscopy studies [31]. Filling is determined by the colloido-osmotic pressure gradient. Prelymphatic or paralymphatic channels are narrow (ranging from 10 to 100 nm in diameter), scattered throughout the inter-stitium (about 1 per μm^2), and lack endothelium. In some anatomical districts, such as the brain, retina, and cerebellar hemispheres, these tissue channels are extended and leave the interstitial fluids of these organs to communicate (by means of adventitial channels of the Willis circle, internal carotid and vertebral artery) with the lymphatic vessels, as well as the neck superficial and deep nodes [32–37]. The actual existence of open pre-lymphatic channels in the interstitium needs to be better confirmed before they can be functionally ascribed to the lymphatic system, even though Casley-Smith [38] considers the group "blood capillary-interstitium-lymphatic vessel" to be a whole system whose components affect each other.

→

Figure 7-1a–7-1d. Lymphatic network (corrosion casts) of mucosa and submucosa from small intestine with an initial fingerlike (Fig. 7-1a — arrows) or ribbonlike (Figs. 7-1b–7-1d — arrows) tract. Arterial blood vessels and their branches are depicted in black. (a) ×30; (b) ×6; (c) ×20; (d) ×70. e: Inner surface of a lacteal vessel (small intestine of land turtle, *Testudo graeca*) with a stabilizing bridge (arrow) between facing luminal surfaces. SEM ×300. f: Enlargement of the black rectangle in Fig. 7-1e. ×3125 and, h: Wavy luminal endothelial surface with grooves and blind-end anfractuoisities (arrows) that correspond to the peripheral edge of an endothelial cell overlapping the adjoining cell. (g) ×1250: (h) ×2500. i: Luminal endothelial surface with a pocket formation (arrow) offering access to a tunnel bounded by two adjoining endothelial cells. In three-dimensional models this morphological aspect corresponds to an "intraendothelial channel," which ensures continuity with the interstitium. ×5000.

85

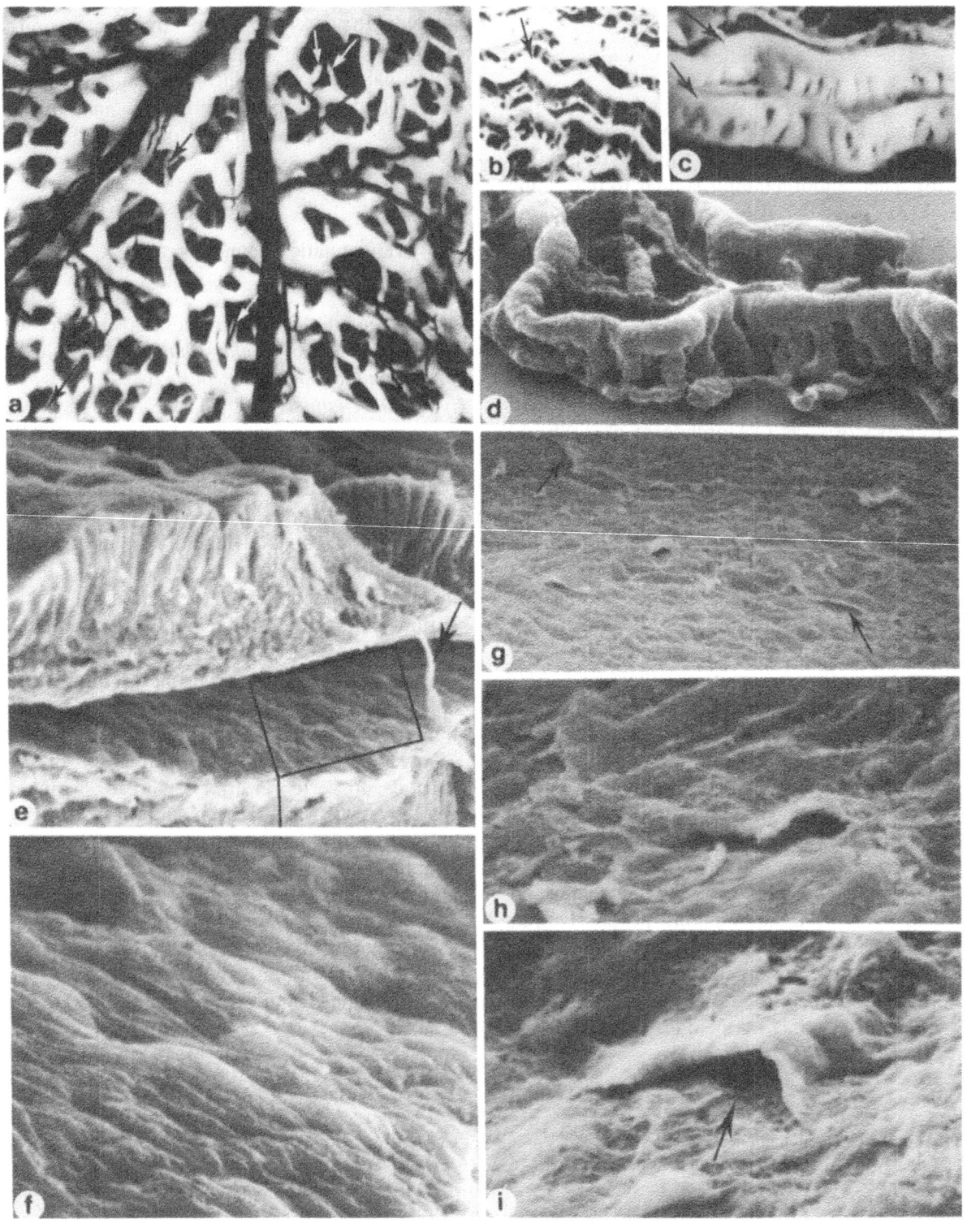

4. The Absorbing Peripheral Lymphatic Vessel (ALPA)

The *ALPA vessel* is the main absorbing segment of the lymphatic vascular system. It rapidly drains water, solutes, corpuscles, and whole cells, and in this way contributes to interstitial homeostasis. It may also be considered to be a carrier viruses, bacteria, and endogenous debris. Its morpho-functional features change according to their anatomical location [9]. Corrosion casts (Figs, 7-1a–7-1d) indicate that the ALPA varies in width (20–60 µm in diameter) and shape; the ALPA can have tubular [39], ribbonlike (Figs. 7-1b–7-1d) [40,41], or saccular [42] aspects, or it can be woven in a mesh (Fig. 7-1a). The initial blind-end tract (Fig. 7-1a arrows) is 70–100 µm long (with the exception of the lacteal vessel, which sometimes reaches 0.5–0.8 mm in length), and it is formed by two, or at most three, endothelial cells. In the "mesh" configuration, the ALPA is formed by five to six endothelial cells arranged in a continuous monolayer and, lacking continuous basal lamina, stomata, pores, and fenestrations.

By SEM and three-dimensional models (TDM), the endothelial cells look mostly rectangular or polyhedral, with a central prominent part. The luminal endothelial surface is slightly wavy (Figs. 7-1e and 7-1f) and has no breaks. Furthermore, simple contact or the overlapping of the peripheral edges of adjoining cells appear as thin grooves and small lumps, respectively. Castenholz [43] asserts that the peripheral edges of the cells are active elements that can affect the intercellular junctions by self-stretching. Often stabilizing elements, such as trabeculae or bridges [41,44], join opposite luminal surfaces (Fig. 7-1e, arrow). Along the border between adjacent cells, the presence of pockets (connected with Casten-holz's branched cells [6]) and small detachments (Figs. 7-1g–7-1i) of limited areas (0.7–1.9 µm high, 1.5–2.7 µm wide) of the endothelial surfaces has been reported; both can vary in number

and shape [15,41,45]. However, the possible continuity of these structures with the inter-stitium cannot be evaluated by SEM, but rather requires three-dimensional reconstruction [45]. The abluminal surface of the lymphatic endothe-lial wall, unlike blood vessels, lacks a basal mem-brane, and therefore has a direct and unlimited relation with the interstitium (Fig. 7-2a). The endothelial wall thickness ranges from 1.9 to 2.4 µm at the cellular body and from 400 to 500 nm in the non-nuclear area. By TEM, endothelial cells show a central bulky portion (O'Morchoe's nuclear area) and a peripheral area (non-nuclear). The former contains the nucleus, which is surrounded by abundant cytoplasm with organelles. The latter represents about two thirds of the total volume and appears as a flat, extended cytoplasmic expansion with a linear profile that, in some experimental conditions (e.g., prolonged fast, lymphatic stasis), becomes very wide (Fig. 7-2b).

The cytoplasmic matrix is poorly electron dense and shows free ribosomes, a few RER tubules and mitochondria, some dense bodies, and many free uncoated vesicles. Thin actinlike filaments (black rectangles of Fig. 7-2c and inset 1), ranging for 3.7 to 4.5 nm in diameter, can also be observed. These filaments are immunocyto-chemically similar to those encountered in smooth muscle cells, are arranged parallel to the longitudinal axis of the cell, and are often grouped in bundles; they prove the existence of an intrinsic contractile system, which accounts for lymphatic endothelium motility and affects lymph propulsion. According to some authors, they could help in preserving the morphological and functional integrity of the vessel [14,46]. The thickness of some areas of the non-nuclear cytoplasm decreases, in particular in physio-logical conditions (such as winter hibernation), as much as 40–60 nm [47,48]. Uncoated vesicles are free, are arranged as rings of a chain, or are grouped [49–51] to form large vacuoles (Fig.

→

Figure 7-2. a: Lacteal vessel (human ileum) with a continuous endothelial wall lacking a basal lamina and fenestrations. ×6500. b: Lacteal vessel (fasting cat ileum) with a markedly wavy endothelial wall. ×8000. c: Endothelial cell of lacteal vessel (adult human). The cytoplasm shows lysosomes, thin actinlike filaments (black rectangles and inset 1), free vesicles, often condensed in large vacuoles (inset 2, v), in channel like structures (small arrow), or arranged in a row (inset 3 — large arrow), draining into an intercellular space bounded by specialized junctional complexes. (c) ×26,600; (inset 1) ×50,000; (inset 2) ×36,900; (inset 3) ×28,700.

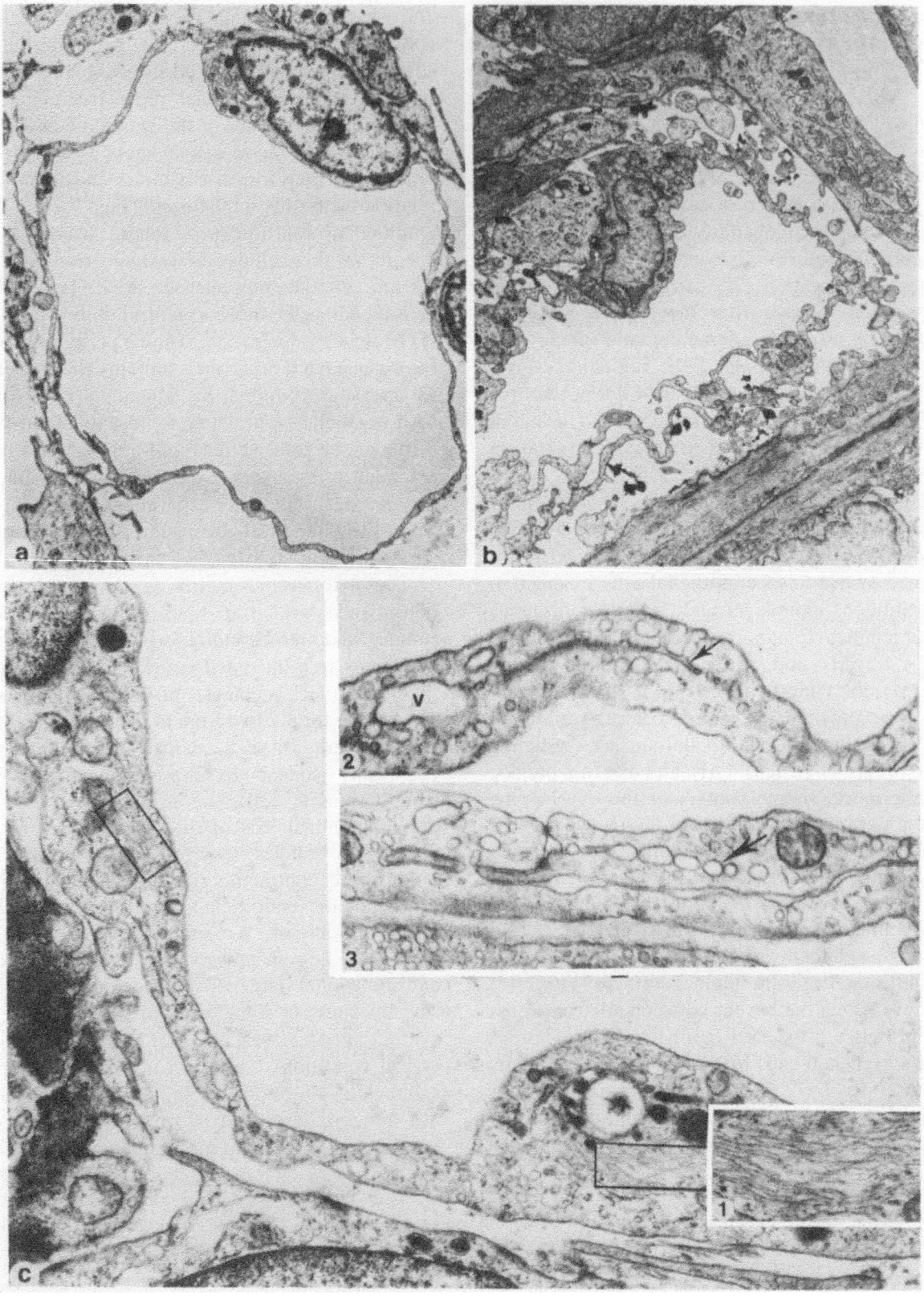

7-2c, inset 2). They also are the origin of tubular structures or transendothelial channels [52–55], which often open into wider intercellular spaces (Fig. 7-2c, inset 3) between two specialized junctional complexes. These spaces drain, as three-dimensional models clearly show, into the lymphatic lumen [41,56]. Jones et al. [19] suggest that most vesicles must not be regarded as isolated units, but rather as rings of chains that bridge endothelial luminal and abluminal surfaces. It is unlikely, nor have we ever observed, that a vesicle arises on a plasma surface, migrates through the cell, and eventually opens on the opposite surface [41,48].

Both luminal and abluminal plasma membranes show many endocytotic and exocytotic invaginations, whose ultrastructural features resemble those of blood vessel endothelial cells [57]; these seem to play an important role in the removal of hydrophilic molecules and inert matter from the interstitium [58–60]. O'Morchoe et al. [18] and Niiro and O'Morchoe [54] have demonstrated, by the injection of colloidal carbon, thorium dioxide, latex spheres, and other tracers, such as ferritin and peroxidase (50–110 Å), that the vesicular route is actually an ultrastructurally evident, fast, and one-way (abluminal-luminal) transport pathway.

In ALPA vessels the peripheral edges of adjacent endothelial cells join with each other via three kinds of contacts: end-to-end, overlapping, and interdigitating [9,49,61,62]. As a rule, a 11–24 nm space divides the plasma membranes of adjoining cells. Specialized junctional complexes, the gap and tight junctions, interrupt this space and fasten the endothelial cells. In kidney vessels, these complexes are 60% fasciae occludentes and 10% fasciae adherentes [62]. Tight junctions prevent the passage of molecules as small as lanthanum and peroxidase [63,64]. In the case of gap junctions, the fusion in the pentalaminar layer of the adjoining membranes does not occur. This allows the passage of plasma protein through a channel of 10–20 nm between opposite plasma membranes [62].

Many authors [60,65–68] assume that the peripheral edges of adjacent cells are loosely joined to each other, and so they can be easily separated in order to allow the fast passage of interstitial fluids, macromolecules, and cells into the lymphatic lumen. According to Leak and Rahil [69], these openings should represent the "stomata" of Recklinghausen [70,71]. During edema states, increased lymphatic flow [72, 73] and protein-losing enteropathies [74], as well as in healthy tissues [29,63,72,75–77], open inter-endothelial junctions (0.1–3 μm) were described as clefts behaving as "inlet valves." This theory is based on the mutual relation between endothelium and basal lamina with the involvement of the interstitial connective tissue collagen, reticular and elastic fibers [29,68]. This opening kind of the endothelial wall has been denied with equal resolution by Roberge et al. [78], Azzali [49,55,79], Azzali et al. [45,48], and others. Albertine and O'Morchoe [62] found only 3 open junctions out of 1000 intercellular contacts in the lymphatics of the dog kidney, while Dobbins and Rollins [80] estimate that only 2% of the open junctions are in the small intestine. According to Albertine and O'Morchoe [62] and O'Morchoe [51], contrary to the opinion of Elhay and Casley-Smith [81], open junctions do not play a significant role in transendothelial transport.

The TEM study of the lymphatic endothelium frequently reveals single luminal and abluminal cytoplasmic projections [54,82–85], which sometimes delimit one or more interstitial spaces between two adjoining cells (Fig. 7-3a). These structures are called *intraendothelial channels* [49]: the lymphatic endothelium organizes these channels according to functional requirements.

The combined analysis of TEM pictures and TDM clarifies the formation and the stereostructure of these channels. They arise from the bifurcation of the cytoplasmic expansion of an endothelial cell. A short luminal branch joins the peripheral edge of an adjacent endothelial cell, while an abluminal branch continues free in the interstitium, bridges an intercellular junction (2.7–3.8 μm) adhering to the abluminal surface of an adjoining endothelial cell (Fig. 7-3a, insets 1–3) with its peripheral edge by means of specialized junctional complexes. This morphological layout delimits a tunnel-shaped interstitial space, (0.7–0.9 μm high, 1.2–1.4 μm wide, 4–7 μm long), with two orifices, luminal and abluminal (Fig. 7-3a, inset 4). The "intra-endothelial channel" is found not only in this, simple structure, but sometimes also in a more involved configuration. However, the only tunnel-shaped portion is complex, whereas the luminal and abluminal orifices are always single [49,55,86]. The number of intraendothelial channels changes with physiological conditions (e.g., hibernation, prolonged experimental fast [45,47,48]), and therefore these channels must be regarded as dynamic, rather than static, structures organized by the lymphatic endothelium according to general metabolic requirements. Furthermore we believe that intraendothelial channels have nothing to share with the open junctions described by many authors [27,39,60,67]. Thus a tridimensional model [45,55] of a lymphatic vessel provided with "open junction", sectioned along different cut surfaces, while preserving the morphological aspects of the vessel, presents such "open junction", on one given cut surface, with the morphological characteristics of an endothelial channel. If the open junction in the endothelial wall was a morphological entity, it would not change according to the cut surface used.

Comparative analysis of the data collected with TEM, SEM, and TDM leads us to believe that the majority of, if not all, open interendothelial junctions should not be regarded as true morphological entities. Exceptions to this assumption are endothelial breaks that follow mishandling during experimental procedures or are the result of histamine release induced by the injection of tracers [20,77,87]. Further and more specific investigations are needed to characterize the factors that induce the formation of the intra-endothelial channels and to determine how they are organized by the lymphatic endothelium. At present the data at our disposal do not enable us to draw conclusions about this topic.

5. Lymph Production

The data collected in the last decade (see also the reviews by Földi and Casley-Smith [88]. Abramson and Dobrin [89], and Johnston [90]) indicate that the transport of fluid and macromolecules across the lymphatic endothelium occurs through
1. Plasma membrane, by both active and passive diffusion processes
2. Endocytoplasmic vesicles [54,91,92] and membrane invaginations [93]
3. Channels of the intercellular space (10–20 nm) between adjoining endothelial cells [62,87,94]
4. Open interendothelial junctions, working both as inlet and outlet valves [20,95]
5. Intraendothelial channels [41,55,96,97], which also allow the migration of large molecules and cells.

At this point the following question arises: Do endothelial cells of the lymphatic vessel play an active role in selective transports? A set of in vivo, TEM, SEM, and TDM studies might be suitable to determine the specificity of the above-mentioned transport pathways. It is generally accepted that lymph derives directly and fully from interstitial contents. Experiments with electron-dense proteic tracers have proven that the absorbing peripheral lymphatic vessel must be regarded as a "selective pathway" devoted to maintaining interstitial homeostasis. Three main mechanisms now considered to be involved in lymph production: are hydrostatic, reticular-oncotic [29,68], and vesicular route [80,91]. However, it is possible that all three of these mechanisms act at the same time.

6. Transendothelial Passage of Cells

The literature concerning cell passage across the endothelium of the lymphatic vessels is rather scanty. In addition, the factors that induce and control this migration are still unknown. Ohtani et al. [98] have published, without any comment, TEM and SEM pictures of lymphocytes in the act of crossing the endothelium of lymphatics located in the thymus-dependent area of the rabbit appendix. Kato [99] published similar findings on the intralobular lymphatics of the thymus cortico-medullary zone, both in normal and hydrocortisone — treated rats. Carr et al. [100], in an experimental model of lymphatic metastasis, reported the passage of tumor cells (Rd/3) and lymphocytes via open intercellular junctions. Reports by van de Velde and Carr [101] and Carr [102] on experimental granuloma confirmed that lymphocytes and macrophages migrate into the lymphatics (as single cells or together with tumoral cells) by the same process, namely, via openings in the interendothelial junctions (intercellular route). The same path has also been postulated for fluids and macro-

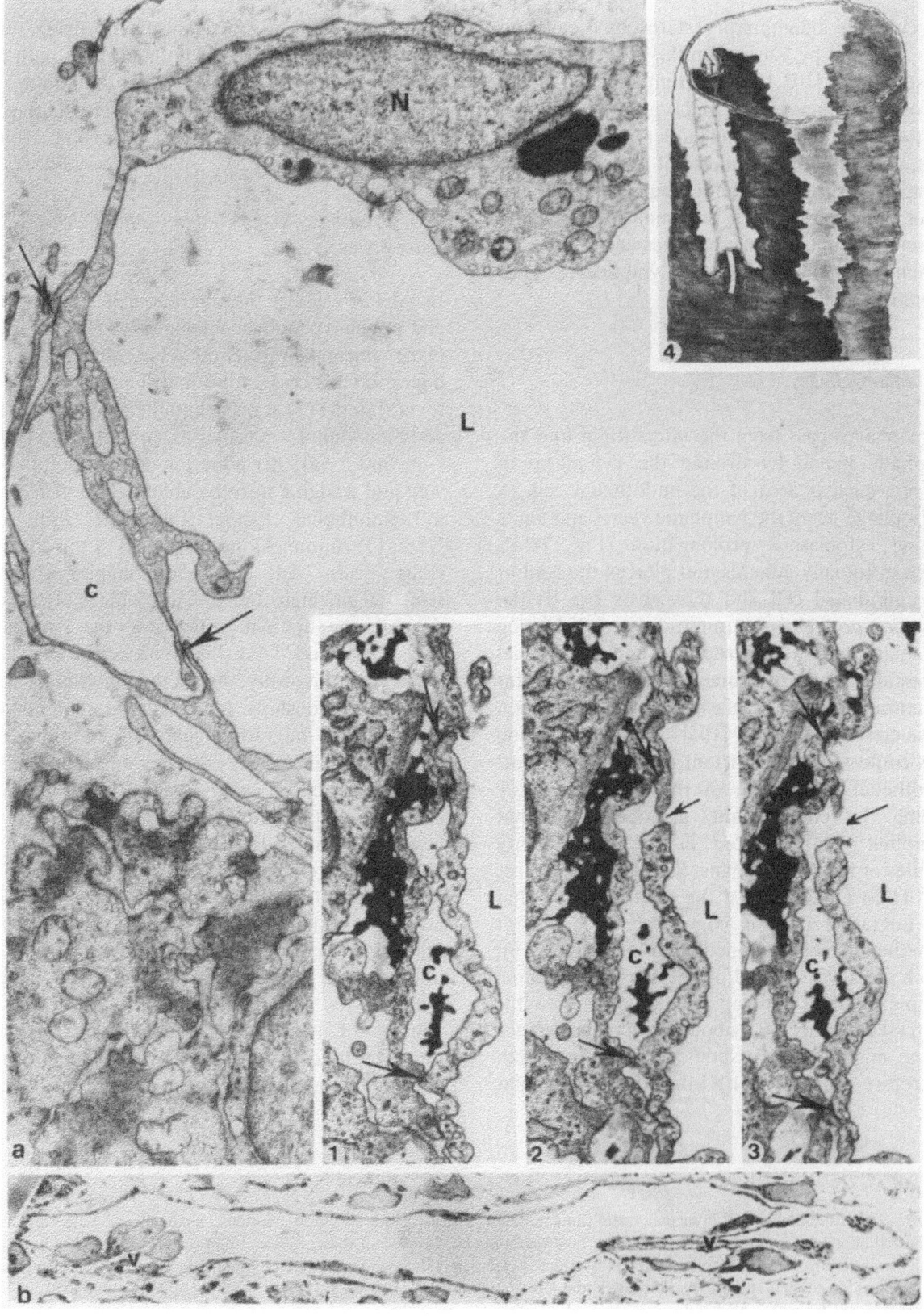

molecules in inflammatory states by Leak and Kato [46] and by Casley-Smith [72]. Our recent investigations [103,104] on humans and some micromammalians (rats and cats) have documented, for the first time under normal conditions, two different modalities of transendothelial passage across the absorbing peripheral lymphatic vessel (lacteal vessel) of the small intestine: (1) transcytoplasmic for macrophages and (2) via the intraendothelial channel for lymphocytes and polymorphonuclear leukocytes.

6.1. Macrophages

Macrophages pass from the interstitium into the lymphatic lumen by drilling the cytoplasm at the non-nuclear area of the endothelial cell. A macrophage nears the lymphatic vessel and emits a long cytoplasmic prolongation (Fig. 7-4a), which eventually adheres and pierces the wall of the endothelial cell and thus gives rise to the *migration pore* (1.3–1.9 μm in diameter). During the migration, the macrophage and endothelial cell establish sites of contact that could represent structural proof of a plasmalemmal interaction in selective migration [105]. These anchoring sites could also be important in the active transendothelial locomotion of the macrophage by folding the endothelium, analogous to what Campbell [106] asserted in the postcapillary venules of lymphoid organs. Serial TEM pictures and TDM [103] enabled the determination that the migration pore is completely independent from the intercellular junctions (Fig. 7-4a inset 1), which are always sealed by specialized junctional complexes (tight and gap junctions). The doubts expressed by Carr [102] about the actual existence of the migration pore should now be removed. Therefore we believe [104] that the transcellular

route, although its mechanisms of origin and control are still obscure, is not a peculiar path for tumor cells or, generally speaking, a pathological event, but rather a physiological event shared by healthy organs and tissues.

6.2. Lymphocytes and Polymorphonuclear Leukocytes

In ALPA vessels, the migration of lymphocytes and polymorphonuclear leukocytes (PMN) takes place through intraendothelial channels. The migration process of both cell types occurs in several steps: (1) approach to the lymphatic vessel and emission of a cytoplasmic prolongation (Figs. 7-4b and 7-5a); (2) adhesion to the endothelial wall and wedging into the abluminal orifice of an intraendothelial channel (Figs. 7-4c, 7-5b, and 7-5c); (3) running along the tunnel of the channel (Figs. 7-4c, 7-5d, and 7-5e), shaping channel size (0.9 μm high and 2–5 μm wide); (4) inflow through the luminal orifice into the lymphatic (Figs. 7-4c and 7-5f). TEM pictures and TDM show unequivocally both the migration and spatial relationships between lymphatic endothelium and migrating cells. In addition, the morphological reality of the intraendothelial channel is clear (Figs. 7-4d). An integrated and complete exploitation of all the different investigative methods (TEM, SEM, TDM) is an absolute prerequisite for a correct interpretation of such delicate and complex morphological aspects as those involved in transendothelial migration. According to many authors [107–111], unlike what happens in the lymphatic vessel, lymphocytes migrate from the postcapillary venules of lymph nodes and Peyer's patches into lymphoid tissue by a transcellular pathway, although Schoefl [112] maintains that a selective

←

Figure 7-3. a: Endothelial wall of lacteal vessel (human ileum) provided with an "intraendothelial channel" (c). Large arrows = specialized intercellular junctions; N = nucleus; L = vessel lumen. Insets 1–3 show, by serial thin sections, the luminal opening (small arrows) of the intraendothelial channel (c), which is completely independent of the interendothelial junctions (large arrows). Black masses are condensed chylomicra located both inside and outside the channel. Inset 4: sketch of an intraendothelial channel. The arrow spans from the abluminal orifice, through the channel, up to the luminal orifice. (a, insets 1–3) ×17,700. b: Absorbing peripheral lymphatic of the subepithelial connective tissue of a bat tongue (anterior $\frac{2}{3}$) with segmental distribution of the valves (v). ×2043.

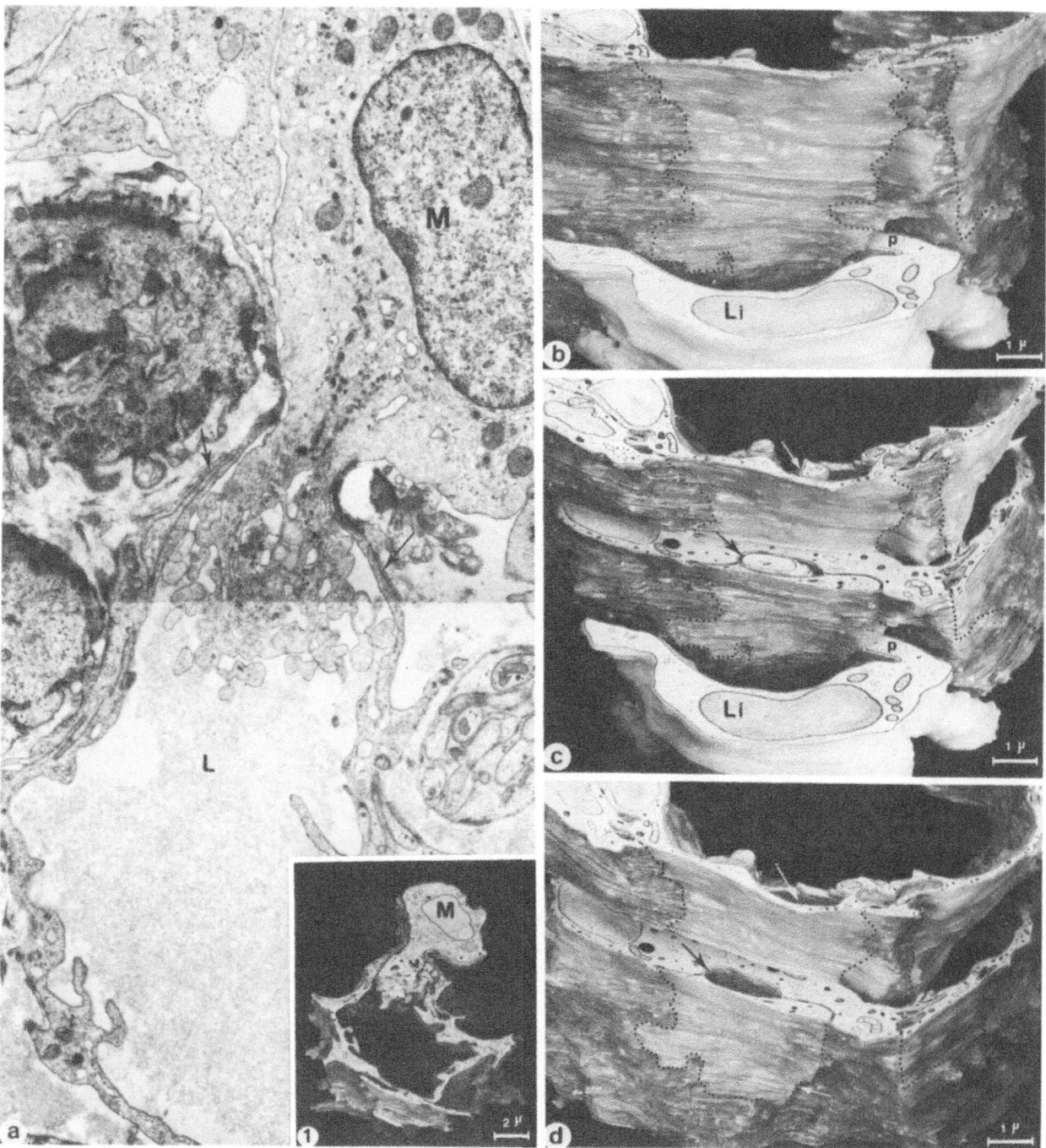

Figure 7-4. a: Lacteal vessel (human ileum) with a macrophage (M) crossing the cytoplasm of an endothelial cell. The area involved in the process is far from the intercellular junctions (arrows), as it stands out in the three-dimensional model (inset 1). L = lymphatic vessel lumen. ×9225. Bar = 2 μm. b, c: Three-dimensional model of lymphatic vessel. The cytoplasmic prolongation (p) of a lymphocyte (Li) is wedged into the abluminal orifice (Fig. 7-4b), and runs through the tunnel portion (Fig 7-4c black arrow) of an "intraendothelial channel" and drains through the luminal orifice into the vessel (white arrow). Dashed lines mark endothelial cell boundaries. Bar = 1 μm. d: Three-dimensional model of the lymphatics depicted in Fig. 7-4b, but devoid of lymphocytes in order to show the stereological aspect of the intraendothelial channel (black arrow) and of its two orifices (white arrows), luminal and abluminal. Dashed lines = cell boundaries. Bar = 1 μm.

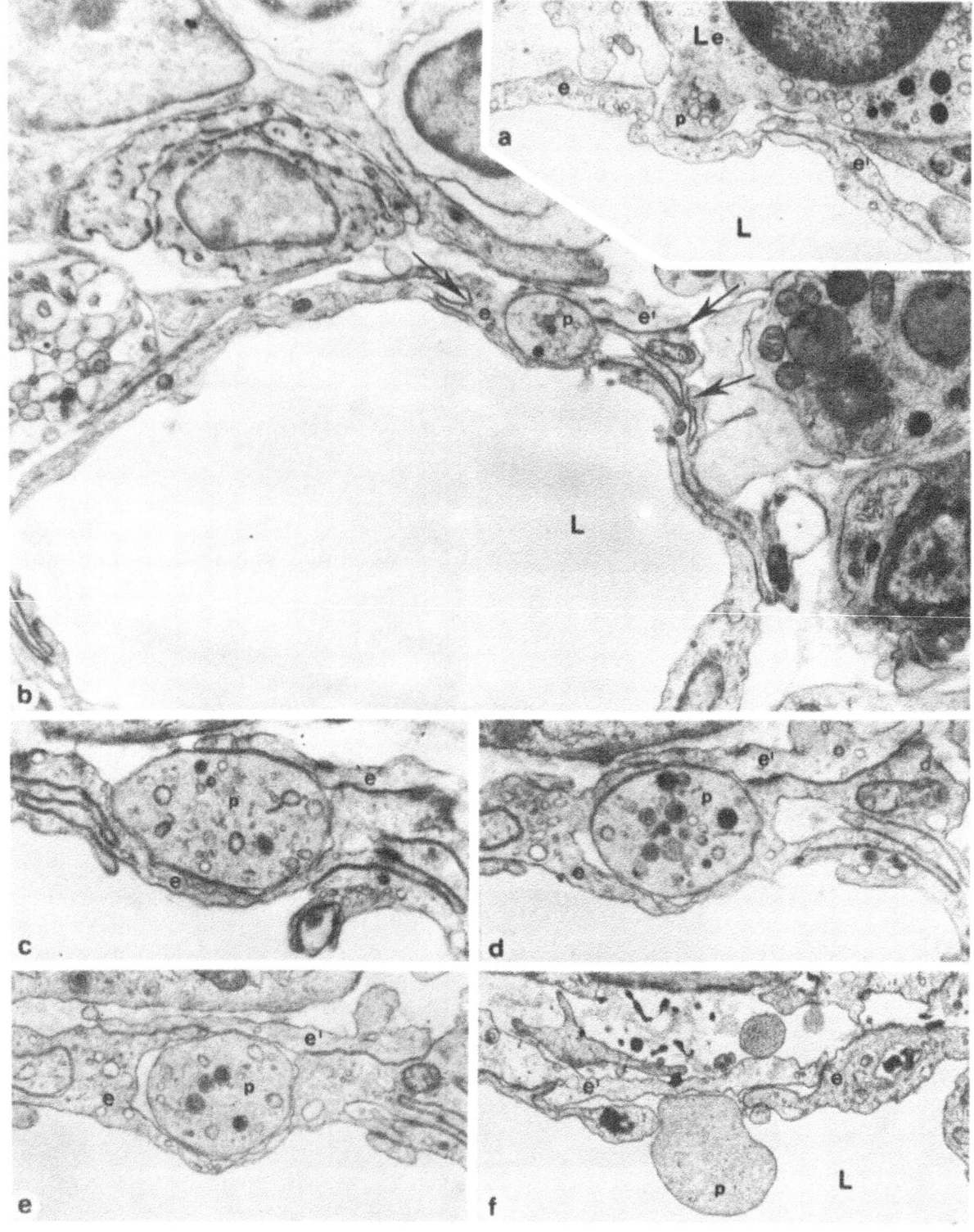

Figure 7-5. a: Polymorphonuclear leukocyte (Le) whose cytoplasmic prolongation (p) abuts the endothelial wall. L = vessel lumen. e, e′ = cytoplasmic expansion of adjoining cells. arrows = intercellular junctions ×17,000. b–f: This series shows sequentially the cytoplasmic prolongation (p) in the abluminal orifice (Fig. 7-5b and 7-5c), in the tunnel portion of the "intraendothelial channel" (Figs. 7-5d and 7-5e), and protruding (Fig. 7-5f) in the lumen (L) of the lymphatic vessel. e and e′ = cytoplasmic expansions of two adjoining endothelial cells that concur in the formation of the above channel. (b) ×11,500; (c–e) ×21,000; (f) ×15,000.

intercellular path exists. In addition, PMN seem to migrate intercellularly [108]. Otherwise, we believe that it is possible that in the peripheral absorbing lymphatic vessels of the small intestine [104] lymphocytes and PMN share the same trans-endothelial migratory path. It is possible that these cells run along the intraendothelial channel by their intrinsic properties, such as proto-plasmatic lability and motility, which Haston et al. [113] have shown in vitro for lymphocytes. Moreover, it is not unlikely that the "channel route" organizes in response to an immunological factor. Besides, the cytoplasmic expansion at the beginning of the migratory phase, could exert with its end portion, the function of a sound in the recognition of the intraendothelial channel; furthermore, it could even represent a motility aspect themselves of the cell [114].

7. Valves

Absorbing peripheral lymphatic vessels of some organs, such as the tongue, diaphragm, lung, and skin [6,39,42,112,115,116], are provided with valves. They are unevenly distributed and lack completely smooth muscle cells, with the exception of those that are arranged in the vessels of the human foot back [117]. The valves are usually bicuspid and funnel shaped. They are formed by thin leaflets (15–18 µm long) arranged as a bishop's miter. Each leaflet is a fold of the endo-thelial wall, with a stroma of connective tissue that differs from the subendothelial tissue. It also contains collagen and elastic fibers and, according to Lauweryns and Boussauw [116], some peculiar connective cells. We did not find such cells in the valves of subepithelial vessel from the tongue and diaphragm [Azzali, unpublished]. Each leaflet is coated with a double endothelial layer, and cellular edges are joined by end-to-end and over-lapping contacts. The subepithelial ALPA vessels in the anterior two thirds of the bat tongue [Azzali, unpublished] have valves arranged at regular intervals of 45–60 µm, and this segmental structure (Fig. 7-3b) resembles that of collect-ing vessels (see Section 2). This peculiar mor-phological structure, in addition to causing a forced one-way lymph flow, could be related to

the contractile function of the organ (tongue), rather than be species linked, in order to prevent lymph backflow and to make easy drainage to-ward precollecting vessels. The valves lack nervous fibers, suggesting a passive function [116].

8. Phagocytosis

Does phagocytosis play an active role in trans-endothelial transport? Nordquist et al. [118] have reported a weak phagocytotic activity. The endo-thelial lymphatic cells can actually phagocytose carbon or latex spheres (0.5–1 µm φ) [6,8,59], but the phagocytosed substances stay inside indefinitely. Ottaviani and Azzali [9] found lipoproteic granules in the cytoplasm of endo-thelial cells of the lacteals, but their location does not allow one to assume that they could be released inside the lymphatic vessel. Dobbins and Rollins [80], by injecting chylomicrons and tracers (ferritin and peroxidase), drew the con-clusion that the transendothelial passage of these substances occurs by phagocytosis rather than via open interendothelial junctions. Castenholz [6] considers endothelial cells of the initial lymphatic to belong to reticuloendothelial system.

9. Interstitium Cells and Lymphatic Endothelium

Smooth muscle cells and lymphatic endothelium are often adjacent in the lacteal vessel. Bubble hyaloplasmic expansions of smooth muscle cells, such as pedunculated pseudopods, often reach the abluminal endothelial surface and adhere to it, but without specialized sites of contact. In these areas the endothelial wall bends as much as to receive the protrusion itself. On this subject, Ottaviani and Azzali [10] and Horstmann and Breucker [119] assume that smooth muscle cells play a role in stirring the interstitial fluid, chang-ing its pressure, and affect transendothelial trans-port modalities. This mechanism would work only under conditions of heavy extracellular fluid flow and only in anatomical districts such as the small and large intestine, but not in the skin, diaphragm, and ear [58,67,120].

10. Innervation

In all the anatomical districts we investigated, ALPA vessels are topographically close to unmyelinated nervous fibers. The latter are located, on the average, 200–600 nm from the wall of the lymphatic vessel [14]. The axoplasm includes mitochondria, neurofilaments, neurotubules, and clear cored vesicles alternating with dense granules. These ultrastructural aspects imply nervous control in the movement of the endothelial wall.

11. Concluding Remarks

The data collected in the last decade about lymphatic canalization using different techniques indicate that light microscopy coupled with direct injection techniques (dyes, epoxy), whether or not followed by corrosion, has not yet exhausted its role. This technique still supplies extremely interesting pictures, unsurpassed in elegance and beauty, of lymphatic angiotectonics and the relation between the lymphatics and blood vessels. Furthermore, electron microscopy (SEM and TEM) has elucidated the ultrastructural features of the lymphatic endothelium, both in the absorbing and conduction sections, and has also improved our knowledge of the mechanisms involved in the transendothelial transport of fluids, micromolecules, macromolecules, and cells. The combined exploitation of these methods, together with three-dimensional models (though the latter technique requires a great expenditure of time), represents the basic tool used to solve the still unsettled points about the mechanisms of lymph formation and the return of cells to the blood stream via the lymphatic system.

Acknowledgments

This paper is dedicated to the memory of Professor Giuseppina Romita Azzali. The author is indebted to Mr. Francesco Azzali for technical assistance. This work was partly supported by a grant from Ministero P.I. — 40%.

References

1. Ottaviani G. Ultrastruttura dei vasi linfatici dell'A.L.P.A. *Mon Zool Ital* 70 (Suppl.):343–353, 1962.
2. Allen L. Lymphatics and lymphoid tissues. *Annu Rev Physiol* 29:197–224, 1967.
3. Casley-Smith JR. Lymph and lymphatics. In: Kaley G, Altura B (eds.), *Microcirculation* Vol. I University Park Press, Baltimore, pp 423–502, 1977.
4. Casley-Smith JR. The fine structure and functioning of tissue channels and lymphatics. *Lymphology* 13:177–183, 1980.
5. Zweifach BW, Prather JW. Micromanipulation of pressure in terminal lymphatics in the mesentery. *Am J Physiol* 228:1326–1335, 1975.
6. Castenholz A. Observations on the structural and functional properties of initial lymphatics. Light and electron microscopic studies of the subepithelial lymphatic plexus in the rat tongue. In Casley-Smith JR, Piller NB (eds.), *Progress in Lymphology*, Proc. Xth Int. Cong. Lymphology, Adelaide, 1985. University of Adelaide Press, Adelaide, pp 20–23, 1985.
7. Wayland H, Silberberg A. Blood to lymph transport. *Microvasc Res* 15:367–374, 1978.
8. Leak LV. Lymphatic endothelial-interstitial interface. *Lymphology* 20:196–204, 1987.
9. Ottaviani G, Azzali G. Ultrastructure des capillaires lymphatiques. In: *Simp. Int. Morphologie et Histochimie de la Paroi Vasculaire — Fribourg 1965*. S. Karger, Basel, Teil II, pp 325–360, 1965.
10. Ottaviani G, Azzali G. Ultrastructure of lymphatic vessels in some functional conditions. *Acta Anat* 73 (Suppl. 56):325–336, 1969.
11. Comparini L, Fruschelli C, Bastianini A, Pompucci G. Morfologia microscopica e struttura del sistema vascolare linfatico. *Arch Ital Anat Embriol* 78 (Suppl.):7–61, 1973.
12. Kubik S. Initial lymphatics in different skin regions. In: Partsch H (ed.), *Progress in Lymphology*, Proc. XIth Int. Cong. Lymphology, Wien, 1987. Elsevier Science, Amsterdam, pp 17–19, 1988.
13. Azzali G, Bucci G, Gatti R, Orlandini G. The lymphatic vascular system of some hibernating mammalians kidney. *Z Mikrosk Anat Forsch* 102:945–961, 1988.
14. Fruschelli C, Gerli R, Alessandrini C, Sacchi G. Il controllo neuroumorale della contrattilità dei vasi linfatici. *Arch Ital Anat Embriol* 88 (Suppl.):49–109, 1983.
15. Berens von Rautenfeld D, Wenzel-Hora BI. The nomenclature of the initial lymphatics from a comparative anatomical point of view. In: Casley-Smith JR, Piller NB (eds.), *Progress in Lymphology*, Proc. Xth Int. Cong. Lymphology, Adelaide, 1985. University of Adelaide Press, Adelaide, pp 17–19, 1985.
16. Comparini L. I precollettori linfatici. *Biologica Latina* 15:479–492, 1962.
17. Nicoll PA, Taylor AE. Lymph formation and flow. *Annu Rev Physiol* 39:73–95, 1977.

18. O'Morchoe CCC, Jarosz HM, Jones WR, O'Morchoe PJ. The study of endothelial vesicles in the isolated perfused lymphatic vessel. In: Hammersen F, Lewis DH (eds.), *Endothelial Cell Vesicles*. Prog. Appl. Microcirc. Vol. 9, S. Karger, Basel, pp 88–99, 1985.

19. Jones WR, O'Morchoe PJ, O'Morchoe CCC. The organization of endocytotic vesicles in lymphatic endothelium. *Microvasc Res* 25:286–299, 1983.

20. Wenzel-Hora BI, Berens von Rautenfeld D, Partsch H. Scanning electron microscopy of the initial lymphatics of the skin after use of the indirect application technique with glutaraldehyde and MERCOX as compared to clinical findings — Part II: The effects on the opening apparatus of the initial lymphatics. *Lymphology* 20:134–144, 1987.

21. Albertine KH, Fox LM, O'Morchoe CCC. The morphology of canine lymphatic valves. *Anat Rec* 202:453–461, 1982.

22. Horstmann E. Über die funktionelle Struktur der mesenterialen Lymphgefässe. *Morphol Jb* 91:483–510, 1951.

23. Mislin H. Active contractility of the lymphangion and coordination of lymphangion chains. *Experientia* 32:820–822, 1976.

24. Mislin H. The lymphangion. In: Földi M, Casley-Smith JR (eds.), *Lymphangiology*. Schattauer Verlag, New York-Stuttgart, pp 165–175, 1983.

25. Johnston MG. Involvement of lymphatic collecting ducts in the physiology and pathophysiology of lymph flow. In: Johnston MG (ed.), *Experimental biology of the Lymphatic Circulation*. Elsevier Science, Amsterdam, pp 81–120, 1985.

26. Achard C. Système lacunaire. In: Roger, Binet (eds.), *Traitè de Physiologie Normale et Patologique*, Vol. 7, Masson T, Paris, pp 351–449, 1934.

27. Collan Y, Kalima TV. Topographic relations of lymphatic endothelial cells in the initial lymphatics of the intestinal villus. *Lymphology* 7:175–184, 1974.

28. Hauck G. Vitalmicroscopical findings and lymph formation. In: Partsch H (ed.), *Progress in Lymphology*, Proc. XIth Int. Cong. Lymphology, Wien, 1987. Elsevier Science, Amsterdam, pp 183–186, 1988.

29. Castenholz A. Histomechanical mechanisms controlling fluid to enter initial lymphatics. In: Partsch H (ed.), *Progress in Lymphology*, Proc. XIth Int. Cong. Lymphology, Wien, 1987. Elsevier Science, Amsterdam, pp 179–182, 1988.

30. Ottaviani G, Gaja C. Il sistema paralinfatico. *Quad Anat Prat* Serie XII:133–152, 1957.

31. Castenholz A, Hauck G. Light and electron microscopic studies on the vital microscopic phenomenon of prelymphatic tissue channels in the mesentery. *Z Lymphol* 11:63–69, 1987.

32. Földi M. Prelymphatic — lymphatic drainage of the brain. *Am Heart J* 93:121–130, 1977.

33. Magari S. The spinal cord and the lymphatic system: Peripheral and autonomic nerves, sympathetic ganglia and the lymphatic system. In: Földi M, Casley-Smith JR (eds.), *Lymphangiology*. Schattauer Verlag, New York-Stuttgart, pp 509–533, 1983.

34. Grüntzig J. The eye and the lymphatic system. In: Földi M, Casley-Smith JR (eds.), *Lymphangiology*. Schattauer Verlag, New York-Stuttgart, pp 535–556, 1983.

35. Csanda E, Obál F, Obál F Jr. Central nervous system and lymphatic system. In: Földi M, Casley-Smith JR (eds.), *Lymphangiology*. Schattauer Verlag, New York-Stuttgart, pp 475–508, 1983.

36. Bradbury MWB, Cserr HF. Drainage of cerebral interstitial fluid and of cerebrospinal fluid into lymphatics. In: Johnston MG (ed.), *Experimental Biology of Lymphatic Circulation*. Elsevier Science, Amsterdam, pp 355–394, 1985.

37. Wang H, Casley-Smith JR. Drainage of the prelymphatics of the brain via the adventitia of the vertebral artery. *Acta Anat* 134:67–71, 1989.

38. Casley-Smith JR. Are the initial lymphatics normally pulled open by the anchoring filaments? *Lymphology* 13:120–129, 1980.

39. Leak LV. Lymphatic capillary ultrastructure and permeability. *Pflügers Arch* 336 (Suppl.):46–55, 1972.

40. Azzali G. Ricerche sul sistema linfatico di piccoli e grossi cheloni (*Testudo graeca, Emys europaea, Thalassochelys caretta*). *Ateneo Parmense*, 29 (Suppl. 4):143–175, 1958.

41. Azzali G. Fine structure of lymphatic vessels of the small intestine in the turtle. *Acta Anat* 130:7–8, 1987.

42. Cliff WJ, Nicoll PA. Structure and function of lymphatic vessels of the bat's wing. *Q J Exp Physiol* 55:112–121, 1970.

43. Castenholz A. Morphological characteristics of initial lymphatics in the tongue as shown by scanning electron microscopy. O'Hare AMF (ed.), *Scanning Electron Microscopy* III: Chicago, pp. 1343–1352, 1984.

44. Wenzel-Hora BI, Berens von Rautenfeld D, Majewski A, Lubach D. Scanning electron microscopy of the initial lymphatics of the skin after use of the indirect application technique with glutaraldehyde and MERCOX as compared to clinical findings. Part I. The nomenclature and microtopography of the initial lymphatics. *Lymphology* 20:126–133, 1987.

45. Azzali G, Orlandini G, Bucci G. Morphological characters of the absorbing peripheral lymphatic vessel by TEM, SEM and three-dimensional models. In: Motta PM (ed.), *Cells and Tissues: A Three-Dimensional Approach by Modern Techniques in Microscopy*, Alan R. Liss, New York pp 487–492, 1989.

46. Leak LV, Kato F. Electron microscopic studies of lymphatic capillaries during early inflammation I. Mild and severe thermal injuries. *Lab Invest* 26:572–588, 1972.

47. Azzali G. Ultrastructural and seasonal aspects of the kidney lymphatic system of hibernating animals. *Experientia* 44:441–444, 1988.

48. Azzali G, Bucci G, Gatti R, Orlandini G. Topography and ultrastructure of kidney lymphatics in some hibernating bats. *Lymphology* 21:212–223, 1988.

49. Azzali G. The ultrastructural basis of lipid transport in the absorbing lymphatic vessel. *J Submicrosc Cytol* 14:45–54, 1982.

50. Romita G, Gatti R. Sulla fine struttura dei vasi linfatici della vescica orinaria di piccoli Cheloni (*Testudo graeca ed Emys europaea*). *Acta Biomed Ateneo Parmense*

54:137–146, 1983.

51. O'Morchoe CCC. Lymphatic drainage of the kidney. In: Johnston MG (ed.), *Experimental Biology of the Lymphatic circulation*, Elsevier Science, Amsterdam, pp 261–304, 1985.

52. Simionescu M, Simionescu N, Palade GE. Permeability of muscle capillaries to small heme-peptides. *J Cell Biol* 64:586–607, 1975.

53. Azzali G, Romita G, Gatti R. Ultrastruttura dei vasi linfatici della vescica orinaria. *Acta Biomed Ateneo Parmense* 54:105–115, 1983.

54. Niiro GK, O'Morchoe CCC. An ultrastructural study of transport pathways across rat hepatic lymph vessels. *Lymphology* 18:98–106, 1985.

55. Azzali G. The "intraendothelial channels" of the peripheral absorbing lymphatic vessel. In: Partsch H (ed.), *Progress in Lymphology*, Proc. XIth Int. Cong. Lymphology, Wien, 1987. Elsevier Science, Amsterdam, pp 187–192, 1988.

56. Azzali G, Bucci G, Gatti R, Romita G. Three-dimensional organization of the gastroenteric lymphatic vessels of the turtle (studied by corrosion casts and SEM). *Quad Anat Prat*, Serie XLIV:16, 1988.

57. Bruns RR, Palade GE. Studies on blood capillaries I. General organization of blood capillaries in muscle. *J Cell Biol* 37:244–276, 1968.

58. French JE, Florey HW, Morris B. The absorption of particles by the lymphatics of the diaphragm. *Q J Exp Physiol* 45:88–103, 1960.

59. Casley-Smith JR. Endothelial permeability; the passage of particles into and out of diaphragmatic lymphatics. *Q J Exp Physiol* 49:365–383, 1964.

60. Leak LV, Burke JF. Fine structure of the lymphatic capillary and the adjoining connective tissue area. *Am J Anat* 118:785–810, 1966.

61. Azzali G. Ultrastructure of small intestine submucosal and serosal-muscular lymphatic vessels. *Lymphology* 15:106–111, 1982.

62. Albertine KH, O'Morchoe CCC. Renal lymphatic ultrastructure and translymphatic transport. *Microvasc Res* 19:338–351, 1980.

63. Leak LV. Studies on the permeability of lymphatic capillaries. *J Cell Biol* 50:300–323, 1971.

64. Leak LV, Burke JF. The passage of electron-opaque tracers across the lymphatic capillary wall. In: *VIth Int. Congr. Electr. Microsc., Kyoto, 1966*, Maruzen Co. Nihonbashi Ltd., Tokyo, pp 731–732, 1966.

65. Leak LV. Physiology of lymphatic system. In: Abramson DI, Dobrin PB (eds.), *Blood Vessels and Lymphatics in Organ Systems*, Academic Press, London, pp 134–171, 1984.

66. Casley-Smith JR. Freeze-substitution observations on the endothelium, and the passage of ions. *Microcirculation* 1:79–109, 1981.

67. Casley-Smith JR. The phylogeny of the fine structure of blood vessels and lymphatics: Similarities and differences. *Lymphology* 20:182–188, 1987.

68. Casley-Smith JR. The two modes of initial lymphatic filling: Colloidal osmotic pressure, hydrostatic pressure, or both? In: Partsch H (ed.), *Progress in Lymphology*,

Proc. XIth Int. Cong. Lymphology, Wien, 1987. Elsevier Science, Amsterdam, pp 173–177, 1988.

69. Leak LV, Rahil KS. Permeability of the diaphragmatic mesothelium. The ultrastructural basis for stomata. *Am J Anat* 151:557–594, 1978.

70. von Recklinghausen FT. Die Lymphgefässe und ihre Beziehungen zum Bindegewebe. Hirschwald, Berlin, 1862.

71. von Recklinghausen FT. Zur Fettresorption. *Virchows Arch Pathol Anat Physiol* 26:172–208, 1863.

72. Casley-Smith JR. Endothelial permeability. II. The passage of particles through the lymphatic endothelium of normal and injured ears. *Br J Exp Pathol* 46:35–49, 1965.

73. Casley-Smith JR. Electron microscopical observations on the dilated lymphatics in oedematous regions and their collapse following hyaluronidase administration. *Br J Exp Pathol* 48:680–686, 1967.

74. Kalima TV, Collan Y, Kalima SH. Variations of lymphatic endothelial cell junctions in experimental conditions. In: Mayall RC, Witte MH (eds.), *Progress in Lymphology*, Proc. Vth Int. Cong. Lymphology, Rio de Janeiro, 1975. Plenum Press, New York, pp 7–12, 1977.

75. Palay SL, Karlin KJ. An electron microscopic study of the intestinal villus II. The pathway of fat absorption. *J Biophys Biochem Cytol* 5:373–383, 1959.

76. Kalima TV. Ultrastructure of the intestinal lymphatics in regards to absorption. *Scand J Gastroenterol* 8:193–196, 1973.

77. O'Morchoe PJ, Yang VV, O'Morchoe CCC. Lymphatic transport pathways during volume expansion. *Microvasc Res* 20:275–294, 1980.

78. Roberge S, Boucher Y, Roy PE. Transient ultrastructural variations of pulmonary lymphatic capillaries during the respiratory cycle of the rat lung. *Anat Rec* 213:551–559, 1985.

79. Azzali G. Sur le passage transendothélial de lipides et de cellules dans les vaisseaux lymphatiques absorbants. *Acta Anat* 120:9, 1984.

80. Dobbins WO, Rollins EL. Intestinal mucosal lymphatic permeability: An electron microscopic study of endothelial vesicles and cell junctions. *J Ultrastruct Res* 33:29–59, 1970.

81. Elhay S, Casley-Smith JR. Mathematical model of the initial lymphatics. *Microvasc Res* 12:121–140, 1976.

82. Azzali G. Ultrastructure du vaisseau lymphatique absorbant pendant les états physiologiques et expérimentaux. *Acta Anat* 111:9, 1981.

83. Azzali G. Transendothelial transport of lipids in the absorbing lymphatic vessel. *Experientia* 38:275–276, 1982.

84. Azzali G, Romita G. Caratteristiche ultrastrutturali dei vasi linfatici dell'intestino tenue dell'uomo. Atti 41° Conv. Naz. Soc. Ital. Anat., Torino. p 97, 1986.

85. Ohkuma M. Electron microscopic observation of the renal lymphatic capillary after injection of ink solution. *Lymphology* 6:175–181, 1973.

86. Azzali G. Il vaso linfatico assorbente del villo intestinale in alcuni stadi funzionali e sperimentali. *Arch Ital Anat Embriol* 85:391–403, 1980.

87. Yang VV, O'Morchoe PJ, O'Morchoe CCC. Transport of protein across lymphatic endothelium in the rat kidney. *Microvasc Res* 21:75–91, 1981.

88. Földi M, Casley-Smith JR. *Lymphangiology*. Schattauer Verlag, New York-Stuttgart, pp 1–810, 1983.

89. Abramson DI, Dobrin PB. *Blood Vessels and Lymphatics in Organ Systems*, Academic Press, London, pp 3–719, 1984.

90. Johnston MG. *Experimental Biology of the Lymphatic Circulation*, Elsevier Science, New York, pp 1–423, 1985.

91. O'Morchoe CCC. Lymphatic system. In: Abramson DI, Dobrin PB (eds.), *Blood Vessels and Lymphatics in Organ Systems*, Academic Press, London, pp 532–543, 1984.

92. O'Morchoe PJ, Han Y, Doyle MD, O'Morchoe CCC. Lymphatic system of the thyroid gland in the rat. *Lymphology* 20:10–19, 1987.

93. Haye R. The capillaries of the middle ear mucosa in the guinea pig. *Z Zellforsch Mikrosk Anat* 143:517–526, 1973.

94. Niiro GK. O'Morchoe CCC. Pattern and distribution of intrahepatic lymph vessels in the rat. *Anat Rec* 215: 351–360, 1986.

95. Kalima TV. Lymphgefässsystem und intestinale Resorption. In: Meyer-Burg J, Arbeiter G (eds.), *Der Abdominelle Lymph-Kreislauf*, Witzstrock G. Verlag, Köln, pp 187–196, 1977.

96. Azzali G. Ultrastruttura del capillare e del vaso linfatico e loro gradiente di permeabilità. *Acta Biomed Ateneo Parmense* 47:14–15, 1976.

97. Azzali G. Il capillare linfatico. *Arch Ital Anat Embriol* 82 (Suppl.):220–221, 1977.

98. Ohtani O, Ohtsuka A, Owen RL. Three-dimensional organization of the lymphatics in the rabbit appendix. A scanning electron and light microscopic study. *Gastroenterology* 91:947–955, 1986.

99. Kato S. Intralobular lymphatic vessels and their relationship to blood vessels in the mouse thymus. Light- and electron-microscopic study. *Cell Tissue Res* 253: 181–187, 1988.

100. Carr I, Norris P, McGinty F. Reverse diapedesis; the mechanism of invasion of lymphatic vessels by neoplastic cells. *Experientia* 31:590–591, 1975.

101. van de Velde C, Carr I. Lymphatic invasion and metastasis. *Experientia* 33:837–843, 1977

102. Carr I. The passage of macrophages across lymphatic walls by reverse diapedesis: An ultrastructural study. *J Reticuloendothel Soc* 21:397–402, 1977.

103. Azzali G, Gatti R, Orlandini G. Macrophage migration through the endothelium in the absorbing peripheral lymphatic vessel of the small intestine. *J Submicrosc Cytol Pathol*, in press, 1990.

104. Azzali G. The passage of the cells across endothelial lymphatic wall. In: *Progress in Lymphology*, Proc. XIIth Int. Cong. Lymphology, Tokyo, 1989. Elsevier Science, Amsterdam, in press, 1990.

105. Cho Y, De Bruyn PPH. Internal structure of the postcapillary high-endothelial venules of rodent lymph nodes and Peyer's patches and the transendothelial lymphocyte passage. *Am J Anat* 177:481–490, 1986.

106. Campbell FR. Intercellular contacts between migrating blood cells and cells of the sinusoidal wall of bone marrow. An ultrastructural study using tannic acid. *Anat Rec* 203:365–374, 1982.

107. Marchesi VT, Gowans JL. The migration of lymphocytes through the endothelium of venules in lymph nodes: An electron microscopic study. *Proc R Soc Lond* (B) 159:283–290, 1964.

108. Marchesi VT, Florey HW. Electron micrographic observations on the emigration of leukocytes. *Q J Exp Physiol* 45:343–348, 1960.

109. Farr AG, De Bruyn PPH. The mode of lymphocyte migration through postcapillary venule endothelium in lymph node. *Am J Anat* 143:59–92, 1975.

110. Cho Y, De Bruyn PPH. The endothelial structure of the postcapillary venules of the lymph node and the passage of lymphocytes across the venule wall. *J Ultrastruct Res* 69:13–21, 1979.

111. Campbell FR. Intercellular contacts of lymphocytes during migration across high-endothelial venules of lymph nodes. An electron microscopic study. *Anat Rec* 207:643–652, 1983.

112. Schoefl GI. The migration of lymphocytes across the vascular endothelium in lymphoid tissue. *J Exp Med* 136:568–584, 1972.

113. Haston WS, Shields JW, Wilkinson PC. Lymphocyte locomotion and attachment on two-dimensional surfaces and three-dimensional matrices. *J Cell Biol* 92:747–752, 1982.

114. Bhalla DK, Braun J, Karnovsky MJ. Lymphocyte surface and cytoplasmic changes associated with translational motility and spontaneous capping of Ig. *J Cell Sci* 39:134–147, 1979.

115. Takada M. The ultrastructure of lymphatic valves in rabbits and mice. *Am J Anat* 132:207–218, 1971.

116. Lauweryns JM, Boussauw L. Striated filamentous bundles associated with centrioles in pulmonary lymphatic endothelial cells. *J Ultrastruct Res* 42:25–28, 1973.

117. Oehmke HJ. Periphere lymphgefasse des menschen und ihre funktionelle struktur. Licht- und elektronenmikroskopische studien. *Z Zellforsch* 90:320–332, 1968.

118. Nordquist RR, Bell RD, Sinclair RJ, Keyl MJ. The distribution and ultrastructural morphology of lymphatic vessels in the canine renal cortex. *Lymphology* 6:13–19, 1973.

119. Horstmann E, Breucker H. Über die Lymphkapillaren in den Darmzotten von Meerschweinchen und Affe. *Z Zellforsch Mikrosk Anat* 133:551–557, 1972.

120. Casley-Smith JR, Florey HW. The structure of normal small lymphatics. *Q J Exp Physiol* 46:101–106, 1961.

Author's address:
Prof. Giacomo Azzali
Institute of Anatomy
University of Parma
Via Gramsci 14
43100 Parma, Italy

Relationship of Blood Microvasculature to Structure and Function in Lymphoid Tissue

AIJI OHTSUKA, ROBERT L. OWEN, & TAKURO MURAKAMI

1. Introduction

In the lymphoid tissues or organs, the blood vascular system allows lymphocyte immigration and recirculation [1]. Thymocytes flow into the blood stream [2], and with lymphocytes originating in other tissues, migrate to lymph nodes [3], Peyer's patches [4], and other mucosal lymphoid organs. This chapter reviews the microvascular architectures of lymph nodes, mucosa-associated lymphoid tissues, thymus, and spleen, as revealed by injection replica SEM methods (scanning electron microscopy of vascular corrosion casts) [5].

2. Methodological Considerations

Target tissues or organs are perfused with physiological solutions through afferent arteries or their parent vessels, followed by methacrylate resin. After complete polymerization of the injected resin, specimens are corroded in NaOH solution, washed in water, air dried, coated with gold, and observed in a scanning electron microscope. Sometimes, 5–10% formalin or 1–2% glutaraldehyde in phosphate buffered saline (pH 7.2–7.4) is perfused prior to resin injection [6]. This prefixation is helpful for stabilizing vascular integrity and making vascular casts with little resin leakage, especially in thymus.

3. Lymph Nodes and Solitary Lymphoid Follicles

3.1. Lymph Nodes

The blood vascular architecture of lymph nodes has been studied by light microscopy of injected samples in rabbits [7], dogs [8], rats [9], mice [10], and guinea pigs [11], by microangiography in rats and dogs [12], and by the injection replica SEM method in mice [13], pigs [14], and rats [15,16].

Almost all the blood vessels destined for the lymph node enter the organ through the hilus (Fig. 8-1), with only occasional small vessels entering through the rest of the capsule, though the latter course is common in pig lymph nodes [14,17]. After entering the lymph node, the arteries initially run within the connective tissue trabecula, where they give off many branches to the medulla (Fig. 8-1). Passing along the medullary cord, these arterial branches give off small medullary arterioles to supply the medullary capillary beds en route, and finally reach the cortex to become cortical arteries. The medullary capillaries are thin, and repeatedly branch and anastomose to form a capillary plexus in the medullary cord.

In the cortex, cortical arteries give off many small arterioles to the parafollicular capillaries, and reach the follicular base, where the terminals of the arteries enter the follicles to give rise to the

Motta, P.M., Murakami, T., and Fujita, H. (eds.), Scanning Electron Microscopy of Vascular Casts: Methods and Applications.

follicular capillaries (Figs. 8-1 and 8-2). In the germinal center, capillaries are scarce in number and form a coarse plexus (Figs. 8-1–8-3). In the interfollicular or parafollicular areas between and around the follicles, capillaries form denser plexuses than in the germinal center. The cortical capillaries drain into postcapillary venules that are located around the follicles (Figs. 8-1 and 8-2). Capillaries at the cortical surface flow into tributaries of superficial venules between the follicles (Fig. 8-3).

Casts of these postcapillary venules have many indentations on their surfaces, which reflect the cobblestone appearance of the lumen of high endothelial venules (HEVs) (Figs. 8-2 and 8-15) (see below). The HEVs do not occur within the follicles but are restricted to the parafollicular (interfollicular) areas (Fig. 8-2). These HEVs are the sites of lymphocyte immigration into lymphoid tissue [3]. At the cortico-medullary boundary, the venules join each other to form thick veins, whose vascular casts have smooth surfaces (Fig. 8-2). These veins travel in the medulla, collecting the venous tributaries from the medullary capillary plexus en route and run in the trabecula to leave the lymph node from the hilus (Fig. 8-1).

Some authors have described arterio-venous anastomoses in the rat lymph node [9,15]. However, in our scanning electron microscopic observation of corrosion casts of rat mesenteric and mediastinal lymph nodes, all vessels that were intercalated between arterioles and venules were capillaries by size. We could not find any such short, thick arterio-venous anastomoses directly joining the arterial and venous systems.

Lymph nodes vary their histological structures according to their state of activity. These variations in structure involve vascular architecture. Experimental studies have shown that antigen challenges produce a significant increase in vascular density and numbers [13,18].

3.2. Solitary Lymphoid Follicles

Solitary lymphoid follicles are seen in the rat mesentery (Figs. 8-4–8-6). The vascular pattern of each follicle is quite similar to that of lymphoid follicles in the lymph node. Afferent arteries that supply the follicle reach the follicular base or hilus to enter the follicle and branch into arterioles that give rise to capillaries. The capillaries repeatedly branch and anastomose to make a follicular capillary plexus. The follicular capillary plexus is drained lateral to the follicle into postcapillary venules, which in turn gather to form efferent venules leaving follicles through the hilus.

In neonatal rats, the solitary lymphoid follicles have a simple vascular architecture (Figs. 8-5 and 8-6). Corrosion casts of venules in some follicles, 0.2–0.3 mm in diameter, show many indentations, as seen in HEVs of mature follicles (Fig. 8-5), while in smaller follicles (0.1–0.2 mm) the venular casts show no such characteristic indentations, but do have rough surfaces (Fig. 8-6). Transmission electron microscopic study of mouse lymph nodes has shown that, at birth, postcapillary venules have an undifferentiated high endothelium [19]. Rough surfaces of the venular casts in the smaller follicles may reflect such undifferentiated or immature postcapillary venules.

→

Figure 8-1. A scanning electron micrograph showing the cut surface of a frozen blood vessel corrosion cast of rat mediastinal lymph node. Arteries (A) and veins (V) enter the lymph node through the hilus (H) and repeatedly branch in the trabecula (T). The cortex (C), including lymphoid follicles (F), are supplied by cortical arteries (a) and veins (v). M = medulla. ×35.

Figure 8-2. Corrosion cast of blood vascular beds of a rat mesenteric lymph node. The coarse capillary plexus of a lymphoid follicle (F) receives the terminals of the cortical artery (a) and is drained by a high endothelial venule (h), which is located at the edge of the follicle in the parafollicular region. The venules join each other to form cortical veins (v), whose vascular casts have smooth surfaces. ×75.

Figure 8-3. Surface view of a vascular cast of a rat mesenteric lymph node. Four follicles (F) are seen. Between the follicles, superficial venous radicals (arrowheads) drain the capillaries of the lymph node surface. ×75.

Figure 8-4. Solitary lymphoid follicle in the rat mesenterium. An afferent arteriole (a) enters the follicle (F) through the follicular base or hilus to supply the follicular capillaries. The capillaries are collected by high endothelial venules (arrowheads), which empty into the efferent venule (v) leaving from the hilus. ×80.

Figure 8-5. Solitary lymphoid follicle (F) in the neonatal rat mesenterium. The venule draining the follicular capillaries is high endothelial (h). a = afferent arteriole. ×120.

Figure 8-6. Mesenteric solitary lymphoid follicle (F) of the rat just after birth. The surface of the efferent venule (v) does not clearly show the indentations that are seen in casts of high endothelial venules in Figs. 8-4, 8-5, and 8-15a. ×270.

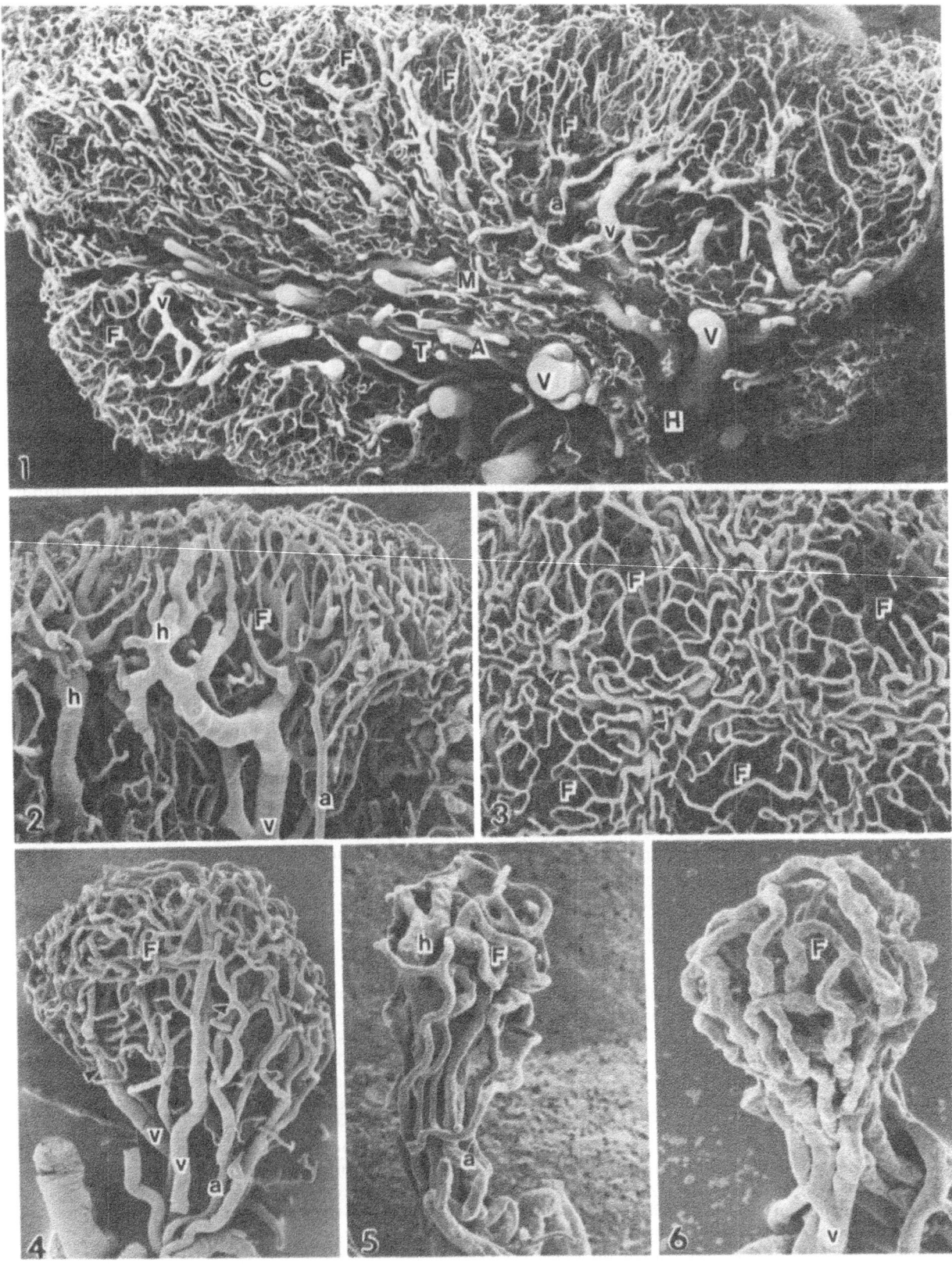

4. Mucosa-Associated Lymphoid Tissues

Mucosa-associated lymphoid tissues consist of solitary and/or aggregated lymphoid follicles in the laminae propriae mucosae with specialized epithelia covering the follicles [20]. We will limit our review of the blood vascular architecture of mucosal lymphoid tissues to the pharynx, small intestine, and large intestine, which have been most extensively investigated. Future studies will be required to determine how basic vascular patterns found in these tissues compare to those of lymphoid follicles in salivary glands, bronchi, conjunctiva, and the avian bursa of Fabricius.

4.1. Lymphoid Tissues in the Pharynx — Palatine Tonsil

The blood vascular architecture of palatine tonsils has been studied in both humans [21] and rabbits [22] by the injection replica SEM method. Arteries that reach the capsule of the palatine tonsil branch into small arteries, which enter the connective tissue septum. Running in the septum, the arteries give off arterioles that course toward the follicle and the interfollicular regions, where the arterioles give rise to the follicular and interfollicular capillary plexuses, respectively. Some of the arterioles reach the subepithelial region, where the arterioles give rise to capillaries, which form dense subepithelial capillary networks. The follicular and interfollicular capillary plexuses are coarser than the subepithelial capillary networks. The subepithelial capillary network overlying the follicle forms a dome protruding towards the tonsillar crypt. In the human tonsil, the subepithelial network consists of many switchback loops of capillaries projecting into the crypt epithelium, around the network located directly over the follicle (Fig. 8-7) [21]. In the rabbit,

hairpin-like capillary loops are observed in the whole subepithelial network, though the loops from the subepithelial network over the follicle are short and those over the interfollicular area are long [22].

The subepithelial capillaries collect into venules, which run in the subepithelial region (Fig. 8-7). These venules connect with the HEVs in the interfollicular region. The HEVs, collecting follicular and interfollicular capillaries en route, course down into the interfollicular region alongside the follicle and ultimately empty into the ordinary veins that run in the septa or in the capsule. Thus, the coarse network of HEVs surround the lateral and basal surfaces of each follicle [21].

4.2. Lymphoid Tissues in Small Intestine — Peyer's Patches

In the small intestine, lymphoid follicles exist singly (solitary lymphoid follicles) and in groups (Peyer's patches). The blood vascular architecture of the Peyer's patch has been studied by light microscopy of injected samples in rats [7], mice [23], guinea pigs [11], rabbits [7], and cats [7], and also by the injection replica SEM method in rats [24,25] and mice [26].

Arteries and veins supply intestines along the mesenteric border and penetrate into the submucosa. In the Peyer's patch, the submucosal arteries and veins run as interfollicular arteries and veins, around and between the lymphoid follicles [24]. The interfollicular arteries give off follicular arterioles to lymphoid follicles, villous arterioles to villi, cryptal branches to crypts, and muscular branches to external muscles.

The follicular arterioles enter the follicle and give branches to the capillary plexus within the follicle (Fig. 8-8). Capillaries within the fol-

Figure 8-7. A luminal view of a corrosion cast of blood vascular beds of human palatine tonsil. Many switchback capillary loops (arrowheads) are seen in the capillary network, around the dome subepithelial network (D) covering the follicle. These subepithelial networks are collected by tributaries of high endothelial venules (h) in the interfollicular region. Courtesy of Dr. O. Ohtani. ×100. Reprinted with permission from Ohtani [20].

Figure 8-8. Vascular cast of rat Peyer's patch. A dome (D) overlying the follicle (F) protrudes among intestinal villi (Vi). A follicular arteriole (a) ascends to the dome to supply the dome subepithelial capillaries. C = crypts around the follicular dome. ×80.

Figure 8-9. Vascular casts of rat cecal patch. A luminal view. Subepithelial capillary network of the follicular dome (D) is connected with the capillary plexus of the crypts (C) around the follicle. Note that arterioles (arrowheads) ascending in the follicles supply the dome capillaries. ×130.

103

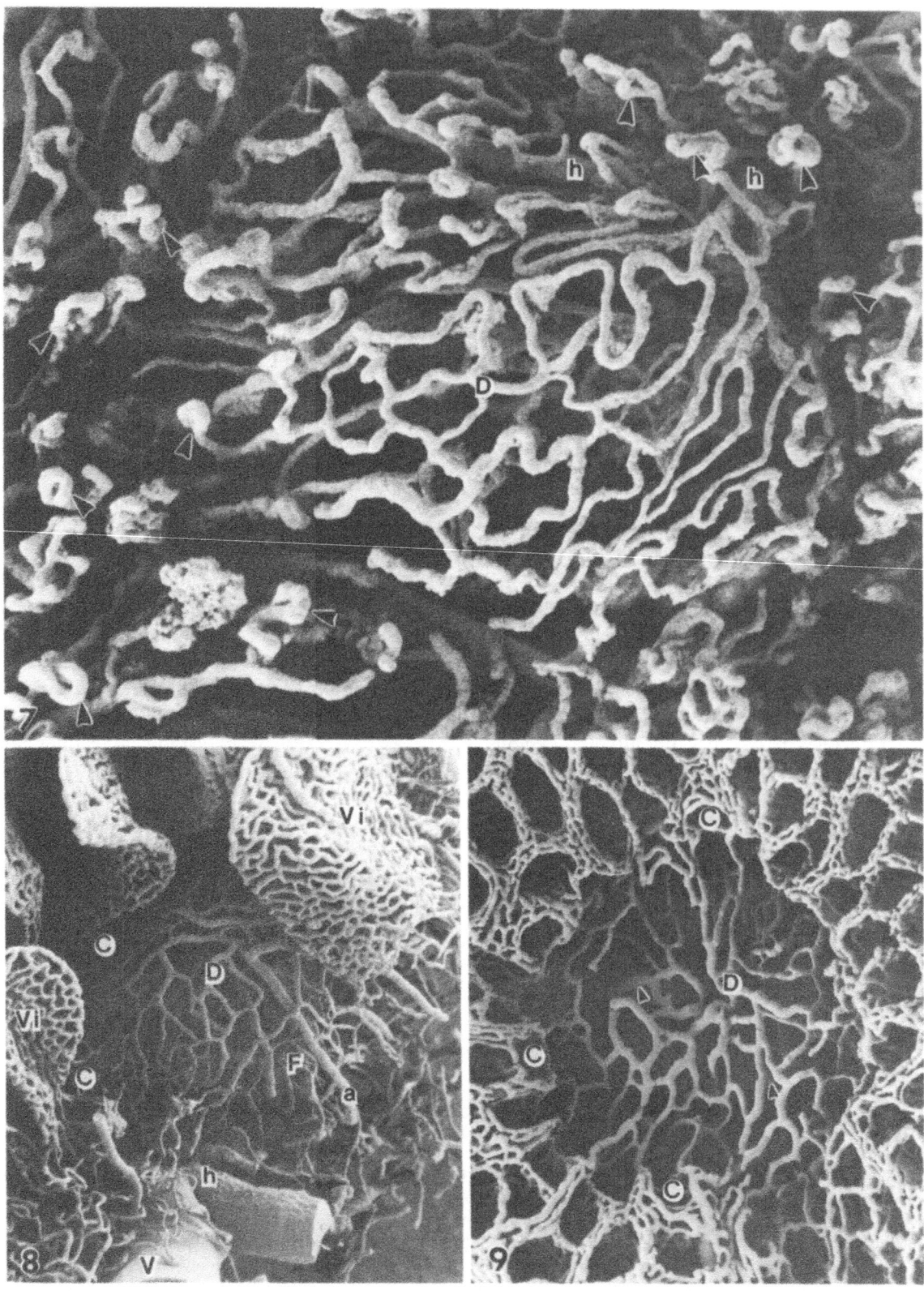

licle are thin and form a coarse plexus. The intrafollicular capillary plexus drains into HEVs basal and lateral to the follicle (parafollicular or interfollicular area). These venules reach the interfollicular space, where they empty into the interfollicular veins.

Some of the follicular arterioles ascend through the follicle and reach the apex area of the dome to give rise to a capillary network beneath the dome epithelium (Fig. 8-8). This capillary network is denser than the intrafollicular capillary plexus, but coarser than the villous capillary network (Fig. 8-8). The dome capillary network drains at its marginal region into the venules, which in turn flow into the HEVs in the interfollicular area. Thus, the microcirculation of the dome area appears to be a modification of the Mall's fountain pattern seen in villi [24]. Some marginal capillaries of the dome network continue into the capillary plexus of the crypts surrounding the follicle (Fig. 8-8). These connections may provide a route for the migration of humoral factors from the follicle apex to the crypts, which may regulate the cells migrating from the crypts to the follicle surface [24].

4.3. Lymphoid Tissues in Large Intestine

Lymphoid follicles occur solitarily and in aggregates in the large intestine of various animals, including humans [16,27–30]. The blood vascular architecture of these lymphoid follicles is essentially similar to that of the Peyer's patch [25], except that follicles are surrounded by flat colonic epithelium, rather than villi, and face crevices in the colonic surface, rather than protrude into the intestinal lumen as domes (Fig. 8-9).

5. Thymus

The blood vascular architecture of the thymus has been studied by light microscopy in guinea pigs

[11], mice [31], and rats [32], and also by the injection replica SEM method in rats [15,16] and guinea pigs [33].

The arteries supplying the thymus originate from the internal thoracic arteries and their mediastinal and pericardiacophrenic branches. They enter this organ penetrating the capsule and run in the interlobular connective tissue, where the arteries ramify into small branches. These branches course in the secondary connective tissue septa toward the cortico-medullary boundary to become cortico-medullary arteries.

In the cortico-medullary boundary, the arteries give off cortical arterioles that ascend into the cortex and give rise to capillaries (Fig. 8-10). These capillaries, ascending to the periphery (outer surface) of the cortex, branch and anastomose to form the cortical capillary plexus (Fig. 8-10). At the surface of the cortex or at the subcapsular region, the cortical capillaries run tortuously and form a superficial network (Fig. 8-11). This superficial network may correspond to the capsular capillary network described by other authors [32]. These superficial capillaries are collected at the cortical surface by superficial venules, which descend in the cortex toward the cortico-medullary boundary, where they flow into thick cortico-medullary veins (Figs. 8-10 and 8-11). Capillaries in the deep cortex are drained by deep venules, which in turn descend to the cortico-medullary boundary, where they empty into the cortico-medullary veins (Fig. 8-10). Three patterns of thymic cortical venous drainage have been proposed, as follows: (1) cortico-medullary drainage [31,34–36]; (2) superficial drainage [11,37]; and (3) dual drainage [32,33, 38]. Our observation of the rat thymus confirms the dual venous drainage at the superficial and cortico-medullary regions.

The medullary capillary plexus receives small arteriolar twigs from the cortico-medullary arteries and drains into the cortico-medullary veins. The cortico-medullary veins drain both the

→

Figure 8-10. Cut frozen vascular casts of rat thymus. A capillary plexus in the cortex (C) receives an arterial branch (a) at the cortico-medullary junction, and is drained by venules (v) at the surface and at the cortico-medullary junction. M = medulla; V = veins. ×80.
Figure 8-11. A surface view of vascular casts of rat thymus. Venules (v) collect subcapsular capillaries. ×100.
Figure 8-12. Rat thymic vascular casts. Note leaked resin (asterisks) around the venules (v) at the cortico-medullary junction. C = cortex; M = medulla; V = vein. ×60.

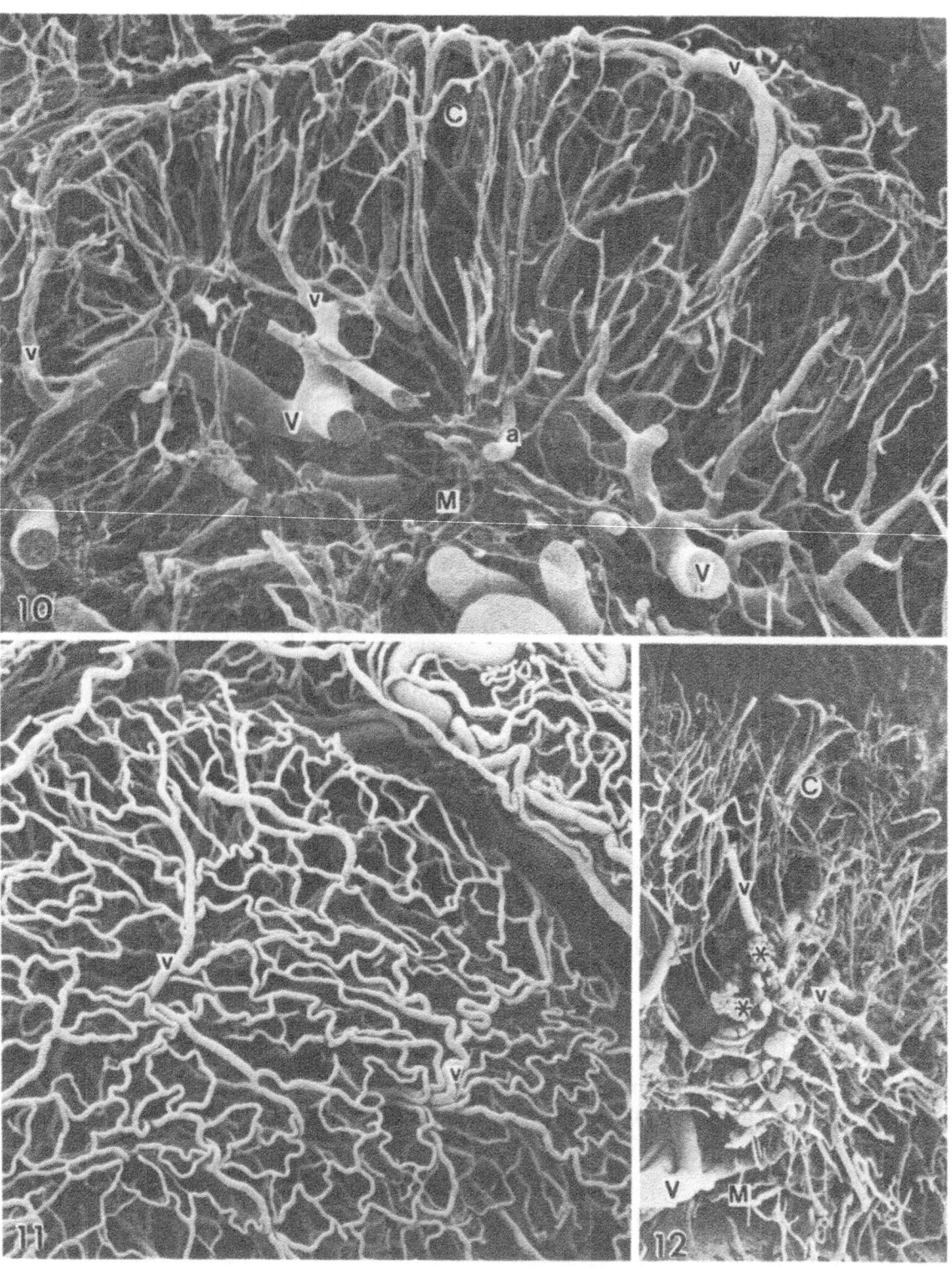

106

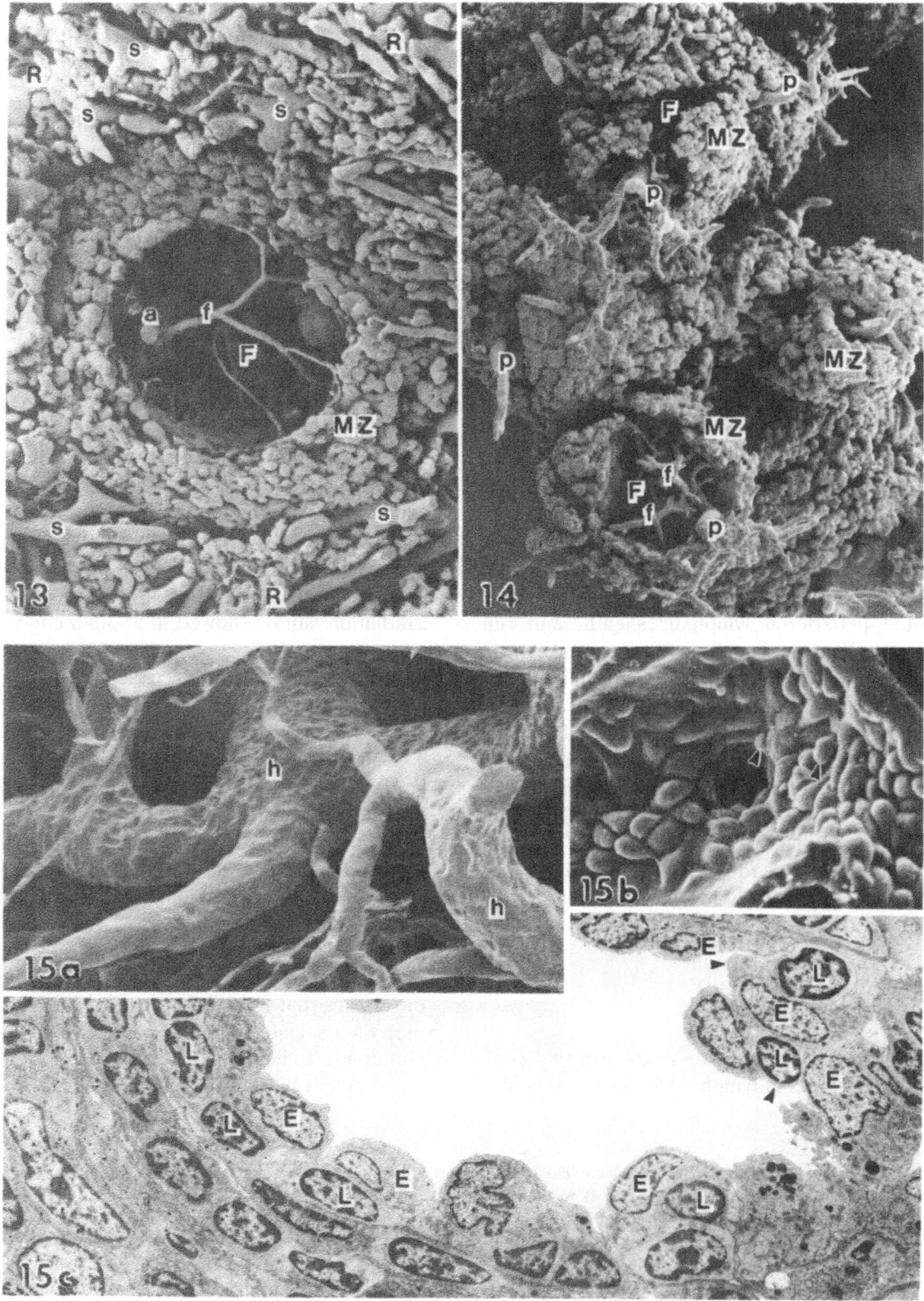

medulla and cortex, and then leave the lobules to empty into the veins in the secondary connective tissues, which run along the corresponding arteries.

The venules at the cortico-medullary boundary have low endothelium with perivascular spaces around the endothelium [33,39]. Transmission electron microscopy has confirmed that thymocytes migrate through these postcapillary venules [2,39]. When perfusion-fixation is not performed prior to resin injection, resin leakage often occurs from the venules in the cortico-medullary boundary (Fig. 8-12). Such resin leakage may reflect the location of endothelial pores that have been demonstrated by scanning electron microscopy [15].

6. White Pulp of the Spleen

The splenic artery divides into the pulp arteries that enter the white pulp. Branches of the pulp arteries, the central arteries, are surrounded by the periarterial lymphatic sheath, and run eccentrically in the lymphoid follicles. The central arteries gives off follicular and penicillar arteries (Figs. 8-13 and 8-14). The follicular arteries branch into capillaries within the follicles, run radially, and terminate at the periphery of the white pulp or at the boundary between the follicle and the marginal zone (Figs. 8-13 and 8-14) [40, 41]. The penicillar arteries penetrate the marginal zone and run toward the red pulp (Fig. 8-14). Other aspects related to the microvascular organization of the spleen are reported in Chapter 9.

7. High Endothelial Venules

Light microscopic studies have shown that postcapillary venules in lymphoid follicles have a high endothelium [42,43] and are sites for lymphocyte migration [44]. Electron microscopic studies have confirmed that lymphocytes migrate through the wall of the postcapillary venules to enter lymph nodes [45,46], Peyer's patches [46,47], and tonsils [48]. Corrosion casts of postcapillary venules in lymphoid follicles have numerous irregular indentation on their surfaces (Fig. 8-15a) [21,26]. These indentations correspond to the cobblestone-like appearance of the luminal surface of the venules (Fig. 8-15b). Transmission electron microscopy of these venules confirmed that there are numerous lymphocytes among the endothelial cells (Fig. 8-15c).

The endothelium of high endothelial venules (HEV) changes its appearance according to the state of activity of the lymphoid tissues. Interruption of afferent lymphatics that reduces the inflow of antigens to lymph nodes results in the disappearance of high-endothelial venules or flattening of endothelial cells, whereas antigen injection into such lymph nodes induces the reappearance of high-endothelial venules [49]. In addition, an irradiation study showed that high-endothelial impressions on the vascular casts become shallow in lymphocyte-depleted mice [13].

8. Concluding Remarks

Thymic lobules are supplied by arteries and veins at the cortico-medullary boundary. Our vascular cast study in the rat showed that the cortex has a dual venous drainage from the surface and the cortico-medullary regions.

The microvascular substructure of the lymphoid follicle is a basic component of those lymphoid tissues that form follicles, i.e., lymph nodes, mucosa-associated lymphoid tissues and solitary lymphoid follicles. The follicular capillaries are supplied by arterioles that enter into the fol-

Figure 8-13. Blood vascular casts of rat spleen. The central artery (a), running in the follicle (F), gives off follicular arteries (f), which terminate at the marginal zone (MZ). R = red pulp; s = splenic sinus. ×100.

Figure 8-14. Arterial partial casts of rat spleen. Red pulp, including splenic sinuses, are not cast. Penicillar arteries (p) penetrate the marginal zone (MZ). F = follicle; f = follicular arteries. ×75.

Figure 8-15. High endothelial venules in the rat cecal patch. (a) Corrosion casts of HEVs (h). Note many indentations on the surface of the casts. ×340. (b) A scanning micrograph of perfusion-fixed HEV. Luminal surface of endothelium resembles cobblestone. Note migrating lymphocytes (arrowheads). ×680. (c) A transmission electron micrograph of perfusion-fixed HEV. Note numerous lymphocytes (L) beneath endothelial cells (E) and also migrating lymphocytes (arrowheads). ×2100.

108

licle from the base region and are drained by venules that dilate into HEVs at the periphery or basolateral region. The capillary plexus in the germinal center, where proliferating lymphoblasts spread vessels apart, is coarser than that in the periphery of the follicle. The corrosion casts of HEVs have irregular indentations on their surfaces, corresponding to sites of adhesions that trap circulating lymphoid cells and facilitate their migration into antigenically stimulated follicles. The HEVs are located in the parafollicular area, but not within the follicle, where lymphocytes proliferate in response to antigen transported through the epithelium over mucosa-associated lymphoid follicles or carried into lymph nodes by afferent lymphatics. Lymphocytes produced in germinal centers and migrating lymphoid cells that enter lymphoid tissues via the HEVs leave by means of lymphatic vessels originating in these tissues [50].

References

1. Raviola E. The immune system. In: Fawcett DW (ed.), *A Textbook of Histology* 11th ed., Saunders, Philadelphia, pp 406–435, 1986.
2. Törö I, Oláh I. Penetration of thymocytes into the blood circulation. *J Ultrastruct Res* 17:439–451, 1967.
3. Gowans JL, Knight EJ. The route of re-circulation of lymphocytes in the rat. *Proc R Soc Lond* (Biol) 159:257–282, 1964.
4. Bhalla DK, Owen RL. Migration of B and T lymphocytes to M cells in Peyer's patch follicle epithelium: An autoradiographic and immunocytochemical study in mice. *Cell Immunol* 81:105–117, 1983.
5. Murakami T. Application of the scanning electron microscope to the study of the fine distribution of the blood vessels. *Arch Histol Jpn* 32:445–454, 1971.
6. Murakami T, Itoshima T, Hitomi K, Ohtsuka A, Jones AL. A monomeric methyl and hydroxypropyl methacrylate injection medium and its utility in casting blood capillaries and liver bile canaliculi for scanning electron microscopy. *Arch Histol Jpn* 47:223–237, 1984.
7. Dabelow A. Die Blutgefässversorgung der lymphatischen Organe. *Verh Anat Ges* 46 (*Erg-H Anat Anz* 87):179–224, 1939.
8. Calvert WJ. The blood-vessels of the lymphatic gland. *Anat Anz* 13:174–180, 1897.
9. Anderson AO, Anderson ND. Studies on the structure and permeability of the microvasculature in normal rat lymph nodes. *Am J Pathol* 80:387–418, 1975.
10. Kowala MC, Schoefl GI. The politeal lymph node of the mouse: Internal architecture, vascular distribution and lymphatic supply. *J Anat* 148:25–46, 1986.
11. Blau JN. A comparative study of the microcirculation in the guinea-pig thymus, lymph nodes and Peyer's patches. *Clin Exp Immunol* 27:340–347, 1977.
12. Herman PG, Ohba S, Mellins HZ. Blood microcirculation in the lymph node. *Radiology* 92:1073–1080, 1969.
13. Steeber DA, Erickson CM, Hodde KC, Albrecht RM. Vascular changes in popliteal lymph nodes due to antigen challenge in normal and lethally irradiated mice. *Scann Microsc* 1:831–839, 1987.
14. Hoshi N, Hashimoto Y, Kitagawa H, Kon Y, Kudo N. Blood supply and microvasculature of the lymph nodes in pigs. *Jpn J Vet Res* 36:15–29, 1988.
15. Irino S, Takasugi N, Murakami T. Vascular architecture of thymus and lymph nodes: Blood vessels, transmural passage of lymphocytes, and cell-interactions. *Scann Electron Microsc* 1981 III:89–98, 1981.
16. Kardon RH, Kessel RG. The microcirculation of lymphoid tissue in three dimensions: Scanning electron microscopy of corrosion casts of the lymph node, thymus, and peri-rectal lymphoid tissue. *Biomed Res* 2 (Suppl):173–179, 1981.
17. Spalding H, Heath T. Blood vessels of lymph nodes in the pig. *Res Vet Sci* 41:196–199, 1986.
18. Herman PG, Yamamoto I, Mellins HZ. Blood microcirculation in the lymph node during the primary immune response. *J Exp Med* 136:697–713, 1972.
19. Van Deurs B, Röpke C. The postnatal development of high-endothelial venules in lymph nodes of mice. *Anat Rec* 181:659–678, 1975.
20. Owen RL, Bhalla DK. Lympho-epithelial organs and lymph nodes. In: Hodges GM, Carr KE (eds.), *Biomedical Research Applications of Scanning Electron Microscopy*, Vol. 3, Academic Press, London, pp 79–169, 1983.
21. Ohtani O, Kikuta A, Terasawa K, Higashikawa T, Yamane T, Taguchi T, Masuda Y, Murakami T. Microvascular organization of human palatine tonsils. *Arch Histol Cytol* 52:493–500, 1989.
22. Terasawa K, Ohtani O, Kikuta A, Taguchi T, Masuda Y, Kawakami S, Ogura Y. Microvascular organization of the rabbit tonsil — a scanning electron microscopic study of corrosion casts. *Jpn J Tonsil* 27:18–22, 1988 (in Japanese).
23. Abe K, Ito T. A qualitative and quantitative morphologic study of Peyer's patches of the mouse. *Arch Histol Jpn* 40:407–420, 1977.
24. Bhalla DK, Murakami T, Owen RL. Microcirculation of intestinal lymphoid follicles in rat Peyer's patches. *Gastroenterology* 81:481–491, 1981.
25. Ohtsuka A, Ohtani O, Murakami T. Microvascularization of the alimentary canal as studied by scanning electron microscopy of corrosion casts. In: Motta PM, Fujita H, Correr S (eds.), *Ultrastructure of the Digestive Tract* Martinus Nijhoff Publishers, Boston, pp 201–212, 1988.
26. Yamaguchi K, Schoefl GI. Blood vessels of the Peyer's patch in the mouse: I. Topographic studies. *Anat Rec* 206:391–401, 1983.
27. Owen RL, Nemanic P. Antigen processing structures of the mammalian intestinal tract: An SEM study of

lymphoepithelial organs. *Scann Electr Microsc* 1978 II: 367–378, 1978.

28. Owen RL, Piazza AJ, Ermak TH. Ultrastructural and cytoarchitectural features of lymphoreticular organs in the colon and rectum of adult BALB/c mice. *Am J Anat* 190: 10–18, 1991.

29. Liebler EM, Pohlenz JF, Woode GN. Gut-associated lymphoid tissue in the large intestine of calves. II. Distribution and histology. *Vet Pathol* 25:503–508, 1988.

30. Langman JM, Rowland R. The number and distribution of lymphoid follicles in the human large intestine. *J Anat* 194:189–194, 1986.

31. Smith C, Thatcher EC, Kraemer DZ, Holt ES. Studies on the thymus of the mammal. VI. The vascular pattern of the thymus of the mouse and its changes during aging. *J Morphol* 91:199–219, 1952.

32. Sainte-Marie G, Peng F-S, Marcoux D. The stroma of the thymus of the rat: Morphology and antigen diffusion, a reconsideration. *Am J Anat* 177:333–352, 1986.

33. Kato S, Schoefl GI. The vasculature of the guinea-pig thymus: Topographic studies by light and electron microscopy. *Arch Histol Jpn* 50:299–314, 1987.

34. Ito T, Hoshino T. Light and electron microscopic observations on the vascular pattern of the thymus of the mouse. *Arch Histol Jpn* 27:351–361, 1966.

35. Raviola E, Karnovsky MJ. Evidence for a blood-thymus barrier using electron-opaque tracers. *J Exp Med* 136: 466–498, 1972.

36. Abe K, Ito T. Vascular permeability in the thymus of the mouse. *Arch Histol Jpn* 36:251–264, 1974.

37. Bargmann W. Der Thymus. In: Möllendorff W (ed.), *Handbuch der Mikroskopischen Anatomie des Menschen*, VI/4, Springer, Berlin, pp 1–172, 1943.

38. Olson IA, Poste ME. The vascular supply of the thymus in the guinea-pig and pig. *Immunology* 24:253–257, 1973.

39. Ushiki T. A scanning electron-microscopic study of the rat thymus with special reference to cell types and migration of lymphocytes into the general circulation. *Cell Tissue Res* 244:285–298, 1986.

40. Yamamoto K, Kobayashi T, Murakami T. Arterial terminals in the rat spleen as demonstrated by scanning electron microscopy of vascular casts. *Scann Electron Microsc* 1982 I:455–458, 1982.

41. Kashimura M, Fujita T. A scanning electron microscopy of human spleen: Relationship between the microcirculation and functions. *Scann Microsc* 1:841–851, 1987.

42. Thomé R. Endothelien als Phagocyten (aus den Lymphdrüsen von *Macacus cynomolgus*). *Arch Mikrosk Anat Entwick-Gesch* 52:820–842, 1898.

43. Schumacher S. Ueber Phagocytose und die Abfuhrwege der Leukocyten in den Lymphdrüsen. *Arch Mikrosk Anat Entw-Gesch* 54:311–328, 1899.

44. Dawson AB, Masur J. Variations in the histological structure of the inguinal lymph nodes of the albino rat. *Anat Rec* 44:143–163, 1929.

45. Marchesi VT, Gowans JL. The migration of lymphocytes through the endothelium of venules in lymph nodes: An electron microscope study. *Proc R Soc Lond* (Biol) 159:283–290, 1964.

46. Cho Y, De Bruyn PPH. Internal structure of the postcapillary high-endothelial venules of rodent lymph nodes and Peyer's patches and the transendothelial lymphocyte passage. *Am J Anat* 177:481–490, 1986.

47. Yamaguchi K, Schoefl GI. Blood vessels of the Peyer's patch in the mouse: III. High-endothelial venules. *Anat Rec* 206:419–438, 1983.

48. Umetani Y. Postcapillary venule in rabbit tonsil and entry of lymphocytes into its endothelium: A scanning and transmission electron microscope study. *Arch Histol Jpn* 40:77–94, 1977.

49. Henderiks HR, Eestermans IL. Disappearance and reappearance of high endothelial venules and immigrating lymphocytes in lymph nodes deprived of afferent lymphatic vessels: A possible regulatory role of macrophages in lymphocyte migration. *Eur J Immunol* 13:663–669, 1983.

50. Ohtani O, Ohtsuka A, Owen RL. Three-dimensional organization of the lymphatics in the rabbit appendix. A scanning electron and light microscopic study. *Gastroenterology* 91:947–955, 1986.

Author's address:
Dr. Aiji Ohtsuka
Department of Anatomy
Okayama University School of Medicine
2-5-1 Shikata-cho
Okayama
700 Japan

CHAPTER 9

Open and Closed Circulation in the Spleen as Evidenced by SEM of Vascular Casts

TSUNEO FUJITA, MAKOTO KASHIMURA, & KAZUO ADACHI

1. Introduction

The spleen is the last large organ left whose microcirculation design is in doubt. The problems to be settled are focused on the open and closed theories. As early as 1861, Billroth [1], the pioneer of spleen histology, noted this problem; his careful observation of human spleen could not determine whether an arterial capillary terminated open into the cord, which is now called after his name, or it passed through the cord to be connected with the sinus. To omit some dispute in the last decades of the 19th century [for literature see 1 and 2], the open theory found its strongest support in Weidenreich [2]. If one reads his treatise of 1901 carefully, however, one will notice that he admits that the arterial capillary either opens to the cord or directly ends in the sinus. Helly [3] opposed Weidenreich, claiming that the closed route was all that existed.

Many researchers later attempted to trace the exact routes of splenic blood flow in different mammalian species, and by using different techniques reached controversial results. Table 9-1 lists only some representative work [for more detailed literature, see 13].

The controversies are partly due to little attention paid by the authors to species differences in the histology of the spleen. Of special importance in this context is the remark by Schmidt et al. [31] that certain animals, including the mouse and cat, favored by some researchers of the spleen belong to the nonsinusal group [37], in which one may easily mistake the reticular spaces of the red pulp to be the sinuses, when reproduced in casts. This problem, however, awaits further investigation, as Hataba et al. [36] demonstrated by SEM that mouse spleen does possess sinuses with slits, permitting the passage of blood cells, though they are conspicuously different in structure from those in human, dog, and other species.

2. Microcirculation of Human Spleen

The present review will deal with the open closed problem and some other topics concerning the microcirculation of the spleen on the basis of our own SEM studies of the tissues and vascular casts of the human spleen. Before dealing with the terminal portions of arteries, the distribution and structure of central arteries and their branches will be discussed.

2.1. Central Arteries and their Branches

Based on reconstructions from serial sections, Snook first pointed out in the spleen in humans [38] and in the monkey [39] that the central artery, after passing through the white pulp, issues a strong branch, turning back into the white pulp and forming a bundle of arterioles and capillaries. This bundle, which is called the *arteriolar-capillary bundle*, and is ensheathed in a connective tissue capsule, opens, with its twigs, into the meshes of the marginal zone. In our SEM

Motta, P.M., Murakami, T., and Fujita, H. (eds.), Scanning Electron Microscopy of Vascular Casts: Methods and Applications.

Table 9-1. Open versus closed circulation proposed in previous studies on mammalian spleen

Author (s)	Year	Result	Species	Method	Ref.
Weidenreich	1901	Open + closed	Human	LM	2
Helly	1902	Closed	Human, cat, rabbit	LM	3
MacNeal	1929	Open	Human	LM	4
McNee	1931	Open + closed	Rabbit	Microsphere injection	5
Knisely	1936	Closed	Rat & mouse	Transillumination microscope	6
MacKenzie et al.	1941	Open	Mouse & rat	Transillumination microscope	7
Björkman	1947	Closed	Rabbit	Starch grain injection	8
Pappart et al.	1955	Open	Mouse	TEM	9
Lewis	1956	Open	Fetal rabbit	LM	10
Snook	1958	Mainly open	Rabbit	LM, also India-ink injection	11
Weiss	1963	Open	Rabbit	TEM	12
Tischendorf	1969	Closed	Human	LM	13
Chen and Weiss	1972	Open	Human	TEM	14
Murakami et al.	1973	Closed	Rat	SEM of cast	15
Fujita	1974	Open	Human	SEM	16
Barnhart & Baechler	1974	Open + closed	Human, dog	SEM	17
Barnhard & Lusher	1976	Open + closed	Human	SEM	18
Irino et al.	1977	Open	Human	SEM of tissue and cast	19
Suzuki et al.	1977	Open	Dog	SEM	20
Chen	1978	Open + closed	Mouse	Microsphere injection	21
Hataba et al.	1981	Open	Mouse	SEM	22
Blue & Weiss	1981	Open	Cat	TEM	23
Fujita et al.	1982	Open	Human	SEM of tissue & cast	24
Schmidt et al.	1982	Closed + open	Dog	SEM of cast	25
Fujita & Kashimura	1983	Open	Human	SEM of tissue & cast	26
Suzuki	1983	Open	Human, dog cat, etc.	SEM of tissue & cast	27
Schmidt et al.	1983	Open	Cat	SEM of cast	28
Schmidt et al.	1983	Open + closed	Dog	SEM of cast	29
Seki and Abe	1984	Open + closed	Rat	SEM (tissue)	30
		Open	Cat, pig, horse, etc.		
Schmidt et al.	1985	Open	Mouse	SEM of cast	31
Schmidt et al.	1985	Open + closed	Rat	SEM of cast	32
Kashimura	1985	Open + closed	Human	SEM (tissue & cast)	33
Kashimura & Fujita	1987	Open + closed	Human	SEM of tissue & cast	34
Schmidt et al.	1988	Open + rarely closed	Human	SEM of cast	35
Hataba & Suzuki	1989	Open	Ferret	SEM	36

specimens of freeze-cracked human spleens, we could confirm occurrence of typical arteriolar-capillary bundles cross fractured in the white pulp [Fig. 13 in 40]. Comparable observations could be gathered from hematoxylin eosin stained sections of perfusion-fixed human spleens.

In their SEM observation of the vascular casts of human spleens, Schmidt et al. [35], as well as our group, demonstrated the arteriolar-capillary bundle. Murakami [Fig. II-27 in 41] obtained SEM images of vascular casts from the spleen of the rhesus monkey that indicated the constant

→

Figure 9-1. Cast of human spleen as seen by SEM. The round space corresponding to a white pulp (W) is surrounded by resin granules, which are the impressions of the reticular spaces of a marginal zone (M). The white pulp is penetrated by an arteriolar-capillary bundle (ACB) whose twigs are connected to the marginal zone (arrows). S = sinuses. ×160. (From Kashimura and Fujita [34], with permission.)

Figure 9-2. Cast of human spleen showing a marginal zone (M) and some twigs of penicillar arteries (PA) terminating in the marginal zone (arrows). S = casts of sinuses. ×200. (From Kashimura and Fujita [34], with permission.)

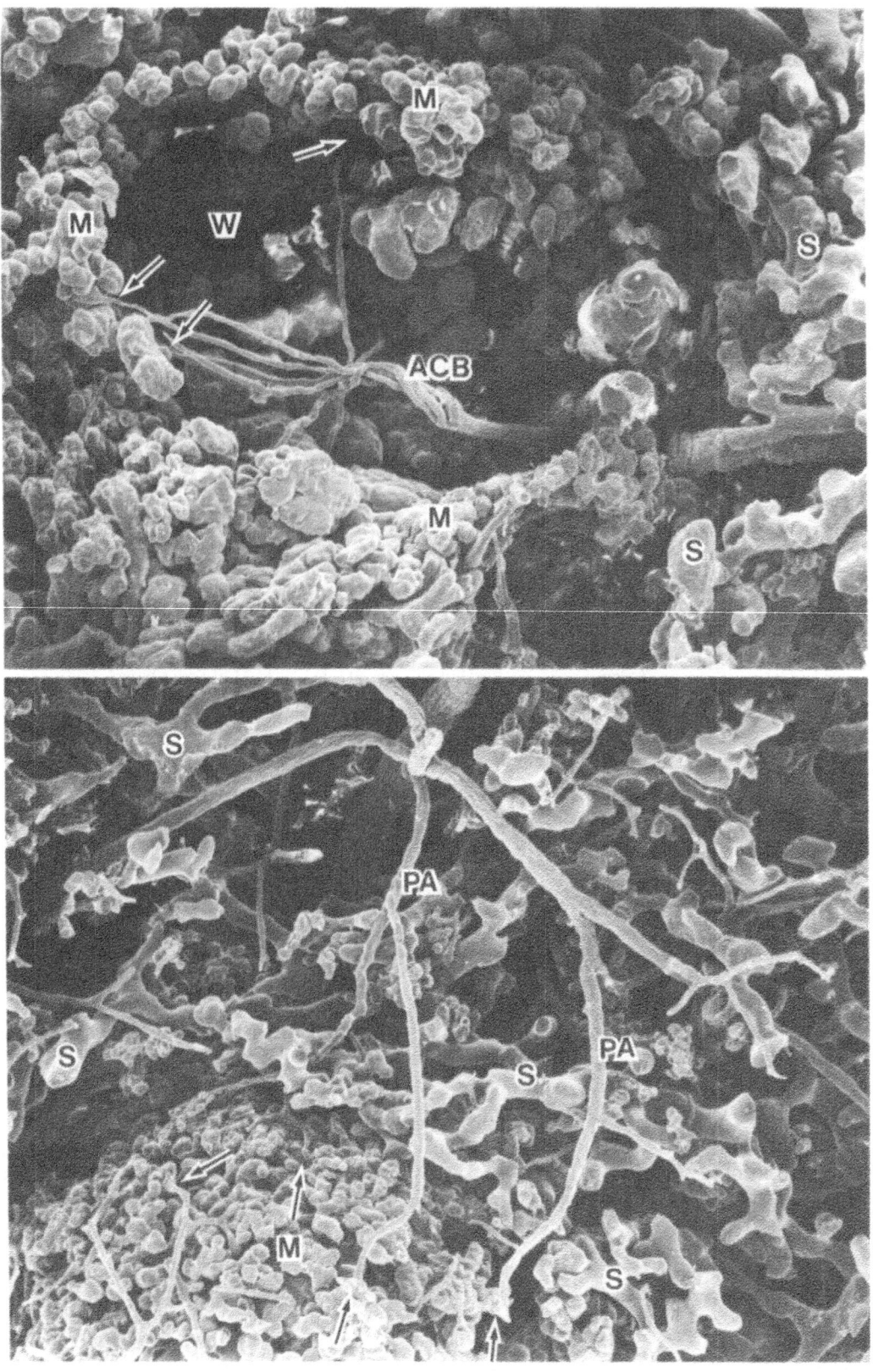

114

occurrence of the arteriolar-capillary bundle (Fig. 9-1) in the white pulp.

The branches of the central artery or the vascular bundle partly terminate in the marginal zone and partly extend into the red pulp. As pointed out by Snook [38] and Schmidt et al. [35], only part of white pulp in humans possess real follicular arteries. The follicular arteries are usually distributed by an arteriolar capillary bundle, which originates from the central artery after it has passed through the white pulp and turns back into the white pulp.

2.2. Marginal Zone

Whatever their origins might be, the terminal branches of the follicular arteries usually end in the marginal zone. As some twigs of penicilli turn back to end in this area [38], the marginal zone is very richly supplied with arterial blood (Fig. 9-2). As depicted in the light microscope by MacNeal [4], arterial capillaries open into the reticular spaces, sometimes through a *terminal ampulla* [4].

SEM images of vascular casts of the spleen usually show the marginal zone to be densely filled with granular masses of resin, which correspond to the reticular spaces of the zone (Figs. 1 and 2). Continuation between the arterial endothelium and the lining of the reticular spaces, occasionally interrupted by a saccular distension of the former, has been demonstrated in tissue specimens [40].

Schmidt et al. [35] produced in human spleen casts of what they identified as the marginal sinus between the white pulp and the marginal zone. The "flattened, anastomosing vascular spaces" [Figs. 4–6 in 35], however, appear to us to be nothing but the innermost layer of the marginal zone, which, as our SEM observation of tissue specimens indicates, comprises several layers of flattened spaces concentrically extending around the white pulp, also supported by concentrically arranged reticular cells and fibers. At any rate,

the term *marginal sinus* seems inadequate for that structure because it has nothing to do with the venous sinus and causes only confusion among researchers.

A more important finding by Schmidt et al. [35] is the outer aspect of the marginal zone, where several channels connecting the granular cast material of the marginal zone and an adjacent red pulp sinus were found (Figs. 9–11 and 16 and 17 in 35). A corresponding observation was reported by the same research group in the dog [29]. These channels are thought to represent very effective blood flow from the marginal zone, which, as emphasized above, receives much arterial blood supply. Yet the route does not deserve to be designated as a closed circulation, as the reticular spaces of the marginal zone are intercalated between the arterial ends and the venous channels.

2.1.1. Penicillar Arteries and their Open Termination. The penicilli or penicillar arteries are those thin arteries repeating branching to terminate in the red pulp. A portion covered by an elipsoid sheath is intercalated on the arteries. This portion, called the *sheathed artery*, is characterized by a thick endothelium with fenestrations that allow the passage of erythrocytes, and by an alternate, concentric arrangement of flattened reticular cells and macrophages, as our SEM observations of the human spleen have confirmed [16].

With regard to the long-disputed terminals of the penicilli, careful examination under the light microscope of sections from perfused human spleens may indicate that the arteries, when hit longitudinally, terminate in the splenic cords. SEM observation of freeze-cracked tissue of the human spleen makes it clear that the arterial ends are located in the cords, assuming either a funnel or a perforated saccule through which blood cells could pass by constricting themselves [16,24, 26,30,42].

→

Figure 9-3. Cast of human red pulp showing sinuses (S) with indentations of rod cells and their nuclei, and some branches of penicillar arteries (PA). The arrow indicates a funnel-like termination of an artery. Granular masses of resin correspond to the spaces in the splenic cords (C). ×230.

Figure 9-4. Cast of penicillar arteries (PA) terminating in saccular swellings (arrows), which continue to the granular resin masses corresponding to the spaces in the cords (C). ×900. (From Kashimura and Fujita [34], with permission.)

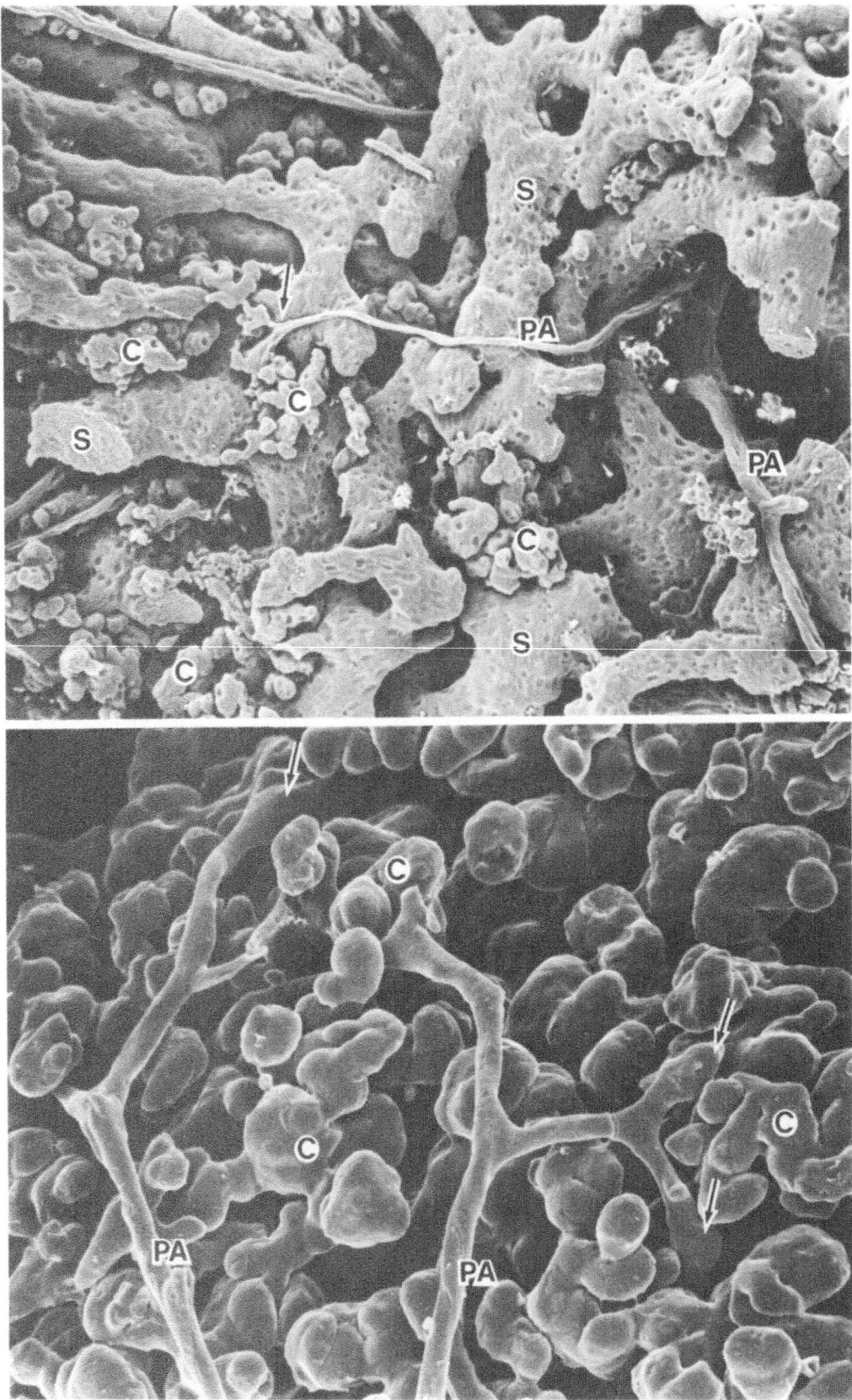

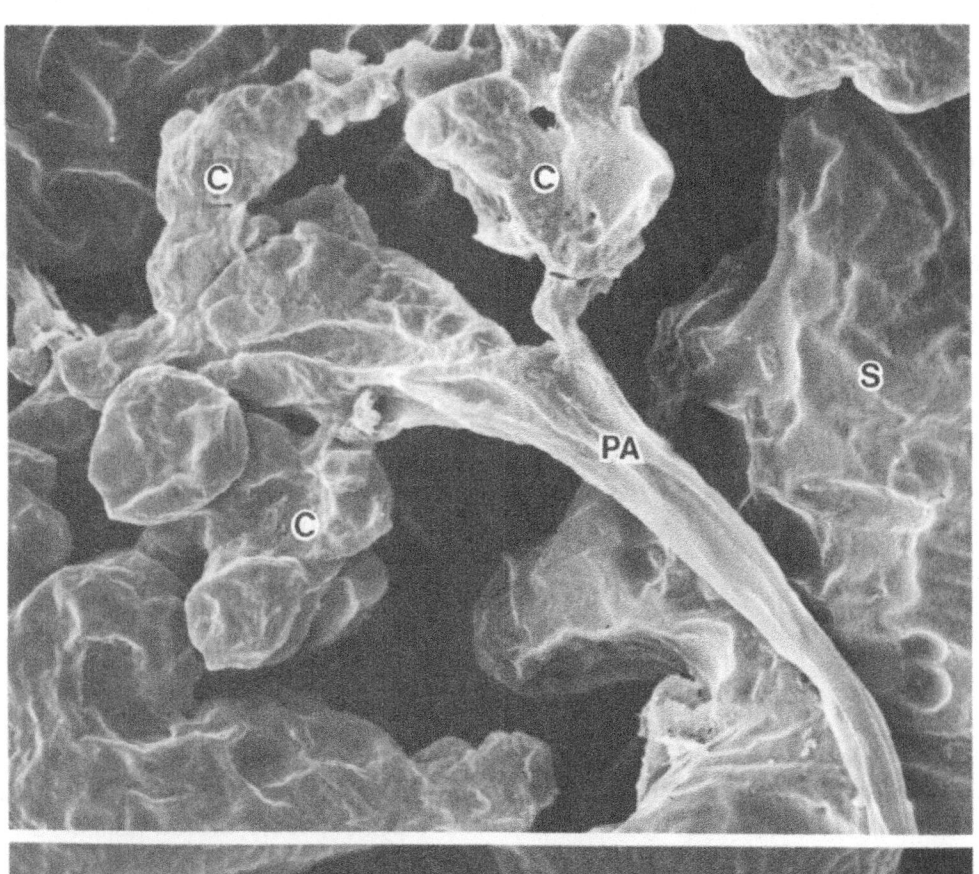

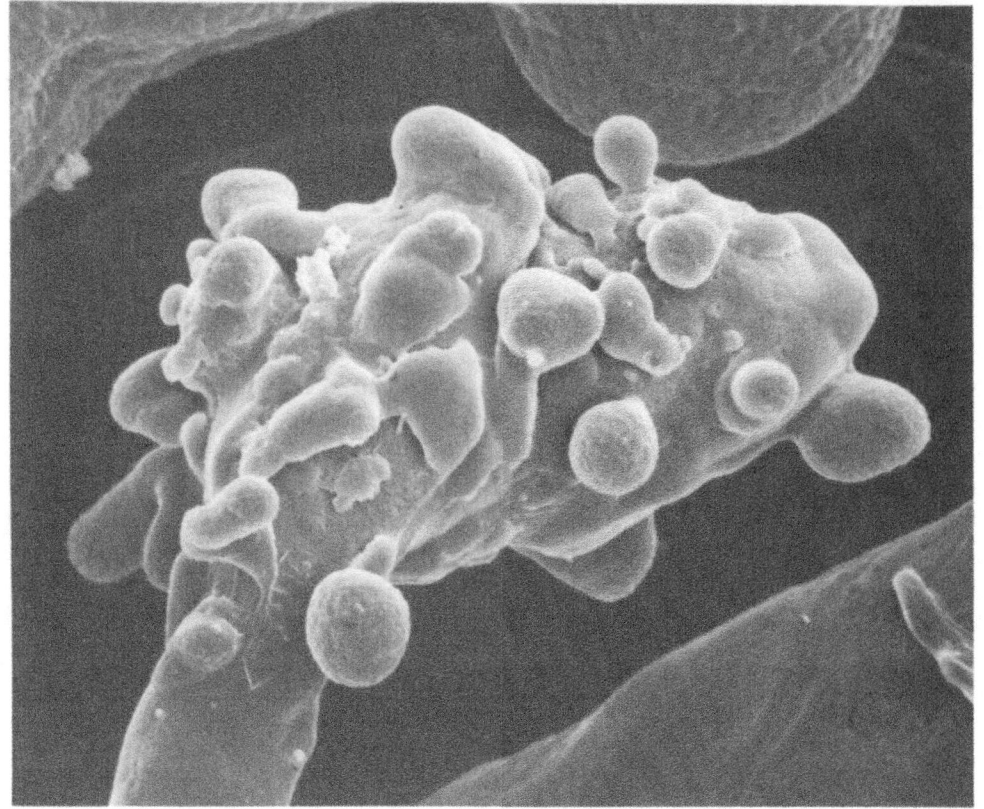

Observation of the vascular casts of human spleens under the SEM supports the notion of an open circulation, as the arterial terminals were shown to not be directly connected with the sinus but to be continuously attached by granular masses of resin corresponding to the spaces of the splenic cord [19] (Figs. 9-3 and 9-4). A closer view of the casts reveals that the arterial termination either fans out in a five-funnel shape or swells into a saccule with perforations through which resin effuses in humplike projections (Figs. 9-5 and 9-6)

Further support of the open circulation theory is provided by light microscopic and SEM images of erythrocytes kept on the lattice of the sinus endothelium, with the cells hanging in a dumbbell shape, with their two heads always projecting into the sinus lumen. This means that the blood flows from the splenic cord towards the sinus. In the spleen of living rats, MacDonald et al. [43] recorded on video the passage of erythrocytes through the sinus wall and concluded that "the direction of flow has been, invariably, from the surrounding reticular meshwork into the venous sinuses."

2.4. Closed Circulation

In 1985 we reported that in restricted portions of the red pulp of human spleen, not far from the white pulp and the peripheral to the marginal zone, arterial terminals expand into a dense labyrinth of vessels [40], which is connected with the tapered ends of the sinuses [33,34]. This finding, obtained both from fractured tissues and from vascular casts of human spleens, provides evidence for a closed circulation.

The cast of a labyrinth resembles coral, composed of thin, winding, and anastomosing vessels (Fig. 9-7). The tissue specimens indicate a smooth and continuous endothelium and, characteristically, there are thin intraluminal trabeculae covered with endothelium (Fig. 9-8). These trabeculae are presumed to prevent overexpansion

of the vessels. The casts reveal pinholes corresponding to the trabeculae. The labyrinth receives a few arteries (branches presumably derived from a central or pulp artery), on the one hand (Fig. 9-9), and, on the other hand, is drained by a considerable number of tapered sinus feet (Fig. 9-10). This arteriovenous transition has been seen in both cast and tissue preparations [33,34].

2.5. Functionally Closed Circulation

Our SEM observations of the fractured tissue of human spleen have occasionally demonstrated structures that seem to allow the rapid passage of blood cells from the arterial ends to the venous sinuses [40]. For example, certain sinuses were found to be loosened in their wall lattice structure at their tapered end and to have a very large aperture at their tapered end [40]. Another example is the perforation in the arterial termination that overlaps a sinus slit with intervention of an attenuated cordal element [24].

The cast observations by Schmidt et al. [32,35] have demonstrated that in human, as well as in rat, spleen many venous sinuses begin as open-ended tubes that are continuous with the marginal zone, which, in turn, receives an ample arterial inflow. As suggested by Schmidt et al., this structure seems to serve "free entry of blood into the venous system and bypassing the red pulp." However, it is necessary to discriminate between this route and a direct connection of the arterial and venous vessels.

3. Conclusions and Physiological Interpretations

The combined SEM observations of vascular casts and spleen tissue itself indicate that the human splenic circulation is, for the most part, open in type, while limited areas possess particular routes of closed circulation (Fig. 9-11).

The open circulation is represented by peni-

←

Figure 9-5. Cast of a funnel-shaped termination of a penicillar artery (PA). Resin masses protrude through openings in the funnel wall into the cordal spaces (C). S = Cast of a sinus that is not connected with the artery. ×1900.

Figure 9-6. Cast of a saccular termination of a penicillar artery. Resin protrudes through its fenestrations, producing globular swellings. ×4275.

118

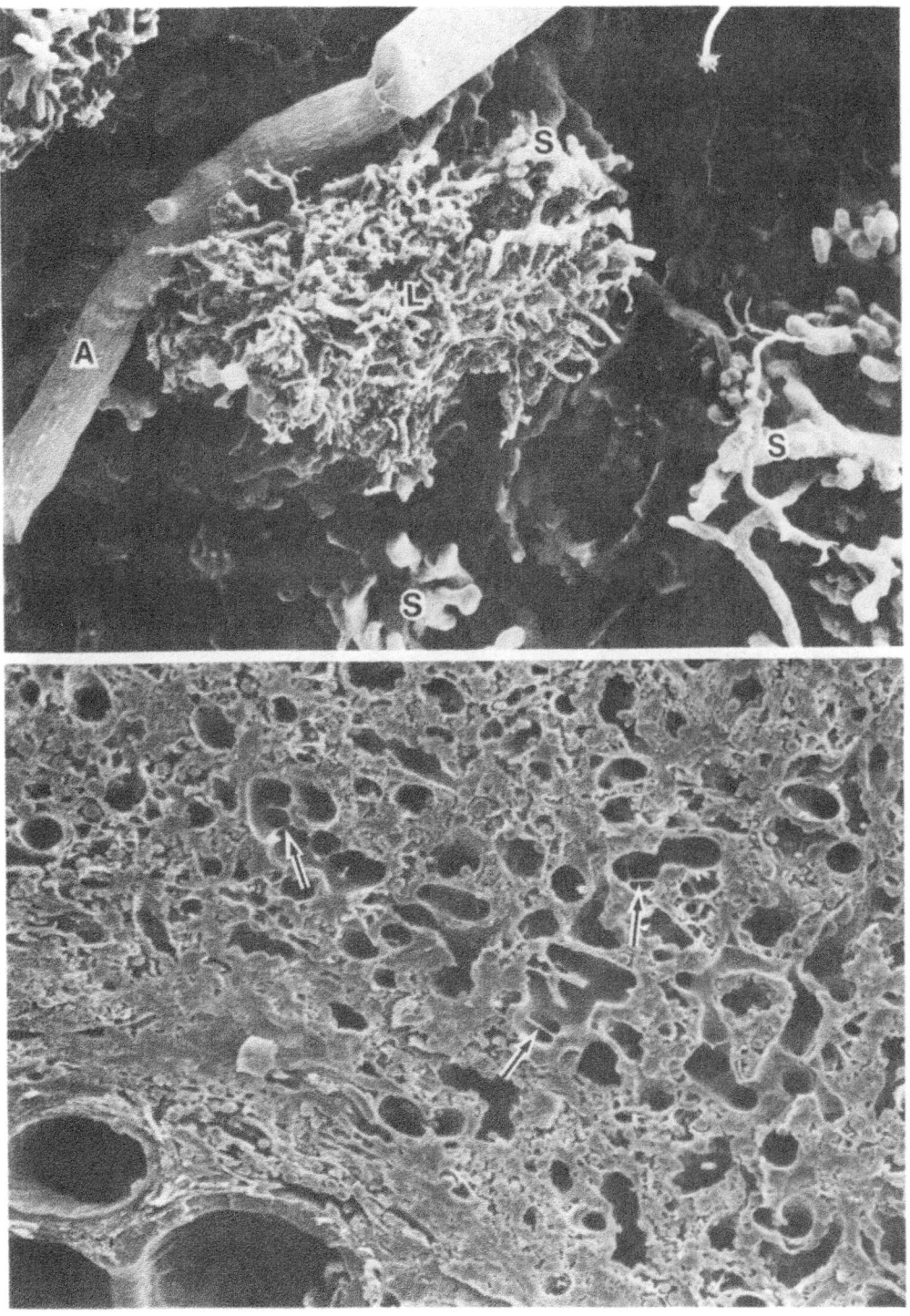

Figure 9-7. Cast of an arterial labyrinth (L) located close to a presumed pulpar artery (A). S = sinuses. ×60.
Figure 9-8. Fractured human spleen showing an arterial labyrinth. Note the irregularly winding channels, which are periodically spanned by thin bars (arrows). ×200.

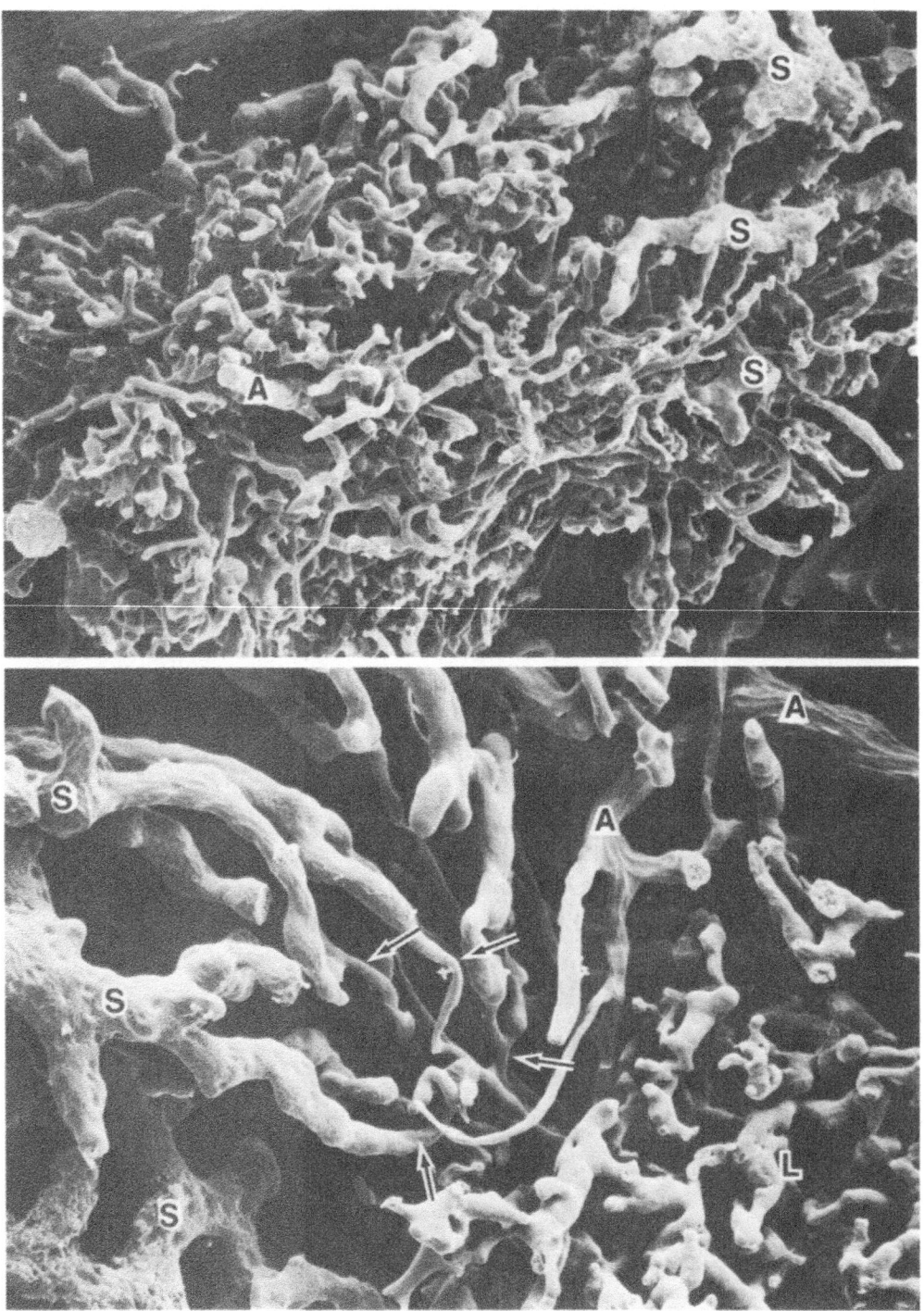

Figure 9-9. Closer view of the arterial labyrinth shown in Fig. 9-7. The labyrinth receives an arteriole (A) and is drained by sinuses (S). ×120.

Figure 9-10. Cast of the peripheral portion of an arterial labyrinth (L) connected with sinuses. Arrows indicate a direct connection between the channels of the labyrinth and the tapered ends of sinuses (S). A = arterioles supplying the labyrinth. ×240.

120

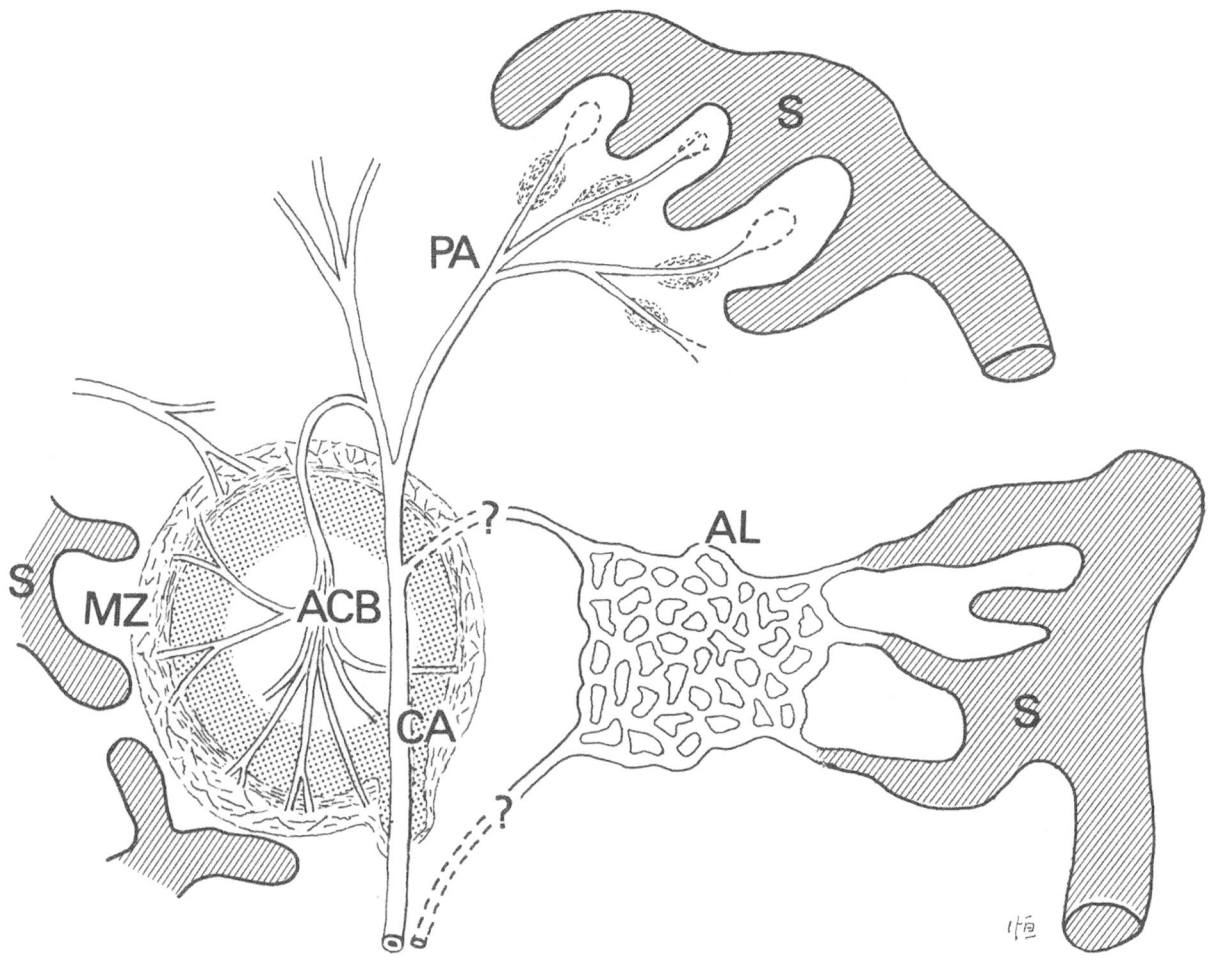

Figure 9-11. Diagram showing our current concept of the microcirculation in the human spleen. CA = central artery; ACB = arteriolar-capillary bundle; MZ = marginal zone; PA = penicillar artery; AL = arterial labyrinth; S = sinus. The derivation of arteries supplying the labyrinth is unclear (marked with ?).

cillar arteries terminating either in a funnel-like fanning out or in a perforated saccule that allows the passage of constricted erythrocytes. Erythrocytes and other cellular elements, as well as foreign bodies, can pass slowly through the arterial terminations into the reticular spaces of the splenic cord, to enter into the venous sinuses through their slits. This slow passage of blood facilitates macrophages, eliminating aged and harmful cells and foreign bodies.

The open circulation also occurs in the marginal zone, receiving large amounts of arterial blood from penicilli and from follicular arteries. The wide-spaced marginal zone is juxtaposed to the perimarginal sinuses and thus apparently

provides a faster and more efficient blood route than other parts of the red pulp.

Small, limited areas in the human red pulp, located rather close to the white pulp and marginal zone, have been demonstrated by our group to be the sites of closed circulation. Here an arterial labyrinth of a specialized structure is directly connected with the venous sinuses. Whether or not corresponding structures occur in various other animal species remains to be studied. Delineation of the mechanism for shunting the blood flow between the open and closed routes seems to be important for a better understanding of the physiology and pathophysiology of the spleen.

References

1. Billroth T. Zur normalen und pathologischen Anatomie der menschlichen Milz. *Arch Pathol Anat Physiol Klin Med* 20:409–425, 1861.
2. Weidenreich F. Das Gefässsystem der menschlichen Milz. *Arch Mikroskop Anat Entw-Gesch* 58:247–376, 1901.
3. Helly K. Zum Nachweise des geschlossenen Gefäss systems des Milz. *Arch Mikroskop Anat Entw-Gesch* 59:93–105, 1902.
4. MacNeal WJ. The circulation of blood through the spleen pulp. *Arch Pathol* 7:215–227, 1929.
5. MacNee JW. The spleen. *Trans Med Soc Lond* 54:185–236, 1931.
6. Knisely MH. Spleen studies. I. Microscopic observations of the circulatory system of living unstimulated mammalian spleens. *Anat Rec* 65:23–50, 1936.
7. MacKenzie DW, Whipple AO, Wintersteiner MP. Studies on the microscopic anatomy and physiology of living transilluminated mammalian spleens. *Am J Anat* 68:397–456, 1941.
8. Björkman SE. The splenic circulation with special reference to the function of the spleen sinus wall. *Acta Med Scand Suppl* 191:1–89, 1947.
9. Pappart AK, Whipple AO, Chang JJ. The microcirculation of the spleen of the mouse. *Angiology* 6:350–362, 1955.
10. Lewis OJ. The development of the circulation in the spleen of the foetal rabbit. *J Anat* 90:282–289, 1956.
11. Snook T. The histology of vascular terminations in the rabbit spleen. *Anat Rec* 130:711–729, 1958.
12. Weiss L. The structure of intermediate vascular pathways in the spleen of rabbits. *Am J Anat* 113:51–91, 1963.
13. Tischendorff F. Die Milz. In: Möllendorff W, Bargmann W (eds.), *Handbuch der Mikroskopischen Anatomie des Menschen*, Vol. 6 Springer-Verlag, Berlin, p 6, 1969.
14. Chen L-T, Weiss L. Electron microscopy of the red pulp of human spleen. *Am J Anat* 134:425–458, 1972.
15. Murakami T, Fujita T, Miyoshi M. Closed circulation in the rat spleen as evidenced by scanning electron microscopy of vascular casts. *Experientia* 29:1374–1375, 1973.
16. Fujita T. A scanning electron microscope study of the human spleen. *Arch Histol Jpn* 37:187–216, 1974.
17. Barnhart MI, Baechler CA. Human and canine splenic vasculature: Structure and function. *Scann Electron Microsc* III:705–712, 1974.
18. Barnhart MI, Lusher JM. The human spleen as revealed by scanning electron microscopy. *Am J Hematol* 1:243–264, 1976.
19. Irino S, Murakami T, Fujita T. Open circulation in human spleen. Dissection scanning electron microscopy of conductive-stained tissue and observation of resin vascular casts. *Arch Histol Jpn* 40:297–304, 1977.
20. Suzuki T, Furusato M, Takasaki S, Shimizu S, Hataba Y. Stereoscopic scanning electron microscopy of the red pulp of dog spleen with special reference to the terminal structure of the cordal capillaries. *Cell Tissue Res* 182:441–453, 1977.
21. Chen LT. Microcirculation of the spleen: An open or closed circulation? *Science* 201:157–159, 1978.
22. Hataba Y, Kirino Y, Suzuki T. Scanning electron microscopic study of the red pulp of mouse spleen. *J Electron Microsc* 30:46–56, 1981.
23. Blue J, Weiss L. Vascular pathways in nonsinusal red pulp — an electron microscopic study of the cat spleen. *Am J Anat* 161:135–168, 1981.
24. Fujita T, Kashimura M, Adachi T. Scanning electron microscopy (SEM) studies of the spleen — normal and pathological. *Scann Electron Microsc* I:435–444, 1982.
25. Schmidt EE, MacDonald IC, Groom AC. Direct arteriovenous connections and the intermediate circulation in dog spleen, studied by scanning electron microscopy of microcorrosion casts. *Cell Tissue Res* 225:543–555, 1982.
26. Fujita T, Kashimura M. Scanning electron microscope studies of human spleen. *Surv Immunol Res* 2:375–384, 1983.
27. Suzuki T. Morphological approaches to the terminal circulation of human and some other mammalian spleens; is it an "open" or "closed"? *Rec Adv Res* 23:3–38, 1983.
28. Schmidt EE, MacDonald IC, Groom AC. The intermediate circulation in the nonsinusal spleen of the cat, studied by scanning electron microscopy of microcorrosion casts. *J Morphol* 178:125–138, 1983.
29. Schmidt EE, MacDonald IC, Groom AG. Circulatory pathways in the sinusal spleen of the dog, studied by scanning electron microscopy of microcorrosion casts. *J Morphol* 178:111–123, 1983.
30. Seki A, Abe M. Scanning electron microscopic studies on the microvascular system of the spleen in the rat, cat, dog, pig, horse and cow. *Jpn J Vet Sci* 47:237–249, 1984.
31. Schmidt EE, MacDonald IC, Groom AC. Microcirculation in mouse spleen (nonsinusal) studied by means of corrosion casts. *J Morphol* 186:17–29, 1985.
32. Schmidt EE, MacDonald IC, Groom AC. Microcirculation in rat spleen (sinusal), studied by means of corrosion casts, with particular reference to the intermediate pathways. *J Morphol* 186:1–16, 1985.
33. Kashimura M. Labyrinthine structure of arterial terminals in the human spleen, with special reference to "closed circulation." A scanning electron microscope study. *Arch Histol Jpn* 48:279–291, 1985.
34. Kashimura M, Fujita T. A scanning electron microscopy study of human spleen: Relationship between the microcirculation and functions. *Scann Microsc* 1:841–851, 1987.
35. Schmidt EE, MacDonald IC, Groom AC. Microcirculatory pathways in normal human spleen, demonstrated by scanning electron microscopy of corrosion casts. *Am J Anat* 181:253–266, 1988.
36. Hataba Y, Suzuki T. Scanning electron microscopic study of the red pulp of ferret spleen. *J Electron Microsc* 38:190–200, 1989.
37. Snook T. A comparative study of the vascular arrangements in mammalian spleens. *Am J Anat* 87:31–61, 1950.
38. Snook T. The origin of the follicular capillaries in the human spleen. *Am J Anat* 144:113–117, 1975.
39. Snook T. The blood supply to the splenic lymphatic

122

nodules in the rhesus monkey. *Anat Rec* 196:461–467, 1980.

40. Fujita T, Kashimura M, Adachi K. Scanning electron microscopy and terminal circulation. *Experientia* 41: 167–178, 1985.

41. Fujita H, Fujita T. *Textbook of Histology*, part 2, 2nd ed., Igaku-Shoin, Tokyo, 1984.

42. Fujita T, Kashimura M. The "reticulo-endothelial system" reviewed by scanning electron microscopy. *Biomed Res* 2 (Suppl.):159–171, 1981.

43. MacDonald IC, Ragan DM, Schmidt EE, Groom AC. Kinetics of red blood cell passage through interendothelial slits into venous sinuses in rat spleen, analyzed by in vivo microscopy. *Microvasc Res* 33:118–134, 1987.

Author's address:
Prof. Tsuneo Fujita
Department of Anatomy
Niigata University School of Medicine
Asahimachi, Niigata, 951
Japan

CHAPTER 10

The Lung Microstructure

DEAN E. SCHRAUFNAGEL

1. Introduction and Historical Background

Casting is an old method of studying the circulation of the lung. Leonardo da Vinci filled the heart and great vessels with wax and described the bronchial circulation. Malpighi, the famous Bolognese anatomist and physiologist, injected mercury and other materials into the blood vessels of the lung [27]. In the 19th century, Guillot injected gelatin into the pulmonary and bronchial circulations, and Miller used Wood's metal to study the alveolar and circulatory structures of the lung [31]. In 1901, Grossner added India ink to gelatin to improve contrast [16]. Schlessinger mixed barium with gelatin to make a radiopaque medium to illustrate lung vascular relationships radiographically [38]. Krahl used vinylite casts of the blood vessels of the lung to show that the branching angles of veins were not as sharp as those of arteries [24].

Nowell and colleagues examined vinylite casts of avian lungs with the scanning electron microscope [35]. They cast the airways and pulmonary arteries with different latexes to show the relationships between the two systems. With Microfil, a silicone rubber, they found that the pulmonary vasculature did not fill as well if the airways were distended. The development of methylmethacrylate by Murakami [33] lead to many scanning electron microscopic studies of casts. This material is well suited for electron microscopy, because it can replicate fine detail and withstand the electron beam.

2. Lung Casting Techniques

Most of the methods of corrosion casting for other organs also apply to the lung. One difference is that only a thin membrane separates the pulmonary capillaries from the alveolar airspace. If the resin is injected with too much pressure, the capillaries will burst. The freed resin will cast alveoli.

2.1. Rinsing

The first step of vascular casting is rinsing the lumina of the blood vessels. Erythrocytes are not usually difficult to clear from normal lungs, but white blood cells may adhere. Adherent intravascular macrophages may be important cells in defending against sepsis and in the regulation of the microvascular blood flow [53]. Cells are trapped more often in casts of damaged lung [43,50], suggesting that the cells cause or contribute to the disease. Warming the rinsing solution may dilate the arteries and cause the resin to harden sooner. Vasodilators, such as papaverine, may improve the casting of systemic vessels, but normal pulmonary arteries are not as muscular as systemic arteries and are usually relaxed, so vasodilators are less beneficial in the lung. If vascular constriction occurs with the ischemia and lactic acidosis of exsanguination, a vasodilator may be valuable. Hypoxia and acidosis, which dilate the systemic circulation, constrict pulmonary arteries.

Motta, P.M., Murakami, T., and Fujita, H. (eds.), Scanning Electron Microscopy of Vascular Casts: Methods and Applications.

Hemodilution may produce pulmonary edema. Large arteries with higher pressure are affected more than small ones. Edema accumulates between the intima and media, and the media and the adventitia [49]. Although the edema may decrease vascular filling in the brain confined in a rigid calvarium, its only affect on lung casts appears to be to increase the space between pulmonary arteries and their adjacent capillaries [49]. Fixing the tissue before rinsing does not prevent edema, but rinsing with 10% dextran-40 in 0.9% saline alleviates the problem.

If an animal dies before the blood is cleared, it may be more difficult to produce a well-filled cast because of the loss of the heart's pumping action, which efficiently removes blood cells. After death the blood coagulates and plasma proteolytic enzymes that are activated may damage the endothelium. Postmortem clots cannot be easily forced through microscopic vessels.

2.2. Fixation

The practice of fixing the lung before casting varies: Hijiya and Okada used 2% glutaraldehyde [15], Koike and associates used 1.2% glutaraldehyde [23], and Nelson used 0.25% glutaraldehyde [34]. Our group used 4% formaldehyde [44], and Ohtani [36] and Caduff and colleagues [4] used no fixative. Formalin and acetic acid given after latex injection help to solidify the cast [22]. Using no fixative might avoid shrinkage caused by the fixative [28], but methylmethacrylate may increase the permeability of unfixed endothelium [9]. However, neither shrinkage nor damage is readily apparent when viewing the casts in the electron microscope, regardless of which fixative is used.

To study the effect of fixation on casting, we compared 10% formalin, 2.5% glutaraldehyde, 0.5% glutaraldehyde, and saline in casting the lungs of normal rats with undiluted Mercox. The major effect of fixation was on leakage of the casting material. No extravasation of resin occurred in the 2.5% glutaraldehyde group, and only a minor amount was seen in the 0.5% glutaraldehyde and formaldehyde groups. However, all but 1 of 8 animals that were not fixed had leakage. Seepage through the vessel wall may

be worse with low-viscosity resins. Increased venous muscular tone may increase permeability. Although good casts can be obtained with many fixation protocols, glutaraldehyde-fixed tissue more often gives better-filled casts than nonfixed tissue. The results of formaldehyde are intermediate between glutaraldehyde and no fixation [48]. This may be because formaldehyde does not crosslink proteins as well as glutaraldehyde, even though it penetrates tissue better. Fixation does not prevent leakage when conditions of increased vascular permeability, such as angiogenesis, are present.

Fixation prevents microbial growth on casts stored in aqueous solutions. Fixation is important if the time between vascular rinsing and casting is prolonged or if studying the lung at a specific inflation size is required. Of course, it is essential for light and transmission electron microscopy.

2.3. Filling

Methylmethacrylate fills all parts of the pulmonary vascular tree well. Although good casts are produced most of the time, there are many reasons for incompletely cast vessels. I usually inject the resin through a cannula that has been inserted into the abdominal vena cava of laboratory animals. This produces a cast of lung vessels at postmortem functional residual capacity. With this method, puncturing the vena cava by the cannula distal to where it was inserted is the most common cause of poor casting. Incomplete rinsing may lead to poor filling if vessels with high resistance shunt blood away from their capillary beds. Focal high resistance can be caused by small clots, vasospasm, local compression, or cells adherent to the endothelium. When casting isolated lungs, the effluent port must be clamped to prevent runoff. The area of interest may be so damaged, or the vessels so distorted by the experimental condition, that it does not fill well. Another reason for incomplete vascular filling is increased alveolar pressure. If pressure is applied to the trachea with the thoracic cavity closed, the lungs will expand and decrease capillary filling [47].

There are several signs that indicate vessels have not been well filled. Well-filled casts usually have impressions caused by the nuclei of endothelial cells that bulge into the blood vessel lumen [16] or alveoli [8]. The resin is applied with pressure that slightly distends the vessels. The nucleus is the least compliant part of the cell, so it is pushed back less than the other areas of the endothelial cell, leaving a hollow on the cast surface. If the injection pressure is low or filling is poor, the nuclear impressions may be absent. Nonconnecting or "dead-end" capillary segments may indicate poor filling, but they also may represent capillary remnants caused by alveolar fragmentation found in emphysema [47] or capillary buds found in angiogenesis. They are increased in experimental lung fibrosis [44]. Nonconnecting segments may be caused by thrombosis [23,50] that occurs in many lung diseases. With poor casting there may be round ending stumps at the branching points of large vessels.

The pressure that must be applied to the syringe to inject the resin is related to the viscosity of the resin and the resistance of the vascular system. For resins with a viscosity near that of blood, the injecting pressure should be near physiologic blood pressure. For high-viscosity resins, high pressures may be needed. The viscosity and the pressure required to fill the vasculature rapidly increase after the catalyst and resin are mixed. Depending on the amount of catalyst added, the resin temperature, and volume, so much pressure may be required that it can be difficult to extrude the plastic from the syringe after a few minutes. Although the syringe pressure may be high, the capillaries may not be exposed to this high pressure because the pressure dissipates as the viscous resin moves through the highly branched pulmonary vessels. Furthermore, the Laplace relationship shows that the capillary wall tension remains small, even when the main pulmonary artery pressure is high [13]. If the capillary wall is breached, nearby alveoli will be cast.

If the viscosity of the resin is reduced to that of blood, a lower injection pressure can be used, but more transudation of casting material results. With increased capillary fragility or angiogenesis, plastic will be found in the alveoli, even with low injection pressures [42]. Filling the pulmonary

vasculature does not require injecting through the pulmonary artery. I recently discovered that I could inject through the aorta and bronchial arteries with nearly as good results [48]. Injecting lung capillaries through the bronchial arteries did not result in a significantly lower capillary density, although there were more deadends and a worse subjective grade of micrographs, compared with specimens cast through the pulmonary arteries [41]. This indicates that the bronchial to pulmonary artery collateral circulation is extensive.

2.4. Corrosion

The agents used to digest the lungs are usually sodium hydroxide or potassium hydroxide, although household bleach, 5.25% sodium hypochlorite, is cheaper and may macerate small specimens. More concentrated bleach, as well as sodium or potassium hydroxide, may corrode. Acids, such as 10–15% hydrochloric, may be useful, especially if digestion of ribs is necessary. Fixation may increase the time required for corrosion [48]. Aldehyde fixatives have another curious effect on casts. Fixation with 2.5% glutaraldehyde and corrosion with concentrated (10 N) sodium hydroxide appears to allow tissue to be permanently preserved [48]. This may be because glutaraldehyde penetrates tissues and polymerizes under alkaline conditions. The polymer derived from the aldehyde is stable in alkali, so it may trap and preserve undigested tissue by impregnating it with plastic.

The temperature may be important for corrosion, but the polymer may soften when warmed (usually above 60°C). Placing specimens in warm water, alone or with proteolytic enzymes, for 1–2 days before alkali corrosion has been recommended [16]. Animals given intratracheal elastase 30 days before their death had lung casts that became clean sooner than animals given saline [47]. To determine the relative importance of corrosion factors, I compared potassium hydroxide, sodium hydroxide, and water as macerating agents. I also tested the size of the tissue sample, prealkali autolysis, detergent, and proteolytic enzymes to improve corrosion. Not unexpectedly, sodium and potassium hydroxide were better than water, and the longer the digestion time, the

126

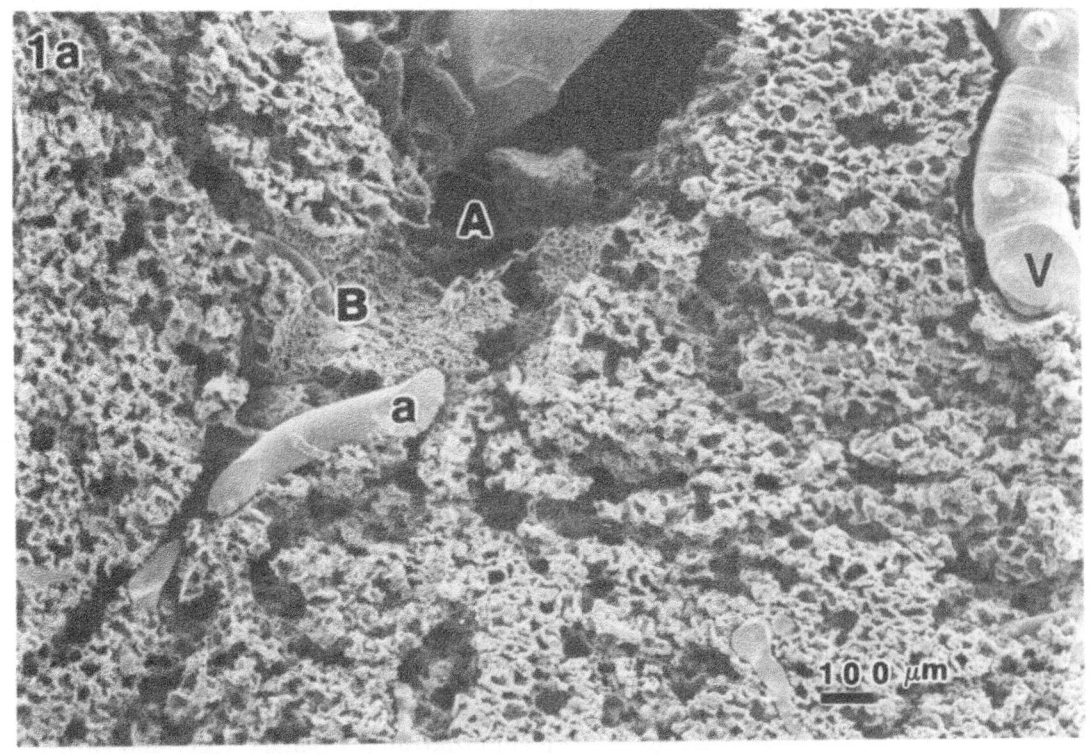

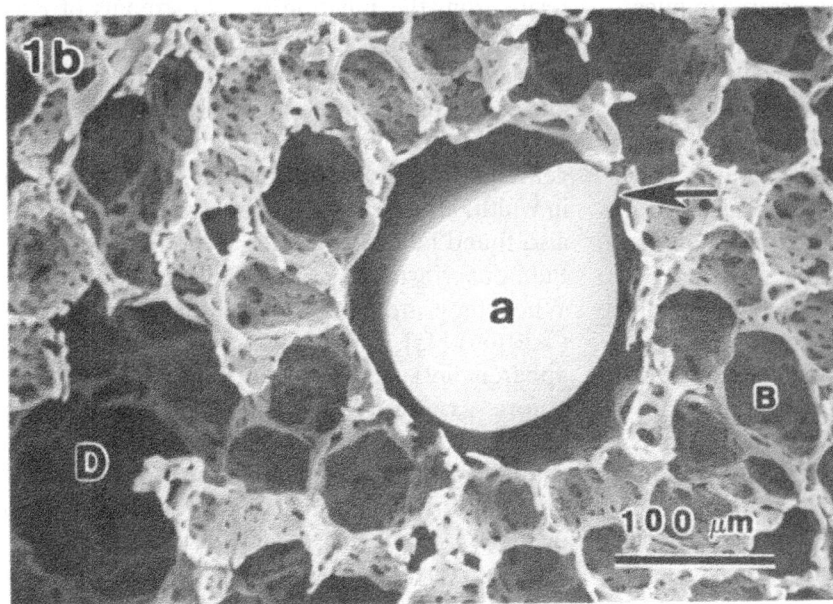

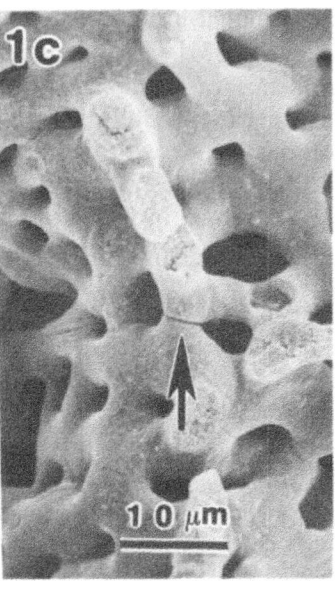

better the corrosion. Although not a controlled factor, higher room temperature was associated with better digestion. One alkali was not better than the other. The use of proteolytic enzymes, detergent, and warm water alone before the alkali treatment did not significantly improve the corrosion [40].

After corrosion, debris and precipitates may stick to the casts, unless they are well rinsed. Detergent can reduce oily residue. Casts of the lung can be cut with razor blades or scissors. Scissors will fracture the cast to give an irregular surface, but this may be better to inspect large vessels lengthwise. Razor cuts provide planar surfaces that are necessary for vascular density comparisons.

3. Lung Vessels

3.1. Arteries and Veins

In most mammals, the pulmonary artery supplies the lung distal to the terminal bronchioles and the bronchial artery supplies the walls of the pulmonary vessels and airways proximal to the terminal bronchioles. The bronchial veins collect blood from the extrapulmonary structures and bronchi, but bronchial artery blood distal to the terminal bronchioles flows into the pulmonary veins [31]. The pleura may be supplied by bronchial or pulmonary arteries, depending on the species [29,30] (Fig. 10-1). The pulmonary arteries course with the airways in the center of the lung lobules, although exceptions are many [17,18]. Small bronchial arteries are in the walls of the airways, arteries, and veins. Veins usually lie between lobules, distal to the arteries. The bronchial circulation of large animals, such as sheep, can be cast by isolating the trachea and bronchi [5,26]. Casts of bronchial vessels larger than capillaries can be difficult to distinguish from pulmonary vessels. The casts of large vessels have been traced down to small vessels to correlate the surface appearance and branching patterns. Particles greater than the diameter of the capillaries have been put into the resin and injected into the pulmonary arteries to distinguish them from veins [1]. We have injected latex microspheres to study large vessel anastomoses. Using a high-viscosity resin that fills only arteries or veins may also be useful for this purpose.

Vascular wall detail can be studied on cast specimens that are completely or partially digested [39]. Like the systemic vessels, casts of arteries have oblong indentations running in the long axis of the vessels; casts of veins have round depressions [32].

Precapillary sphincters and arterial cushions that occur at bifurcations in the systemic circulation have not been described in the lung. Isolated thin-ring indentations are found in the highly collateralized lung capillary casts and probably result from the junction of two streams of resin (Fig. 10-1c). On vessels larger than capillaries, cast indentations have more physiologic importance. Circumferential narrowing occurs differently in arteries and veins. Casts of the pulmonary veins have narrow constrictions, usually $1-3\,\mu m$ in width, about every $30\,\mu m$ (Fig. 10-2). They are also found before and after accepting tributaries. Pulmonary arteries usually have no constriction. When they are found, the infolding is broader ($>30\,\mu m$) [45]. The bands on the veins are more apparent in specimens that have not been fixed before casting [48]. Pulmonary veins do not have leaflet valves, but the narrow bands may be sphincters with a valvelike function [45]. The indentations of the venous sphincters can be deepened by massive neural discharge, implying that they are controlled by the central nervous

---

Figure 10-1. a: A low-magnification view of the cut surface of a rat lung shows a pulmonary artery (a) next to an airway (A). Note the pattern of capillaries of the bronchovascular space (B). A vein (v) with circular constrictions lies distally. b: The cast of a small artery (a) surrounded by alveolar capillaries arranged in baskets (B). The baskets are more prominent in the lung of this rat given intratracheal elastase. Note the artery first gives off a short segment (arrow) before branching into capillaries. Veins of this size will accept blood directly from capillaries. Alveolar ducts (D) are also more prominent in this lung treated with elastase. c: A more highly magnified view of capillary baskets. Note the thin ring (arrow) that is seen in alveolar casts. This artifact results from two streams of resin meeting in the highly collateralized lung.

128

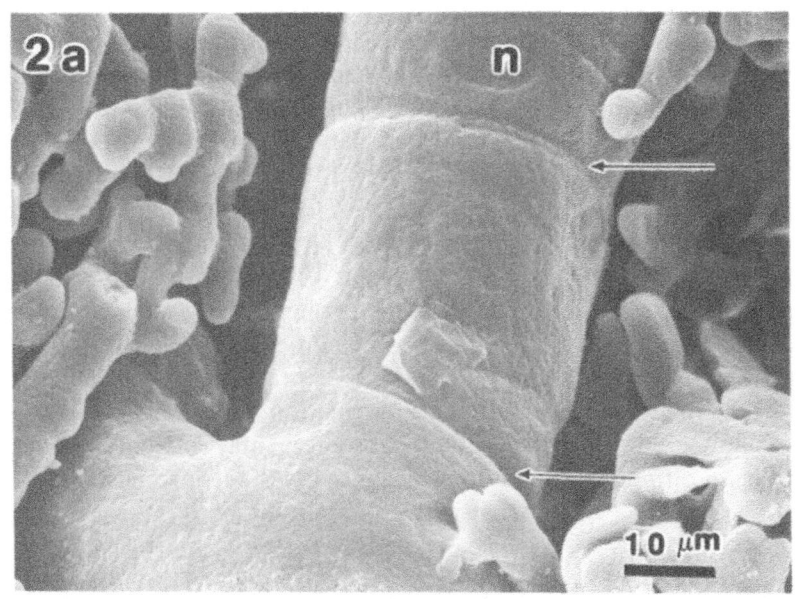

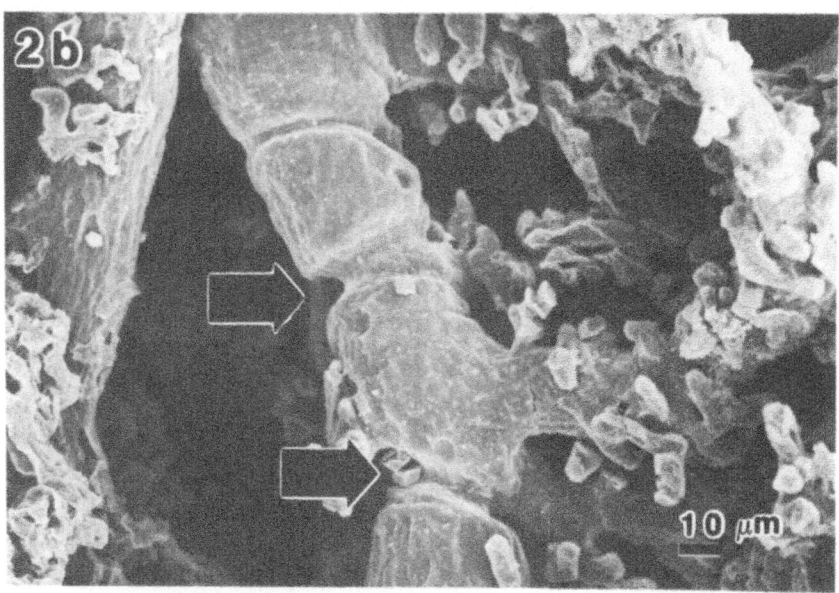

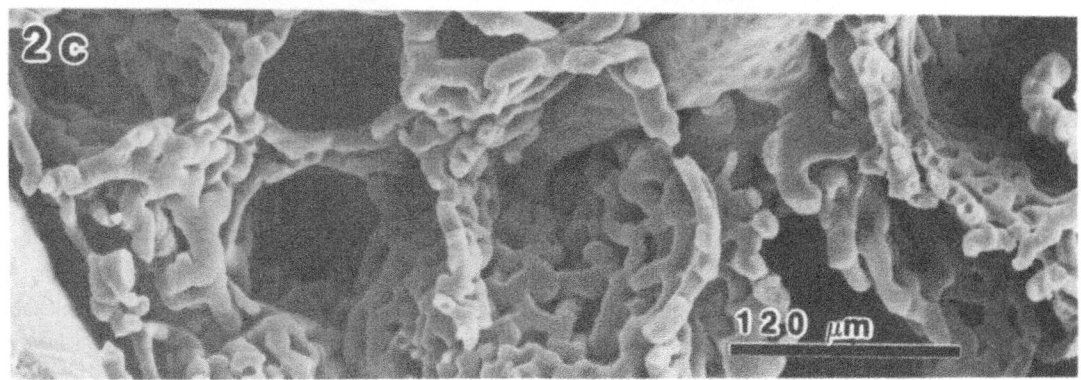

system [45]. They may be particularly important in neurogenic pulmonary edema. Bronchial veins have leaflet valves similar to other systemic veins. The presence of sphincters in the pulmonary veins could regulate blood flow in the low-resistance pulmonary circulation and keep blood from pooling in the lung bases. Sphincters could provide a mechanism for valve function without adding the impedance of leaflet valves. Valves are not under neural or humoral control, but muscular sphincters may be. Kay described a rhythmic valvelike action of striated muscle of the pulmonary venous wall during systole in rats [20]. The presence of sphincters would allow erythrocytes to better pass through capillaries at different velocities.

3.2. Branching

Microvascular casting can display three-dimensional branching of blood vessels well, but it is difficult to measure branching angles. The terms 'conducting' and 'distributing' vessels, often used to describe systemic vessels [46], are less useful in the lung. Using stereomicroscopy, Koike and colleagues [23] measured the diameters and lengths of pulmonary artery casts and found that large vessels give off single branches with diameters about the same size as the stem. Arteries larger than 1 mm have about 75% of their branches single. Distributing arteries, 200–300 μm, have only about 50% of their branches single. The average branch to stem-diameter ratio is about 0.8 for the larger branches and 0.5 for the smaller ones. Lane and coworkers showed that obtuse pulmonary artery branches have the same concentric muscle layer structure as parent vessels, but right-angle branches have spiral muscle bundles that lead to nonmuscular precapillary vessels [25]. Capillaries branch differently than arteries and veins, which makes them easy to identify. They divide at right angles and in Y shapes [46]. One capillary supplies more than one alveolus, and capillary branching is extensive within an alveolus. In the rat, normal alveolar capillaries have about 16 branches per 100 μm length. Pleural capillaries have about 11 branches per 100 μm [44]. Capillaries may attach directly to large veins, but equal-sized arteries first give off short precapillary segments (Fig. 10-1) [39].

Microvascular branching has important hemodynamic implications. It may cause a variation in local hematocrit [7]. Right-angle branching will conserve pressure in the low-pressure pulmonary circulation. If blood flow is laminar, cells and substances may remain in a stream for long distances without mixing and will preferentially go to certain lobes or organs.

3.3. Capillaries

Capillaries account for most of the cast vessels observed by scanning electron microscopy. Capillaries do not change in size with branching, but may vary their diameter at bends and other places [46]. Capillaries branch sharply and cover most of the alveolar wall. They are found on the surface of terminal and respiratory bronchioles. Casts show that lung capillaries are arranged in two patterns [4,12,44]. On the pleural surface the capillaries are less dense, planar, and branch less often (Fig. 10-3). On the cut surface the capillaries form baskets around alveoli [36] (Fig. 10-1). The three-dimensional pattern of capillaries in the bronchovascular bundle is similar to that of the pleural surface [44]. To account for the low resistance of the pulmonary vascular bed, Fung and Sobin [10] proposed that blood flows through lung capillaries in sheets. Knowing the alveolar blood volume and transit time, they described the pulmonary capillaries as a vascular space with the endothelial walls for the upper and lower boundaries and cellular posts to hold them apart. Guntheroth and colleagues [12] disputed this sheet-and-post theory, because the

Figure 10-2. a: Veins have round nuclear impressions (n) and narrow bands that represent sphincters. (Reprinted with permission, from Schraufnagel and Patel [45].) b: This vein has deeply constricted areas (arrows). This rat received a blow on the head after the pulmonary vascular tree was cast, but before the resin hardened. Constriction is a response to head injury. c: In this rat that received bleomycin to produce lung fibrosis, the alveolar capillaries are separated more widely than normal. Light microscopy shows that the space is occupied by collagen and cells. (Reprinted with permission from Schraufnagel and Mehta [44].)

130

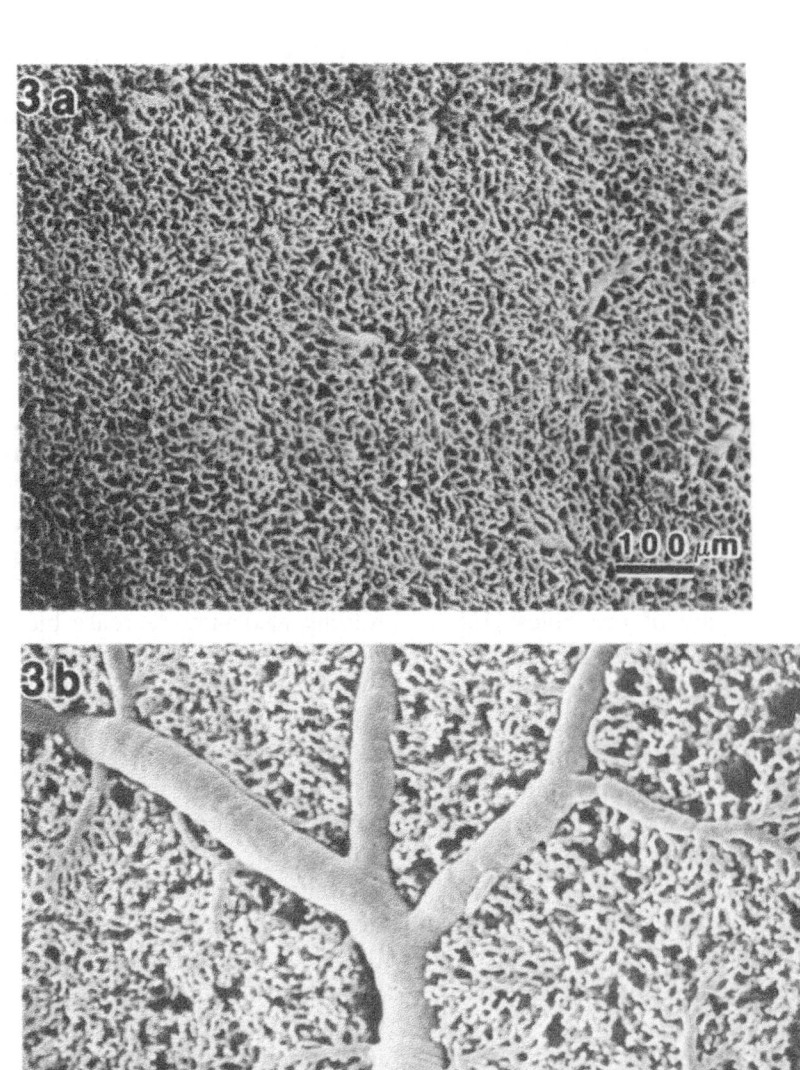

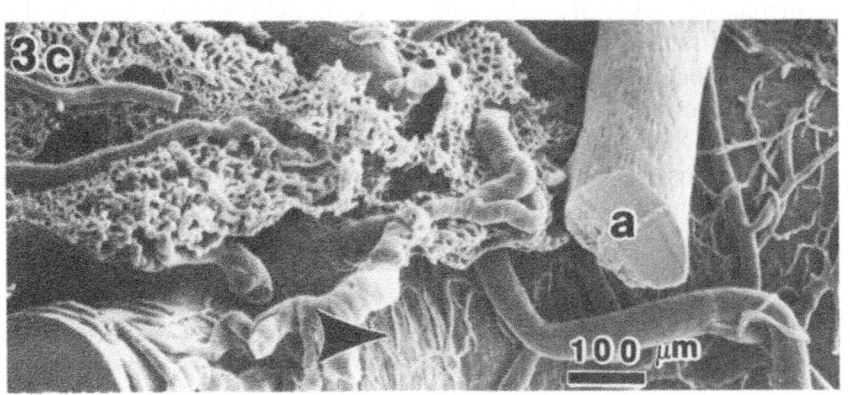

capillaries' diameter was not greater than their distance between branches. Scanning electron microscopic images of cast capillaries show alveolar capillaries branch every 1.07 diameters distance and pleural capillaries branch every 1.6 diameters length [44]. The casts also show that alveolar capillaries are tubular, with flattened and widened areas, but capillaries on the pleural surface are only tubular. Alveolar capillaries are arrranged in pentagonal rings [15] (Fig. 10-4), with spokes radiating out from them. The number of spokes does not appear to change with experimental fibrosis [44] or emphysema [47]. The average diameter of the central holes is 4.2 μm in normal rats at low lung volumes [44], which is compatible with the size of a single cell. Hijiya and Okada showed that alveolar brush cells and type II cells occupied some of the holes [15]. Caduff and coworkers considered the holes to be the pillars of tissue that were the posts for the sheet-flow model of lung capillaries [4]. Others have suggested that the holes are sites of interstitial contractile cells. When situated here, those myofibroblasts that contract with hypoxia could control local blood flow [19]. Adler and associates showed that myofibroblasts are increased in the lungs of rats with bleomycin-induced fibrosis [2], and we have found that the size of the holes is also increased in this condition [44].

Casting studies can be helpful to understand the pulmonary vascular physiology. Blind ends on casts occur more often in the subpleural areas, indicating that these capillaries fill less readily. Distal capillaries have less supporting structure than central ones, so that they are squeezed more by the increasing alveolar size with inspiration. Krahl [24] showed that central vessels have greater change of their branching angles than peripheral vessels because central vessels encompass more alveoli. The greater branching change with inspiration may mean more resistance centrally and less blood flow peripherally.

Shape change in capillaries may affect the perfusion and ventilation matching by a similar explanation. During tidal breathing the alveoli that expand the most may receive the most blood. Alveoli at the top of the lung that are already expanded at the beginning of inspiration may have capillaries narrowed by the large air volume and have little blood flow. Collapsed alveoli that are small at the beginning of inspiration and do not expand during the breathing cycle may have capillaries that remain folded and allow little blood flow. Alveoli that are small at the beginning of inspiration and expand greatly may have the least resistance to capillary blood flow. This would be a purely mechanical explanation for matching ventilation with perfusion, which is critical for maximal blood oxygenation. Capillary folding also may decrease blood flow in expiration, although its importance compared to the rising intrathoracic pressure that occurs during this time is unknown.

Charan and colleagues cast the bronchial arteries of sheep and found that they filled pulmonary capillaries and veins, but not pulmonary arteries [5], indicating that most communications between the bronchial and pulmonary circulations are at the capillary level. They suggested that increased alveolar pressure might impede bronchial blood flow. Although it is widely held that only about 1% of the normal cardiac output is shunted through the bronchial circulation to bypass the alveoli, these investigators suggested that because the bronchopulmonary communication is mostly capillary, bronchial venous blood may be oxygenated and, therefore, may not contribute to the physiologic shunt measured. This group also showed that the dense plexus of bronchial vessels in the peribronchial space drained into pulmonary capillaries. They hypothesized that increased hydrostatic pressure in the pulmonary veins may contribute to the submucosal and peribronchial edema seen radiographically as "hilar haze" and "peribronchial cuffing" in early pulmonary edema.

←

Figure 10-3. a: This is the pleural surface of a normal rat. Normally there are no large blood vessels away from the hila. b: Animals given monocrotaline grow large arteries and veins on their pleural surface. Alveolar capillaries appear to not undergo angiogenesis. c: An artery (a) has typical elliptical nuclear impressions. The large vessel at the bottom of the micrograph has many small vessels (vasa vasorum, arrowhead) on its surface. This rat received 100 mg/kg of monocrotaline 20 days before casting and had prominent bronchial arteries.

132

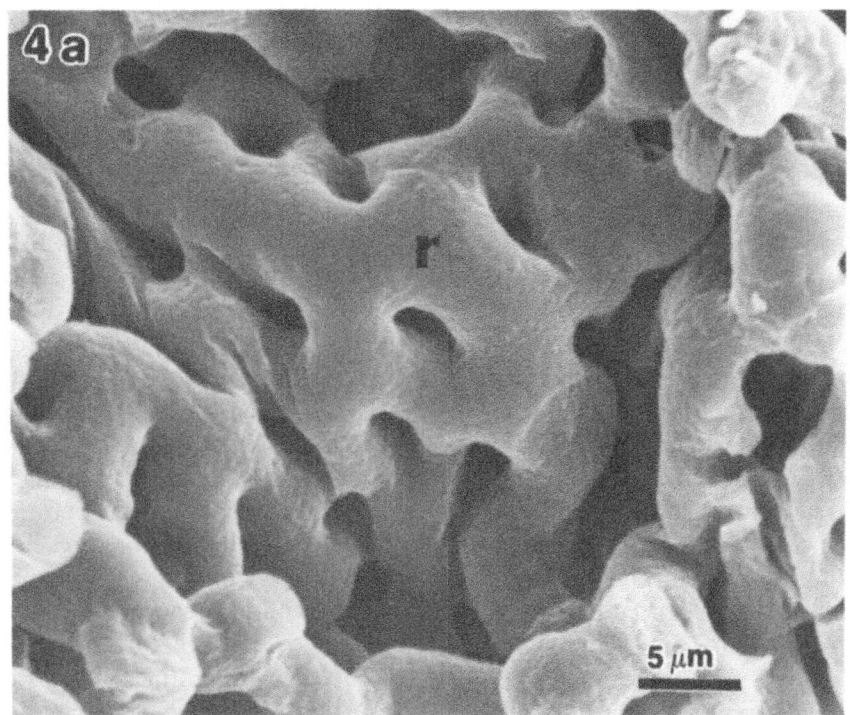

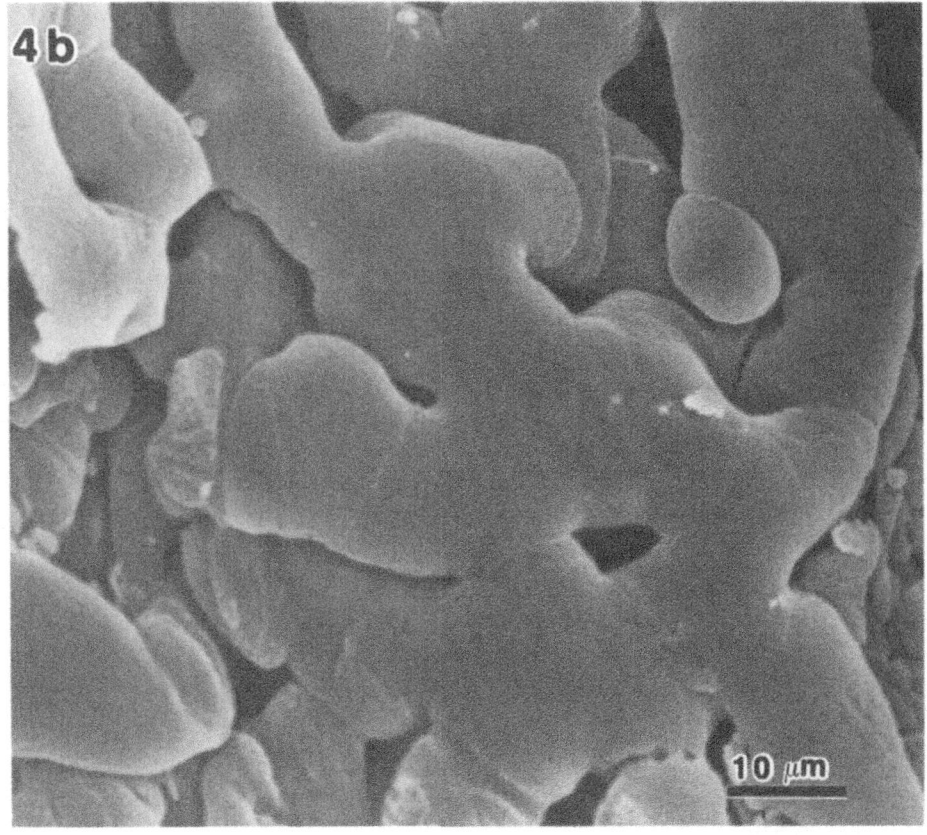

4. Lung Disease

Casting shows that alveolar capillaries have limited responses to injury. They can enlarge, exemplified by adult respiratory distress syndrome, hepatic cirrhosis, and lung fibrosis [44] (Fig. 10-4b). The enlargement does not increase the alveolar-capillary surface area [44]. Only a few casting studies have been done on autopsied human or experimentally injured mammalian lungs. Focal missing capillary casts result from many perturbations [47,50]. Capillary casts have different densities, diameters, and branching distances in different disease states [39,42]. Casts of lungs made emphysematous by giving elastase intratracheally to rats have large alveolar baskets and vessels that end abruptly near areas of destruction [47]. The size of these baskets depends on lung inflation and are larger in emphysema because of increased lung compliance [47]. In lung fibrosis, inflammation and collagen deposition in the interstitium disturb the capillary-alveolar relationship. By casting animals with bleomycin-induced fibrosis, Hijiya showed strictures in the pulmonary arteries [14]. We found irregularly enlarged capillaries with decreased branching [44]. In monocrotaline-induced pulmonary hypertension in rats, the microvascular density is decreased, probably from microthrombi that occlude capillaries. Most lung diseases and experimental conditions are heterogenous, with patchy capillary involvement.

Angiogenesis is a topic that may be studied by casting [52]. The lung undergoes new blood-vessel formation in growth and many disorders, such as neoplasia, bronchiectasis, and thromboembolism. In thrombus-occluded large pulmonary arteries, recanalization comes from the vasa vasorum, which arises from bronchial vessels. In pulmonary infarction and lung abscess, new capillaries come from the bronchial circulation. Remodeled or new vessels occur in lung fibrosis around terminal bronchioles where the bronchial circulation ends.

The exact form that new capillaries in the lung take is not yet established. Some years ago we found miniature pulmonary capillaries, only 2–3 µm in diameter, in normal rat [44], but we have been unable to find similar structures since that time. The pleural surface normally has few large vessels away from the hilum, but extensive new vessel growth occurs in neoplastic and inflammatory states, such as in monocrotaline-induced pulmonary injury (Fig. 1). Around pulmonary arteries extensive remodeling and probably new capillary growth occurs [43]. This is intriguing because this is the same place that Kay and colleagues [21] found increased mast cells. Mast cells contain heparin, which is known to bind to angiogenic factors. The reason the bronchial and not the pulmonary circulation forms new vessels is not clear. Metastatic cancers to the lung [52] and lung cancer transplanted into other sites have been studied with corrosion casts [11]. A problem with casting new vessels is that they are leaky because of incompletely formed basement membranes. Extravasation and alveolar filling may obscure the capillary surface.

5. Airway Casting

Casting has also been used to demonstrate the conducting airways and alveoli, and to measure their branching. This is more difficult than vascular casting because there is no exit port for the air. Injecting resin into the trachea may result in good lung casting, but usually leaves air bubbles in many parts of the cast. Part of the air that is in the lung is pushed out of alveoli through the interalveolar pores of Kohn. Good casts of alveoli can be obtained because the resin adheres to alveolar walls and will cast even blind sacs, leaving the air in the middle. For quantitative casting the lungs must be degassed. This has been done by ventilating animals with carbon dioxide and then adding degassed saline to absorb the carbon dioxide. A slow-setting resin, such as latex, may

←

Figure 10-4. a: Closeup of an alveolar basket in normal rat lung. Here the ring (r) appears square at the low lung volume at which the lung was cast. The rings have five spokes. (Reprinted with permission from Schraufnagel DE et al. [44].) b: Another change that occurs with lung fibrosis is ballooning of alveolar capillaries. These were as large as 19µm in diameter. (Reprinted with permission from Schraufnagel DE et al. [44].)

then been added at a constant pressure as the saline seeps out through the lung wall. These techniques show the regular branching pattern of human bronchi. Silicone casting has been used to define the pulmonary acinus and to measure its volume [37]. Methylmethacrylate does not cast fluid-filled airways well because the resin mixes with the aqueous solution present and the cast does not harden. Casts of the bronchi and bronchioles, with their ridges, gland openings, and ciliated surfaces, appear different than the scanning electron microscopy of tissue. Casting air spaces is hindered, as the fluid layer that covers their surface may obscure epithelial detail. Free-standing clusters of cast alveoli are delicate and easily charge if electrons cannot escape.

Schreider and Raabe casted human alveoli and found the average diameter to be 250 μm and that there are about 7000 alveoli per acinus [51]. Andersen and Jespersen found intraacinar bronchial communication with a diameter of 80–130 μm in postmortem lungs cast with methylmethacrylate [3]. These were different from the pores of Kohn, which have diameters of 3–13 μm, and the channels of Lambert, which have diameters of about 30 μm. Clark and colleagues successfully cast embryonic chick airways and respiratory blood vessels with Batson's Compound No. 17 thinned with glacial acetic acid [6]. Dilly cast the lungs of human fetuses from 19 weeks gestation to term, a 5-year-old, and two adults [8]. She used Tensol cement No. 70, which has a viscosity of 130 cPs at room temperature, and found that it did not penetrate beyond the small bronchi unless they were prefixed. On the other hand, Batson's penetrated into the interstitium and created a solid plastic mass. The best casts were made with Tensol and formaldehyde fixation. Epithelial cell impressions were clearly visible. Tall columnar epithelium had regular, deep nuclear impressions, whereas the cuboidal epithelium made irregular, shallow indentations. She found that only two or three generations of respiratory bronchioles were present at 19 weeks. The number of generations of intraacinar airways continually increased and the most peripheral airspaces expanded to form saccules. Measurements of the maximum diameter of the most peripheral airspaces showed

a doubling of size between 19 weeks and birth, and a further doubling by 5 years [8].

6. Quantitative Casting

An important area that needs more work is obtaining quantitative information from casts. One problem is how trustworthy the plastic is in portraying real-life dimensions. The volume or weight of the casts may be a measure of the adequacy of filling, but no standard exists. Morphometric measurements have been made, but they are fraught with problems. First, filling the vessels at an unknown lung volume makes it difficult to compare data to previous published morphometric work from other laboratories. Most morphometric measurements have been made at the total lung capacity, but lung capillaries cannot be cast at that lung volume. We cast the lungs of rats at the functional residual capacity at death, with the thorax closed for density measurements [44,50]. We also inflated the lung after casting [47]. However, filling the vessels with methylmethacrylate leads to decreased lung compliance, an effect that is different in normal and emphysematous lungs [47].

How the lungs are fixed may affect the density of blood vessels. Fixation through the airway may give a lower capillary density measurement than fixation through the vessels, because alveoli are distended and capillaries are compressed. Cast arteries of the elderly may be less distensible and may appear smaller. Postmortem tissue may have higher vascular density because the blood-vessel wall does not resist the distension of perfusion fixation or casting [13]. On the other hand, cast postmortem organs may have lower vascular density because casting after death may not fill the vessels as well.

Morphometry can be carried out on cast specimens, as on paraffin sections of tissue, although cast vessels may be more completely filled, so the density measured may be greater. Photogrammetric techniques have been used to measure distances and branching angles, but stereoscopic measurements are time consuming and are not easily reproduced.

Three-dimensional angle measurements cannot be made from two-dimensional scanning electron micrographs, but branching frequency and patterns may be compared [46]. Image analysis is commonly used for two-dimensional pictures, where morphometric estimates of density are valid. The pleural surface is planar and the area of the surface capillary layer can be reasonably estimated. For the alveolar surface, the depth of focus can be calculated by knowing the final aperture, the accelerating voltage, and the working distance. A relative index of capillary density can be obtained by counting the points in the areas within the focal plane [44]. Gray-level image analysis gives a more reproducible relative measurement of the vascular density on micrographs of casts, but is still imperfect for the alveolar surface. The optimal gray level to separate the regions in focus — and to cause them to appear white — from regions out of focus — and to cause them to appear black — must be established and kept constant during the study. The ratio of the white area to the total area is an index of vascular density [47]. Image analysis is sensitive to photographic variation in contrast and brightness, and the electron microscope settings must be kept constant during the study. If capillary density is measured at low magnification, an estimate of the general vascularity is obtained. The conducting airways, central areas of the alveolar baskets, and focally destroyed areas may appear black. At higher magnification, only the capillaries are white and the space between them is black. Therefore, in emphysema a greater change occurs with the size of the alveolar basket than the space between capillaries, which is reflected by using a lower magnification. Normally, at high magnification (>1000X) the cut surface has a higher vascular density than the pleural surface, because there is less distance between capillaries, but at low magnification (about 100X) the alveolar baskets are black, so the pleural surface has a greater density. Although Nelson [34] found a good correlation between quantitative casting data and freeze-fractured, quick-frozen lung specimens; the problem of great depth of field should prompt a search for better methods. Casts could be made two dimensional by slicing and carrying out image analysis on the cut surfaces that are perpendicular to the electron collector. Placing casts in polyethylene glycol or ice, and cutting the embedded cast, may give even surfaces. The polyethylene glycol or ice can be melted so that the three-dimensional structure changes can be studied on the same images on which the vascular densities are measured. Actual volumes can then be obtained by using the volume proportion assumption of standard morphometry. Cast lungs can be embedded in paraffin for light microscopy with only occasional problems — the hard methylmethacrylate in the soft paraffin may cause uneven sectioning.

Acknowledgements

I thank the Electron Microscopy Facility of the Research Resources Center at the University of Illinois at Chicago for providing equipment and technical help.

References

1. Abdalla MA, King AS. The functional anatomy of the pulmonary circulation in the fowl. *Respir Physiol* 23:267–290, 1975.
2. Adler KB, Callahan LM, Evans JN. Cellular alterations in the alveolar wall in bleomycin-induced pulmonary fibrosis in rats. *Am Rev Respir Dis* 133:1043–1048, 1986.
3. Andersen JB, Jespersen W. Demonstration of intersegmental respiratory bronchioles in normal human lungs. *Eur J Respir Dis* 61:337–341, 1980.
4. Caduff JH, Fischer LC, Burri PH. Scanning electron microscope study of the developing microvasculature in the postnatal rat lung. *Anat Rec* 216:154–164, 1986.
5. Charan NB, Turk GM, Dhand R. Gross and subgross anatomy of bronchial circulation in sheep. *J Appl Physiol* 57:658–664, 1984.
6. Clark EB, Rooney PR, Martini DR, Rosenquist GC. Plastic casts of embryonic respiratory and cardiovascular system: A technique. *Teratology* 19:357–360, 1979.
7. Dellimore JW, Dunlop MJ, Cannham PB. Ratio of cells and plasma in blood flowing past branches in small plastic channels. *Am J Physiol* 244:H635–H643, 1983.
8. Dilly SA. Microcorrosion casting of the human respiratory acinus. *Scann Electron Microsc* 3:1095–1101, 1986.
9. Fairman RP, Morrow C, Glauser FL. Methylmethacrylate induces pulmonary hypertension and increases lung vascular permeability in sheep. *Am Rev Respir Dis* 130:92–95, 1984.
10. Fung YC, Sobin SS. Theory of sheet flow in lung alveoli. *J Appl Physiol* 26:472–490, 1969.

136

11. Grunt TW, Lametschwandtner A, Karrer K. The characteristic structural features of the blood vessels of the Lewis lung carcinoma. *Scann Electron Microsc* 2:575–589, 1986.

12. Guntheroth WG, Luchtel DL, Kawabori I. Pulmonary microcirculation: Tubules rather than sheets and posts. *J Appl Physiol* 53:510–515, 1982.

13. Harris P, Heath D. *The Human Pulmonary Circulation. Its Form and Function in Health and Disease.* Churchill Livingstone, Edinburgh, 1986.

14. Hijiya K. Ultrastructural study of lung injury induced by bleomycin sulfate in rats. *J Clin Electron Microsc* 11:245–292, 1978.

15. Hijiya K, Okada Y. Scanning electron microscopy of the pulmonary capillary vessels in rats. *J Electron Microsc* 1:49–53, 1978.

16. Hodde KC. Cephalic Vascular Patterns in the Rat. Doctoral Thesis. Amsterdam University, Amsterdam, 1981.

17. Hojo T. A reexamination of making anatomical corrosion casts, especially the lung. *Sapporo Med J* 43:1–4, 1974.

18. Hojo T. An anatomical study of trachea, bronchi and pulmonary vessels of the harbor seal (*Phoca vitulina*) with a corrosion cast. *Acta Anat Nippon* 50:229–235, 1976.

19. Kapanci Y, Assimacopoulos A, Irle C, Zwahlen A, Gabbiani G. "Contractile interstitial cells" in pulmonary alveolar septa: A possible regulator of ventilation/perfusion ratio? Ultrastructural fluorescence and *in vitro* studies. *J Cell Biol* 60:375–392, 1974.

20. Kay JM. Pulmonary vasculature and nerves. Comparative morphologic features of the pulmonary vasculature in mammals. *Am Rev Respir Dis* 128:S53–S57, 1983.

21. Kay JM, Gillund TD, Heath D. Mast cells in the lungs of rats fed on *Crotalaria spectabilis* seeds. *Am J Pathol* 51:1031–1044, 1967.

22. Kendall MW, Eissman E. Scanning electron microscopic examination of human pulmonary capillaries using latex replication method. *Anat Rec* 197:275–283, 1980.

23. Koike K, Ohnuki T, Ohkuda K, Nitta S, Nakada T. Branching architecture of canine pulmonary arteries: A quantitative cast study. *Tohoku J Exp Med* 149:293–305, 1986.

24. Krahl VE. Relationships of peripheral pulmonary vessels to the respiratory areas of the lung. *Med Thorac* 19:194–207, 1962.

25. Lane BP, Zeidler M, Weinhold C, Drummond E. Organization and structure of branches in the rat pulmonary arterial bed. *Anat Rec* 205:272–279, 1983.

26. Magno MG, Fishman AP. Origin, distribution and blood flow of bronchial circulation in anesthetized sheep. *J Appl Physiol Respir* 53:272–279, 1982.

27. Malpighi M. De Pulmonibus, 1661. Translated by Young J. *Proc R Soc Med* 23:1–14, 1930.

28. Mazzone RW, Kornblau S, Durand CM. Shrinkage of lung after chemical fixation for analysis of pulmonary structure-function relations. *J Appl Physiol Environ Exercise Physiol* 48:382–385, 1980.

29. McLaughlin RF Jr. Bronchial artery distribution in various mammals and in humans. *Am Rev Respir Dis* 128:S57–S58, 1983.

30. Miller WS. Vascular supply of the pleura pulmonalis. *Am J Anat* 7:389–407, 1907.

31. Miller WS. *The Lung*, 2nd ed. Charles C Thomas, Springfield, IL, 1947.

32. Miodonski A, Hodde KC, Bakker C. Rasterelektronenmikroskopie von Plastik-Korrosions-Praparaten: morphologische Unterschiede zwischen Arterien und Venen. *Beitr Elektronenmikroskop Direktabb Oberfl* 9:435–442, 1976.

33. Murakami T. Application of the scanning electron microscope to the study of fine distribution of blood vessels. *Arch Histol Jpn* 32:445–454, 1971.

34. Nelson AC. Study of rat lung alveoli using corrosion casting and freeze fracture methods coupled with digital image analysis. *Scann Microsc* 1:817–822, 1987.

35. Nowell JA, Pangborn J, Tyler WS. SEM of avian lung. *Scann Electron Microsc* 249–256, 1970.

36. Ohtani O. Microvasculature of the rat lung as revealed by scanning electron microscopy of corrosion casts. *Scann Electron Microsc* III:349–356, 1980.

37. Rodriguez M, Bur S, Favre A, Weibel ER. Pulmonary acinus: Geometry and morphometry of the peripheral airway system in rat and rabbit. *Am J Anat* 180:143–155, 1987.

38. Schlessinger MJ. New radio-opaque material for vascular injections. *Lab Invest* 6:1–11, 1957.

39. Schraufnagel DE. Microvascular corrosion casting of the lung. A state-of-the-art review. *Scann Microsc* 1:1733–1747, 1987.

40. Schraufnagel DE. Ranking corrosion efficiency: A Latin square study on rat lung microvascular corrosion casts. *Scann Microsc* 3:299–304, 1989.

41. Schraufnagel DE. Microvascular casting of the lung: Bronchial versus pulmonary filling. *Scann Microsc* 3:575–578, 1989.

42. Schraufnagel DE. Corrosion casting of the lung. In: Schraufnagel DE (ed.), *Electron Microscopy of the Lung.* Lung Biology in Health and Disease Series. Marcel Dekker, New York, 1990.

43. Schraufnagel DE. Monocrotaline-induced angiogenesis. Differences in the bronchial and pulmonary vasculature. *Am J Pathol* 137:1083–1090, 1990.

44. Schraufnagel DE, Mehta D, Harshbarger R, Treviranus K, Wang NS. Capillary remodeling in bleomycin-induced pulmonary fibrosis. *Am J Pathol* 125:97–106, 1986.

45. Schraufnagel DE, Patel KR. Sphincters in pulmonary veins: A anatomic study in rats. *Am Rev Respir Dis* 141:721–726, 1990.

46. Schraufnagel DE, Roussos C, Macklem PT, Wang NS. The geometry of the microvascular bed of the diaphragm: Comparison to intercostals and triceps. *Microvasc Res* 26:291–306, 1983.

47. Schraufnagel DE, Schmid A. Capillary structure in elastase-induced emphysema. *Am J Pathol* 130:126–135, 1988.

48. Schraufnagel DE, Schmid A. Microvascular casting of the lung: Effects of various fixation protocols. *J Electron Microsc Techn* 8:185–191, 1988.

49. Schraufnagel DE, Schmid A. Microvascular casting of the lung: Vascular lavage. *Scann Microsc* 2:1017–1020, 1988.

50. Schraufnagel DE, Schmid A. Pulmonary capillary density in monocrotaline-induced pulmonary hypertension: A cast corrosion study. *Am Rev Respir Dis* 140:1405–1409, 1989.
51. Schreider JP, Raabe OG. Structure of the human acinus. *Am J Anat* 162:221–232, 1981.
52. Wang ZW, Song XB, Huang WW, Tao LX, Xu JY. Investigation of experimental metastases in the lung by cast-scanning electron microscopy. *Shih Yen Sheng Wu Hsueh Pao* 19:68–79, 1986.
53. Winkler GC, Pulmonary intravascular macrophages in domestic animal species: A review of structure and functional properties. *Am J Anat* 181:217–234, 1988.

Author's address:
Dr. Dean E. Schraufnagel
Section of Respiratory and
    Critical Care Medicine
Department of Medicine M/C 787
University of Illinois at Chicago
P.O. Box 6998
Chicago, IL 60680-6998
USA

CHAPTER 11

Postnatal Development and Growth of the Pulmonary Microvasculature

PETER H. BURRI

1. Introduction

In the current literature (although not officially sanctioned by the third edition of *Nomina Embryologica*, 1989) lung development is generally divided into four successive, but partially overlapping, phases, known as the pseudoglandular, canalicular, saccular, and alveolar stages. Based on recent structural observations, we have advocated the idea that alveolization does not represent the final step in lung development and have proposed the introduction into the nomenclature of a stage of microvascular maturation [1]. As will become evident in this chapter, the process of alveoli formation depends upon the presence of a specific morphology of the parenchymal[1] capillaries, i.e., on the presence of a double capillary network within the intersaccular and interalveolar walls [2–5]. The adult mature interalveolar septa, however, are slender and contain only a single capillary system, allowing gas exchange on both of its faces. The transformation of the postalveolar to the adult pulmonary capillary system therefore requires a stage of microvascular maturation before pulmonary development may be considered to be complete.

2. Lung Development and its Stages (with Special Reference to Capillary Morphology)

2.1. Embryonic Period (1–7 Weeks)

The embryonic period is not a developmental phase specific to the lung. Rather, it encompasses the first phase of development of the new organism, during the course of which almost all organs are laid down.

The human lung arises around day 26 after fertilization as a ventral outpouching of the foregut. Two longitudinal furrows, the laryngotracheal grooves, appear laterally, deepen, and separate the lung bud from the foregut in a caudo-cranial direction; a connection is maintained at the site of the prospective entrance into the larynx (hypopharynx). The bud elongates, divides, and grows into the surrounding mesenchyme. At an age of 4.5 weeks, the lung primordium already shows five tiny saccules corresponding to the future lobar bronchi.

During the course of a few more days, the branching process proceeds rapidly, and in light microscopic sections the organ consists of a number of tubular cross sections embedded in a loose mesenchyme. The lung now has the appearance of a small primitive gland and has entered the pseudoglandular stage of development.

Motta, P.M., Murakami, T., and Fujita, H. (eds.), Scanning Electron Microscopy of Vascular Casts: Methods and Applications.

140

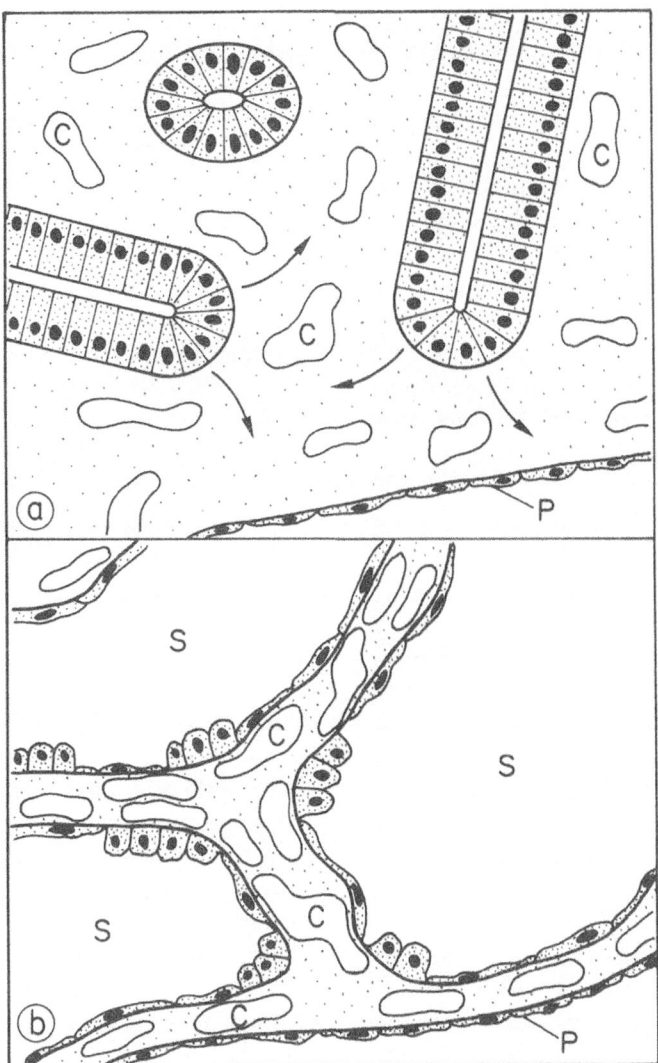

Figure 11-1. Illustration of relationship between capillaries and air spaces during lung development. C = capillaries; S = prospective air spaces; P = pleural epithelium; dotted area = mesenchyme. a: Pseudoglandular stage. Epithelial tubes grow and branch into the mesenchyme containing a loose three-dimensional network of capillaries. Arrows show direction of growth and branching. b: Canalicular stage. Through growth and expansion of the future air spaces, the mass of intervening mesenchyme has decreased and the epithelial tubes have acquired an envelope of capillaries. Wherever the capillaries are in close contact with the epithelium, the originally cuboidal epithelial cells develop thin cytoplasmic extensions and differentiate into the squamous type I cells, thus preforming the future air-blood barrier. The remaining cuboidal cells represent either undifferentiated or type II epithelial cells. Where two canaliculi come to lie against each other, the intercanalicular wall contains a capillary bilayer. The two capillary layers are interconnected because they are both derived from the same three-dimensional mesenchymal capillary network.

2.2. *Pseudoglandular Stage (5–17 Weeks)*

This stage could also be referred to as the "bronchial phase" of lung development, since by the end of this stage the complete set of generations of conducting airways of the future lung is present [6]. However, cells have been identified at the periphery of the airway tree that are capable of differentiating into gas-exchanging structures, as shown by immunohistochemical techniques [7]. According to Boyden [8], the prospective acinus is formed by week 17.

The tubules exhibit an epithelium of varying height and differentiation according to their age and topographical location. They are embedded in a primitive and pluripotent type of connective tissue, the mesenchyme, consisting of a loose network of cells and abundant interstitial substance, but with only a small fibrous component. Within the mesenchyme, primitive vessels develop; they form a wide three-dimensional network connected to the heart and to larger neighboring vessels, such as the aorta and the cranial cardinal veins.

At the end of this stage, at the uppermost extremity of the airway tree, the future gas-exchange region develops, a process that leads into the next phase.

2.3. Canalicular Stage (16–26 Weeks)

This phase comprises the early development of the pulmonary parenchyma. From a small cluster of buds delineated by a rarified mesenchymal envelope and representing a future acinus, the gas-exchange region develops by further peripheral branching and by lengthening of each tubular branch. These distal airspaces widen at the expense of the intervening mesenchyme. The new airways are called *canaliculi*; with their presence, the lung assumes a spongy appearance for the first time. Capillaries continue to develop within the mesenchyme, some of which begin to surround the canaliculi. As a consequence of this process, parts of the capillary network come into close association with the epithelial lining of the airways. At these contact sites, the epithelial cells become more flattened and develop attenuated processes (Fig. 11-1); as such, the future structure of the air-blood barrier is realized. The close apposition of the capillaries to the epithelial lining and the differentiation of the latter into type I cells appear to be closely synchronized occurrences, but the precise interrelationship existing between these processes is not known.

In parallel with the appearance of type I cells, glycogen-rich cuboidal cells, the future type II epithelial cells, begin to accumulate lamellar bodies within their cytoplasm. These membrane-bound structures, consisting of multiple osmiophilic lamellae arranged in an onionlike or concentric fashion, represent the

intracellular storage form of the surface-active material [9], also found subsequently in the prospective air spaces, as secretion begins.

Respecting differentiation of the lung parenchyma, the canalicular stage is an eminently important period in lung development, since survival of a prematurely born infant becomes possible if two conditions are fulfilled: a thin air-blood barrier large enough to allow sufficient O_2 uptake, and a sufficient supply of surfactant (adequately composed) for the decrease of surface tension at the air-tissue interface.

2.4. Saccular Stage (24 Weeks to Term)

This stage may be defined as a period of rapid volume growth of the prospective lung parenchyma. At the transition from the previous stage to this one, the peripheral airways form typical terminal clusters of rounded saccules, hence the nomenclature of the saccular or terminal sac stage. Within the next few months, the terminal sacs undergo several successive series of lengthening and dichotomous division, the consequence of which is that a terminal sac gives rise to two new terminal sacs and is itself transformed into a tube. The tube again becomes an alveolar duct when, later, alveoli are formed. These structures have therefore been termed *transitory* because of their changing morphology [5]: A transitory saccule becomes a transitory channel or duct, which is definitively transformed into an alveolar duct when alveolar formation occurs.

The term *saccule* hints at the relatively wide dimensions of these peripheral airways. Furthermore, with their tremendous multiplication, the interposed interstitial tissue is markedly "compressed" and decreases in terms of volume proportion. This process has a profound effect on the three-dimensional structure of the parenchymal capillary network; as the airways push aside the intervening interstitium, the capillary networks surrounding each airway approach one another. This process results in the formation of a thick intersaccular septum with a double capillary layer (Fig. 11-1). These septa, which are typical of the saccular stage, have been termed *primary septa* [10].

142

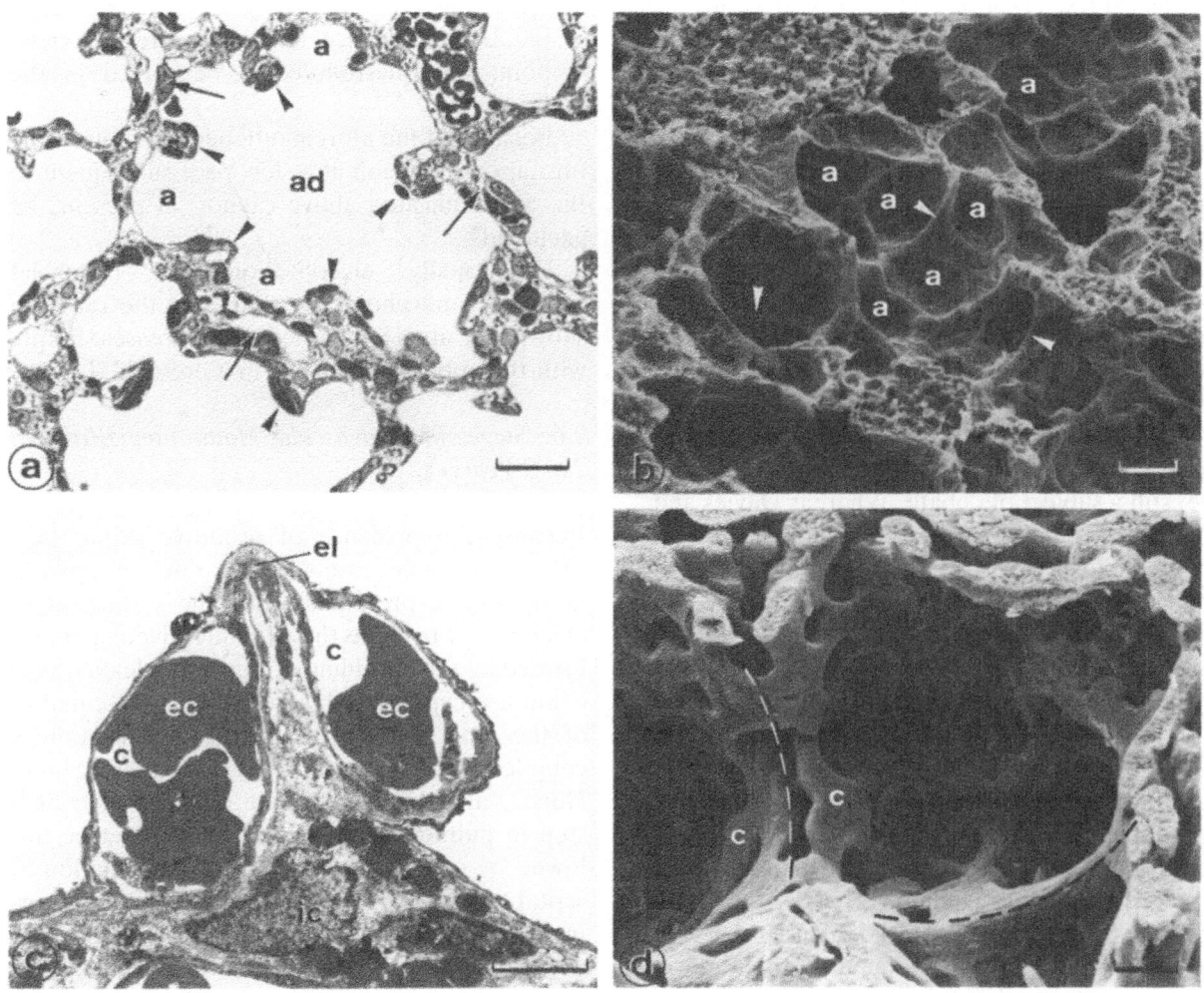

Figure 11-2. Illustration of alveolar formation by four different imaging techniques. a: Light micrograph (rat lung, age 7 days) showing numerous secondary septa (arrowheads) arising from primary septa (arrows). The secondary septa delineate shallow depressions, the newly formed alveoli (a), transforming the original saccule into an alveolar duct (ad). Bar = 50 µm. b: Scanning electron micrograph (rat lung, age 8 days) illustrating the formation of alveoli (a) by secondary septa (arrowheads). Bar = 20 µm. c: Electron micrograph (rat lung, age 7 days) of a transversely cut secondary septum. Note central axis of connective tissue flanked on both sides by capillaries (c) containing erythrocytes (ec). Elastin (el) is often located at the tip of the crest. ic = nucleus of interstitial cell. Bar = 2 µm. d: Scanning electron micrograph of Mercox cast of lung capillaries (rat lung, age 7 days). Secondary septa are formed along the dotted lines by an upfolding of one of the two capillary layers of the primary septum. The septum on the left with the two capillaries running in parallel (c) corresponds to the cross section of Fig. 11-2c.

2.5. *Alveolar Stage (36 Weeks to 18 Months Postnatal)*

At the beginning of this stage, there still exists a relative paucity of connective tissue fibres. Elastin is, however, deposited beneath the epithelium in regions created by cytoplasmic infoldings of interstitial cells. It has long been recognized that elastin formation plays an important role in the

further development of the lung, since inter-alveolar septa appear where strands of elastin are laid down in the intersaccular walls [11].

Alveolization begins with the appearance of low ridges on the saccular wall (Figs. 11-2a–11-2c). The ridges incompletely subdivide the saccular space into a number of smaller units, a kind of shallow outpouchings arranged around the central lumen of the transitory channels or saccules (Figs. 11-2a and 11-2b). Also known as secondary septa, these ridges arise by an upfolding of one of the two capillary layers of the primary septa and delineate the alveoli, as it is illustrated in Figs. 11-2a, 11-2c, and 11-2d.

At this time, the primary and secondary septa contain two capillary layers, i.e., they are still of the so-called primitive type, as opposed to the mature septa of the adult lung. Considering the mechanism of secondary septa formation discussed above, it appears that new alveoli can be formed only if primitive septa still exist in the lung.

In humans the onset, duration, and termination of alveolar formation have long been, and are, still a subject of debate. Whereas Davies and Reid [12] counted about 20 million alveoli at birth, Boyden [8] postulated that alveolization begins postnatally; in his opinion the air spaces present at birth were still saccules.

In an investigation of over 50 fetal and postnatal lungs, Langston and coworkers [13] described the first alveoli to be present as early as week 32 in some instances, and counted on average 50 million alveoli at birth. These authors also found that alveolization was completed at about 2 years of age, which is much earlier than had been previously assumed: Dunnill [14] had set the limit at 8 years, Emery and Wilcock [15] at 20 years. In our own investigations of human postnatal development and growth, we found alveolar formation to be well advanced at 1 month after birth and likely to be completed at about 18 months [1, 16]. Based solely upon morphological criteria (maturity of interalveolar septa, see below), it even appeared that bulk alveolar formation was completed by about 6 months of age.

For several reasons, termination of the alveolization process cannot be precisely determined:

1. The final number of alveoli in the adult lung varies considerably (an average 300 million [17], or 375 × 106, with a range of 212–605 × 106 according to body size [18]).
2. The clear definition of an alveolus as it appears in sections is difficult [19], and this problem is exasperated during alveolar formation.
3. The stereological counting techniques currently used for alveoli in an attempt to solve the problem are not bias free. However, new

stereological approaches have been proposed [20,21], and hence from a theoretical viewpoint the question could be solved in the future.

Because of the aforementioned difficulties, the formation of alveoli at a slow pace subsequent to the ages indicated above cannot, at present, be excluded.

Functionally, alveolization is a beneficial event. As morphometric studies in the rat have shown, the alveolar surface area increases sharply with the appearance of the first crests [22].

2.6. Stage of Microvascular Maturation (Birth to 2 or 3 Years)

Because the presence of primitive septa, i.e., of inter-air space walls with a double capillary network is a prerequisite for the formation of alveoli, and because the new interalveolar septa, formed by the upfolding of a capillary leaflet, also contain a capillary bilayer, the microvasculature of the lung parenchyma during alveolization is completely different from that in the mature lung. Hence, alveolization cannot represent the final step in pulmonary development. It must be followed by a stage of structural transformation in septal morphology, involving marked alterations in microvascular architecture [1]. Although for wide areas of the lung this maturation proceeds swiftly, the process may last 2 or 3 years. As with alveolar formation, the end of this process can hardly be clearly determined, because limited immature areas, especially at the base and the tips of the septa, can even be found in adult lungs. The maturation process occurring at the level of the capillary network is ideally suited for investigation by the casting technique and will be described in the following sections of this chapter.

3. The Pulmonary Capillary System During Alveolization

3.1. Technical Information Concerning Pulmonary Microvascular Casting[2]

The experimental work detailed here relates to the rat. The lung development in this rodent species differs from that in the human in that

144

no alveoli are present at birth. Alveolization occurs within a very short period of time between postnatal day 4 and the age of about 2 weeks. Capillary maturation then occurs during the third postnatal week. Nonetheless, since the structural alterations are almost identical to those occurring in the human lung, the rat lung represents an appropriate model for studying and understanding the postnatal phase of lung development in general. The compression of events into just $2\frac{1}{2}$ weeks is also a positive aspect.

Following numerous attempts with various casting materials, the following approach using Mercox was found to yield the best results in our hands [4]. Rats aged 1,4,7,10,13,17,21, and 44 days, and 4 and 9 months, were premedicated with Valium® and deeply anesthetized with Ketalar® combined with Hypnorm® or Nembutal®. Rats were maintained in the supine position and the skin was incised from mandible to symphysis; the trachea was exposed and the abdomen opened. The lungs were collapsed by perforating the diaphragm, and the anterior chest wall was removed. The trachea was cannulated and a ligature was placed around the pulmonary artery. The lungs were expanded to the midrespiratory level under visual control by intratracheal instillation of Ringer's solution. By avoiding lung atelectasis, this procedure ensures easier assessment of lung and capillary structure. The instillation of fluid into the lungs removes the surfactant at the air-tissue interface and may produce, depending upon the applied pressures, an artificial bulging of capillaries into the air-space lumina [23]. In our study, however, this artifact was judged beneficial, since by improving the deployment of the capillaries, pleating of the septa could be avoided. The pulmonary artery was cannulated through the right ventricle and, after incision of the left auricle, the vascular bed was flushed with Ringer's solution, containing 1% procaine and 1% Liquemin®, at ambient temperature. Using a fresh syringe, Mercox was perfused through the same catheter, carefully

avoiding air-bubble formation in the system. This procedure was performed continuously by hand, with no monitoring of the applied pressure, until viscosity increased and impeded flow. The secret of successful casting resides in the critical selection of the Mercox polymerization times. After numerous trials, we found that a freshly prepared 50:1 mixture of Mercox base resin (no monomer added) and catalyst was optimal. Immediately after mixing, the viscosity of the resin is low enough to permit access to the smallest capillaries; within 5 minutes it increases sufficiently to impede flow completely under the same pressure conditions. After 20 minutes, polymerization of the Mercox is virtually complete. This rapid setting time avoids cast deformation, which is consistently found with desiccation of the specimens (tissue shrinkage). Tissue dissolution was achieved by immersing the excised lungs in 15% KOH for 1–3 weeks with regular changes of the caustic solution.

After complete dissolution of the tissue (microscopically controlled), the casts were rinsed in tap water, placed in plastic cups, and frozen in distilled water degassed under vacuum. The ice blocks were cut into small pieces with a fret saw, and the surface to be examined was polished with a cryomicrotome. The specimens were then dehydrated in an ascending ethanol series, critical-point dried, glued onto stubs with carbon, and sputtered with gold before viewing in a Philips PSEM 500 scanning electron microscope.

3.2. Morphology of Capillary Networks

The study of Mercox casts of postnatal rat lung vasculature by scanning electron microscopy yielded some new insights into the structural alterations produced within the capillary bed.

On days 1 and 4, corresponding to a phase of lung expansion at the end of the saccular stage, the intersaccular wall appears as a triple-layered structure consisting of two capillary networks

→

Figure 11-3. Differences in capillary network structure between immature and mature lungs in scanning electron micrographs of Mercox casts. a: Rat lung, age 7 days. Normal inter-air space walls contain a double capillary system (arrows), while capillary networks abutting the wall of a larger vessel (v) form only a single layer (arrowheads). Bar = 20 μm. b: Rat lung, age 44 days. Surface view of the pleural capillary network with small collecting veins (ve) and side view of the capillaries of the gas-exchange region. Note the small meshes of the interalveolar capillaries, in contrast to the coarse pleural network. Bar = 50 μm.

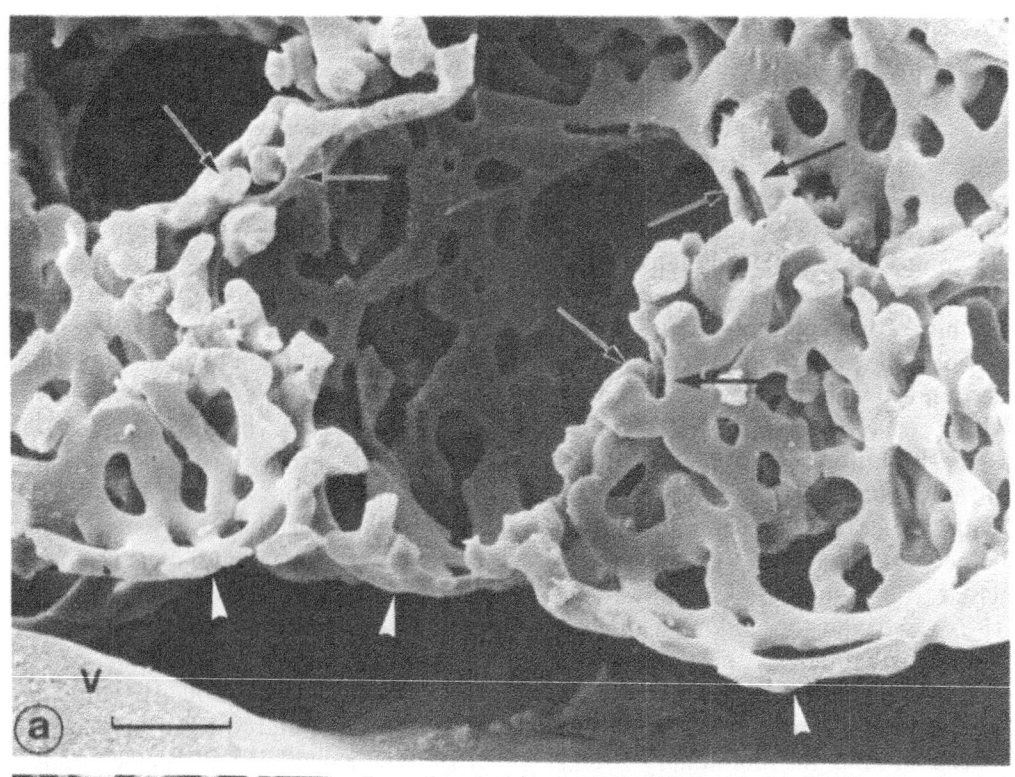

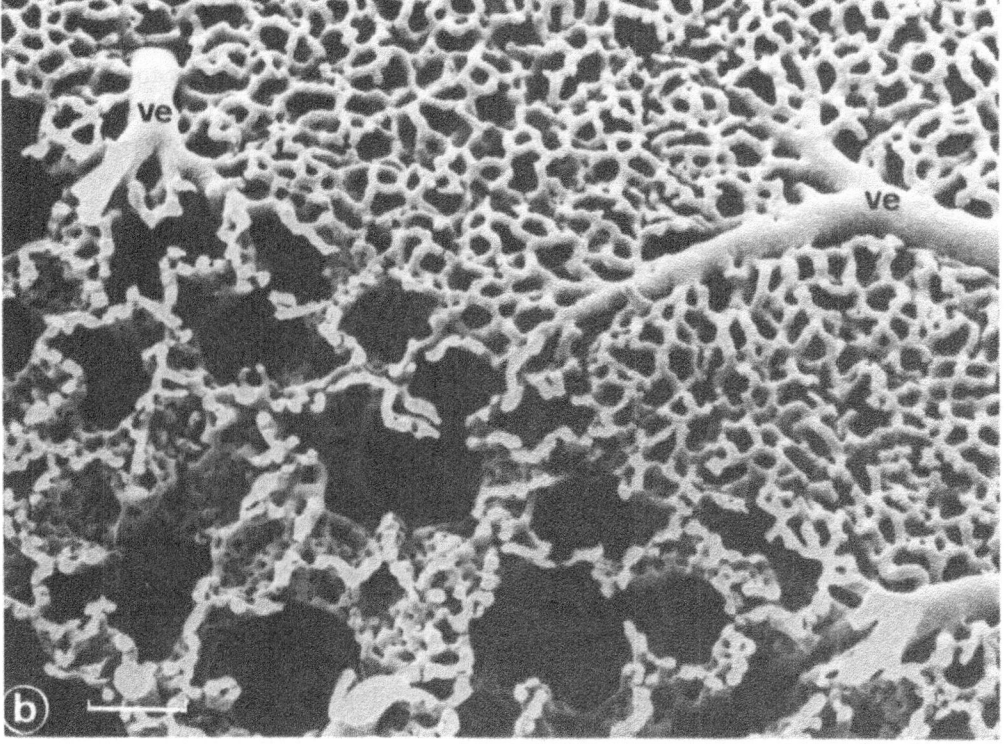

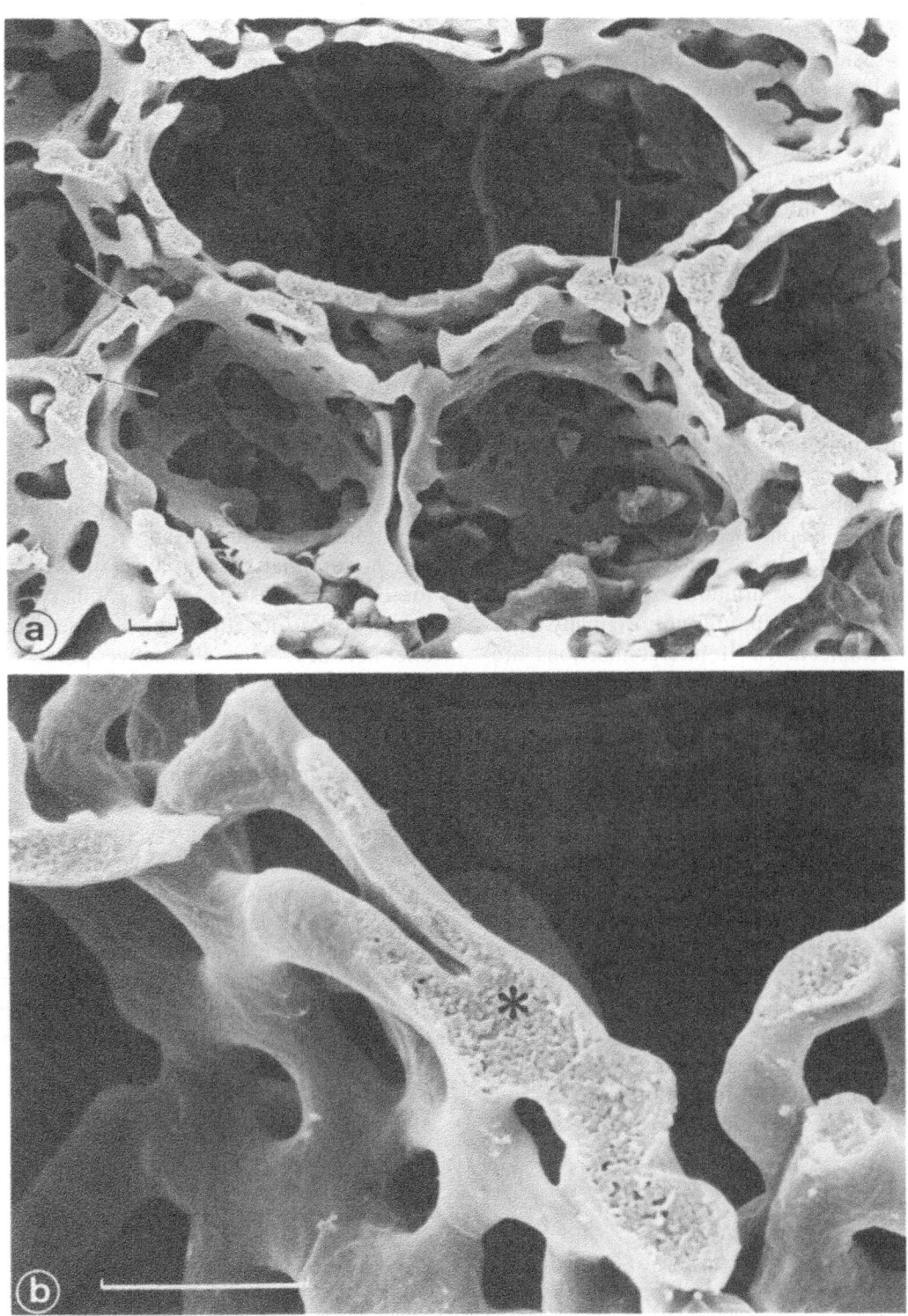

interposed by a wide gap corresponding to the central sheet of connective tissue digested away by KOH treatment (Figs. 11-3a and 11-4a). Connecting vessels between the two capillary layers are often seen. This organization is easily explained by the ontogenetic development of these networks: They arise as a consequence of the growth of the transitory airways and saccules that compress the intercalated mesenchyme and its capillary system into an interconnected bilayer (Figs. 11-1 and 11-3a). Interestingly, however, only one capillary layer can be distinguished around the larger nonparenchymal structures, such as arteries and veins, conducting airways, and the pleura.

Three days later alveolization has begun. Typically, low and curved ridges of capillaries appear on primary septa (Fig. 11-2d); these correspond to the secondary crests illustrated in Fig. 11-2c. The double capillary system of these arises as an upfolding of one of the capillary layers of the primary septa. In secondary septa the two networks are often continuous over the edge of the crest (Fig. 11-2d), but do not originally show other interconnections. If interconnections are found, they are the result of capillary fusions.

Within a few days, the width of the slit between the networks is markedly narrowed and, on primary and higher secondary septa, interconnections become more frequent. The observation of a decreasing amount of interstitial tissue is corroborated by previous morphometric measurements [22].

During the third week, the vascular pattern of the interalveolar septa is profoundly transformed (compare Figs. 11-4a and 11-4b). Large areas of the septal microvasculature consist of a single network. Some septa, do however, retain an immature morphology at their base and tip, where capillary segments form loops over the free edge of the crests (as illustrated for the human lung in Fig. 11-5a) or where two capillaries often run in parallel along the septal border.

At 6 weeks and thereafter, the rat lung appears to be mature in every respect. This is confirmed by the microvascular morphology of the casts: The interalveolar walls contain a single-layered dense network of capillaries meandering along an imaginary central axis (Fig. 11-5b), an image that fits well with the description of a septal axis formed by connective tissue fibers [24].

4. Stage of Microvascular Maturation — The Last Step in Lung Development

On the basis of observations made in the transmission electron microscope [2], we have postulated that the morphological alterations related to the microvascular maturation are brought about by two processes that run in parallel and are of about equal importance: capillary fusions and preferential growth. We found no evidence for extensive destruction of capillary segments, in particular, for the complete degradation of one network. This alternative process for the reduction of the double network was therefore dismissed.

4.1. Intercapillary Fusions

A process of intercapillary fusion is difficult to prove unless observed in vivo. Unfortunately, continuous microscopy of lung capillaries in vivo for hours and days cannot, at present, be achieved. Hence, evidence in favor of this process is derived from static transmission and scanning electron micrographs, which, nonetheless, permit a logical derivation or construction of the sequence of events involved.

By serial sectioning of the primitive interalveolar wall of rats aged between 7 and 13 days,

←

Figure 11-4. Maturation of the pulmonary capillary network by intercapillary fusions. Scanning electron micrographs of Mercox casts. a: Rat lung, age 7 days. The two capillary layers in primitive septa show partly wide and partly narrow gaps between them. The gaps correspond to the interstitial tissue digested away by KOH treatment. Note the numerous interconnections between the two capillary layers (arrows). Bar = 10 μm. b: Rat lung, age 21 days. Mature capillary network of interalveolar wall with local remnants of the immature stage. In view of the network morphology observed on day 7 (Fig. 11-4a), the two capillary segments merging at the asterisk are suggestive of a fusion process. Bar = 10 μm.

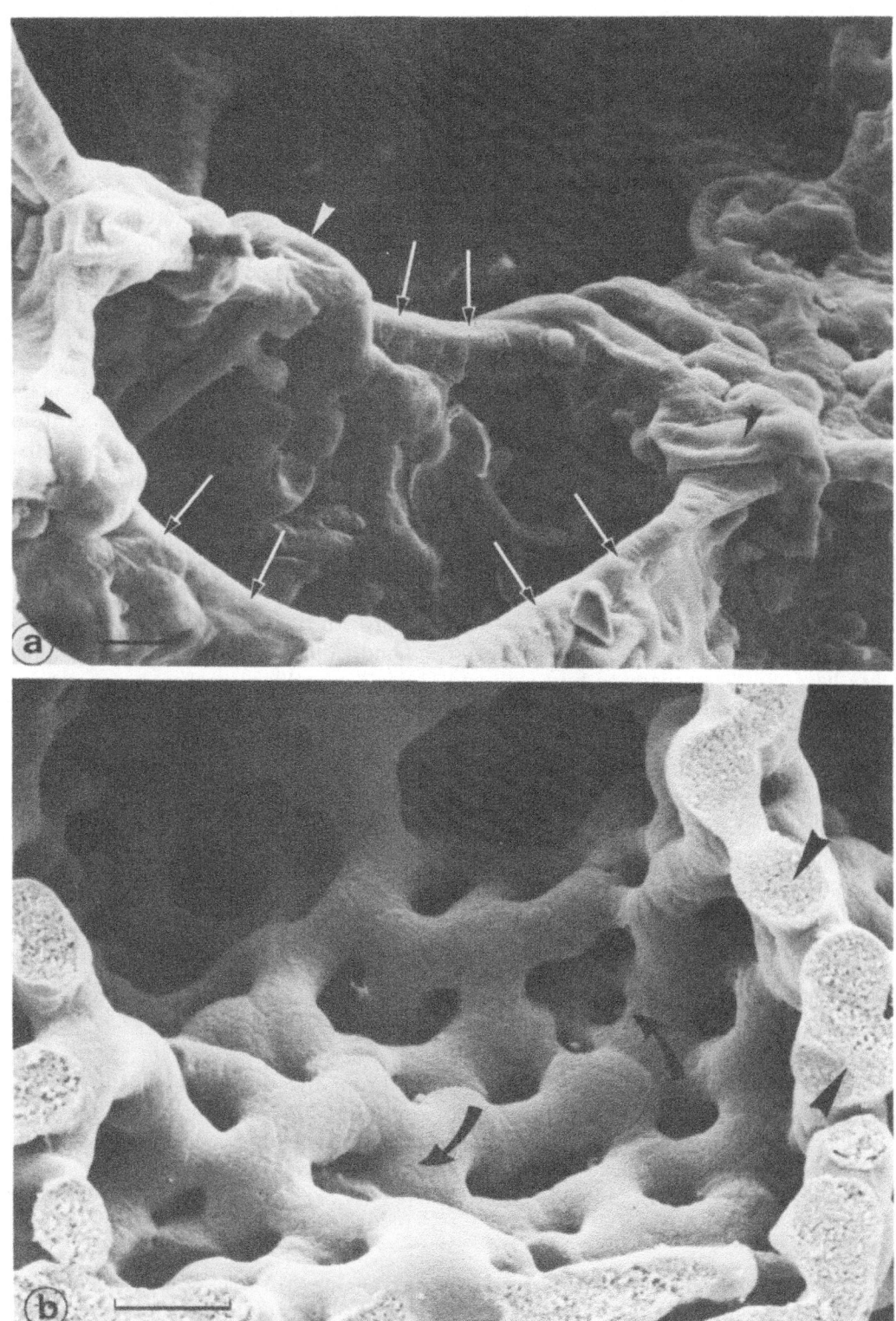

we were able to demonstrate that the two capillary layers were sometimes so closely apposed that they were separated merely by a cytoplasmic extension of a single endothelial cell without any interconnection of the respective lumina [Burri and Tarek, in preparation]. Such a situation is clearly a necessary intermediate step in the proposed fusion process of the capillary bilayer. Scanning pictures of the vascular casts also suggest that a merging of the capillary systems occurs (Figs. 11-4a, 11-4b, and 11-5b).

Further arguments in support of this mechanism include the following:

1. The capillary network of the alveoli is much coarser when adjacent to nonparenchymal structures, such as the pleura or peribronchial and perivascular sheaths. This peculiarity, already mentioned by Miller in 1947 [25], and illustrated in Fig. 11-3b, may be explained if we assume that the tight interalveolar network is the result of the final merging of the two networks shown in Fig. 11-1. The capillary networks of the peripheral air spaces abutting the peribronchial and perivascular sheaths and pleura have no counterparts to fuse with and thereby remain coarse.

2. The development of the lung per se represents an argument in favor of a fusion process: The mass of intervening mesenchyme surrounding the peripheral air spaces during fetal development undergoes continuous reduction. As stated above, the decrease has been demonstrated by morphometric measurements to continue after birth. Obviously, the ultimate logical consequence of such a process would be the contact of endothelial cells from the two adjacent capillary layers, followed by merging of the lumina. There would be no need for actual cell-to-cell fusions: Adjacent, originally unconnected capillary lumina can become continuous by rearrangement of the interendothelial junctional complexes.

We may therefore summarize that, in addition to the simple and static morphological observations, the logical interpretation of additional, partially indirect, evidence is consistent with the fusion concept.

4.2. Preferential Growth

In embryology, preferential growth is a well-recognized phenomenon, by means of which considerable morphological changes can be effected, as, for example, in heart development. Although it is very difficult to prove, preferential growth of certain areas of the capillary network is likely to play a role in microvascular maturation. If one assumes that the areas of fusion grow at a faster rate than the rest of the capillary bed, then the appearance of the interalveolar septa would be expected to alter rapidly (Fig. 11-6). Even if only one of the two networks expanded rapidly without the involvement of any fusions, septal morphology would be greatly altered. Capillary growth does not imply merely a lengthening of individual capillary segments, thus producing wider capillary meshes, but mainly an increase in the number of capillaries. From quantitative data in the rat [22] we know that between birth and 4 months of age, the capillary volume increases 35-fold and the capillary surface area over 20-fold. This implies intensive growth processes at the capillary level, providing ample latitude for a differential growth process to drastically influence septal morphology.

5. Postnatal Intussusceptive Growth of the Capillary System

From the quantitative data and morphological observations that capillary mesh size does not increase with age, we may deduce that new capillary vessels are continually added, as long as

←

Figure 11-5. a: Scanning electron micrograph of a human lung, age 18 months. The mouth of an alveolus is delineated by ropelike reinforcements of the interalveolar septa corresponding to the alveolar entrance ring (arrows). At places, capillary loops extend over the edge of the septa (arrowheads). Bar = 10 µm. b: Detail of capillary network of the alveolar wall. Rat lung, age 44 days; Mercox cast. The cross section illustrates the mature septal morphology with a single capillary network (arrowheads); the top view shows patches of prominent and patches of recessed capillary network areas, suggestive of a process of fusion (arrows). Bar = 10 µm.

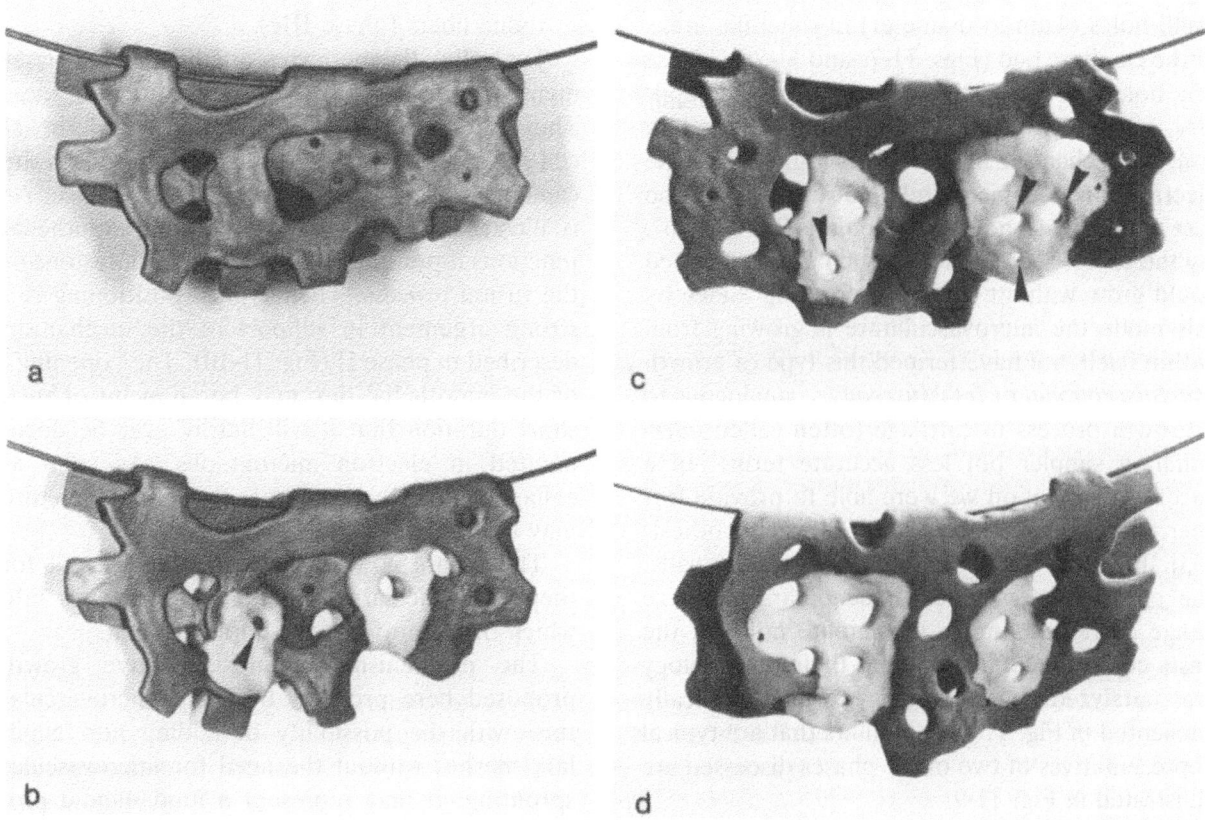

Figure 11-6. Plasticine model of microvascular maturation of the interalveolar septa (original size 35 × 20 cm). The capillary bilayer is suspended on a wire representing bundles of elastic tissue in the alveolar entrance ring (compare with Fig. 11-5a). a: The septum primarily contains a double capillary network with meshes of various sizes. b: The areas shaded in light gray indicate zones of fusion between the two layers. c: The septum has grown and its capillary network has expanded, particularly the fused areas. New capillary meshes have been added (arrowheads) according to the principle of intussusceptive capillary growth illustrated in Fig. 11-7. d: Rear view of Fig. 11-6c showing that the process of preferential growth of the fused areas (light gray) is capable of transforming the internal septal structure.

growth proceeds. Indeed, the ratio of alveolar to capillary surface area remains more or less constant throughout life [16,22,26]. So far, one has tacitly assumed that de novo capillary formation in the lung proceeds by mechanisms similar to those detailed for other systems, both in vivo and in vitro [27–29]. According to this concept, capillaries would form solid endothelial sprouts or hollow protrusions, which grow into the connective tissue, eventually contact existing capillary segments, interconnect, and open up to the circulation. In casts, such vessels commonly appear as abruptly ending vascular segments, sometimes with a bulbous tip [30–32]. Since incomplete filling of capillaries can produce identical pic-

tures, differentiation between the process occurring in vivo and artifacts is not always easy. Blindly ending vessels were observed in all of our casts. However, since they mostly pointed towards each other in pairs, and serial sectioning at the electron microscopic level yielded no evidence of sprout formation, the vascular stumps of our casts most likely represent filling defects, rather than capillary sprouts.

The observations made in our study suggest a completely different mechanism for capillary formation [4]. Almost irrespective of age, the intercapillary meshes varied greatly in size, not only in the interalveolar walls, but also in the perivascular, peribronchial, and subpleural lung regions. In particular, we observed numerous small holes (1 μm in diameter) in sheetlike areas of the capillary bed (Figs. 11-7c and 11-7d). These tiny holes are suggestive of newly formed capillary meshes, which are destined to expand in size, a mechanism illustrated in Fig. 11-7a. The electron microscopic correlate of the holes in the cast would be slender tissue pillars (Fig. 11-7b). By the successive addition of such pillars, the bed could grow without capillary sprouting. Since by this mode the microvasculature is growing from within itself, we have termed this type of growth *intussusceptional* or *intussusceptive*,[3] analogous to a growth process in cartilage (often called *interstitial*, a simpler but less accurate term). In a recent investigation we were able to provide further evidence in favor of the above hypothesis [33]. By serial sectioning of interalveolar walls, the existence of slender tissue pillars in a size range corresponding to the minute holes of the casts could be confirmed, and their morphology was analyzed. The findings are schematically presented in Fig. 11-8, and pillars that are typical representatives of two of the phases discussed are illustrated in Fig. 11-9.

According to the new concept of intussusceptive capillary growth, the formation of a new capillary mesh would proceed as follows:

1. Formation of a transcapillary interendothelial disc-like zone of contact (= endothelial pillar; phase I).
2. Reorganization of the intercellular junctions with a sealing of the capillary lumen around the disc, followed by invasion of the endothelial pillar by interstitial components (phase II; Fig. 11-10).
3. The central core of the interstitial pillar is invaded by a cylindrical cytoplasmic extension of a myofibroblast (phase IIIa). The cell process interconnects the opposite layers of the epithelium by attaching to their respective basement membrane. Bundles of actin filaments run across the cytoplasm and terminate in adhesion plaques at the cell membrane, at sites of attachment to the basement membrane. The interstitial pillar is surrounded by the capillary endothelium.
4. The pillar structure is then successively completed by the addition of pericyte processes covering the lateral walls of the capillaries

(phase IIIb) and by laying down connective tissue fibers (phase IIIc).

Typically, the pericytes are lined by a basement membrane on their contraluminal face. They often "cover" the lateral interendothelial cell junctions, which, as a rule, are present in the capillary wall around the pillar circumference. As is illustrated in Fig. 11-10, the interendothelial junctions represent, so to speak, the tombstone of the fusion process. Their typical positioning is a strong argument in support of the mechanism described in phase II (Fig. 11-10). The "opening" of the endothelial disc may be an event of such short duration that it will hardly ever be documented in electron micrographs. As soon as collagen fibrils are present, the pillar structure may be considered to be stabilized.

The pillars now closely resemble, except for their size, the larger intercapillary meshes into which they transform by simple growth.

The mechanism of intussusceptive growth proposed here provides the lung microvasculature with the possibility of adding new capillary meshes without the need for microvascular sprouting. It may represent a fundamental process of morphogenesis in the pulmonary vascular bed and, perhaps also in other mainly "two-dimensionally" or "surface"-oriented capillary systems. Its discovery can clearly be attributed to the simple observations made in corrosion casts.

Acknowledgments

This work was supported by grant no. 3.492-0.86 from the Swiss National Science Foundation. I express my thanks to Mrs. R.M. Fankhauser, Mrs. B. Krieger, Mr. B. Haenni, and Mr. K. Babl for their excellent technical collaboration.

References

1. Zeltner TB, Burri PH. The postnatal development and growth of the human lung. II. Morphology. *Respir Physiol* 67:269–282, 1987.
2. Burri PH. The postnatal growth of the rat lung. III. Morphology. *Anat Rec* 180:77–98, 1974.
3. Amy RWM, Bowes D, Burri PH, Haines J, Thurlbeck WM. Postnatal growth of the mouse lung. *J Anat* 124: 131–151, 1977.

152

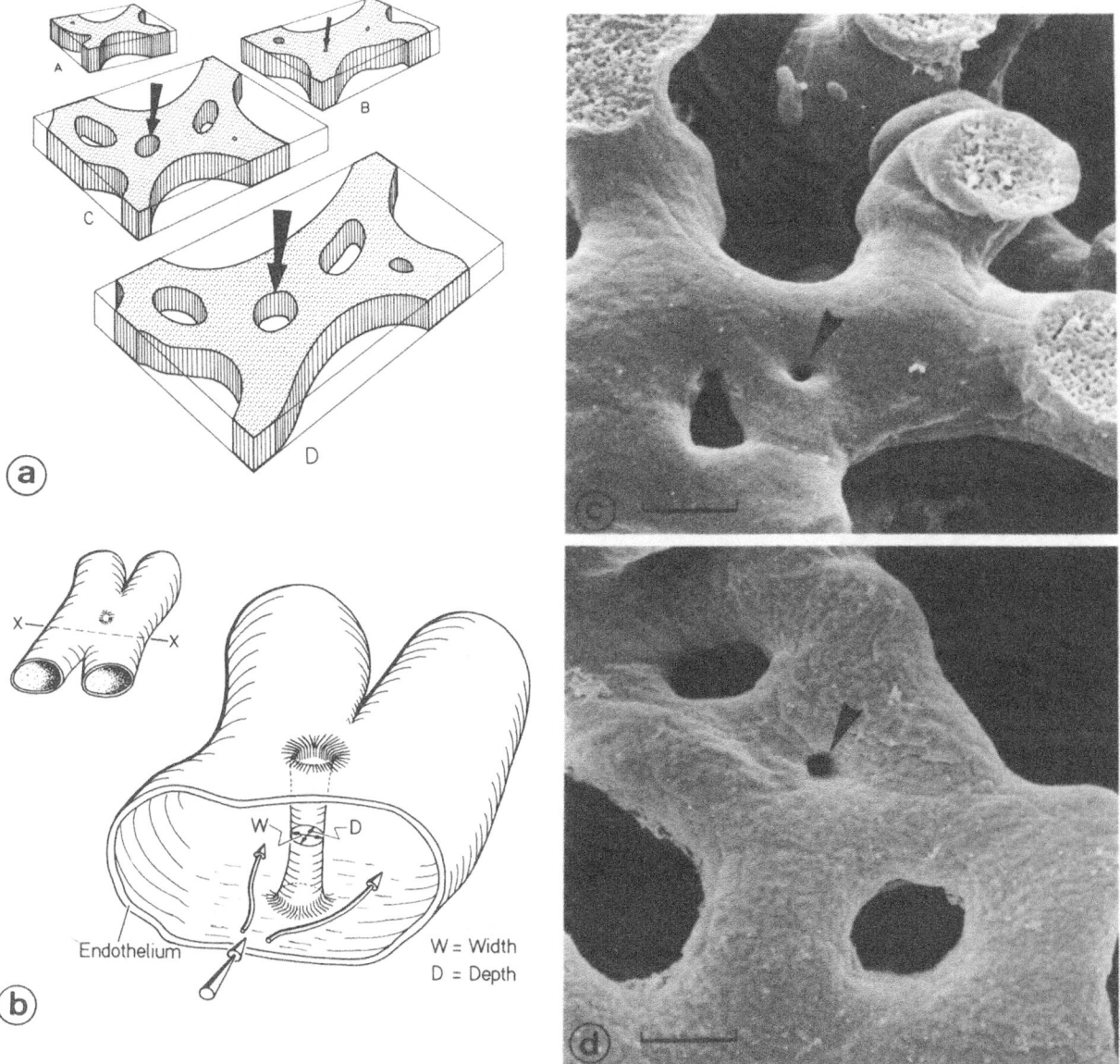

Figure 11-7. Concept of intussusceptive microvascular growth. a: Schematic drawing of casts illustrating the different stages of a growing capillary segment. In an enlarged capillary segment, a new capillary mesh appears as a small hole (arrow in B). The new hole enlarges to form a normal capillary mesh (arrow in C and D). b*: The holes of the casts correspond to tissue pillars through the capillary lumen. With increasing diameter, the pillar will become a capillary mesh. We have arbitrarily defined holes with diameters <2.5 μm as pillars and holes >2.5 μm as capillary meshes**. c,d: Scanning electron micrographs of casts showing extremely small holes (diameter 1 μm; arrowheads) corresponding to newly formed transcapillary tissue pillars. c: Rat lung, 4 days, Bar = 5 μm. d: Rat lung, 44 days. Bar = 5 μm.

* (Reproduced with permission from Caduff et al. [4]).

** (Reproduced with permission from Burri and Tarek [33]).

4. Caduff JH, Fischer LC, Burri PH. Scanning electron microscopic study of the developing microvasculature in the postnatal rat lung. *Anat Rec* 216:154–164, 1986.

5. Burri PH. Development and growth of the human lung. In: Fishman, AP, Fisher AB (ed.), *Handbook of Physiology, Section 3: The Respiratory System*, American Physiological Society, Bethesda, MD, pp 1–46, 1985.

6. Bucher U, Reid L. Development of the intrasegmental bronchial tree: The pattern of branching and development of cartilage at various stages of intra-uterine life. *Thorax* 16:207–218, 1961.

7. Ten Have-Opbroek AAW. The development of the lung in mammals: An analysis of concepts and findings. *Am J Anat* 162:201–219, 1981.

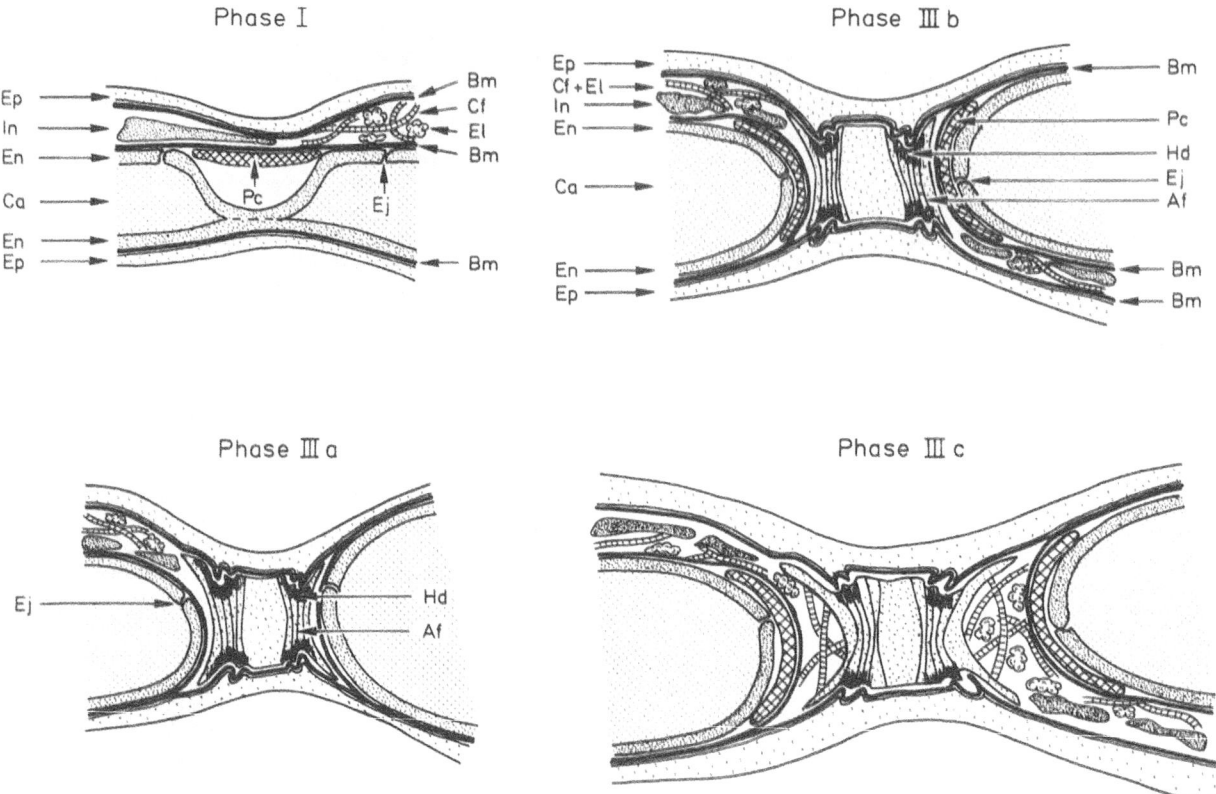

Figure 11-8. This schematical illustration of the intussusceptive capillary growth concept is based on the four phases (phases I, IIIa–IIIc) documented so far by electron microscopic investigation of serial sections. Following the formation of a transcapillary interendothelial zone of contact (phase I), the interendothelial junctions are reorganized and the endothelium gives way to the components of the interstitium (= phase II, not documented thus far by electron microscopy and not illustrated here). The tissue post is then consecutively invaded by a cytoplasmic extension of a myofibroblast (phase IIIa), by pericytic processes (phase IIIb), and finally by collagen fibrils (phase IIIc). The myofibroblast attaches to the basement membranes of the epithelium and shows bundles of actin filaments running across its cytoplasm. The pericytes may play a role in phase I, where they are roofing over the endothelial infolding, and in phases IIIb and IIIc, where they are often found covering the lateral interendothelial junctions. Af = actin filaments; Bm = basement membrane; Ca = capillary lumen; Cf = collagen fibrils; Ec = erythrocyte; Ej = endothelial cell junction; El = elastin; En = endothelium; Ep = epithelium; Hd = hemidesmosomelike structure or attachment body; In = interstitium; Pc = pericyte. (Reproduced with permission from Burri and Tarek [33]).

8. Boyden EA. Development and growth of the airways. In: Hodson WA (ed.), *Lung Biology in Health and Disease. Development of the Lung.* Marcel Dekker, New York, pp 3–35, 1977.

9. Gil J, Reiss OK. Isolation and characterization of lamellar bodies and tubular myelin from rat lung homogenates. *J Cell Biol* 58:152–171, 1973.

10. Burri PH. Fetal and postnatal development of the lung. *Ann Rev Physiol* 46:617–628, 1984.

11. Dubreuil G, Lacoste A, Raymond R. Observations sur le développement du poumon humain. *Bull Histol Tech Microsc* 13:235–245, 1936.

12. Davies G, Reid L. Growth of the alveoli and pulmonary arteries in childhood. *Thorax* 25:669–681, 1970.

13. Langston C, Kida K, Reed M, Thurlbeck WM. Human lung growth in late gestation and in the neonate. *Am Rev Respir Dis* 129:607–613, 1984.

14. Dunnill MS. Postnatal growth of the lung. *Thorax* 17:329–333, 1962.

15. Emery JL, Wilcock PF. The postnatal development of the lung. *Acta Anat* 65:10–29, 1966.

16. Zeltner TB, Caduff JH, Gehr P, Pfenninger, J, Burri PH. The postnatal development and growth of the human lung. I. Morphometry. *Respir Physiol* 67:247–267, 1987.

17. Weibel ER. *Morphometry of the Human Lung*, Springer-Verlag, Heidelberg, 1963.

18. Angus GE, Thurlbeck WM. Number of alveoli in the human lung. *J Appl Physiol* 32:483–485, 1972.

19. Hansen JE, Ampaya EP. Lung morphometry: A fallacy in the use of the counting principle. *J Physiol* 37:951–954, 1974.

20. Sterio DC. The unbiased estimation of number and sizes of arbitrary particles using the dissector. *J Microsc* 134:127–136, 1984.

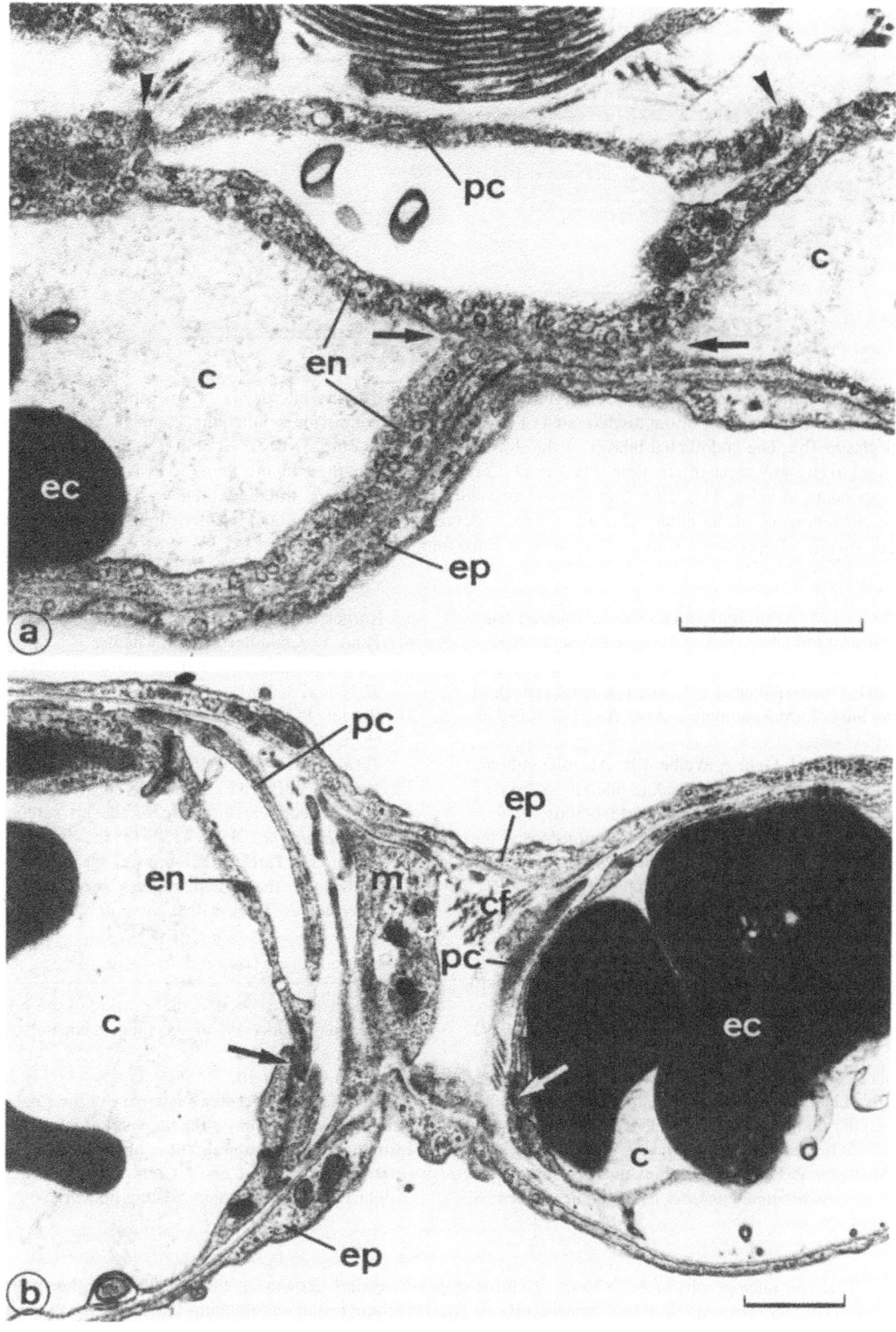

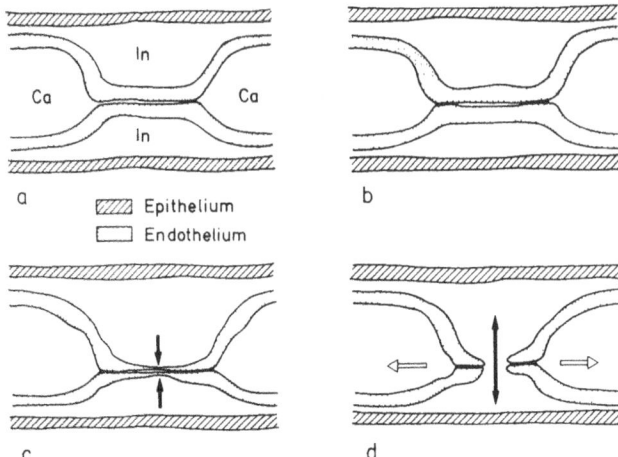

Figure 11-10. Schematic presentation of phase II events starting from phase I morphology (a). Ca = capillary; In = interstitium. In preparing the central perforation of the disclike area of interendothelial contact, new junctional complexes are formed to prevent leakage of plasma (b). The endothelial bilayer of the disc is then thinned out (arrows in c), and it finally breaks up and retracts (white arrows) to give way to the interstitium (black double arrow). The interstitial pillar is then successively invaded by cellular and fibrillar components, as in Fig. 11-8. Note that with the mechanism presented here, interendothelial cell junctions must be found all around the circumference of the pillar. This was indeed observed in all the pillars and smaller capillary meshes investigated by electron microscopy. (Reproduced with permission from Burri and Tarek [33].)

21. Cruz-Orive LM. Arbitrary particles can be counted using a dissector of unknown thickness: the selector. *J Microsc* 145:121–142, 1986.

22. Burri PH, Dbaly J, Weibel ER. The postnatal growth of the rat lung. I. Morphometry. *Anat Rec* 178:711–730, 1974.

23. Gil J, Bachofen H, Gehr P, Weibel ER. Alveolar volume-surface area relation in air- and saline-filled lungs fixed by vascular perfusion. *J Appl Physiol* 47:990–1001, 1979.

24. Weibel ER, Gil J. Structure-function relationships at the alveolar level. In: West JB (ed.), *Bioengineering Aspects of the Lung*, Marcel Dekker, New York, pp 1–81, 1977.

25. Miller WS. *The Lung*, 2nd ed., Charles C Thomas, Springfield, IL, pp 74–88, 1947.

26. Weibel ER. *The Pathway for Oxygen. Structure and Function in the Mammalian Respiratory System*. Harvard University Press, Cambridge, MA, pp 1–425, 1984.

27. Folkman J. Angiogenesis: Initiation and control. *Ann NY Acad Sci* 401:212–227, 1982.

28. Nicosia RF, Tchao R, Leighton J. Histotypic angiogenesis in vitro: Light microscopic, ultrastructural, and radio-autographic studies. *In Vitro* 18:538–549, 1982.

29. Simpson JG, Fraser RA, Thompson WD. Angiogenese und Angiogenese-Faktoren. In: Messmer K, Hammersen F (eds.), *Struktur und Funktion endothelialer Zellen*. S.

Karger, Basel, pp 80–96, 1983.

30. Tano Y, Chandler DB, Machemer R. Vascular casts of experimental retinal neovascularization. *Am J Ophtalmol* 92:110–120, 1981.

31. Burger PC, Chandler DB, Klintworth GK. Scanning electron microscopy of vascular casts. *J Electron Microsc Tech* 1:341–348, 1984.

32. Yoshida Y, Ikura F, Watabe K, Nagata T. Developmental microvascular architecture of the rat cerebellar cortex. *Anat Embryol* 171:129–138, 1985.

33. Burri PH, Tarek MR. A novel mechanism of capillary growth in the rat pulmonary microcirculation. *Anat Record*, 228:35–45, 1990.

---

[1] Parenchyma is defined as that portion of the organ supporting the main function; in the lung it encompasses the gas-exchanging structures.

[2] See also Chapter 10.

[3] In the Merriam Webster Dictionary of the English language, *intussusception* means "the deposition of new particles of formative material among those already embodied in a tissue or structure." In our case, it refers to the formation of new capillary meshes within an existing network.

←

Figure 11-9. Electron micrographs of two phases of intussusceptive capillary growth. a: Phase I: Disclike area of transcapillary interendothelial contact (arrows). Note the pericytic process (pc) attaching to the endothelium (arrowheads). c = capillary; ec = erythrocyte; en = endothelium; ep = epithelium of type I. Bar = 1 μm. b: Phase IIIc: Fully developed transcapillary interstitial tissue pillar containing axially a cell process of a myofibroblast (m) interconnecting the opposite epithelial layers (ep), pericytes (pc) adjacent to the capillary walls, and a few collagen fibrils (cf). The pillar diameter is still below 2.5 μm. Note the location of the interendothelial cell junctions (arrows), which is typical for all the phase III pillars. c = capillary; ec = erythrocyte; en = endothelium. Bar = 1 μm.

156

Author's address:
Prof. Peter H. Burri
Institute of Anatomy
Department of Developmental Biology
University of Bern
Bühlstrasse 26
Postfach 139
CH-3000 Bern 9
Switzerland

CHAPTER 12

Microvasculature of Bone and Bone Marrow

FUMIHIKO IWAKU

1. Introduction

The blood vascular architecture of bone, until lately, has mainly been investigated by various methods using long bones. These methods have been the India-ink injection method, the latex injection replica technique, angiography, and others. Many features of the blood circulation and bone vascular architecture have been clarified, but the fine three-dimensional vascular architecture remains inadequately understood [1–12].

In 1952, a technique involving the injection of a proper volume of methacrylate resin solution into blood vessels to target organs was improved by Taniguchi et al. to study the microvascular architecture of various tissues [13,14]. Using this technique, corrosion cast preparations of the fine three-dimensional blood vascular architecture in bone were produced.

Recently, an improved injection technique [15–17], using the newly developed methacrylate resin solution Mercox resin, has provided new information about the bone blood vascular architecture as observed by SEM [18–25].

The present study was carried out to elucidate (1) the three-dimensional architecture of the microvascular system of long and flat bone surfaces, (2) the relationship between the microvascular distribution of the bone surface and bone

remodeling, and (3) the microvascular architecture of the bone marrow of the long bone.

2. Materials and Methods

Wistar-strain mature male rats weighing 260–350 g were used. Thoracotomy and laparotomy were conducted at the same time under ether anaesthesia and the right atrium was opened [21, 24,25]. The blood vessels were perfused with saline (36°C) through the left ventricle and/or abdominal aorta for 1 or 2 minutes. To facilitate the perfusion, a vasodilator (Sunar agent: Oken Chemical Co., Ltd.) was added to the saline (3/100 mg in 500 ml). A suitable volume, about 5 ml, of Mercox resin solution (Dainippon Ink Chemical, Inc.) was subsequently injected into the blood vascular system (21,24,25). After 15 minutes, when the resin solution was fully polymerized at room temperature, the adhering soft tissues were carefully removed using a magnifying glass. Immediately thereafter, the extracted specimens were dipped into a 10.0–30.0% NaOCL solution for about 2 hours or soaked in a 10.0–15.0% NaOH or KOH solution for 24 hours to remove any remaining minute soft tissue by corrosion. They were then thoroughly washed in tap water to obtain a vascular cast preparation and, in some cases, they were

Motta, P.M., Murakami, T., and Fujita, H. (eds.), Scanning Electron Microscopy of Vascular Casts: Methods and Applications.

158

cleaned in distilled water by ultrasonic washing.

To prepare bone medullary microvascular casts, preparations including bone material were dipped again in a 5.0% trichloroacetic acid solution for 7 days to completely decalcify the bone material, followed by washing in tap water and, soaking in the corrosive solution for about 2 or 3 hours to remove the decalcified bone material [25].

After air drying, the specimens were coated with gold and observed at 20 kV using a Hitachi SSM II type SEM.

3. Results

3.1. Identification of Cast Blood Vessels

The criteria for classification of cast microvascular vessels were based upon (1) the inside diameter of a branch, (2) a replica picture of an endothelial cellular nucleus, (3) the impression of the vascular smooth muscle, and (4) the vascular running form [18,19,21,24,25]. The following vessels were identified in the bone surface of long and flat bones: (1) small, round arteries of about 30–50 μm in diameter, resembling smooth muscle; (2) precapillary arteries, about 20–30 μm in diameter, similar to (1) but finer; (3) capillaries, about 5–12 μm in diameter, closely resembling an endothelial cellular nucleus and very fine; (4) postcapillary venules about 20–50 μm in diameter, with indistinct weak impressions of smooth muscle and appearing to be flat. The large, round vessels that strongly resembled vascular smooth muscle ran a straight couse and belonged to the arterial system. In contrast, the small or elliptical branches were generally strong and/or weak serpiginous running and slightly resembled vascular smooth muscle, and were classified as capillaries and/or postcapillary venules.

The nomenclature that has been used for bone medullary vessels has varied among different authors. In this chapter, Ono's nomenclature is most generally used [18,19].

3.2. Microvasculature of Long Bone

It has been known that a very complex network, consisting of small arteries, precapillary arteries, capillaries, and postcapillary venules, lies in the periosteum of long bone. This vascular network is supplied with branches of nutrient vessels, vessels ramifying from those in muscle attachments, diaphyseal and metaphyseal vessels, and/or vessels arising from various vascular foramina of bone, such as Volkman's canal [26–29].

Two vascular layers of blood vessels have been found in the diaphyseal periosteum [21]. The outer layer lay on the fibrous coat is composed of ramified large vessels and their branches. These vessels derived from either nutrient vessels or vessels originating from muscle attachments (Figs. 12-1b, 12-1c, and 12-2a). As in the case of tibia, these vessels run like cutaneous veins to form a rough network of anastomosing branches. Some of them run into Volkman's canals. Postcapillary venules and a large number of small branches originate from a network of small arteries and veins present in the outer layer. Many of these branches ramify with the capillaries in the adjacent inner layer. The arterioles and precapillary arteries run straight everywhere and/or curve sharply (Figs. 12-1b, 12-1c, and 12-2a), while many venous branches, including the postcapillary venules, show various running patterns (Figs. 12-1b, 12-1c, and 12-2a).

The inner vascular layer is in the osteogenetic zone [21]. Its vessels form a capillary bed consisting of an irregular fine network of nutrient capillaries reaching the adjacent osteoblasts (Figs. 12-1b and 12-1c). In the area corresponding to the posterior mesial bone surface of the diaphysis of rat tibia, complex and irregular vascular networks are present.

Figure 12-1. Two layers of microvascular architecture are evident on the bone surface of the posterior tibial (b, c). The inner layer mainly consists of capillaries (CA), adhering to the bone surface (B) (b, c), and capillaries running longitudinally along the long axis of the tibia (b, c). The outer layer consists of precapillary arteries (PCA), postcapillary venules (PCV), and small veins (SV) (c). The meandering capillaries (CA) appear on the bone surface (B) (b, c). Arrowheads show capillaries embedded in the bone substance (b, c). The large and shallow forming lacunae (FL) are scattered on the bone-forming surface (FS) (a, b).

159

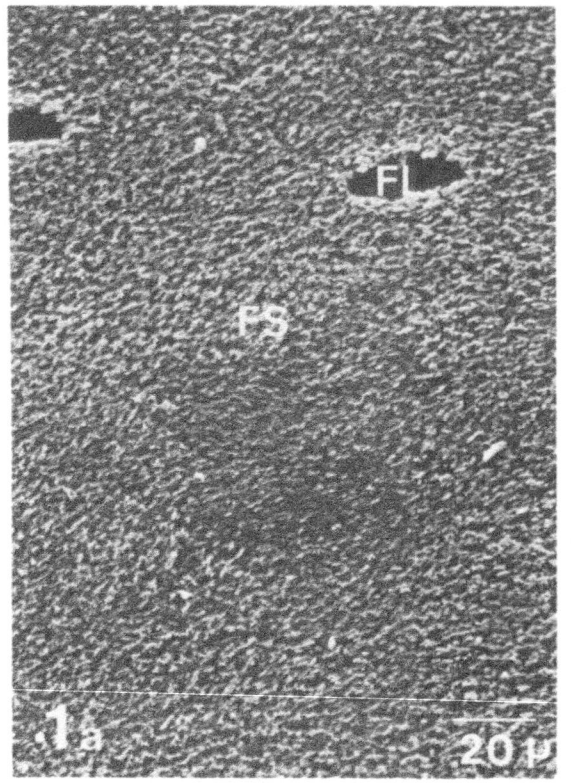

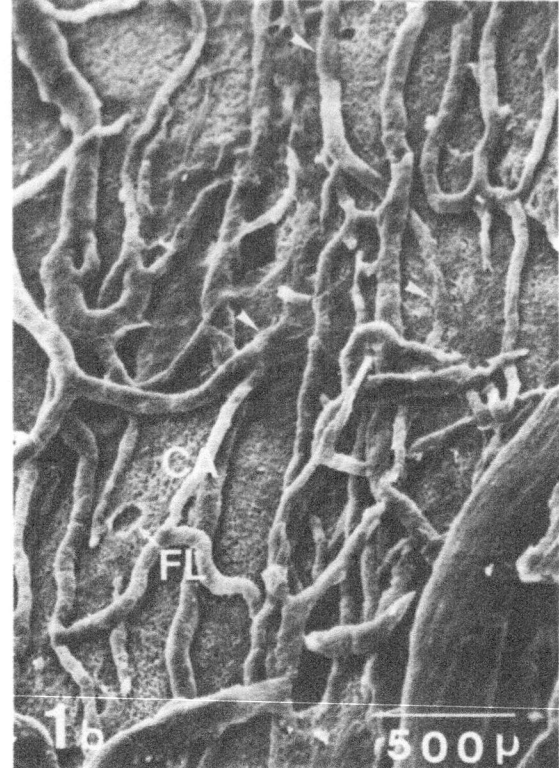

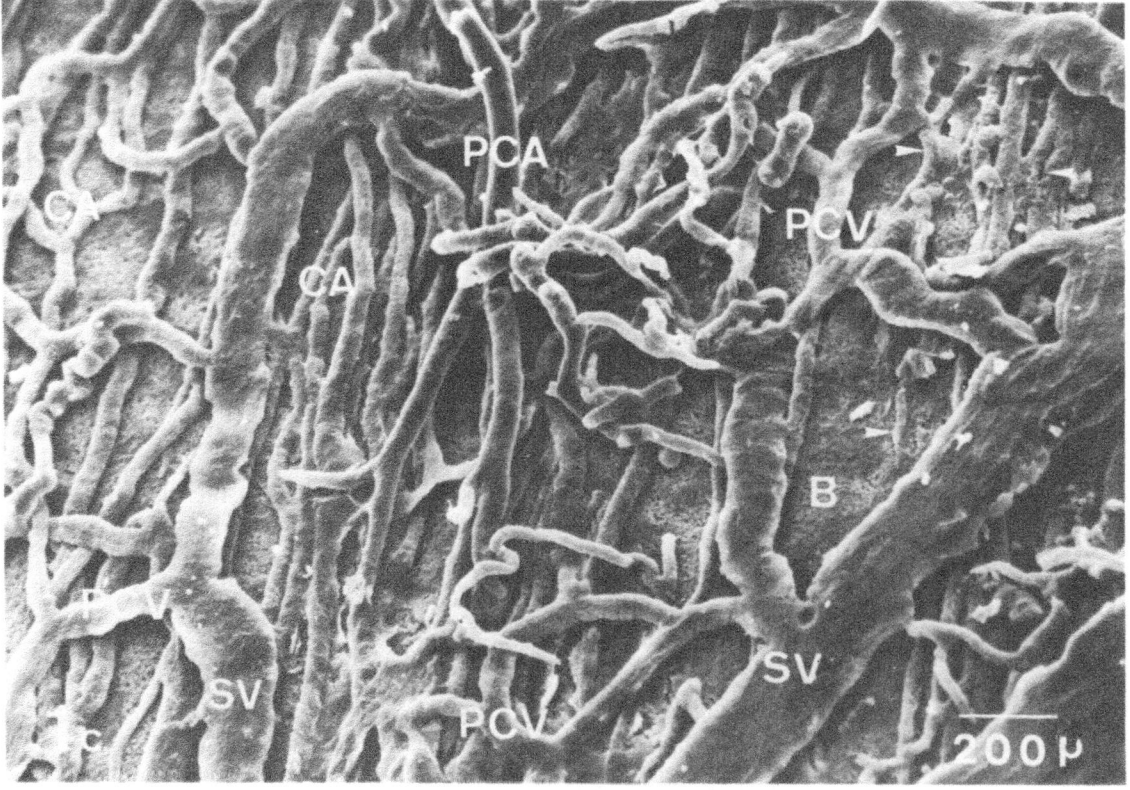

160

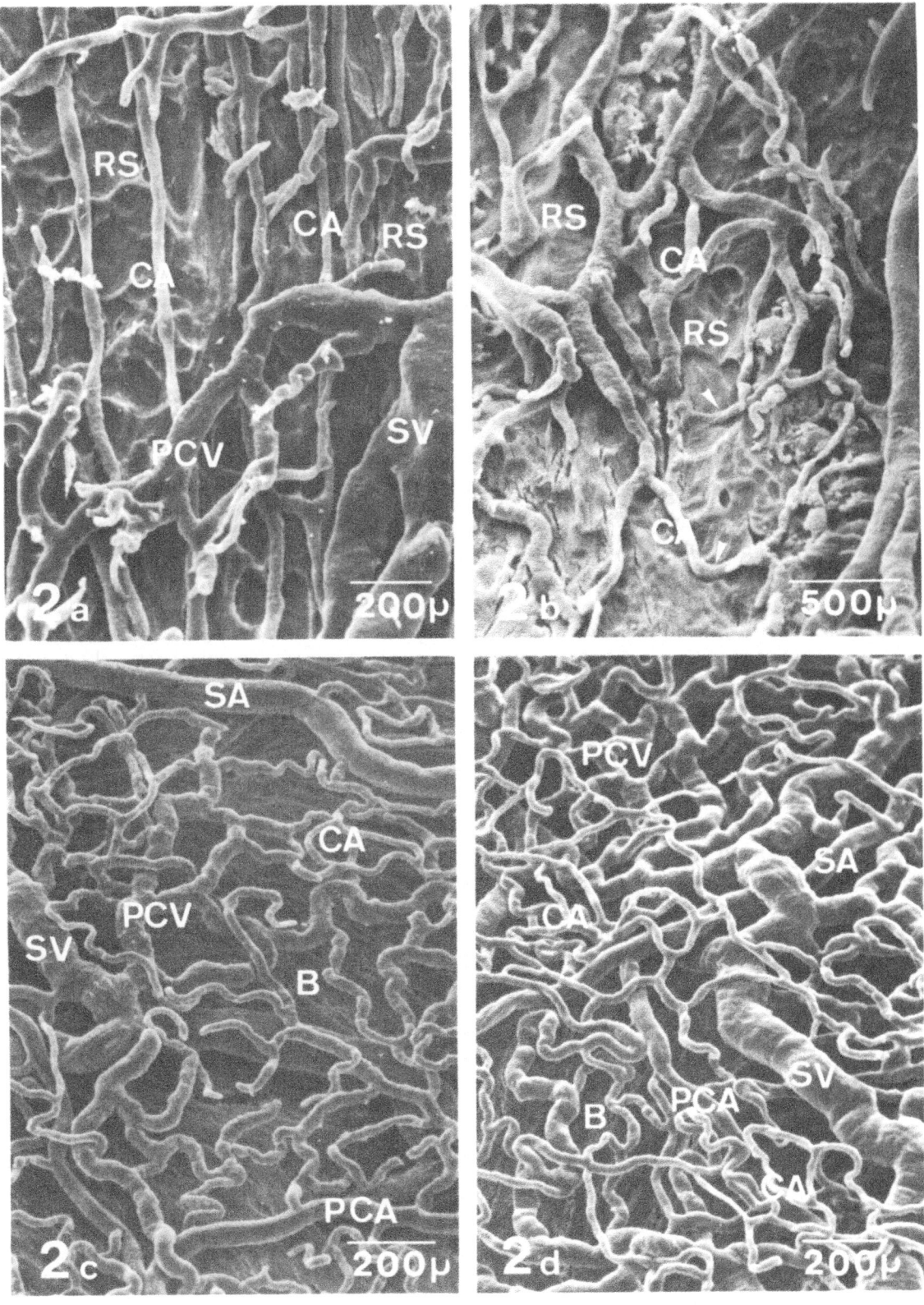

These outer and inner vascular layers consist of many branches arising directly from nutrient blood vessels and from other muscle branches of tibia blood vessels. Among these blood vessels there are found (1) small arteries, running in a straight course, with a diameter of about 30–40 μm round in shape, and resembling endothelial cells; (2) precapillary arteries, similar to small arteries, round, and about 10–30 μm in diameter; (3) capillaries of about 5–10 μm in diameter, sometimes running in a line and at other times curved; and (4) postcapillary venules with curved patterns and a diameter of about 10–30 μm having a flat replica. Small veins, which run the same as postcapillary venules, are about 30–60 μm in diameter.

The network of the inner layer is mostly composed of many capillaries, and sometimes precapillary arteries and postcapillary venules. The features of the vascular architecture of the inner layer are discussed below for flat bone and rough bone surface [21].

3.3. Microvasculature of Flat Long-Bone Surface

The vascular network, consisting mainly of capillaries running along the long axis of the bone, but in some places sharply curved, is composed of irregular, fine, and rough networks (Fig. 12-1c). This network is distributed close to the bone surface (Fig. 12-1b). In several zones of the bone surface, the capillaries are embedded in the bone substance (Figs. 12-1b and 12-1c). The bone surface under this network is rather flat everywhere, and scattered bone-forming lacunae of osteoblasts [31,32,34] are seen on this surface (Figs. 12-1a and 12-1b). Thus the network in the active stage of bone formation is formed by irregular and varied networks of capillaries adhering closely to the bone surface (Fig. 12-1).

3.4. Microvasculature of Rough Long-Bone Surface

The vascular network consists mainly of capillaries and in some places has postcapillary venules, which lay apart from the bone surface (Fig. 12-2a). The network composed by rough meshes of longitudinally running capillaries along the long axis of the bone is a single layer and is distributed adjacent to the rough bone surface, where many Howship's lacunae of various sizes and shapes are found (Figs. 12-2a and 12-2b). In several Howship's lacunae capillaries from bone substance during bone resorption are seen (Fig. 12-2b).

3.5. Microvasculature of Flat Bone

The microvascular architecture of the sagittal lateral region beyond the temporal muscle line of the parietal bone has been observed in adult Wistar strain rats. The bone surface is almost smooth, and collagenous bundles running almost parallel to each other are seen everywhere. The single layer of vascular network is composed of the following elements [24]: (1) precapillary arteries, about 15–30 μm in diameter, branching from peripheral arterioles of the superficial temporal artery; (2) capillaries, about 6–12 μm in diameter, originating from these precapillary arteries; (3) postcapillary venules, about 15–40 μm in diameter, bifurcating from the peripheral branches of the superficial temporal vein; (4) branches, about 30–60 μm in diameter, of small veins that emerge outward from and/or enter into the vascular foramen of the bone, and drain into the superficial temporal vein. These vascular vessels show curved running patterns and lay apart from the bone surface. The meshes of this network are irregular and vary in size,

Figure 12-2. The straight capillaries (CA) run longitudinally apart from the bone resorbing surface (RS) along the long axis of the tibia (a) and enter the postcapillary venules (PCV) (a). Arrowheads indicate that the meandering capillaries, facing the bone-resorbing surface (RS), are working their way out of the resorbing lacunae (b). SV = small vein (a). A single layer of microvascular architecture faces the lateral surface of the parietal bone (c, d). The considerably and slightly curved running capillaries (CA), precapillary arteries (PCA), and postcapillary venules (PCV) constitute an irregular rough network (c) and a fine network (d). Both are distributed away from the flat bone surface (B) (c, d).

being rough in some places and fine in others (Figs. 12-2c and 12-2d).

3.6. Vascular Network and Bone Remodeling

The relationship between the bone surface and the stage of bone remodeling has already been reported [30–32, 34]. The following characteristics of the bone surface in the active stage of bone formation have been described: (1) a large number of shallow and large osteoblast lacunae with rough margins, named *forming lacunae* by Boyde; (2) an almost flat and coarse surface having no bone-absorptive cavities, like *Howship's lacunae*; and (3) bundles of collagen fibers running in various directions. In the bone-absorptive stage of bone remodeling a concave surface having many bone-absorbing cavities of various sizes, called *Howship's lacunae*, were noted. Finally, in the resting stage of bone remodeling, the following structures were observed: (1) small and deep or shallow osteocytic lacunae possessing rough and/or smooth margins, (2) flat and smooth bone surface having no bone-absorbing cavities, and (3) bundles of collagen fibers running almost parallel. The parallel and annular lamellar arrangements of collagenous bundles noted around the vascular foramen may possibly be an indication of advanced calcification.

Iwaku and Ozawa [21] described a single network in the absorptive bone surface of tibia. This consists of rough meshes of longitudinally running capillaries that are distinct from the surface. The network of bone surface in the active stage of bone formation, consisting of irregular, rather fine, and in some places, rough, meshes adhers closely to the bone surface [21]. The network of the bone surface in the resting stage of bone remodeling, and/or in the further advanced stage of calcification, is a single structure consisting of irregular rough and sometimes fine meshes, which lay apart from the bone surface, as described for smooth bone surface of the parietal bone [24].

It would thus appear that the form and position of the vascular network shows a remarkable variability depending on the metabolic activity of bone tissue. It has also been reported that the vascularization and vascular disappearance of alveolar bone are always dependent on physical external forces [20,22,23]. In fact, the disappearance of the vascular network was observed on the bone surface due to corrective dental force wherein bone resorption had occurred. Furthermore, vascularization and the formation of a vascular network were found on the bone surface at the stage of bone formation. These findings indicate that the disappearance of the vascular network and its rough distribution lead to bone resorption, while vascularization and an irregular dense distribution lead to osteogenesis.

3.7. Microvasculature of Bone Marrow

In this section we describe the microvascular architecture of bone marrow, which has been investigated in the central part of femoral diaphysis of Wistar strain rats by SEM (Figs. 12-3–12-5). The following vascular features were noted [18,19,25,33]: (1) In the arterial system, the ascending and descending branch arising from the nutrient artery in the central part of diaphyseal bone marrow are the main trunks. These supply many small arteries and precapillary arteries, which branch from small arteries and pass among many sinusoids to peripheral regions of the bone marrow. (2) In the venous system are found many central veins formed by the fusion of sinusoidal veins, which flow into large central veins and finally enter the largest ascending and descending veins, which continue to nutrient veins in the central part of bone marrow. (3) Many sinusoids form a more complex three-dimensional vascular architecture by anastomosing with each other. After small arteries, about 30–40 μm in diameter, branch from the main ascending branch, they run obliquely in a tortuous path among the

→

Figure 12-3. The blood vascular architecture of the bone marrow is shown in the longitudinal (a) and cross section (b and Fig. 12-4c) preparations of the femur diaphysis. The ascending artery (AA) from the nutrient artery runs spirally and longitudinally in the central region of the bone marrow (a, b), and winding small arteries (SA) branching from the (AA) run obliquely to the cortical bone (B) among the sinusoids (a, b). Numerous sinusoids (SiV) constitute a coral-like unit (large arrowhead) and enter the central veins (CV) (a, b). The small arrowhead points to the vascular network facing the bone surface (a).

163

164

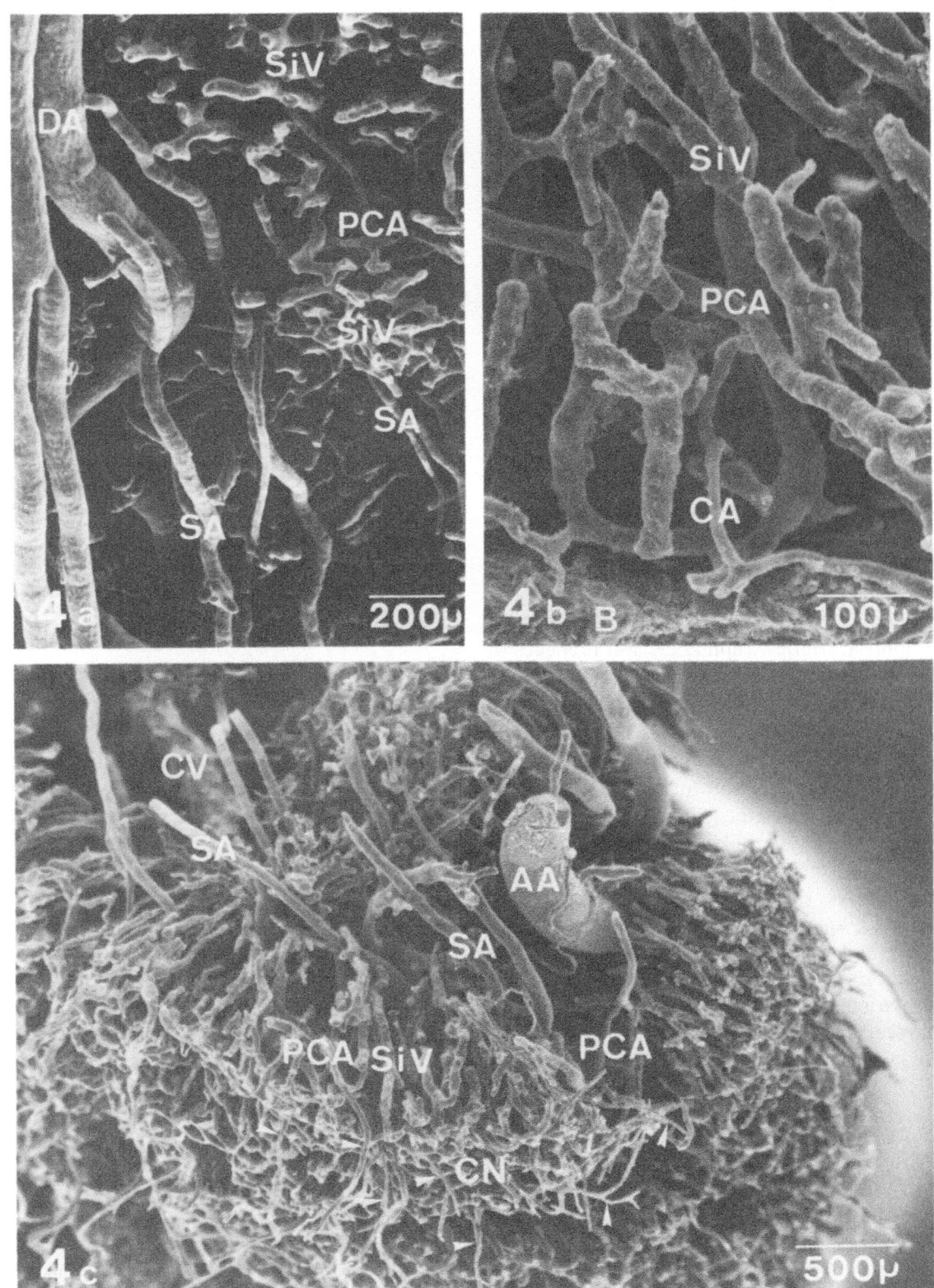

sinusoids toward the peripheral region of the bone marrow (Figs. 12-3 and 12-4). Precapillary arteries, about 20–30 μm in diameter, arise arborescently from the small arteries and run toward the cortical bone surface among the sinusoids (Fig. 12-4). They give rise to capillaries of about 7–12 μm in diameter near the cortical bone surface (Figs. 12-4 and 12-5b). The capillaries anastomose with peripheral branches of sinusoids of about 20–30 μm in diameter, and these vascular branches form a complex coral-like vascular architecture, with rough and varied meshes (Figs. 12-4c, 12-5a, and 12-5b). Near the cortical bone surface, this architecture becomes a complex network of irregular, rough and fine meshes consisting of capillaries and sinusoids anastomosing with each other (Figs. 12-4b, 12-4c, and 12-5b). In some places, this network protrudes into the central region of the bone marrow along trabeculae, in accordance with the shape of the cortical bone surface (Figs. 12-3a and 12-5a). No great differences in the meshes of the network could be found for the flat, concave, or absorbent surfaces (Fig. 12-5). Small arteries, precapillary arteries, capillaries, and branches of sinusoids enter Volkman's canals after running obliquely, tortuously, and/or in parallel near the cortical bone surface (Figs. 12-4c, 12-5, and 12-6b). The sinusoids combine with each other to form a coral-like structure of branches of about 30–40 μm in diameter, show a conic three-dimensional labyrinthlike structure, and grow gradually to the center of the bone marrow to form venous sinuses of about 40–120 μm in diameter, and finally, become main venous sinuses and then the central vein (Figs. 12-3a and 12-3b). These sinusoidal veins enter various concave regions delimited by various trabeculae and recessed areas of large cortical bones to form units, which drain into these regions and areas.

3.8. Microvasculature of Metaphysis and its Distal Region

The microvasculature of these areas consists of many vascular branches of small arteries and veins, sinusoids, precapillary arteries, capillaries, and postcapillary venules, and is similar to the architecture of diaphysis, without sinusoid running. The small arteries, arising like the twigs of a tree from the descending branch in the central region of the bone marrow, run longitudinally and obliquely to the epiphyseal part, and while running among the sinusoids they branch into several precapillary arteries near the distal epiphyseal regions (Figs. 12-6a and 12-6c). In the epiphyseal regions, numerous sinusoids are arranged longitudinally, and their distal parts become enlarged (Figs. 12-6c and 12-6d). The form of the sinusoids and their distribution appear to promote the blood circulation of the bone marrow for the growth of cartilaginous bone, but the microvascular architecture of this region is still not well known (Figs. 12-6a, 12-6c, and 12-6d).

4. Concluding Remarks

The three-dimensional microvascular architecture of long and flat bone has been classified based on SEM observations of microvascular casts produced by corrosion casting in which Mercox resin solution is injected into blood vessels. The irregular and complex network possessing fine, and in some places rough, meshes of capillaries is closely distributed on the bone surface of the diaphysis of rat tibia. On the outside of this network, a rough network consisting of many vascular branches of small arteries and veins, precapillary arteries, and postcapillary venules

←

Figure 12-4. The microvascular architecture of the bone marrow of the diaphysis is shown in the femur preparation. The bone material of the cast preparation of Figs. 12-4a and 12-4c was previously removed by the NaOCl-corrosion technique. A small artery (SA in a) from the descending artery (DA) passes among the sinusoids (SiV) and approaches the peripheral region of the bone marrow (a). The precapillary artery (PCA in b) bifurcates from the small artery and runs distally through numerous sinusoids, finally approaching the surface of the cortical bone (B) (b). Small arrowheads indicate blood vessels entering the cortical bone (c and Fig. 12-5).

166

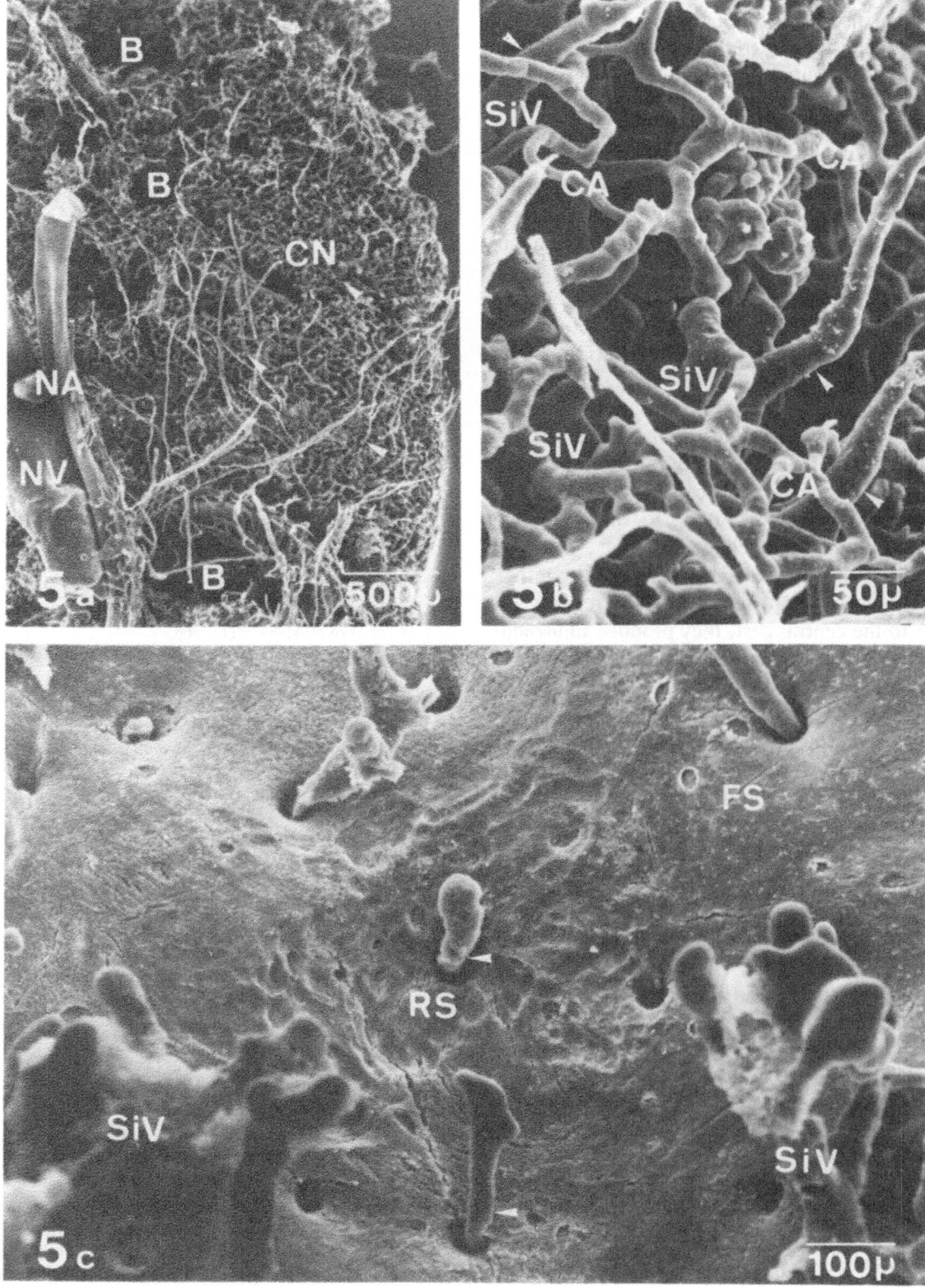

is present. In the meshes of the inner capillary network are observed more changes, determined by the condition of the bone surface. For example, on the bone formative surface, capillary meshes are irregular, finer, and/or in some places rough and adhere to the surface. On the bone-absorptive surface, the meshes become more rough and are distributed apart from the surface. However, the capillaries that produced these meshes run mainly longitudinally along the long axis of the bone. The microvasculature of the flat bone surface, such as rat parietal bone, is formed by a single layer of network possessing irregular, rough and/or finer meshes, consisting of small arteries and veins, precapillary arteries, capillaries, and postcapillary venules, lying apart from the surface. This bone surface is smooth and has no bone-absorptive cavities on its furface. These results indicate that the capillary bed on the bone surface may be remarkably capable of adapting to the nutrient supply according to the metabolic activity in bone remodeling.

In the bone marrow, sinusoids gather together their branches supplying peripheral regions. In route to the central part, they produce sinusoidal units having a conical three-dimensional labyrinthlike structure that perfuses each area of various concave regions delimited by many trabeculae and grows gradually to the central part of the bone marrow. Several trunks of the sinusoids draining into this constituent unit combine to form a large venous sinus. By adding trunks from other regions, this becomes the main larger venous sinus and as such enters the central vein. Near the cortical bone surface, capillaries and branches of sinusoids anastomose with each other and form an irregular complex network having various rough and/or fine meshes. However, remarkable differences in the meshes could not be found on flat or absorptive bone surfaces.

References

1. Nelson GE Jr, Kelly PJ, Peterson LFA, Janes JM. Blood supply of the human tibia. *J Bone Joint Surg* 42A: 625–636, 1960.
2. Vasciaveo F, Bartoli E. Vascular channels and resorption cavities in the long bone cortex of the bovine bone. *Acta Anat* 47:1–33, 1961.
3. Morgan JD. Blood supply of growing rabbit's tibia. *J Bone Joint Surg* 41B:185–203, 1959.
4. Simpson AHRW. The blood supply of the periosteum. *J Anat* 140:697–704, 1985.
5. Cohen J, Harris WH. The three dimensional anatomy of Haversian system. *J Bone Joint Surg* 40A:419–434, 1958.
6. Treuta J, Harrison MHH. The normal vascular anatomy of the femoral head in adult man. *J Bone Joint Surg* 35B:442–461, 1953.
7. Brookes M. Femoral growth after occlusion of the principal nutrient canal in day-old rabbits. *J Bone Joint Surg* 39B:563–571, 1957.
8. Treuta J, Cavadias AX. A study of the blood supply of the long bones. *Surg Gynecol Obstet* 118:485–498, 1964.
9. Brookes M. The vascular architecture of tubular bone in the rat. *Anat Rec* 132:25–47, 1958.
10. Brookes M, Elkin AC, Harrison GR. A new concept of capillary circulation in bone cortex. *Lancet*:1078–1081, 1961.
11. Rhinelander FW. The normal microcirculation of diaphyseal cortex and its response to fracture. *J Bone Joint Surg* 50A:784–811, 1968.
12. Kelly PJ. Anatomy, physiology and pathology of the blood supply of bones. *J Bone Joint Surg* 50A:766–776, 1968.
13. Taniguchi Y, Ohta Y, Tajiri S. New improved method for injection of acrylic resin. *Okajimas Fol Anat Jpn* 24:259–267, 1952.
14. Taniguchi Y, Ohta Y, Tajiri S. Supplement to new improved method for injection of acrylic resin. *Okajimas Fol Anat Jpn* 27:401–406, 1955.
15. Naito I. The development of glomerular capillary tufts of the bullfrog kidney from a straight interstitial vessel to an anastomosed capillary network. A scanning electron microscopic study of vascular casts. *Arch Histol Jpn* 47:411–456, 1984.
16. Ohtani O, Ohtsuka A. Three-dimensional organization of lymphatics and their relationship to blood vessels in rabbit small intestine. A scanning electron microscopic study of corrosion casts. *Arch Histol Jpn* 48:255–268, 1985.

Figure 12-5. Cast preparations (a, b) of the microvascular system of the bone marrow can be seen from the cortical bone side in the central part of the femur diaphysis. Following decalcification of the bone, this material was removed by the same corrosion technique (a, b). Arrowheads indicate blood vessels stemming not only from precapillary arteries, but also peripheral branches of sinusoids (SiV in b and c and in Fig. 12-4c), that are entering vascular canals of the cortical bone. The arrowheads indicate branches passing through meshes of the network (CN in a and Fig. 12-4c) composed of capillaries (CA) and sinusoids (SiV) (a, b). NV is nutrient vein and NA is nutrient artery. The void places (B in a) are areas of bones of the trabeculae that were previously removed (a). Flat (FS) and resorbed (RS) surfaces of the cortical bone (c) are evident in the bonc area facing the vascular network consisting of capillaries (CA in b) and peripheral branches of sinusoids (SiV in b) which will subsequently be removed (c). Arrowheads indicate peripheral branches of sinusoids entering the cortical bone (c).

168

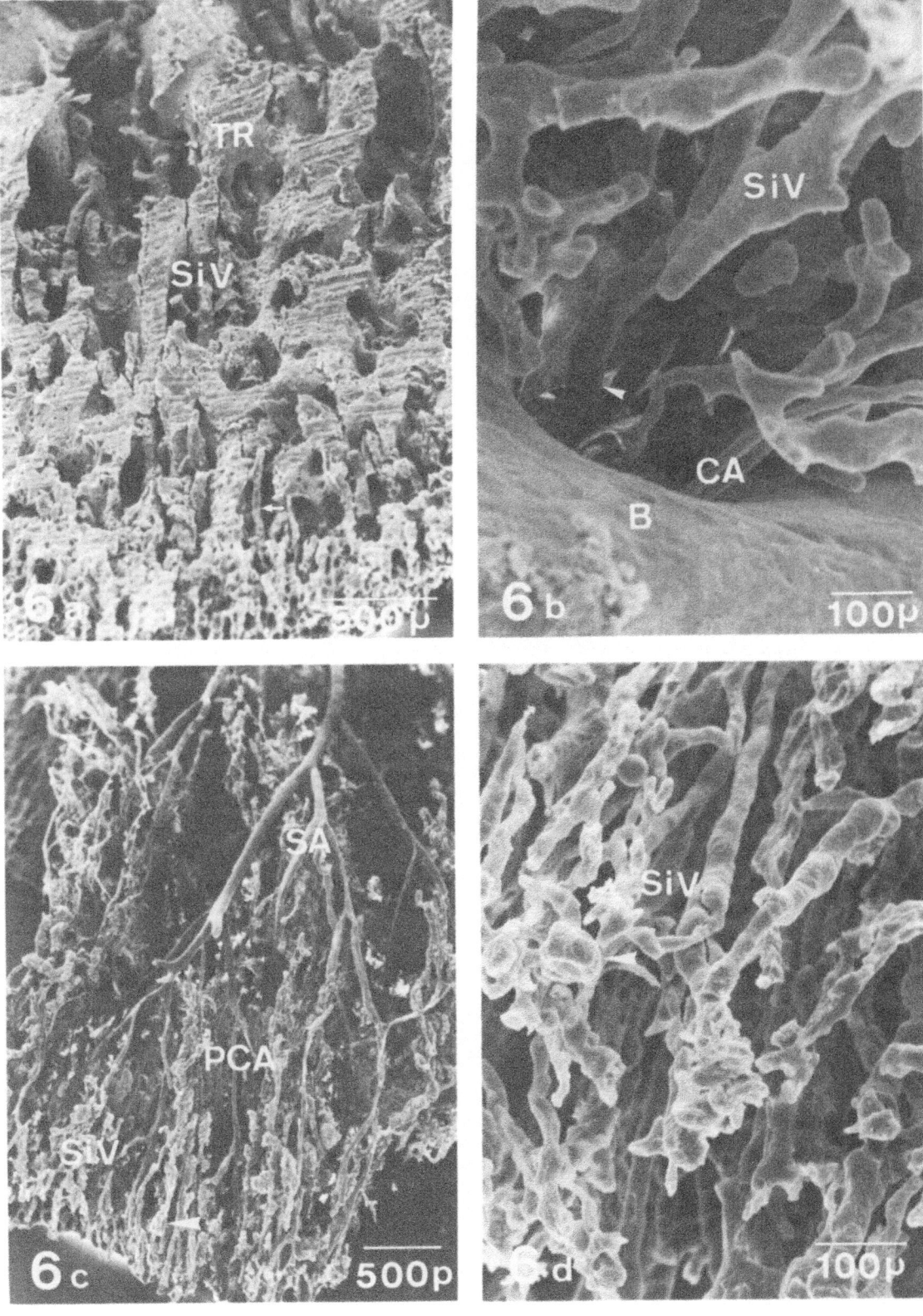

17. Ohtani O, Ushiki T, Kanazawa H, Fujita T. Microcirculation of the pancreas in the rat and rabbit with special reference to the insulo-acinar portal system and emissary vein of the islet. *Arch Histol Jpn* 49:45–60, 1986.
18. Ono T. Scanning electron microscope studies on microvascular architecture of bone marrow. Report 1. Injection replica scanning electron microscope study on arterial system of the rat femoral marrow. *Okayama Igakkai Zasshi* 90:1–16, 1978.
19. Ono T. Scanning electron microscope studies on microvascular architecture of bone marrow. Report 2. Injection replica scanning electron microscope study on venous system of the rat femoral marrow. *Okayama Igakkai Zasshi* 90:17–31, 1978.
20. Takahashi K. Changes in the vasculature of the alveolar bone and its resorption. *J Jpn Assoc Dent Scie* 2:76–109, 1983.
21. Iwaku F, Ozawa H. Microvasculature of bone. I. The three dimensional corrosion cast microvasculature of the periosteum of the diaphysis of the long bone. *Niigata Dent J* 15:11–18, 1985.
22. Matsuo M. Morphological changes of the vascular network of the periodontal membrane and alveolar bone incident to orthodontics tooth movement. *J Kanagawa Odont Soc* 21:·21–48, 1986.
23. Matsuo M, Kishi Y, Takahashi K. The periodontal vascular changes and bone resorption incident to experimental teeth movement using corrosion resin cast with SEM. *J Jpn Orthodont Soc* 46:217–229, 1987.
24. Iwaku F. Microvasculature of bone. II. The three dimensional corrosion cast microvasculature of the outer surface of the rat's parietal bone. *J Bone Miner Metab* 6:7–11, 1989.
25. Iwaku F. Microvasculature of bone. III. Microvasculature of the bone marrow of the rat's femur by means of vascular corrosioncast SEM method. *J Bone Min Metab* 7:85–89, 1989.
26. Vaughan JM. *The Physiology of Bone, 2nd ed. 4. The Blood Supply of Bone*. Oxford University Press, London, pp 17–22, 1975.
27. Warwich R, Williams PL (ed.). *Gray's Anatomy*, 35th ed. Longman, London, pp 215–232, 1973.
28. Williams EA, Fitzgerald RH Jr, Kelly PJ. Microcirculation of bone. In: Mortillaro NA (ed.), *The Physiology and Pharmacology of the Microcirculation*, Vol. 2 Academic Press, New York, pp 267–279, 1985.
29. Treuta J. The role of the vessels in osteogenesis. *J Bone Joint Surg* 45B:402–418, 1963.
30. Boyde A, Lester KS. Electron microscopy of resorbing surface of dental hard tissues. *Zeit für Zellforsch* 83: 538–548, 1967.
31. Boyde A, Hobdell MH. Scanning electron microscopy of primary membrane bone. *Zeit für Zellforsch* 99:98–108, 1969.
32. Boyde A, Hobdell MH. Scanning electron microscopy of lamellar bone. *Zeit für Zellforsch* 93:213–231, 1969.
33. Doan CA. The circulation of the bone-marrow. *Contrib Embtryol* 67:29–47, 1922.
34. Boyde A. Scanning electron microscope studies of bone. IV. Adult bone. In: Bourne GH (ed.), *The Biochemistry and Physiology of Bone*, 2nd ed., Vol. 1, Academic Press, New York, pp 268–274, 1972.

Author's address:
Prof. Fumihiko Iwaku
First Department of Oral Anatomy
Asahi University
School of Dentistry
1851-1 Hozumi Hozumi-Cho
Motosu-Gun, Gifu Pref. 501–02
Japan

Figure 12-6. Blood vessels, such as sinusoids (SiV) and precapillary arteries (arrow), pass among many trabeculae (TR) (a). Capillaries (CA) and peripheral branches of sinusoids (arrowhead) enter together in great numbers into the large vascular foramen of the bone (B) (b). In the distal region of the epiphysis, the precapillary arteries (PCA) bifurcate from the small arteries (SA) and run distally, resembling tree twigs (c). Numerous sinusoids (SiV) are arranged in parallel in the epiphyseal distal area (d). Their distal parts (arrowheads) become more complex and enlarged (c, d).

CHAPTER 13

Pericyte Topography of the Microvasculature of Skeletal Muscle: Correlated Analysis of Corrosion Casts and KOH-Digested Specimens

EUGENIO GAUDIO, LUIGI PANNARALE, ALBERTO CAGGIATI,
ANDREA MAGGIONI, GIULIO MARINOZZI & PIETRO M. MOTTA

1. Introduction

During the past few years, different approaches to the study of pericytes in the skeletal muscle microvasculature have been made through the use of different techniques, including corrosion casts.

In our own studies [1] the occurence of certain imprints noticed on the surface of the capillary casts has been correlated with the presence of pericytes on the basis of results obtained by other researchers who used different techniques [2–5].

Previously, Castenholz [6–9] discussed the presence of pericytes, based on his observations of "plastic strips" on the surface of corrosion casts. But these structures, most probably due to a casual extravasation of the casting media [10], were observed only rarely at the level of capillaries, so that limited information was provided concerning pericyte topography.

On the other hand, imprints or concave traces [1] of the presence of pericytes are obtained with the same mechanism used in casting endothelial nuclei [11–15]. The injected resin tends to homogeneously distend the wall of the vessels. This is more rigid and thick at the level of pericyte attachment [3,4,16], and consequently, it is less stretched at that level than in the neighboring areas. The higher degree of distension of neighboring areas outlines the pericyte shape.

In any case, what corrosion casts show is only a trace of perivascular cells at definite sites in the microvascular bed. These traces, in order to be interpreted, need to be compared with the possible tridimensional arrangement of the perivascular cell around the vessels.

The observation of chemically digested tissue specimens can provide this type of information [17–21]. With this technique it is possible to evaluate the extension of the perivascular cells around and along the vessel, but, on the other hand, the overall pattern of the microvascular bed cannot be followed as thoroughly as when using corrosion casts.

2. Materials and Methods

The technique we employed has been reported earlier [22].

In our studies on skeletal muscle microcirculation, we used the rat tibialis anterior (TA) and soleus (S) muscles, because they appeared to be well studied, both from the structural and physiological points of view [23,24]. Moreover, the TA and S represent two functionally different groups of muscles, the twitch and the tonic muscles, respectively. From the microvascular point of view, our studies have demonstrated that they present structural differences, both at the

Motta, P.M., Murakami, T., and Fujita, H. (eds.), Scanning Electron Microscopy of Vascular Casts: Methods and Applications.

172

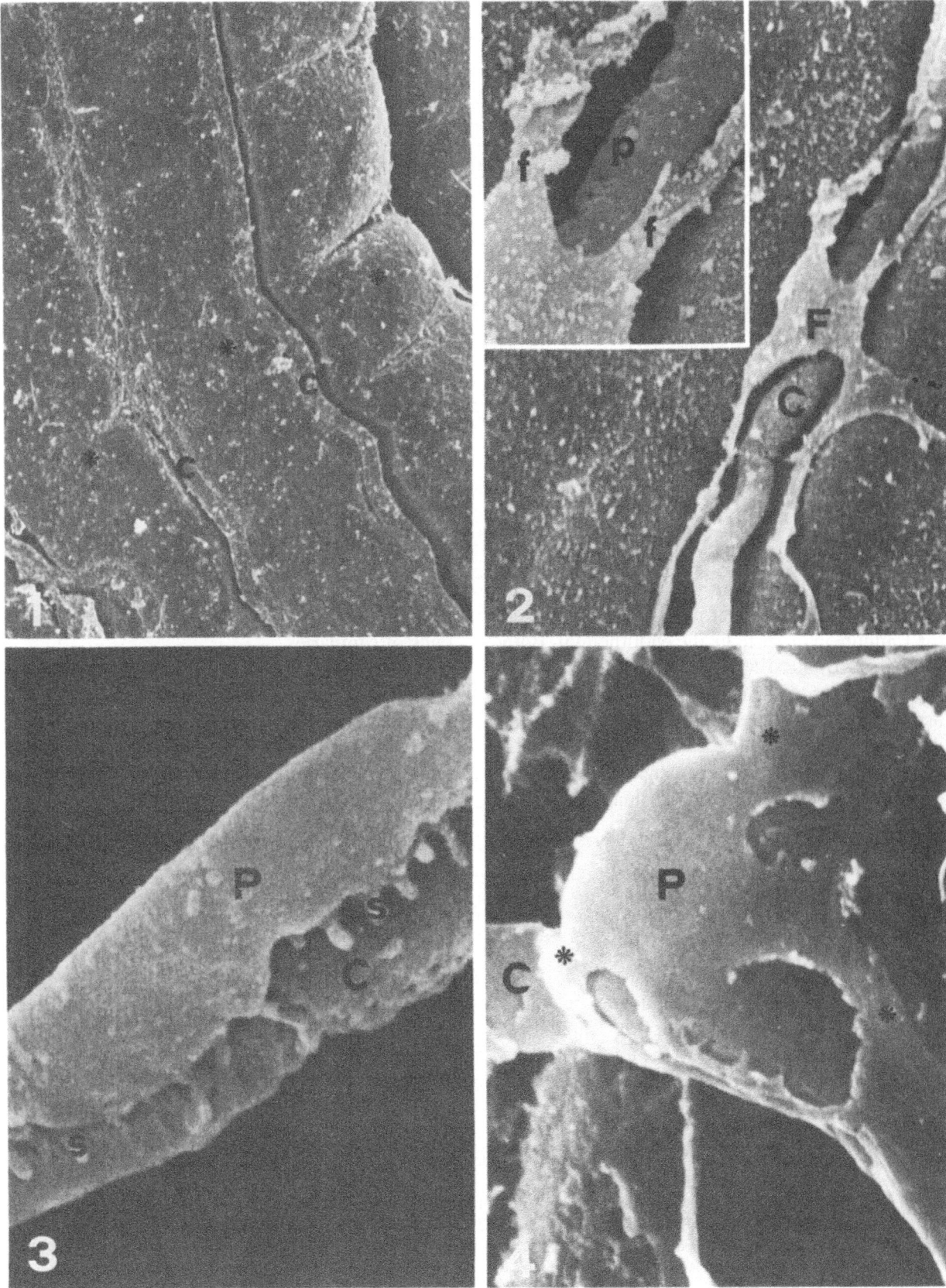

levels of precapillary and capillary sections of the microvascular bed [24].

2.1. KOH Digestion Technique

Samples for KOH digestion were prepared following the procedure devised by Maggioni et al. [21], which has proved to be very reliable and simpler than the techniques employed by other authors [25].

The muscles were left free to contract under the fixative solution effect. Critical-point drying was used for dehydration [26].

2.2. Corrosion Cast Technique

Corrosion casts were prepared following a technique described previously [22]. The vascular bed was rinsed using heparinized 0.9% saline solution [11–27] with 1% carbocaine [28]. This washing, in our experience, results in more consistent filling of the capillary bed.

"Prefixation" [29] with 1% glutaraldehyde in 0.1 M cacodylate buffer at pH 7.3 was performed for 1 minute in order to fix the endothelium. Both a Cambridge 150 SEM and a Hitachi S 4000 field emission SEM were used for observation. The use of the field emission SEM provided high-definition images of the cast surface. Thus, it allowed us to follow every single wave of the cast surface and made tissue-remnant detection easier.

3. Observations

3.1. Pericytes

Observation of KOH-digested tissue specimens shows the surface of myofibers and capillaries. Both structures run parallel to the long axis of the muscle (Fig. 13-1). In the zones where the muscle

fibers are well preserved, the capillaries are placed very close to the fibers and are often "embedded" between two myofibers. In places where the muscle fibers are more corroded and washed out, one can observe the whole circumference of the vessels and follow their course more extensively. At these levels it is easier to observe transverse vessels connecting the longitudinal capillaries.

Most of the connective tissue appears to be removed. Nevertheless, a few fibroblasts appear lying in the neighborhood of the vessels (Fig. 13-2). These cells show a central body that gives rise to three or four tapering main processes with a regular outline. Capillaries show a smooth surface, a variable tortuous trend, and a mostly constant diameter. On the surface of the capillaries, it is possible to observe pericytes (Figs. 13-2–13-5). These perivascular cells can be observed in different positions. Most often we observed at least one pericyte in the middle of the tract between two transverse anastomoses. Other sites where pericytes are most likely to be encountered are branching points and venous ends of capillaries.

These cells appear to differ from fibroblasts in some morphological features (Fig. 13-2). The pericyte central body always lies in close contact with the capillary wall and gives rise, in most cases, to only two primary processes. Pericyte primary processes are, in most cases, equally attached onto the wall of the vessels. The borders of these cells are not smooth and do not have irregular secondary digitations, as some authors have observed in fibroblasts [19,25]. In fact, many secondary processes arise from the edge of the primary processes, or of the cell body, and tend to encircle the vessel.

Morphological differences can be observed whether the pericyte is placed at the level of a longitudinal capillary or at a branching point. At the level of a longitudinal capillary, the pericyte

←

Figure 13-1. Digested tissue specimen of tibialis anterior. Parallel capillaries (C) can be observed in the spaces between myofibers (asterisk). SEM. 1000×.

Figure 13-2. Digested tissue specimen of tibialis anterior: Fibroblast (F) overlying a capillary (C). SEM. 2000×. Insert: At higher magnification, it is possible to distinguish fibroblast processes (f) from pericyte processes (p). SEM. 5000×.

Figure 13-3. Digested tissue specimen of soleus. P = pericyte body; C = capillary; S = secondary processes SEM. 10,000×.

Figure 13-4. Digested tissue specimen of soleus. A capillary branching point is shown. C = capillary; P = pericyte body; asterisk = pericyte processes. SEM. 7000×

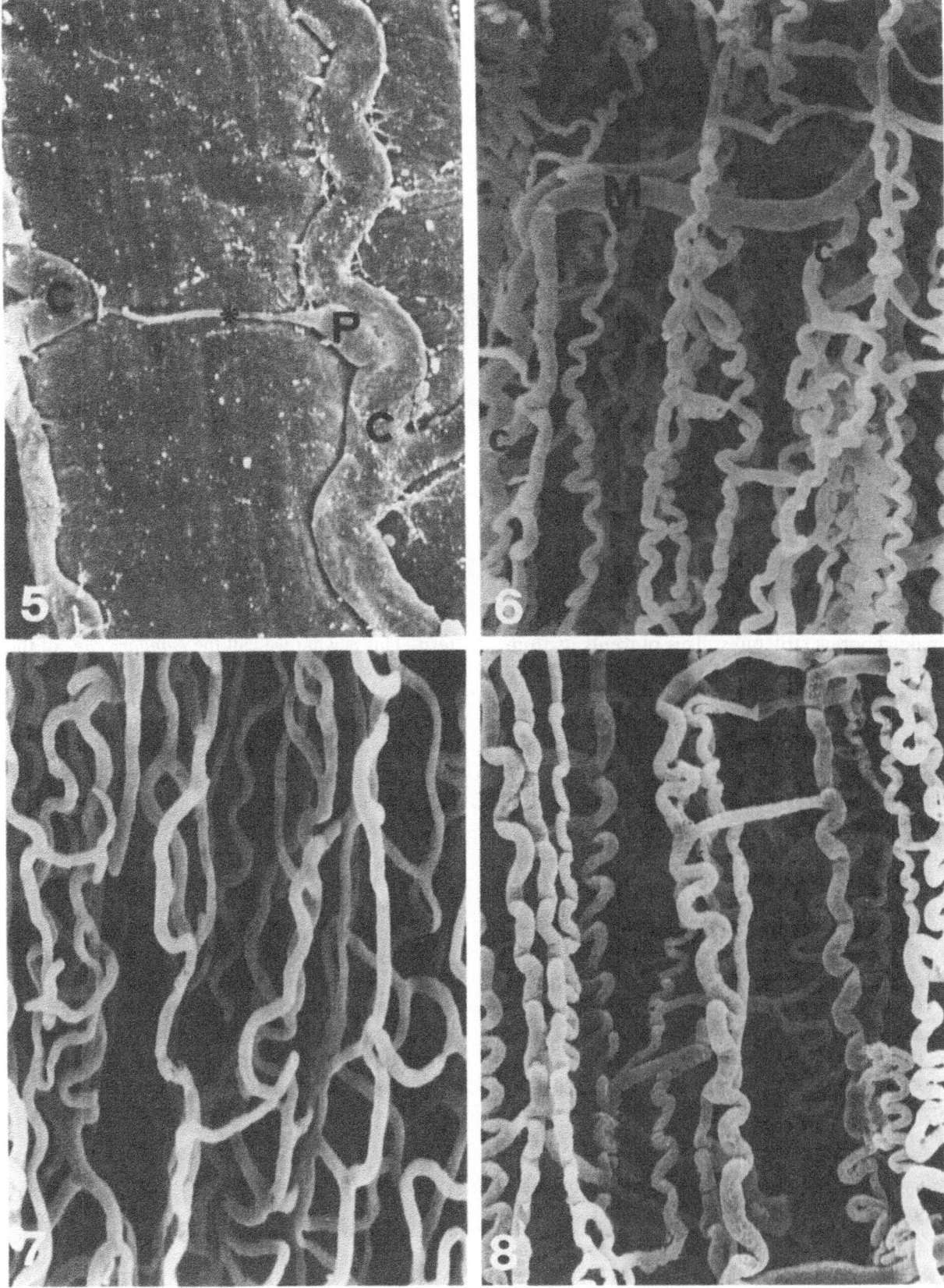

body gives rise to two longitudinal primary processes that run along the capillary, while giving rise to secondary processes.

At the level of branching points (Fig. 13-4), one or more pericytes are often present. The cell body can give rise to more numerous, wider, and thicker primary processes that run along the different branches or encircle almost completely one branch at the level of the pericyte body.

Sometimes, a pericyte process extends itself from one capillary to another, bridging two parallel vessels (Fig. 13-5). This same arrangement has been observed by Williamson et al. using fluorescent microscopy in the muscle, but not in the retina [5]. These data, obtained by the technique of tissue digestion devised by Maggioni et al. [21], are very similar to those obtained by other authors by means of different tissue-digestion techniques [18,19,25,30].

Pericytes seem to be satisfactorily preserved by this technique, and their relation to vessels does not change in different samples. However, a certain degree of pericyte loss must be taken into account, and some perivascular cells are hidden by neighboring structures. This prevents us from making any statistical calculations of pericyte frequency.

3.2. Corrosion Casts

The corrosion casts showed, both in TA and in the S muscles, primary, secondary, and precapillary arterioles. These vessels allow segmentary control of blood flow through capillaries [23, 31,32]. In muscles characterized by oxidative metabolism, metarterioles showing two different types of courses could take origin from the secondary arteriole [33].

These vessels can follow a longer oblique course, running from an arteriole to a venule; in this case, they give rise to capillaries that a rise from one side of the vessel (Fig. 13-6). At the origin of the capillaries we have observed

imprints that could be attributed to the presence of pericytes. This type of vessel represents the only precapillary selective control of blood flow through the single vessel.

Otherwise, metarterioles can run parallel to the capillaries and anastomose with them on both sides. Metarterioles get progressively thinner toward their venular end, where they may be confused with capillaries. This type of arrangement makes the metarteriole a preferential channel that is able to share its blood flow selectively with additional vessels.

In both muscles the capillary bed develops from subsequent tuning-fork divisions. Both in the contracted and in the extended positions, the overall capillary patterns of the two muscles appear to be different.

In the tibialis anterior we recognize longitudinal capillaries with a straight or slightly tortuous trend (Fig. 13-7). Transverse anastomoses of different lengths connect capillaries at different distances. They are more frequent in the venous halves of vessels.

In the soleus we recognize a more complex pattern. This is what we call the *capillary cage* of the muscle fiber (Fig. 13-8). The cage is composed of three or more tortuous longitudinal capillaries (which, due to their possible function, have also been called *main capillaries*) and their side branches. The side branches are transverse anastomoses and capillary loops connected at both ends with the same longitudinal capillary.

At higher magnification the surface of the cast appears to be regularly wavy. On the surface of longitudinal tracts of capillary casts, far from the transverse branches, it is often possible to observe the presence of a very shallow groove, about $3\,\mu m$ wide, running along the cast (Fig. 13-9). Inside the groove the superficial waves of the cast change somewhat in orientation. This feature could be followed, sometimes, for up to $60\,\mu m$ along the cast. Sometimes, at the middle of its course, the groove widens in a roundish pattern

←

Figure 13-5. Digested tissue specimen of soleus. A pericyte process bridges two parallel capillaries. C = capillary; P = pericyte body; asterisk = pericyte process. SEM. 2000×.

Figure 13-6. Corrosion cast of tibialis anterior. A metarteriole follows an oblique course. Metarteriole (M) and capillaries arise from the metarteriole. SEM. 500×.

Figure 13-7. Corrosion cast of a contracted tibialis anterior: capillary bed. SEM. 600×.

Figure 13-8. Corrosion cast of an extended soleus. Capillary bed. SEM. 600×.

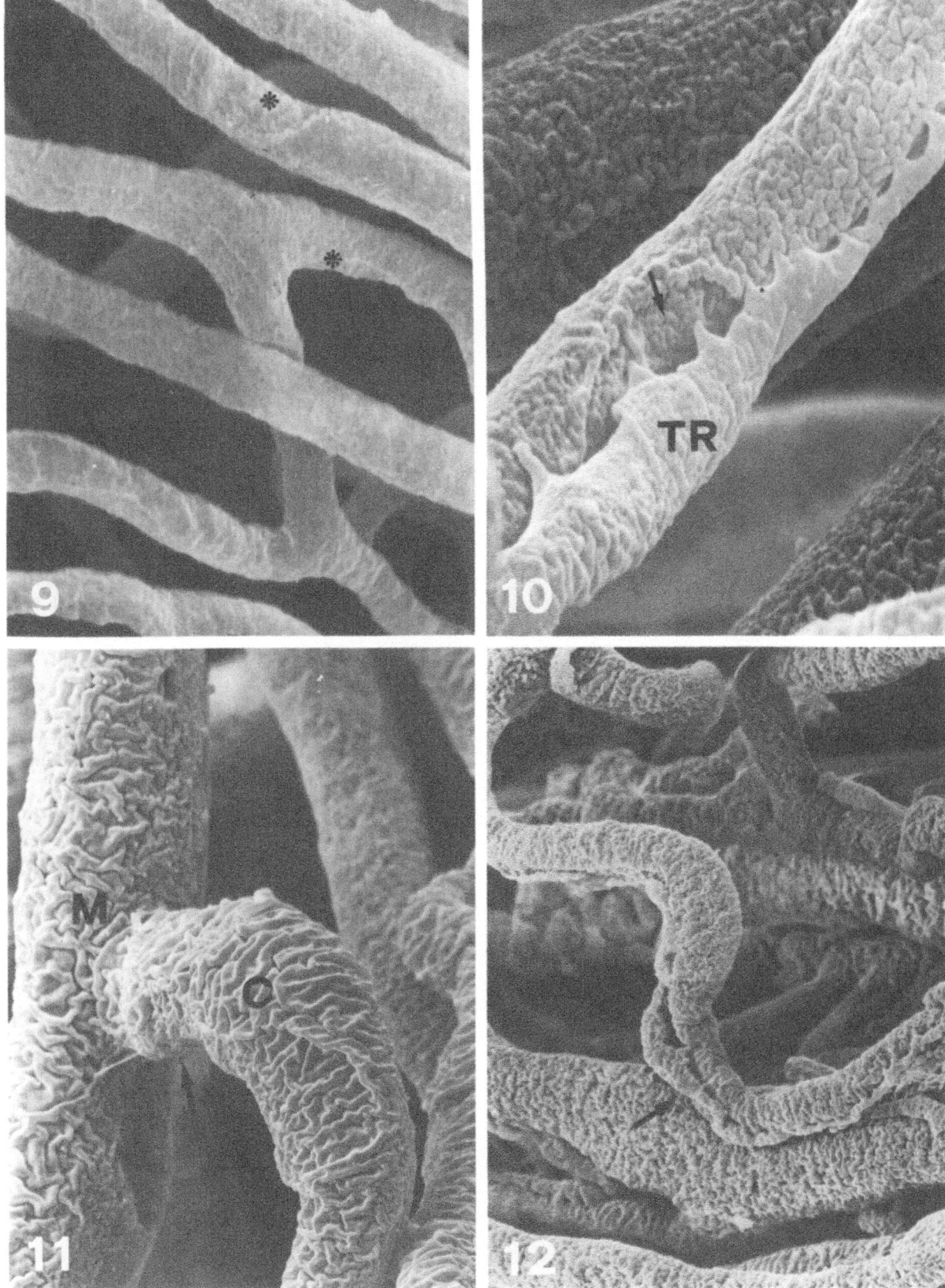

that extends around one-half the circumference of the cast (Fig. 13-10). The depth of the groove is so small that a proper combination of observation parameters is essential for clear observation of their morphology. For the same reason, we can argue that even a slightly excessive corrosion can make them not viable for study.

3.3. Vascular Imprints

Different kinds of imprints can be observed at the level of capillary branching points, more specifically at the level of side branches, and at the venular and metarteriolar end of capillaries.

These imprints are often grooves that are similar in the width and type of surface to those encountered on the longitudinal capillaries. They generally extend from the longitudinal capillary to the side branch. At times, more than one groove is present on the same tract of capillary, at the opposite sides of the cast.

Round imprints can be seen at branching points. They can be as shallow as the grooves are or deeper, with a punched-out concave shape.

In some cases, instead of showing the mentioned imprint, the cast appears tapered or restricted by a circumferential groove at branching points (Fig. 13-11). This is almost constantly seen at the metarteriolar end of capillary casts.

The mentioned surface features correspond in their anatomical situation and their extension around and along the cast to the arrangement of pericytes in digested tissue specimens. On the other hand, we have not found any evidence for attributing those traces to endothelium.

The traces that we found do not share the shape, dimensions, or frequency with endothelial nuclei imprints. Moreover, imprints of endothelial nuclei at the capillary level have seldom been described.

Some of the round imprints at branching points are reminiscent of the shape of endothelial nuclei, but if they were due to endothelial nuclei we should have found them regularly spaced all along the capillary cast.

Further evidence of a direct correlation between the pericytes and cast imprints can be obtained from the study of those parts of some specimens in which "mummified" tissue remnants are present around the cast (Figs. 13-10–13-12).

These less-corroded zones can often be easily recognized from the presence of "threadlike structures" connecting vessels and partially besmeared waves of the cast surface.

Within such zones, it is possible to see some structures that show a specific form and relation with certain sites of the cast.

At the level of longitudinal capillaries, the remnants measure about 4 µm in width and up to 120 µm in length. Most of these structures show regularly serrated borders (Fig. 13-12).

In places where the structures appear to be partially detached from the cast, it is always possible to see a corresponding imprint on the surface of the cast itself (Fig. 13-10). At many points the remnants change into the material with which the cast was smeared.

At the level of capillary cast bifurcations, the remnants show a thicker central part, which gives rise to two or three thinner processes that run along or around the cast. Sometimes, these structures appear receded from capillary nook and are taut between the two branches.

None of the mentioned structures showed the same type of densely wavy surface as the rest of the cast.

4. Comments

In order to intepret pericyte function, both the already discussed attachment sites [3–5,21] and their contractility, as demonstrated by Tilton [4], must be taken into account.

First, pericytes attached at capillary bifurcations, with their cell bodies strategically placed

←

Figure 13-9. Corrosion cast of tibialis anterior. Shallow imprints (asterisk) can be observed along the cast. SEM. 2200×.

Figure 13-10. Corrosion cast of tibialis anterior. The roundish (arrow) imprint of the cast is partially covered by mummified tissue remnant (TR). SEM. 4000×.

Figure 13-11. Corrosion cast of soleus. Tapering (arrow) of capillary cast at its "metarteriolar" origin is seen. M = metarteriole; C = capillary. Field emission (FE) SEM. 3600×.

Figure 13-12. Corrosion cast of soleus. A mummified tissue remnant (arrow) overlying a capillary cast is seen. FE SEM. 2000×.

and primary processes running around the vessel, can play a role in the regulation of the extremely variable blood cell flow in the capillary bed of skeletal muscles [31,32,34]. The same role can be attributed to any pericyte enwrapping the vessel with its thicker processes, as demonstrated by Tilton et al. [3,4]. Such a role is even more important where many secondary paths for blood cells are available, that is, at the level of capillary cage and of the "metarteriolar" origin capillaries in the soleus muscle. When the peculiar microvascular pattern of this slow-twitch muscle and the pericyte contractility are taken into account, it becomes easy to understand that the flow of blood cells can be differentially directed in different metabolic situations. This agrees with the high variability of blood flow in this muscle [35] as compared with that in other skeletal muscles.

In a recent work [22], we considered the importance of pericyte actions, other than their concentric contraction around the vessel. Their strong connection with the endothelium [36–42] and their resistance against stretching seem to suggest that they are a mechanical support for the vessel wall. But, of course, this cannot be considered to be their only role, as in Cogan's opinion [16] pericyte contractility must also be taken into consideration. In fact, the contraction of pericyte processes along and between the vessels strengthens their supporting role for their vascular function. Such a role could be even more desirable in contractile organs, where the capillaries could undergo excessive binding. In this respect a role in limiting excessive binding can certainly be attributed to the bridging processes of pericytes. It must be noted that such processes have not been observed in noncontractile organs, such as the retina [5].

Other authors [2,19] have hypothesized that the contraction of longitudinal pericytes themselves can bind the capillary wall, thus producing a buckling of the wall towards the lumen and, consequently, affecting blood flow. This hypothesis implies an inhomogeneous adhesion of pericytes along the vessel, but this fact has never been demonstrated.

A further hypothesis on pericyte function [43] is that they would stretch endothelial intercellular junctions, thus influencing vascular permeability. As we have noticed in KOH-digested tissue spec-imens, the length of pericytes in the skeletal muscle microvascular bed can exceed 100 µm. These dimensions suggest that one pericyte bridges different endothelial cells. Thus, longitudinal contraction of the pericyte as a whole is quite unlikely to stretch intercellular junctions.

5. Concluding Remarks

The study of imprints represent a means of trying to acquire the maximum amount of information possible from vascular casts. Further research is needed in order to enhance the traces of perivascular cells on the corrosion cast surface. A comparison with observations made through the use of digestion technique remains essential.

Casual observation of tissue remnants have provided unexpected support for the correct interpretation of imprints. In the future a technique should be devised for semidigested cast specimens, as well as a progressive corrosion technique, in order to allow the observation of the digested tissue in the same specimen and, subsequently, the observation of the cast.

Any development of this approach can prove itself even more useful in the study of perivascular cells morphology, localization, and their role in organs where this needs to be elucidated.

References

1. Gaudio E, Pannarale L, Marinozzi G. An SEM corrosion cast study on pericyte localization and role in the microcirculation of skeletal muscle. *Angiology* 36:458–464, 1985.
2. Courtoy P, Bovles J. Fibronectin in the microvasculature: Localization in the pericyte-endothelial interstitium. *J Ultrastruct Res* 83:258–273, 1983.
3. Tilton RG, Kilo C, Williamson JR. Pericyte-endothelial relationships in cardiac and skeletal muscle capillaries. *Microvasc Res* 18:325–335, 1979a.
4. Tilton RG, Kilo C, Williamson JR, Murch DW. Differences in pericyte contractile function in rat cardiac and skeletal muscle microvasculatures. *Microvasc Res* 18:336–352, 1979b.
5. Williamson JR, Tilton RG, Kilo C, Yu S. Immunofluorescent imaging of capillaries and pericytes in human skeletal muscle and retina. *Microvasc Res* 20:233–241, 1980.
6. Castenholz A. Scanning electron microscopy of myocytes and pericytes in terminal blood vessels of the rat. *Microvasc Res* 19:395, 1980.

7. Castenholz A, Zoltzer H, Erhardt H. Structures imitating myocytes and pericytes in corrosion casts of terminal blood vessels. A methodical approach to the sphenomenon of "plastic strips" in SEM. *Mikroskopie* 39:95–106, 1982.

8. Castenholz A. Visualization of periendothelial cells in arterioles and capillaries by scanning electron microscopy of ultrasound treated and plastoid injected brains in rats. *SEM* 161–170, 1983.

9. Castenholz A. The outer surface morphology of blood vessels as revealed in scanning electron microscopy in resin cast, non-corroded tissue specimens. *SEM* 1955–1962, 1983.

10. Castenholz A. Interpretation of structural patterns appearing on corrosion casts of small blood and initial lymphatic vessels. *Scanning Microscopy* 315–325 1989.

11. Hodde KC, Miodonski A, Bakker C, Veltman WAM. SEM of microcorrosion casts with special attention on arterio-venous differences and application to the rat's cochlea. *Scann Electron Microsc* II:477–484, 1977.

12. Miodonski AJ, Hodde CK, Backer C. Rasterelektronenmikroskopie von Plastik-korrosion-praparaten: Morphologische Unterschiede zwischen Arterien und Venen. *Beitr Elektronenmikroskop Direktabb Oberfl* (Munchen) 9:435–442, 1976.

13. Miodonski AJ, Bar T. Arterial supply of the choriocapillaris of anuran amphibians (*Rana temporaria, Rana esculenta*). Scanning electron microscopic study of microcorrosion casts. *Cell Tissue Res* 249:101–109, 1987.

14. Nowell JA, Lohse CL. Injection replication of the microvasculature for SEM. *Scann Electron Microsc* 267–274, 1974.

15. Kendall MW. Eissman E. Scanning electron microscopic examination of human pulmonary capillaries using a latex replication method. *Anat Rec* 196:275–283, 1980.

16. Cogan DG, Toussaint D, Kuwabara T. Retinal vascular patterns. IV. Diabetic retinopathy. *Arch Ophthalmol* 66:366–378, 1961.

17. Mazanet R, Reese BF, Franzini-Armstrong C. Reese TS. SEM of satellite cells and their response to muscle fiber injury, *J Cell Biol* 83, 383a, 1979.

18. Mazanet R, Franzini-Armstrong C. SEM of vascular pericyte in rat red muscle. *J Cell Biol* 87:259a, 1980.

19. Mazanet R, Franzini-Armstrong C. Scanning electron microscopy of pericytes in rat red muscle. *Microvasc Res* 23:361–369, 1982.

20. Shotton DM, Heuser JE, Reese BF, Reese TS. Postsynaptic membrane folds of the frog neuromuscular junction visualized by scanning electron microscopy. *Neuroscience* 4:427–435, 1977.

21. Maggioni A, Caggiati A, Macchiarelli G. Scanning electron microscopy of microvessels and perivascular cells in different organs after KOH digestion. In: Motta P (ed.), *Cells and Tissues: A Three-Dimensional Approach by Modern Techniques in Microscopy*, Alan R. Liss, New York, pp 469–474, 1989

22. Gaudio E, Pannarale L, Caggiati A, Marinozzi G. A three-dimensional study of the morphology and topography of pericytes in the microvascular bed of skeletal muscles. *Scann Microsc*, 4(2):491–500, 1990.

23. Gaudio E, Pannarale L, Marinozzi G. A tridimensional study of microcirculation in skeletal muscle. *Vasc Surg* 18:372–381, 1984.

24. Pannarale L, Gaudio E, Marinozzi G. Microcorrosion casts in the microcirculation of skeletal muscle, *Scann Electron Microsc 1986* III:1103–1108, 1986.

25. Miller BG, Woods RI, Bohlen HG, Evan AP. A new morphological procedure for viewing microvessels: A scanning electron microscopic study of the vasculature of the small intestine. *Anat Rec* 203:493–503, 1982.

26. Boyde A. *Pros and Cons of Critical Point Drying and Freeze Drying for SEM.* SEM/1987/II, SEM, Inc., AMF O'Hare, IL, pp 303–314, 1978.

27. Miodonski A, Jasinski A. SEM study of microcorrosion casts of the vascular bed in the skin of the spotted salamander, *Salamandra salamandra* L. *Cell Tissue Res* 196:153–162, 1979.

28. Gaudio E, Pappalardo G, Pannarale L, Pintucci S, Resta S, Ripani M. Three-dimensional observations (by SEM) on the microcirculation of adenocarcinoma of the rectosigmoid junction in man. *It J Surg Sci* 12:221–225, 1982.

29. Lametschwandtner L, Lametschwandtner U, Weiger T. Scanning electron microscopy of vascular corrosion casts — techniques and application. *Scann Electron Microsc. 1984* II:663–695, 1984.

30. Holley JA, Fahim MA. Scanning electron microscopy of mouse muscle microvasculature. *Anat Rec* 205:109–117, 1983.

31. Stingl J. Fine structure of precapillary arterioles of skeletal muscle in the rat. *Acta Anat* 96:196–205, 1976.

32. Zweifach BW, Metz DB. Selective distribution of blood through the terminal vascular bed of the mesenteric structure and skeletal muscle. *Angiology* 6:282–290, 1955.

33. Gaudio E, Pannarale L, Marinozzi G. Corrosion casts in the microcirculation of skeletal muscle. In: Motta PM (ed.), *Cells and Tissues: A Three Dimensional Approach by Modern Techniques in Microscopy*, Alan R. Liss, New York, pp 443–450, 1989.

34. Myrhage R, Hudlicka O. The microvascular bed and capillary surface area in rat extensor hallucis propius muscle (EHP). *Microvasc Res* 11:315–323, 1976.

35. Reis DJ, et al. Differential regulation of blood flow to red and white muscle in sleep and defense behavior. *Am J Physiol* 217:541–546, 1969.

36. Rhodin JAG. Ultrastructure of mammalian venous capillaries, venules, and small collecting veins. *J Ultrastruct Res* 25:452–500, 1968.

37. Rhodin JAG. *Histology: A Text and Atlas*, Oxford University Press, New York, pp 331–370, 1974.

38. Weibel ER. On pericytes, particularly their existence on lung capillaries. *Microvasc Res* 8:218–235, 1974.

39. Matsusaka T. Tridimensional views of the relationship of pericytes to endothelial cells of capillaries in the human choroid and retina. *J Electron Microsc* 24:13–18, 1975.

40. Forbes MS, Rennels MC, Nelson E. Ultrastructure of pericyte in mouse heart. *Am J Anat* 149:47–70, 1977.

41. Ryan US, Ryan JW, Whitaker C. How do kinins affect vascular tone? *Adv Exp Med Biol* 102A:237–291, 1979.

180

42. Wallow IH, Burnside B. Actin filaments in retinal pericytes and endothelial cells. *Invest Ophthalmol Vis Sci* 19:1433–1441, 1980.
43. Miller FN, Sims DE. Contractile elements in the regulation of macromolecular permeability. *Fed Proc* 45:84–88, 1986.

Author's address:
Dr. L. Pannarale
Department of Human Anatomy
Faculty of Medicine
University "La Sapienza"
via A. Borelli 50
00161 Rome
Italy

CHAPTER 14

Blood Vascular Casting in Cardiac and Skeletal Muscle

RICHARD F. POTTER & ALAN C. GROOM

1. Introduction

Cardiac and skeletal muscle depend upon an adequate blood vascular system for the delivery of nutrients and the removal of metabolites. The functional unit for blood-tissue exchange is a network of vessels visible only with the aid of a microscope. The geometric complexity of this microvascular network reflects, in large measure, the functional requirements of the tissue it serves. Thus, in order to fully appreciate the importance of this network in cardiac and skeletal muscle, a knowledge of the three-dimensional geometry of the microvasculature is required.

Light microscopy has the disadvantage of a very restricted depth of focus, limiting the field of view to two dimensions. Because of this, attempts have been made to study cleared and/or serial sectioned tissue whose vasculature has been injected with a contrasting medium. Although superior to standard histological techniques, these methods are essentially still restricted to two dimensions. Corrosion casting techniques offer a distinct advantage for the study of vascular geometry, since the surrounding tissue is removed, leaving an intact replica of the vascular network, which can be examined under the scanning electron microscope (SEM). Because of the enormous depth of focus of this instrument, SEM images of corrosion casts possess a three-

dimensional quality, allowing unobstructed views of otherwise hidden parts of the vasculature. Vascular corrosion casting methods have been classified under various designations [1], but in this chapter we shall use the terms *vascular corrosion casting* and *microcorrosion casting* to refer to casts of the arterial or venous network, and of the microvasculature, respectively.

The purpose of this chapter is to describe recent advances in our understanding of vascular geometry in cardiac and skeletal muscle, brought about by the study of corrosion casts. First, some methods and materials for the preparation of corrosion casts are described, since the value of a cast depends on the accuracy and detail with which in vivo conditions can be replicated. Second, we review what is presently understood of the coronary arterial vessels, coronary microvasculature, and lastly, the microvasculature in skeletal muscle, extended versus shortened, in situ.

2. Methods of Blood Vascular Casting

The evolution of corrosion casting techniques has been extensively reviewed in the literature [1,2]. Nevertheless, a brief comment here on some methods and compounds commonly used in the preparation of blood vascular casts is in order.

Motta, P.M., Murakami, T., and Fujita, H. (eds.), Scanning Electron Microscopy of Vascular Casts: Methods and Applications.

Replication of the microvasculature necessitates the introduction of foreign material (i.e., casting compound) into the vascular compartment. All casting compounds are, to some extent, toxic to the vascular wall, and their introduction may cause contraction of vascular smooth muscle cells (SMC), resulting in luminal narrowing and increased resistance to flow. If such changes are sufficiently severe, incomplete filling of the microvasculature may result. There are many variations in methodology that attempt to reduce or eliminate this effect, such as exsanguination by perfusion with saline [3–5] or physiological solutions [6–8], and/or the administration of vasoactive agents such as 0.2% procaine [8] or 1% carbocaine [3]. Some investigators [9] have found pretreatment of the vascular network generally unnecessary and claim that such pretreatment may exacerbate the occurrence of incompletely filled areas within the casts.

The tip of the cannula through which the casting material is infused is generally placed as close as possible to the organ or tissue of interest. This limits infusion of the compound into other vascular beds and lessens the risk of incomplete filling of the microvascular network of interest. Under some circumstances it may not be physically possible or convenient to place the cannula into the artery supplying the organ of interest. In these situations, it may be desirable to pretreat the microvasculature with either a vasoactive or paralytic agent. However, it should be emphasized that the effect of such pretreatment on the quality of the cast has as yet to be studied [1].

Infusion of the casting compound into the microvascular network has been carried out under either constant pressure [4,6,7,10–12] or constant flow [9,13]. No studies to date have supplied conclusive evidence in support of one method over the other. However, progressive polymerization of the infused compound causes the viscosity to increase with time, giving rise to problems with both methods of infusion. Under constant-pressure conditions, incomplete filling of the microvasculature may result. Under constant-flow conditions, high intracapillary pressures may cause exudation of the casting material into the tissue spaces. The degree of exudation has been reported to be minimal, however, and capillary diameters measured from such casts are comparable to those obtained in vivo [9]. These findings suggest that the pressures generated at the capillary level, during infusion of the casting material under constant flow conditions, may not be greatly above the normal range.

The use of microcorrosion casts to examine microvascular geometry in cardiac and skeletal muscle depends critically on the detail with which the microvascular network can be replicated. This, in turn, depends on the characteristics of the casting compound itself. Compounds formerly used were neoprene, nylon, vinylite or vinyl acetate, marco resin, and polyester. However, all of these compounds have major disadvantages that limit their use for blood vascular casting. Latex, nylon, and vinyl acetate are extremely viscous and therefore are suitable only for the replication of large vessels (arteries and veins). Latex has the additional disadvantage of being extremely flexible, often requiring structural support by immersion in a water bath. Mercox resin has the major disadvantage of penetrating vessel walls, and even entering the surrounding tissue and preventing proper maceration. The use of polyester is restricted to the replication of large vessels, although its viscosity appears to be lower than those of latex or nylon; use of this compound produces, at best, incomplete replicas of the capillary network.

Modern methods of microcorrosion casting employ one of the polymer resins. Although it is unlikely that any material would meet all of the criteria listed for the ideal casting compound [2], the newer polymer resins come close. The first report describing their use in the production of vascular casts was by Murakami [14]. Murakami's plastic underwent further modification by Ganon [15], who successfully lowered the viscosity to 2.5–5 mPa.s (approx). Today, many prepolymerized methacrylates have become commercially available, such as Batson's No. 17 and Mercox, which were first used by Nowell et al. [16] and Matsusaka and Fujibashi [17], respectively. Modification of the Batson's No. 17 compound by addition of a dental acrylic, Sevriton, reduces the viscosity from 260 mPa.s (approx.) to 20–22 mPa.s [18]. It also produces an extremely hard cast, which is less fragile than casts obtained with Mercox.

Modifications to existing polymer resins, or the introduction of new compounds may further enhance the quality of microvascular casts in terms of their durability, reproducibility, and the detail with which in vivo conditions are replicated. A new compound, L.R. White, having a low viscosity (8–10 mPa.s) and being highly durable, may find a use in the preparation of such casts. To our knowledge, this material has been used only in the myocardium [8], without being subjected to the rigors of the corrosion process. Further tests of this compound may uncover yet another valuable tool for the preparation of microcorrosion casts.

After infusion and polymerization of the casting compound, maceration of the surrounding tissue in sodium hydroxide solution (10–40%) yields a free-standing, three-dimensional replica of the blood vascular network. This replica can then be viewed using the SEM. The remarkable depth of focus of the SEM and the possibility of generating stereopaired photographs of the casts make it possible to obtain information about the three-dimensional structure of the blood vascular system. However, certain cautions are necessary if a quantitative assessment of the vascular network is to be made using individual photographs taken from the SEM. Because of possible errors due to parallax caused by the large depth of field of this instrument, the accuracy of any absolute measurement must be questioned. Nevertheless, conclusions based on general trends indicated by mean values or frequency distributions of the measurements can be trusted. It is significant that measurements of capillary luminal diameter taken from SEM photographs of microcorrosion casts are in close agreement with published data obtained using other techniques [9,13].

3. Coronary Arterial Vessels

Interest in the vascular geometry of the heart has been sparked by the varied clinical implications of altered blood flow to this highly aerobic tissue. This interest has focused on the course and branching pattern of the main coronary arteries, since their relatively large diameter and exposed position on the surface of the myocardium allows their observation, unobstructed by the surrounding muscle tissue. However, the geometry of the coronary arterial network as a whole has remained elusive, since the smaller branches enter the myocardium and direct visual inspection is then no longer possible.

The introduction of corrosion casting methods made possible more complete descriptions of the geometry of the coronary arterial tree. The use of compounds such as neoprene and vinyl acetate make possible a qualitative description of the arrangement and appearance of the main coronary vessels. However, since inconsistencies in either the position or anatomy of coronary arteries within individual hearts are known to occur, such qualitative descriptions have largely failed to relieve the confusion in the literature.

The aim of this section is to review what is currently known about the geometry of the coronary arteries. This will be accomplished, first, by reviewing the classification of the vascular network; second, by providing a description of the coronary network based on quantitative measures; and, third, by presenting a description of "neovascular meshes" in the wall of a coronary artery. By reviewing the literature in this way, it is hoped the highly complex geometry of the coronary arterial system may be better appreciated.

3.1. Classification of the Vascular Network

Much of the information regarding coronary arterial geometry in humans has been derived from angiography, a procedure most often applied to hearts suffering pathological change and restricted to a more-or-less two-dimensional field of view. Such descriptions of vascular geometry rely on the use of gross anatomical features of the myocardium as landmarks for distinguishing various vessels. However, because of the wide variations in coronary artery geometry among individuals, there is confusion in the literature regarding the position, course, and area of myocardium served by any particular artery. Vascular corrosion casts of animal and human hearts obtained at autopsy provide an unparalleled opportunity to study the three-dimensional geometry of the arterial network in normal subjects.

Investigations based on the use of vinyl acetate corrosion casts in various animal models have described the origin, course, diameters, and lengths of the left (L.) coronary, circumflex, and accessory arteries and the right (R.) coronary arteries and their branches. In addition, evidence for the presence of anastomoses in the coronary arterial network has been obtained. A simplified description of the vascular supply to the posterior aspect of the human heart was proposed in 1978 by Hadziselimovic [19], based on careful studies of vinyl acetate corrosion casts. He described three basic types of vascularization: (1) "symmetrical," in which the L. coronary artery served only the L. ventricle and the R. coronary artery served only the R. ventricle; (2) "right type," in which the R. coronary artery vascularized the L. ventricle; (3) "left type," in which the L. coronary artery vascularized the entire posterior aspect of the R. ventricle. These three types of vascularization were found in 24%, 63%, and 13%, respectively, of the hearts examined. Although this system of classification simplified earlier anatomical descriptions, there still remained a continuum of variations from heart to heart that could not adequately be described in this way.

An entirely new approach to the morpho-functional anatomy of the human coronary arteries was presented in 1985 by Zamir and Silver [20], based on observations of corrosion casts of 44 human hearts prepared by the use of Batson's No. 17 casting compound. The hearts were obtained from autopsy, where the cause of death was not related to cardiovascular disease. This investigation showed that, in spite of the marked variability in vessel distribution from one heart to another, the coronary arterial branches invariably divided the myocardium into the same six distinct zones, each having a (presumably) different demand for blood supply. Each zone of the myocardium was demarcated by arteries that brought the blood supply to its borders (i.e., "distributing" arteries). While most borders were always supplied by the same arteries, others were not, although the borders themselves did not change. Whereas distributing arteries remained at the border of each zone but did not enter it, other arteries branched from these distributing vessels and entered the zones, actually delivering the blood to the tissue (i.e., "delivering" arteries). Thus, when the zones and vessels are classified in this way a comparable network structure appears in every heart. The zones and the distributing and delivering roles of coronary arteries are permanent features of the network,

whereas the exact course, size and identity of each vessel are incidental features [20].

3.2. Quantitative Description of the Coronary Arterial network

In most instances the information derived from vascular corrosion casts, whatever the tissue involved, has been of an entirely qualitative nature. Recently, however, the branching angles, diameters, and lengths of coronary arteries in rats and humans have been measured from corrosion casts [21–23]. Based on these data, Zamir and his colleagues have been able to compare the branching characteristics of the coronary network with those predicted theoretically for optimal efficiency as a blood conveying system.

At arterial branches, the angle between a daughter vessel and the flow direction in the parent is usually less than 90°. In this way the flow in the parent vessel is not required to reverse its direction prior to entering the daughter, as would be necessary if the angle were greater than 90°. However, the coronary-aortic junctions do not conform to this general pattern [24], supporting the possibility that flow at the base of the ascending aorta comes to rest before entering the coronary arteries. The coronary-aortic branching angles would not then be important from a fluid dynamics standpoint and would be influenced mainly by the gross anatomy of the heart. These authors point out that atherosclerotic disease often affects the coronary arteries in the vicinity of these junctions, and that their unusual structure and the unusual flow conditions under which they function may help to explain why these sites are particularly prone to atherosclerosis.

Optimal branching of a bifurcation is such that four factors are at or near a minimum [23]:
1. Power required to drive blood through the bifurcation
2. Drag force exerted on endothelial tissue
3. Volume of blood required to fill the lumen of the bifurcation
4. Surface area of luminal tissue required to form the bifurcation

Studies from corrosion casts of various vascular beds show that branching angles and branch diameters in arterial networks are interrelated, being determined by the above optimality criteria. However, similar measurements in rat heart show the existence of significant departures from optimality, suggesting that the branching pattern of coronary arteries is governed by other factors, in addition to those listed above [23]. Interestingly, in human hearts the corresponding measurements reveal a much closer agreement with the values predicted from optimality considerations, the additional energy required as a result of the departures from optimality being 5% or less in most cases [22].

The optimality of the branching pattern depends critically on the vessel diameters, not on their lengths. For this reason, the rate of branching of a coronary artery is best defined as the rate at which the diameter decreases as a function of the number of branching sites along the vessel, not as a function of distance. In fact, a wide variation of segment lengths is found at all levels of the network. When the branching rate is measured in this way, it is clear that "delivering" arteries (Section 3.1) branch at a much greater rate than the "distributing" arteries [21]. This finding provides additional support for the classification of coronary arteries into these two distinct groups [20], for if the role of "distributing" arteries is to convey blood to the borders of the various myocardial zones, they should not branch as profusely as "delivering" arteries, and their caliber should not diminish as rapidly along their course [21].

3.3. Neovascular Meshes in the Walls of Coronary Arteries

Based on observations from angiograms, the presence of *neovascularization* in the region of atherosclerotic plaques in the walls of human coronary arteries was reported [25] in 1984. One year later, corrosion casts of fine vascular meshes at similar sites were produced using Batson's No. 17 casting compound [26]. These meshes were present not only in association with calcified masses in arterial walls, but also in regions where no evidence of calcification was found. The fine vessels had diameters ranging from 20 to 50 μm, considerably larger than the diameters of blood capillaries in the myocardium (2–9 μm in rat L. ventricle: [9]). The density of fine vessels in these meshworks far exceeded that of normal

vasa vasorum, and each meshwork was linked to the lumen of the host artery by as few as only one or two very fine branches.

The evidence indicates that these vascular meshworks are not uniformly distributed along the walls of coronary arteries, but form almost exclusively in association with sites of atherosclerotic plaque formation. Their localized occurrence would support the view that these meshworks represent the formation of new blood vessels. Zamir and Silver [26] suggested a possible association of this neovasculature with a potential for intramural hemorrhage, inasmuch as casting material was sometimes found to have leaked out from the fine vessels into the calcified masses.

4. Coronary Microvasculature

When corrosion casting compounds of high viscosity are employed, incomplete filling of the microvasculature will result. Areas totally devoid of capillary and venular filling will be present in such casts, permitting clear views of the coronary arterial network. It can then be seen that casts of the arterial and arteriolar lumens, all exhibit characteristically elongated endothelial nuclear impressions, similar to those reported for arteries in other tissues and in marked contrast to the rounded impressions seen in venous vessels [27].

When the viscosity of the casting compound is reduced sufficiently, the material will penetrate into even the smallest capillaries, and virtually complete filling of the microvasculature can be obtained [9]. Under these circumstances, corrosion casts of the epicardial and endocardial surfaces of the heart demonstrate clearly the course of venular channels and capillaries that lie in the surface plane. However, the locations of the arterioles that ascend toward the surface from deeper layers of the myocardium can be inferred only on the basis of the branching pattern of the capillaries overlying these points.

4.1. Microvascular Network Geometry

A microcorrosion cast of the surface of the canine myocardium [28] showing a capillary network with its arteriolar origin and venular drainage is shown in Fig. 14-1a. The capillaries originate

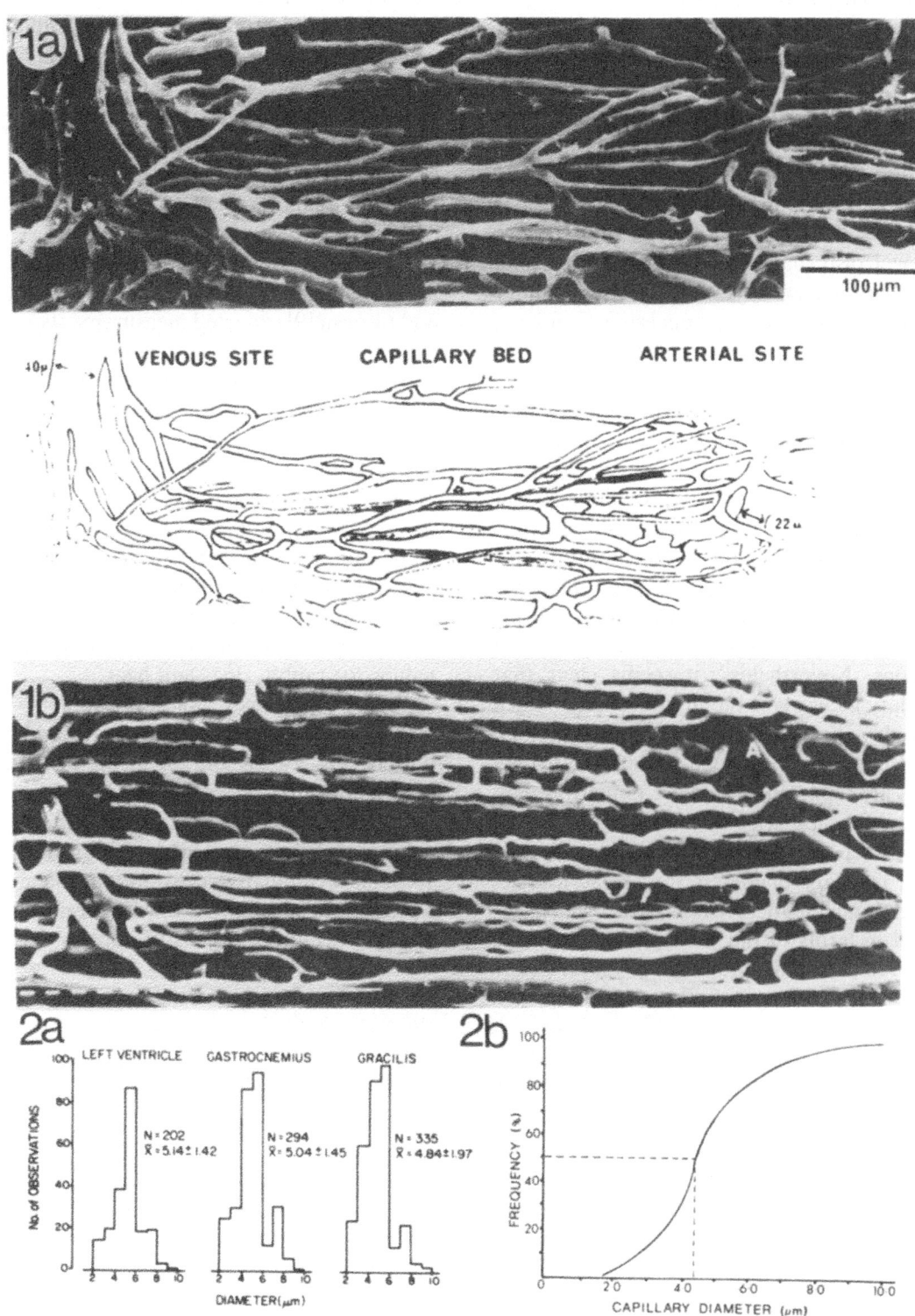

VENOUS SITE   CAPILLARY BED   ARTERIAL SITE

from an arteriole that ascends toward the surface from below, and then travel a distance of 400 μm (approx.) to unite with tributaries of a common venular outflow. The axes of these tributaries begin roughly parallel to the capillaries and then turn through 90° as the vessels merge to form a venule 40–50 μm in diameter. Acid-digestion studies show that most of the capillaries lie parallel to the surrounding myocardial fibers [11], although the branched nature of these fibers causes local disturbances in this pattern. Thus, capillaries in different layers of the myocardium run in different directions, appearing in microcorrosion casts to cross over one another [7: Fig. 3].

The measurement of the distribution of capillary interbranch distances and total path lengths from microcorrosion casts of the heart is made difficult by the complex three-dimensional geometry of the network. For this reason descriptions of the network found in the literature are based on qualitative observations, with quantitation of capillary diameters and/or intercapillary distances mentioned only as a corollary of such descriptions. Nevertheless, it seems clear that the capillary network of the myocardium is much more highly branched than that of skeletal muscle. Connections between capillary networks and veins are reported to be more numerous than those between capillaries and arteries [7].

Several reports in the literature cite the values of capillary diameters in cardiac muscle, measured from microcorrosion casts, which are expressed in terms of the extremes of the range encountered. Although the range of values differs in each case, all the ranges reported lie between the limits of 2.5 and 10 μm (rabbit [7], dog [11], bear [5]). More insight into the caliber of capillaries is obtained when the measurements are expressed in terms of the population mean and standard deviation. Such data have been reported [12] for rat epicardium and endocardium (diameters 5.7 μm ± 1.7 (SD), n = 190, and 5.5 μm ± 2.0 (SD), n = 164, respectively). Although the statistical significance of the difference between these two population means was not reported, it seems unlikely that there exists any difference in capillary diameter between the epicardium and endocardium.

Capillary diameters in the myocardium appear to be similar to those in skeletal muscle. Potter and Groom [9] reported measurements from casts of rat L. ventricle [5.14 μm ± 1.42 (SD), n = 202] and found the values to be no different, statistically, from those in the gastrocnemius and gracilis muscles [5.04 μm ± 1.45 (SD), n = 294, and 4.84 μm ± 1.97 (SD), n = 335, respectively]. At present, there is no satisfactory explanation for the apparent difference between the above values for rat heart and those of Hossler et al. [12]. The extreme values for capillary diameters in rat (Fig. 14-2a) agree closely with the range of 2.5–10 μm cited above, which included the values for rabbit, dog, and bear.

Since the range of values for capillary diameter was practically identical in heart and skeletal muscle (Fig. 14-2a) and the mean diameters were not significantly different, the three sets of data may be grouped together to yield an overall frequency distribution. This has been done in Fig. 14-2b as a cumulative distribution, the analysis of which reveals a skewing toward larger diameters. The lower limit of the distribution is of particular interest, since it lies below 2.45 μm, which represents the smallest diameter of cylindrical tubes through which red cells of the rat can easily pass. From a population of 831 capillaries in rats, 2.5% were less than 2.5 μm and 1.1% were less than

←

Figure 14-1. SEM micrographs (composite) taken at the surface of microcorrosion casts from (a) canine myocardium and (b) rat gastrocnemius fully extended in situ. A portion of the capillary network from each is shown, together with its associated arteriolar and venular trees. In a the arteriole ascends to the surface from beneath, whereas in b it lies within the plane of view. The capillary network is more highly branched in a than in b. The short white bar in b represents 10 μm. (Figures from: (a) Shozawa et al. [28] and (b) Groom et al. [41], with permission.)

Figure 14-2. a: Distributions of capillary diameters measured from microcorrosion casts of L. ventricle, gastrocnemius, and gracilis of the rat. All histograms appear similar, and the mean diameters are not significantly different (p > 0.2). b: Cumulative frequency distribution of capillary diameters, obtained by combining the data for all muscles. The overall mean capillary diameter is 4.97 μm ± 1.44 (SD); 10.3% of diameters are below 3.0 μm, 2.5% are below 2.5 μm and 1.1% are below 2.3 μm. (Figures from Potter and Groom [9], with permission.)

2.3 µm in diameter [9]. In the microcirculation the pressure difference between the arteriole and venule is less than 10 mmHg, whereas in order to aspirate rat red cells into micropipets 100 µm in length and 2.3 µm in diameter, pressure differences of 30–40 mmHg are required. Therefore, the data suggest that the smallest capillaries (1–2% of the total population) must be channels for plasma flow alone. It is possible that these may be newly formed capillaries, for in rat skeletal muscle the union of opposing capillary sprouts has been observed in vivo to yield a vessel 0.5–1.5 µm in diameter [29].

4.2. "Anomalous" Vascular Routes

The focus of this section is on routes other than those of the typical arteriolar-capillary-venular network shown in Fig. 14-1a; these routes are thus designated as *anomalous*. It is of interest that whereas arteriovenous anastomoses (AVAs) have never been described in skeletal muscle, at least two groups of investigators have reported their existence in microcorrosion casts of cardiac muscle from a variety of wild and domesticated animals [30], rabbits [7], and humans [4]. Two different types of AVAs have been observed, the first lying just beneath the epicardial surface (diameters are 50–140 µm in rabbit and 150 µm in humans) and the second of much smaller diameter (20–35 µm in rabbit) in the subendocardium and ventricular septum. In the epicardium several AVAs converge on a single vein without communication with the capillary network, whereas in the subendocardium the vessels communicate with the capillary network as well. To our knowledge, the functional significance of AVAs in cardiac muscle has not been addressed.

The intraarterial anastomoses described in cardiac muscle [30] have their counterpart in the arterial "arcades," which are well known in skeletal muscle. Venous-venous anastomoses have been described in the human heart [4]. In the epicardium a coarse network of such vessels, 30–75 µm in diameter, forms interconnections with capillaries, whereas in the myocardium similar interconnections have not yet been reported. Small arterial vessels from the ventricular free wall and septum are reported to terminate in the R. ventricular cavity, but never in the L. ventricular cavity [7]. However, it is difficult to determine, from the micrographs presented, whether the "terminations" were not actually the result of breakage of the arterial casts. The functional significance of arterial terminations in the R. ventricular cavity is difficult to assess. Venous terminations in the ventricular cavities [7] presumably correspond to Thebesian veins.

4.3. Evidences of Localized "Vascular" Contraction

The innermost layer of smooth muscle in the walls of small arterioles may be arranged circularly, helically, or both. It is intriguing that replicas of perivascular structures, resembling the description of smooth muscle cells, have been reported around small arterioles in microcorrosion casts [for instance, see 31]. These replicas are caused, presumably, by exudation of the casting compound between adjacent endothelial cells. Whether these structures reflect the true replication of smooth muscle cells remains speculative, at best.

Dramatically altered luminal configurations of arterioles have been reported in microcorrosion

→

Figure 14-3. The tortuous configuration of capillaries is seen occasionally in microcorrosion casts of myocardium. a: Contracture of cat myocardium results in a profusion of extremely tortuous capillaries. Bar = 100 m. (From Phillips et al. [6], with permission.) b: Under conditions thought to represent diastole, most capillaries in rat myocardium are reasonably straight, but a few are tortuous. The capillary network is predominantly parallel in nature, with many loops and anastomoses. Bar = 20 µm. (From Potter and Groom [9], with permission.)

Figure 14-4. a, b:Microcorrosion casts of rat soleus muscle. In extended muscle [a: sarcomere length, l_o, 2.55 µm ± 0.19 (SE)] the capillaries show slight undulations, whereas in shortened muscle [b: l_o, 1.62 µm ± 0.02 (SE)] the capillaries are very tortuous. Insets: Light micrographs of portions of capillaries and adjacent muscle fibers from the same cast muscles, taken before the remaining tissue was digested. Bars = 10 µm. c: The fraction of capillary length density (i.e., capillary length/fiber volume) contributed by tortuosity and branching, c(k, o), increases as l_o is reduced below 2.0 µm. Filled circles = cast muscles; open circles and squares = perfusion-fixed muscles; solid line = theoretical curve. (Figures from Mathieu-Costello et al. [13], with permission.)

189

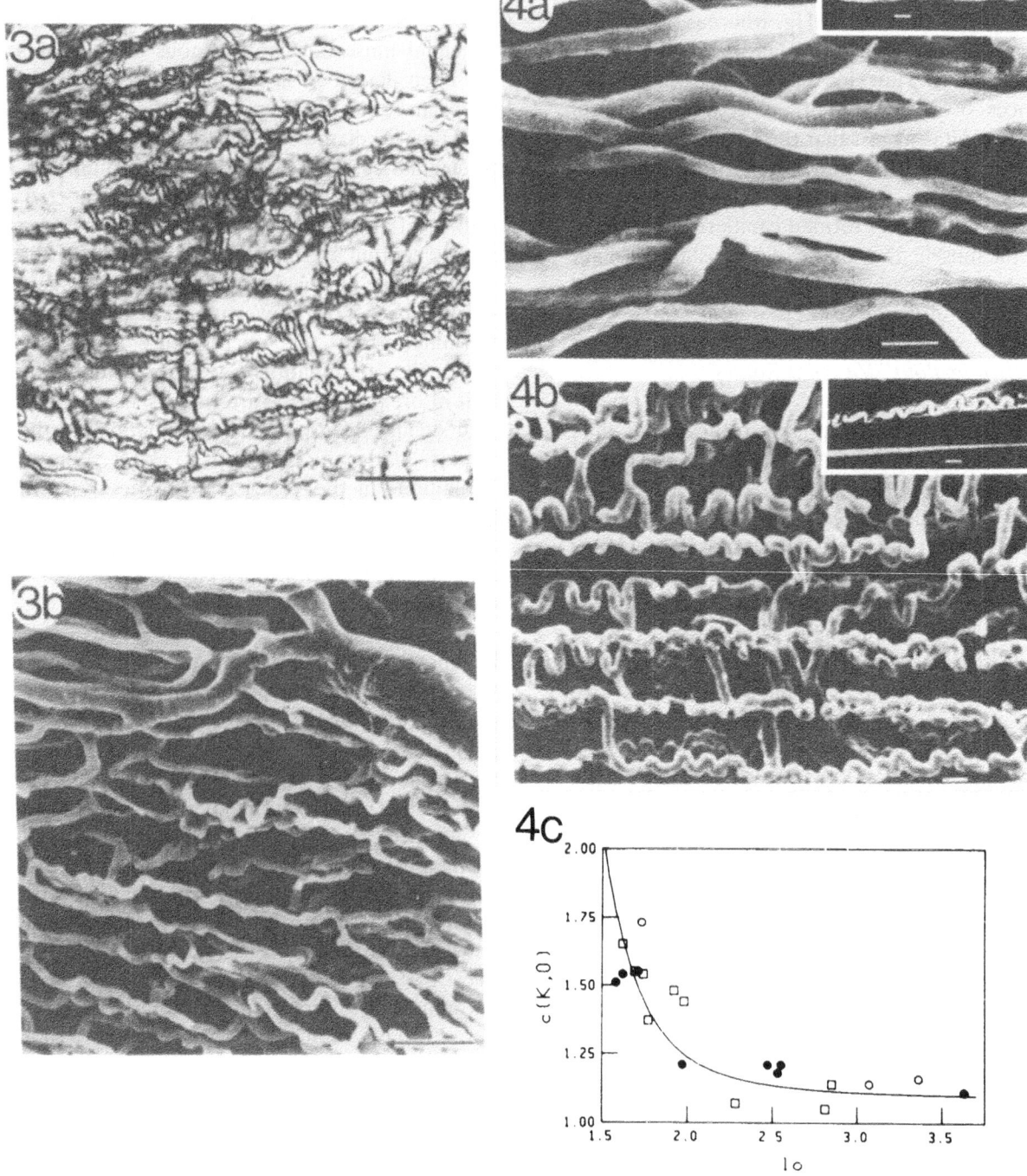

casts of dog myocardium [10: Fig. 3]. Assuming that no localized myocyte contraction had occurred, the fact that the precapillary arterioles became highly tortuous, whereas the capillaries remained relatively straight, suggests that contraction of vascular smooth muscle must have been the cause. Variations in the degree of smooth muscle contraction may result in a less dramatic conformational change, such as the torsion of arterioles in cat myocardium [6].

Evidence from microcorrosion casts for the existence of vascular sphincters in myocardium is limited to local narrowing of arteriolar lumens. However, in our opinion most of the micrographs presented in the literature are at too low a magnification for the reader to be fully convinced. In view of the importance of vascular sphincters for flow redistribution within the microcirculation, there is need for high-resolution images of areas of the casts that are purported to represent sphincters.

4.4. Capillary Geometry in Relation to the State of Cardiac Muscle

SEM micrographs published in the literature suggest that in most microcorrosion casts of heart the capillaries are comparatively straight. However, an extremely tortuous (serpentine) capillary configuration is seen when the myocardium is in a state of contracture (Fig. 14-3a). Taken together these findings might suggest that in normal hearts the capillaries are straight and that only during contracture do they become tortuous. This interpretation may not be correct, since it does not account for the presence of both straight and tortuous capillaries in the same field of view (Fig. 14-3b). An alternative possibility might be that changes in capillary configuration occur as a continuous function of sarcomere length, rather than as a sudden change when the myocardium passes from its normal state into a state of contracture. (In skeletal muscle, the degree of capillary tortuosity has indeed been shown to be a consequence of muscle shortening; see next section.) On this basis it can be inferred that the presence of straight capillaries, exclusively, in vascular casts of the heart signifies that casting must have taken place during diastole. There remains a challenge for researchers to prepare

casts of the microvasculature from hearts in various degrees of contraction, in order to confirm or refute the above possibilities.

An unusual feature seen in microcorrosion casts of human myocardium is that in areas of focal necrosis, the arterioles and capillaries are grossly dilated and tortuous [4]. The diameters of these capillaries appear to be two to three times those of the normal capillaries with which they communicate. Histological sections of such necrotic areas show fewer muscle bundles than normal, but many dilated capillaries. After their origin from short dilated arterial segments that branch at 90° from 100–150 µm arteries, these capillaries form coarse networks. The size and tortuous appearance of capillaries in areas of focal myocardial necrosis are strongly reminiscent of the neovasculature found in tumors. This raises the interesting possibility that similar angiogenic factors may be involved in both pathologic situations.

5. Skeletal Muscle Microvasculature

The classical description of the microvascular network in skeletal muscle was given by Spälteholz [32] over a century ago. It reads as follows:

The arteries supplying a muscle branch freely, and between the branches there are very numerous anastomoses forming a primary network. Into the meshes of this net small arteries are given off at regular intervals, and these anastomose freely, forming a secondary cuboidal net of great regularity. From the threads of this network the arterioles branch off, generally at right angles to the muscle fibres and at very regular intervals (of about 1 mm in the warm blooded animal) and these arterioles split up into a large number of capillaries running along the muscle fibres and in the main parallel to them, but with numerous anastomoses, forming long narrow meshes about the fibres. The capillaries unite into venules intercalated regularly between the arterioles, and the whole system of veins reproduces and follows almost exactly that of the arteries.

(Translation by Plyley [33].)

Much of this description has been confirmed from later investigations, based on methods such as histological staining, intravascular injection of contrast material, and intravital microscopy. Since each of these techniques possesses unique advantages as well as limitations, it follows that no one technique is sufficient, by itself, for a comprehensive study of the microvasculature. However, when results obtained from several different methods are combined, the achievement of this goal becomes feasible. The purpose of this section is to review the contributions of microcorrosion casting to our knowledge of skeletal muscle microvasculature.

5.1. Microvascular Network Geometry

Scanning electron microscopy of microcorrosion casts provides an unparalleled three-dimensional visual perspective of the microvascular network. In Fig. 14-1b, photomicrographs of adjacent fields of view from skeletal muscle are combined, so as to display entire capillary pathways from arterioles to venules. This cast was prepared with the muscle in its fully extended position in situ, which ensured that the capillaries were predominantly straight and parallel to the muscle fibers. Measurements of capillary path lengths from such views are difficult, because of the large depth of field afforded by SEM. More accurate measures may be obtained using light microscopic techniques to examine the muscle in vivo, or after perfusion with contrast material (silicone elastomer [34], fluorescein-labelled albumin [35]). Microcorrosion casts may offer the possibility of evaluating the concept of discrete microvascular units in skeletal muscle, described from fluorescein studies of hamster tibialis in vivo [35].

Corrosion casts provide a good basis for the measurement of capillary luminal diameters (see Section 4.1). Use of the modified Batson's casting compound described by Nopanitaya et al. [18] yielded a similar range of capillary diameters in cast as in perfusion-fixed noncast samples, from muscles of the rat hindlimb [13]. This suggests that systematic errors due to swelling or shrinkage of the material had not occurred during polymerization. Furthermore, capillary luminal diameters measured from such casts are in good agreement with in vivo measurements [9: Fig. 2].

Stereo-paired SEM micrographs of corrosion casts offer the possibility of quantitative spatial analysis (stereogrammetry) of the microvascular network. However, in view of recent rapid advances in digital image analysis, such an application of stereogrammetry to vascular networks has been superseded by computer digitization of serial sections followed by three-dimensional reconstruction. This technique also offers the opportunity of viewing the network from any desired direction, or of viewing any desired plane within the network.

5.2. Changes in Capillary Configuration as a Consequence of Muscle Shortening

The increased capillary sinuosity in contracted muscle, recognized many years ago by Ranvier [36] and Krogh [37], remained almost unnoticed in later investigations until very recently. The advent of microcorrosion casting, by providing "3-D snapshot" views of the microvasculature at any instant in time, offered a unique opportunity to investigate changes in capillary orientation as a consequence of muscle shortening.

In 1983, Potter and Groom [9] demonstrated a dramatic change in capillary configuration between the extended and passively shortened gastrocnemius and gracilis muscles of the rat. In extended muscle the capillary network was comprised of long, straight vessels, running parallel to the fibers, and joined at intervals by short anastomotic channels. This arrangement corresponded closely to the description given by Spälteholz (Section 5). In passively shortened muscle, however, the capillaries formed a tightly meshed network of convoluted vessels around the fibers. The fact that tortuous capillaries were found only in shortened muscles suggested that the change in capillary configuration from straight to tortuous must have occurred as a result of the process of muscle shortening.

Later reports, although confirming the existence of tortuous capillaries in skeletal muscle, attributed their origin to factors other than (or in addition to) muscle shortening. It has been proposed that the presence or absence of tortuous vessels is determined primarily by fiber type [3], whereas another report has suggested that capillary tortuosity represents a chronic response to endurance training [38]. The existence of tortuous capillaries has been demonstrated in chronically stimulated muscle [39]. Appell [38] considered that "contraction is probably not responsible for the formation of tortuous capillaries." A weakness common to all the reports cited in Section 5.2 thus far is that the degree of extension or shortening of the muscles (i.e., sarcomere length) was not taken into account. In the absence of measurements of sarcomere length, all the foregoing conclusions remain tentative, at best.

There has been, to date, only one study designed to investigate capillary tortuosity from microcorrosion casts on a quantitative basis [13]. In this study transmission electron microscopy (TEM), SEM, and stereological methods were combined in the same cast muscle to provide information on (1) the degree of muscle shortening, by measuring the sarcomere lengths (l_o) of the muscle fibers, and (2) the fraction of capillary length density (i.e., capillary length/fiber volume) contributed by tortuosity and branching, c(k,o). In casts of extended muscle [$l_o = 2.55\,\mu m \pm 0.19$ (SE)] the capillaries were found in a relatively straight configuration, lying parallel to the direction of the muscle fibers (Fig. 14-4a). In casts of shortened muscle [$l_o = 1.62\,\mu m \pm 0.02$ (SE)], however, all capillaries running in a direction parallel to the muscle fibers now showed an amazingly tortuous configuration (Fig. 14-4b). Interconnections between neighboring capillaries were formed by anastomoses passing around the perimeter of adjacent empty spaces originally occupied by muscle fibers. The stereological results leave little doubt that the degree of capillary tortuosity was related to the degree of muscle shortening (Fig. 14-4c).

In microcorrosion casts of the heart, the presence of tortuous capillaries had been assumed to be an artefact, the result of muscle contracture induced (presumably) by the casting material [4,6,7]. In the case of skeletal muscle, however, this possibility was excluded by TEM studies of cast muscles that had been treated with

glutaradehyde, rather than by the conventional digestion process [13]. This investigation established that, following microcorrosion casting in passively shortened or extended muscle in situ, sarcomere lengths lay within the normal range of $1.6-3.5\,\mu m$, and no evidence of contracture was found [13: Fig. 3]. In the same study, stereological estimates of capillary anisotropy as a function of sarcomere length were found to be the same in perfusion-fixed as in cast muscles, indicating that capillary configuration had not been altered by either the casting compound or the method of infusion.

When microcorrosion casts of shortened muscles were transected [9] and viewed end-on (Fig. 14-5a), the tortuous capillaries, joined by anastomotic vessels, were seen to form envelopes of polygonal (usually pentagonal or hexagonal) shape. The diameters of these envelopes were $20-60\,\mu m$, comparable in magnitude to the diameters of the muscle fibers themselves. Thus, a single muscle fiber could be contained within each of these envelopes, the capillary configuration suggesting the possibility of a roughly uniform oxygen field around each fiber. At the surface of the muscle, however, the capillary convolutions and anastomotic vessels formed troughs (Fig. 14-5b), each presumably having contained one muscle fiber. From such views, it would appear that the superficial muscle fibers are not "enclosed" by the capillary network when muscle shortens.

5.3. Implications of Capillary Network Geometry for Oxygen Transport to Tissue, in Extended Versus Shortened Muscle

The capillary network is the major functional unit for exchange of diffusible substances between blood and tissue. In skeletal muscle, therefore, capillary/fiber geometry is of crucial importance in determining both the rate and uniformity of oxygen transport to tissue. For modelling oxygen transport, the distribution of capillaries and

→

Figure 14-5. a: Cross-sectional view of microcorrosion cast from rat gastrocnemius muscle at its shortest in situ length. Note polygonal envelopes, formed by convoluted capillaries and their anastomotic vessels. A single muscle fiber could be contained within each core. Bar = 20 μm. (From Ellis et al. [42], with permission.) b: View at surface of microcorrosion cast from rat gastrocnemius muscle at its shortest in situ length. U-shaped troughs, formed by tortuous capillaries and their anastomotic vessels, reveal locations previously occupied by individual muscle fibers. Bar = 25 μm. (From Groom et al. [43], with permission.)

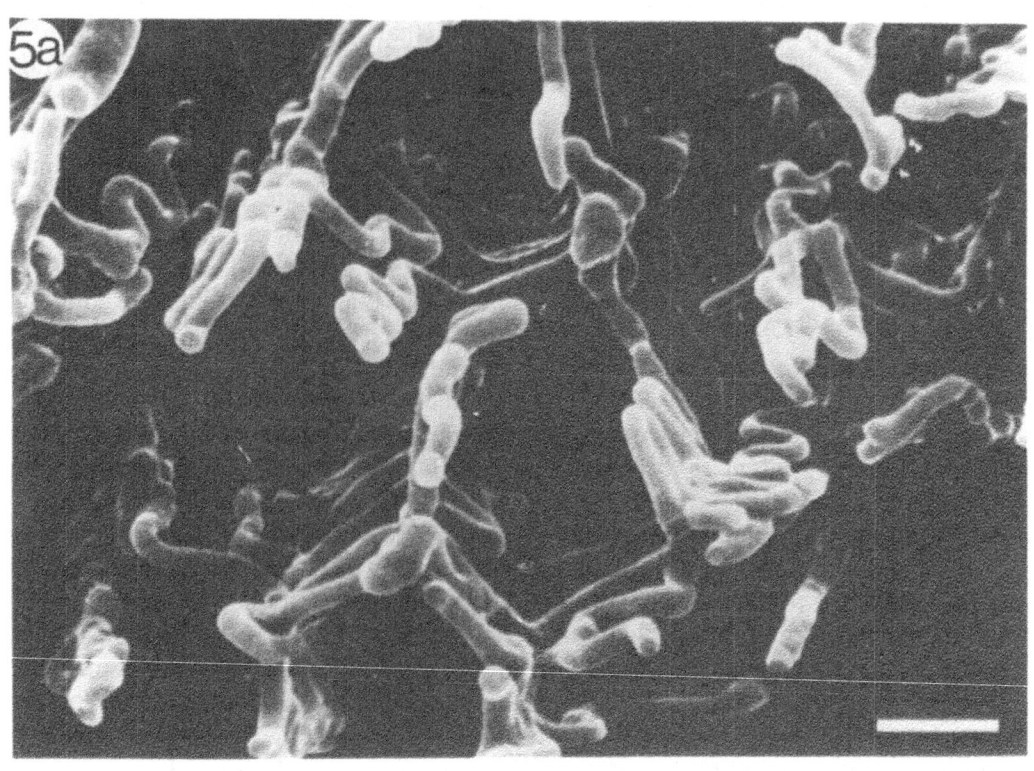

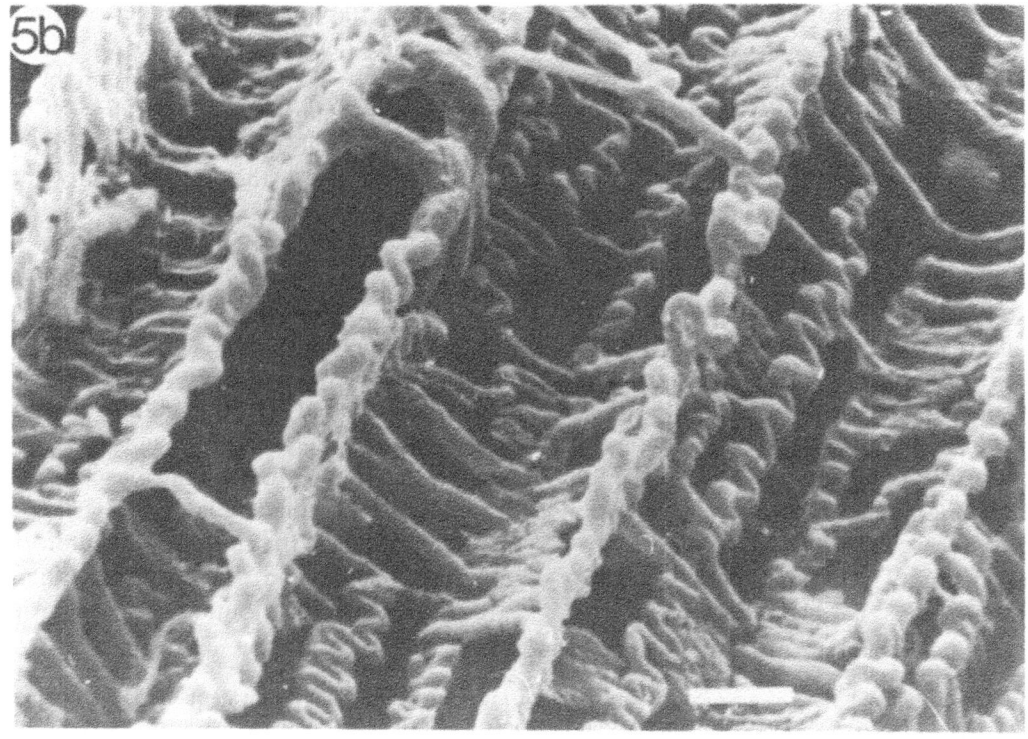

194

the oxygen consumption rate are assumed to be uniform throughout each tissue cross section. In reality, however, capillaries are located solely in the spaces between muscle fibers, the maximum diffusion distances (fiber radii) being governed by fiber type and oxidative capacity. Increasing the number of capillaries at the perimeter of a fiber (i.e., increased capillary density per mm^2 in transverse section) will not diminish the maximum diffusion distance at all, except when intercapillary distances are greater than the fiber diameter.

The fact that differences in oxidative capacity exist among muscle fibers suggests the possibility of a heterogeneous distribution of RBC flow within the capillary network. Such heterogeneity could be brought about by the physical design of the capillary network, as well as by the response of contractile elements at the arteriolar level to local tissue needs. The presence of bifurcations and anastomotic vessels must of itself result in differences in RBC perfusion among different segments of the capillary network. In addition, a fourfold range of luminal diameters among capillaries [9] and differences in total path length [34] will augment the nonuniformity of capillary RBC perfusion. Consequently, substantial differences in blood oxygen tension must exist among capillaries in any cross section of the tissue, leading in turn to diffusional interactions between capillaries. This raises doubts concerning the fundamental assumptions on which the classical "Krogh cylinder" model for oxygen diffusion from blood to tissue is based [40].

The tortuous configuration of capillaries seen in microcorrosion casts from shortened muscles appears to provide two major advantages for oxygen exchange [13]. First, the capillary surface area per unit length of muscle fiber is greater than predicted, on the assumption of straight capillaries, from the capillary density per mm^2 in transverse sections of muscle. Second, the geometry for oxygen transfer is altered. The sinuous folding of capillaries about each muscle fiber provides for axial diffusion of oxygen between the folds of each capillary, as well as enhanced diffusional interactions between adjacent capillaries. This will tend to create a relatively uniform oxygen tension around the periphery of each fiber cross section [40].

For purposes of modelling oxygen transport to tissue, the models of Krogh and Hill represent limiting cases at opposite extremes [40]. The Krogh cylinder model is based on the outward diffusion of oxygen from a straight capillary into the surrounding tissue, whereas the Hill model is based on the inward diffusion of oxygen into a cylinder of tissue (in this case, a muscle fiber) from a uniform peripheral oxygen supply. In extended muscle, the situation corresponds to some position intermediate between these two extremes. When muscle shortening occurs, however, the new situation will correspond closely to that represented by the Hill model. Analysis shows that as a result of the tortuous capillary configuration seen in microcorrosion casts of shortened muscle, there will be improved oxygen transfer from blood to tissue. Indeed, at high rates of oxygen consumption during exercise, this improvement is essential to prevent a fraction of each muscle fiber from becoming anoxic [40].

6. Concluding Remarks

Blood vascular corrosion casting has made possible significant recent advances in the understanding of vascular network geometry in cardiac and skeletal muscle. In normal human hearts the arterial network always divides the myocardium into the same six distinct zones, in spite of variabilities in vessel location from heart to heart [20]. Arterial branching patterns agree closely with theoretical predictions based on optimality principles, the energy costs of departures from optimality being <5% in most cases. In the epicardium, capillaries, originating from an arteriole that ascends toward the surface from below, travel a distance of 400 µm (approx.) to unite with tributaries of a common venular outflow. Most capillaries lie parallel to the adjacent myocardial fibers, forming a network that is more highly branched than and one-half the path length of that in skeletal muscle. Capillary diameters in the myocardium [rat: 5.14 µm ± 1.42 (SD)] are not significantly different from those in skeletal muscle. A dramatic change in capillary configuration, from straight to tortuous, occurs during shortening of skeletal muscle. The degree of tortuosity is related to the sarcomere length of the muscle fibers. A tortuous capillary configuration may create a relatively uniform oxygen tension around any fiber cross section, enhancing oxygen exchange severalfold.

The full potential of blood vascular casting in heart and skeletal muscle has yet to be realized. The focus of most studies to date has been the description, in qualitative terms, of vascular pathways at the surface of the epicardium or skeletal muscle. Investigators are now faced with several important new challenges. First, the extraction of quantitative information from vascular corrosion casts promises to provide new insights on structure/function relationships. This possibility becomes even greater when morphometric analysis of tissue and SEM analysis of corrosion casts are carried out side by side on aliquots of the same tissue specimen. Second, there is a need to follow microvascular pathways three dimensionally in deeper layers of the cast and, indeed, throughout the entire thickness. In this connection, the value of incomplete casts needs to be appreciated. Because of the paucity of vessels filled with casting material, one is able to see in considerable detail the structure of those vessels that have filled. Third, the interpretation of fine details of microcorrosion casts is enhanced immeasurably when SEM of tissue and SEM of casts are combined.

Using these approaches to study the heart, it will be possible to reveal in detail the arteriolar portion of the microvasculature, the distribution of arteriolar-to-venular distances, and the capillary diameters and branching patterns at the venular versus arteriolar ends of the network. Such studies are needed for a comparison of the microvasculature in the endocardium and epicardium. Does capillary tortuosity occur in the heart solely as a result of contracture, or is it a physiological phenomenon, as in skeletal muscle? The combination of stereological measurements with microcorrosion casting may provide an answer to this question. Further studies of "anomalous" microvascular pathways are needed, together with high-resolution images, to support claims for the existence of vascular sphincters.

In skeletal muscle, there is need for definitive identification of the discrete microvascular units proposed recently on the basis of observations in vivo. Moreover, are there characteristic differences in microvascular network geometry between oxidative and glycolytic muscles? Are special adaptations to be found in highly oxidative muscles, e.g., flight muscles in migratory birds? Much might be learned from a comparison of microcorrosion casts of pectoralis versus hindlimb muscles in these species. The beauty of such studies is that we may discern from them the operation of basic biophysical principles that govern the design of microvascular networks under various situations.

Acknowledgments

This work was supported by a grant from the Heart and Stroke Foundation of Ontario to A.C. Groom. The authors wish to thank Mrs. Barbara Anderson for typing the manuscript.

References

1. Lametschwandtner A, Lametschwandtner U, Weiger T. Scanning electron microscopy of vascular corrosion casts — technique and applications. *Scann Electron Microsc* II:663–695, 1984.
2. Hodde KC, Nowell JA. SEM of micro-corrosion casts. *Scann Electron Microsc* II:89–106, 1980.
3. Gaudio E, Pannarale L, Marinozzi G. An S.E.M. corrosion cast study on pericyte localization and role in microcirculation of skeletal muscle. *Angiology* 36:458–464, 1985.
4. Ono T, Shimohara Y, Okada K, Irino S. Scanning electron microscopic studies on microvascular architecture of human coronary vessels by corrosion casts: Normal and focal necrosis. *Scann Electron Microsc* I:263–270, 1986.
5. Anderson WD, Anderson BG, Seguin RJ. Microvasculature of the bear heart demonstrated by scanning electron microscopy. *Acta Anat* 131:305–313, 1988.
6. Phillips SJ, Rosenberg A, Meir-Levi D, Pappas E. Visualization of the coronary microvascular bed by light and scanning electron microscopy, and x-ray in the mammalian heart. *Scann Electron Microsc* III:735–742, 1979.
7. Irino S, Ono T, Shimohara Y. Microvascular architecture of the rabbit ventricular walls: A scanning electron microscopic study of corrosion casts. *Scann Electron Microsc* IV:1785–1792, 1982.
8. Sage MD, Gavin JB. Morphological identification of functional capillaries in the myocardium. *Anat Rec* 208:283–289, 1984.

9. Potter RF, Groom AC. Capillary diameter and geometry in cardiac and skeletal muscle studied by means of corrosion casts. *Microvasc Res* 25:68–84, 1983.

10. Anderson BA, Anderson WD. Microvasculature of the canine heart demonstrated by scanning electron microscopy. *Am J Anat* 158:217–227, 1980.

11. Anderson BA, Anderson WD. Myocardial microvasculature studied by microcorrosion casts. *Biomed Res* 2 (Suppl.):209–217, 1981.

12. Hossler FE, Douglas JE, Douglas LE. Anatomy and morphometry of myocardial capillaries studied with vascular corrosion casting and scanning electron microscopy: A method for rat heart. *Scann Electron Microsc* IV:1469–1475, 1986.

13. Mathieu-Costello O, Potter RF, Ellis CG, Groom AC. Capillary configuration and fiber shortening in muscles of the rat hindlimb: Correlation between corrosion casts and stereological measurements. *Microvasc Res* 36:40–55, 1988.

14. Murakami T. Application of the scanning electron microscope to the study of the fine distribution of the blood vessels. *Arch Histol Jpn* 32:445–454, 1971.

15. Gannon BJ. Pre-polymerization of methacrylate corrosion casting media for microvascular replication using ultra-violet light. *J Anat* 120:665, 1979.

16. Nowell JA, Pangborn J, Tyler WS. Replication of internal biological surfaces. In: Arceneaux CJ (ed.), *Ann. Proc. Electron Microscopy Soc. Amer.*, pp 308–309, 1972.

17. Matsusaka T, Fujibashi I. Ultrastructural specialization appearing on the plastic cast of ocular blood vessels. *J Clin Electron Microsc* 7:227–228, 1974.

18. Nopanitaya W, Aghajanian JG, Gray LD. An improved plastic mixture for corrosion casting of the gastrointestinal microvascular system. *Scann Electron Microsc* III:751–755, 1979.

19. Hadziselimovic H. Vascularization of the conducting system in the human heart. *Acta Anat* 102:105–110, 1978.

20. Zamir M, Silver MD. Morpho-functional anatomy of the human coronary arteries with reference to myocardial ischemia. *Can J Cardiol* 1:363–372, 1985.

21. Zamir M. Distributing and delivering vessels of the human heart. *J Gen Physiol* 91:725–735, 1988.

22. Zamir M, Chee H. Branching characteristics of human coronary arteries. *Can J Physiol Pharmacol* 64:661–668, 1986.

23. Zamir M, Phipps S, Langille BL, Wonnacott TH. Branching characteristics of coronary arteries in rats. *Can J Physiol Pharmacol* 62:1453–1459, 1984.

24. Zamir M, Sinclair P. Roots and calibers of the human coronary arteries. *Am J Anat* 183:226–234, 1988.

25. Barger AC, Beeuwkes R III, Lainey LL. Hypothesis: Vasa vasorum and neovascularization of human coronary arteries. *N Engl J Med* 310:175–177, 1984.

26. Zamir M, Silver MD. Vasculature in the walls of human coronary arteries. *Arch Pathol Lab Med* 109:659–662, 1985.

27. Gnepp DR, and Green, FHY. SEM of collecting lymphatic vessels and their comparison to arteries and veins. *Scann Electron Microsc* III:756–762, 1979.

28. Shozawa T, Kawamura K, Okada E. Study of intramyocardial microangioarchitecture with respect to pathogenesis of focal myocardial necrosis. *Bibl Anat* 20:511–516, 1981.

29. Myrhage R, Hudlicka O. Capillary growth in chronically stimulated adult skeletal muscle as studied by intravital microscopy and histological methods in rabbits and rats. *Microvasc Res* 16:73–90, 1978.

30. Hadziselimovic H, Secerov D, Gmaz-Nikulin E. Comparative anatomical investigations on coronary arteries in wild and domestic animals. *Acta Anat* 90:16–35, 1974.

31. Castenholz A. Visualization of periendothelial cells in arterioles and capillaries by scanning electron microscopy of ultrasound treated and plastoid® injected brains in rats. *Scann Electron Microsc*, 1983.

32. Spälteholz W. Die Vertheilung der Blutgefässe in Muskel. *Abh Sachs Ges Eiss Math Phys* 14:509–528, 1888.

33. Plyley MJ. The Organization of the Capillary Network in Skeletal Muscle. Ph.D. Thesis, University of Western Ontario, Canada, pp 20, 1977.

34. Plyley MJ, Sutherland GJ, Groom AC. Geometry of the capillary network in skeletal muscle. *Microvasc Res* 11:161–173, 1976.

35. Lund N, Damon DH, Damon DN, Duling BR. Capillary grouping in hamster tibialis anterior muscles: Flow patterns and physiological significance. *Int J Microcirc Clin Exp* 5:359–372, 1987.

36. Ranvier L. De quelques faits relatifs à l'histologie et à la physiologie des muscles striés. *Arch Physiol Norm Pathol* 2e série 1:5–15, 1874.

37. Krogh A. The number and distribution of capillaries in muscles with calculations of the oxygen pressure head necessary for supplying the tissue. *J Physiol* (London) 52:409–415, 1919.

38. Appell HL. Variability in microvascular pattern dependent upon muscle fiber composition. *Prog Appl Microcirc* 5:15–29, 1984.

39. Dawson JM, Hudlicka O. The effects of long-term activity on the microvasculature of rat glycolytic skeletal muscle. *Int J Microcirc Clin Exp* 8:53–69, 1989.

40. Groom AC, Ellis CG, Potter RF. Microvascular architecture and red cell perfusion in skeletal muscle. *Prog Appl Microcirc* 5:64–83, 1984.

41. Groom AC, Ellis CG, Wrigley SM, Potter RF. Architecture and flow patterns in capillary networks of skeletal muscle in frog and rat. In: Popel AS, Johnson PC (eds.), *Microvascular Networks: Experimental and Theoretical Studies*, S. Karger, Basel, pp 61–76, 1986.

42. Ellis CG, Potter RF, Groom AC. The Krogh cylinder geometry is not appropriate for modelling O_2 transport in contracted skeletal muscle. *Adv Exp Med Biol* 159:253–268, 1983.

43. Groom AC, Ellis CG, Potter RF. Microvascular geometry in relation to modeling oxygen transport in contracted skeletal muscle. *Am Rev Respir Dis* 129 (Suppl.):S6–S9, 1984.

Authors' address:
Dr. Alan C. Groom and Dr. Richard F. Potter
Department Medical Biophysics
Health Sciences Centre
University of Western Ontario
London, Ontario
Canada   N6A 5Cl

Vasculature of the Carotid Body as Observed by Scanning Electron Microscopy of Vascular Casts

TAKEHITO TAGUCHI, TAKURO MURAKAMI, & AIJI OHTSUKA

1. Introduction

The carotid body is a chemoreceptor sensitive to the partial pressure of oxygen (pO_2) and carbon dioxide (pCO_2), pH, osmolarity, and temperature of the blood [1,2]. In this organ, capillary flow can be regulated independently of total flow [3,4]. It is generally believed that capillaries [5] and arteriovenous anastomoses [3,4] are involved in such regulation in the organ. Many individuals have studied the blood flow or vascular channels in the mammalian carotid body [5–27]. This chapter introduces the scanning electron microscope findings on the cast carotid vasculature, with some comments on previous light or transmission electron microscopic findings.

2. Preparation and Scanning Electron Microscopy of Vascular Casts

Male Wistar rats weighing 150–200 g were used. Vascular casts were prepared by the method of Murakami [28]. Animals were thoroughly perfused with physiological saline through the ascending aorta. Then a small amount (0.8–1.0 ml) of a commercially available methacrylate medium (Mercox; Japan Vilene Inc., Tokyo) was injected through the aorta. The carotid body and its surrounding tissues were isolated as a block, immersed in a hot water bath (60°C) for 2–3 hours, corroded in a 20% NaOH solution overnight, and washed several times in hot water (60–80°C). Vascular casts thus prepared were mounted on metal stubs, coated with gold, and observed with a scanning electron microscope. Some casts were frozen and cut with razor blades to expose the inner vessels of interest.

3. Vascular Organization of the Carotid Body

3.1. Carotid Body Artery

The rat carotid body receives a single artery, the carotid body artery, which arises from the external carotid artery or from the occipital artery [5,26] (Fig. 15-1). Before entering the carotid body, the carotid body artery also gives off two or more branches [26] (Fig. 15-1), which supply the vagus nerve, superior cervical ganglion, nodosa ganglion, and carotid sinus [26]. A marked constriction indicative of an intraarterial cushion is constantly imprinted at the origin of the carotid body artery [5,26] (Fig. 15-1).

Previous light microscopic studies of tissue sections or dye-injected specimens have shown that in dogs [8,9], cats [8,9,18], and rabbits [8,27], the carotid body receives one or more arteries that arise from the ascending pharyngeal, occipi-

Motta, P.M., Murakami, T., and Fujita, H. (eds.), Scanning Electron Microscopy of Vascular Casts: Methods and Applications.

tal, external carotid, or internal carotid arteries. The carotid body arteries of these animals also have additional branches that supply the tissues or organs surrounding the carotid body [8,18,26]. Previous studies have further shown that the cushion acts as a valve, rather than a sphincter [20,25], since it contains few muscle fibers [5] and receives few nerve terminals [20]. Some species differences in occurrence have been reported in the cushion; remarkable cushions are observed in rats [5,19,20,23,25,27], dogs [9], and mice [15], but not in cats [20] and rabbits [20,27].

3.2. Arteries and Arterioles

Scanning electron microscopy of vascular casts clearly shows that in the rat, the carotid body artery enters the carotid body at the caudal pole and immediately divides at this pole into three or more branches (Fig. 15-1). These branches further divide into finer branches. Some of these branches terminate as the proper carotid branches within the carotid body, while the remaining branches pass as the pseudo-carotid branches through the body (Fig. 15-1). The terminal arterioles of the proper branches are randomly distributed throughout the body. The proper carotid branches always show shallow constrictions at their origins, suggesting the existence of an intraarterial cushion; the pseudo-branches show no constrictions [26] (Fig. 15-1).

Acker and his associates indicated, by measuring hydrogen partial pressure (pH_2) in the cat and rabbit carotid bodies, that dissociation between capillary flow and total flow occurs during changes of pO_2 and blood pressure: in the normoxic condition, changes in blood pressure correlatively induce an alteration of total flow, but not of capillary flow; in the hypoxic state, capillary flow significantly decreases, regardless of the increase in total flow [3,4]. They considered that these dissociated alterations are due to the existence of blood flow through shunt vessels, such as arteriovenous anastomoses (see below) [3,4]. Some other authors contended that in the rat carotid body, the capillary flow is regulated by the terminal arterioles, precapillary sphincters, and the intraarterial cushions of the proper carotid branches [5,21,24]. Noteworthy is that most branches of the carotid body artery or pseudo-carotid branches pass through the body. These branches may be involved in eliminating excessive blood flow into the body [26].

Measurements of tissue pO_2 in the cat and rabbit carotid bodies have shown that the carotid body is possibly perfused with plasma alone or plasma containing very few red blood cells [4,29,30]. It has been indicated that small arteries, which arise at a relatively large angle from their parent vessel [31], would carry a larger volume fraction of plasma. In the cast samples, the proper carotid branches with original constrictions or intraarterial cushions resemble such arteries. It may be well exemplified in the rat kidney that the intraarterial cushion skims the plasma [32].

3.3. Capillaries

Two types of capillaries are observed in vascular casts of the rat carotid body: thick capillaries with an average luminal diameter of 8 μm, and thin capillaries with 4 μm in diameter (Fig. 15-2). Terminal arterioles of the proper branches give rise to both types of capillaries, with few anastomoses between them (Fig. 15-2). Thus, our

→

Figure 15-1. Scanning electron micrographs of the vascular beds of the rat carotid bodies cast with methacrylate. A: A dorsal view of the vascular cast of the left carotid body and its surrounding tissues. The carotid body artery (CBA) arises from the external carotid artery (ECA) and gives off additional branches (Ab) supplying tissues or organs around the carotid body (CB). The carotid body artery shows a marked constriction (arrow) at its origin. In the carotid body, the carotid body artery divides into the proper carotid branches (as) supplying the body and the pseudo-carotid branches (ap) passing through the body. Vessels of the carotid body connect with the efferent veins (V), receive veins from surrounding tissues, and finally drain into the internal or external jugular veins. ICA = internal carotid artery; OA = occipital artery; CS =blood vessels of the carotid sinus. B: A ventral view of a vascular cast of the carotid body. The elaborate venous plexus (Vp) is situated in the superficial layer of the body and gathers into efferent veins (V). as = proper carotid branches. C: A cut-surface view of the vascular cast of the carotid body. Proper carotid branches (as) arise at a large angle from the parent vessels and show shallow constrictions (arrowheads) at their origins. ap = pseudo-carotid branches; CBA = carotid body artery; vc= interior collecting vein; V = efferent veins. A: ×120; B: ×250; C: ×400.

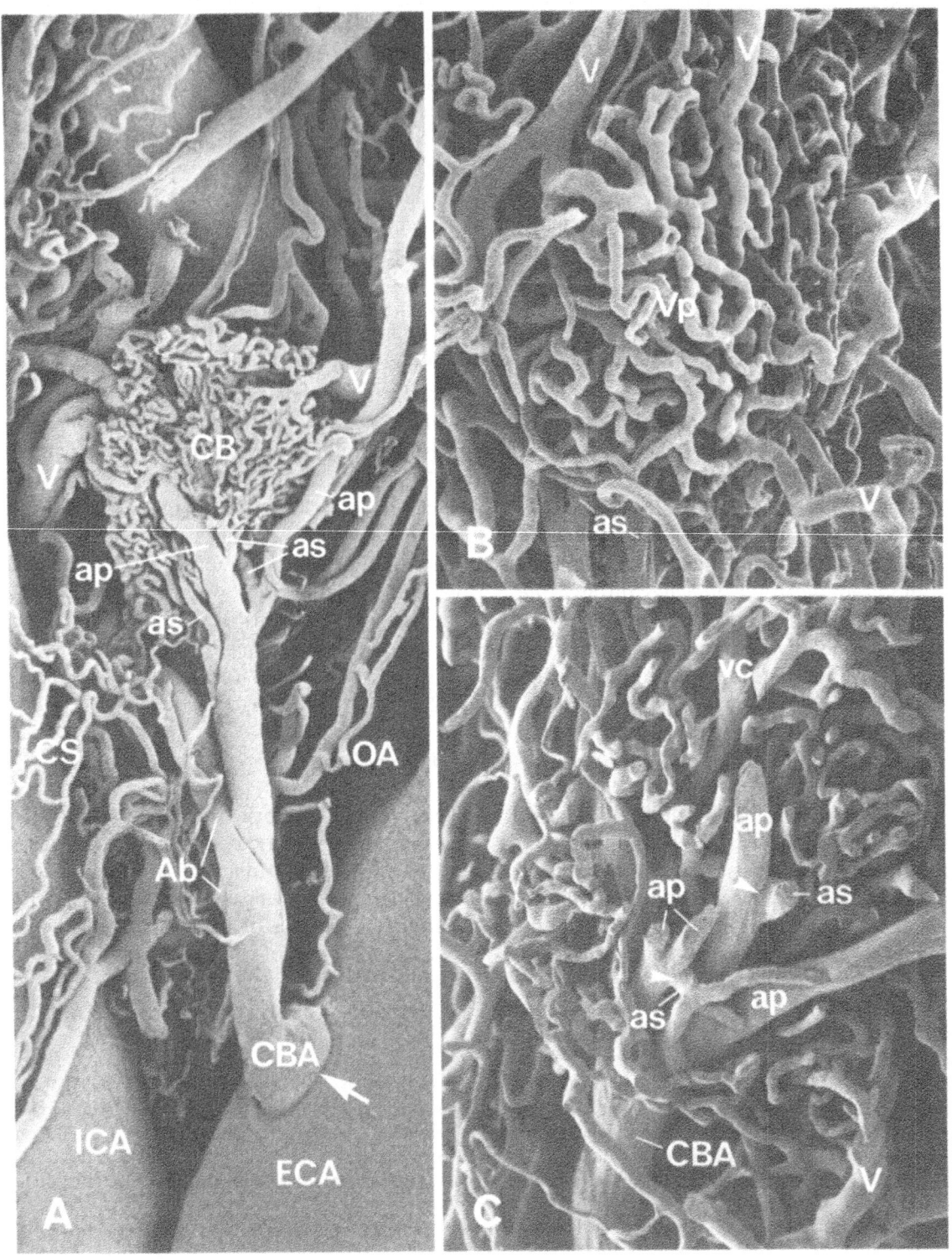

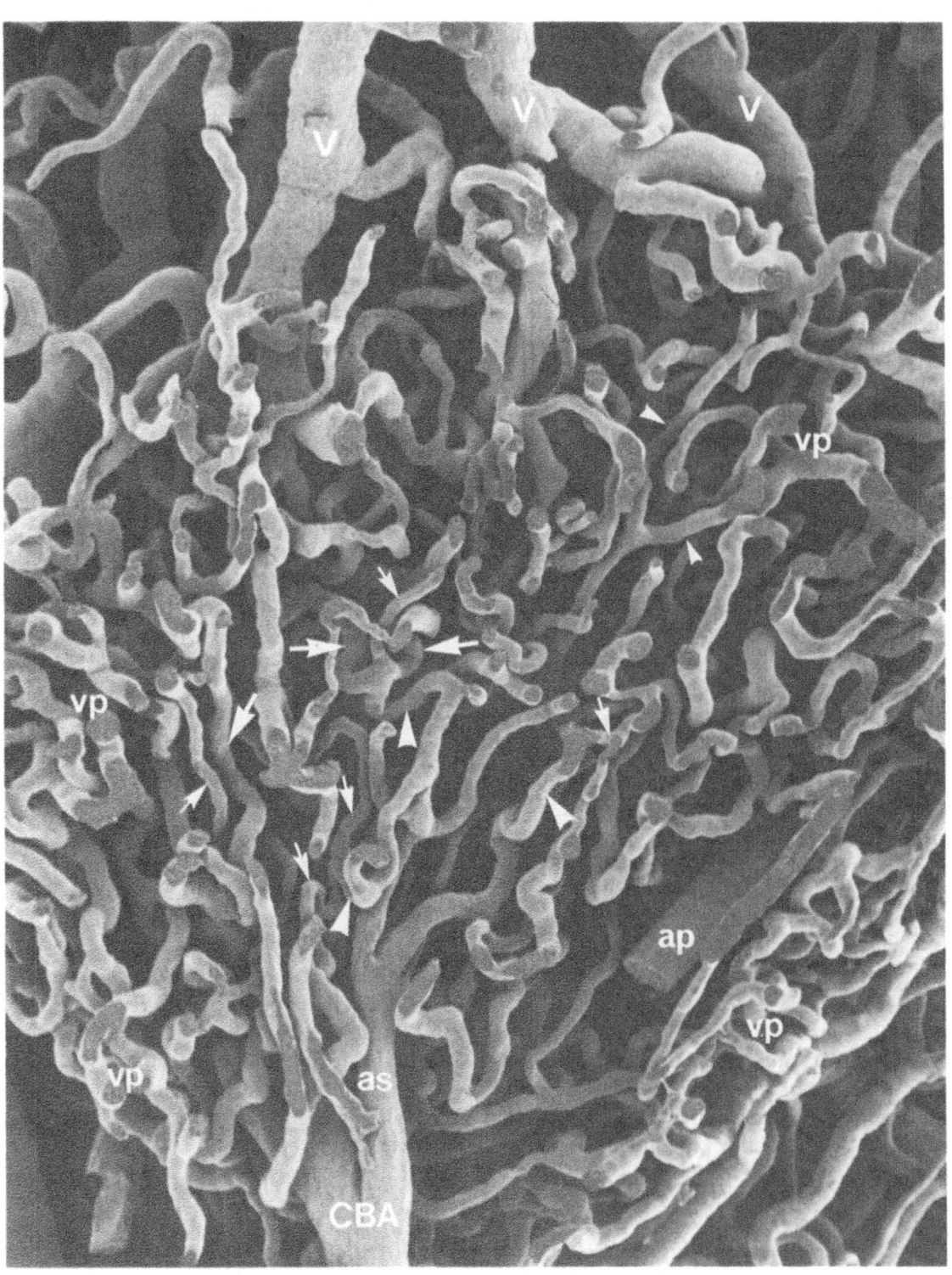

scanning observations of casts coincide with the previous light and transmission electron microscopic findings.

McDonald and his associates described in the rat that Type I capillaries have a luminal diameter ranging from 8 to over 20 μm, vary in caliber along their length, and follow a winding course. Type II capillaries have a uniform luminal diameter of about 7 μm and follow a course that has both straight and curved regions [5,24]. Type I capillary penetrates a single cluster of glomus cells and is associated with this cluster over its entire length, whereas Type II capillary does not penetrate the cluster [5,24]. Furthermore, it has been shown that Type I capillary has numerous endothelial fenestrations, while Type II capillary located in connective tissue has solely endothelial fenestrations near its venous end [5,24]. Similar findings have been obtained by many authors in the human [13], monkey [10], dog [12], seal [14], mouse [15], and rabbit [7].

3.4. Venules and Veins

Venules of the rat carotid body are interconnected with each other to form an elaborate venous plexus (Fig. 15-1). This plexus is conspicuous in the superficial layer of the body [26] (Fig. 15-1). Both types of capillaries drain, at various levels in the body, into the collecting vessels and finally continue into the venous plexus (Fig. 15-1). The plexus gathers into the rostral and caudal efferent veins (Fig. 15-1). The efferent veins receive some veins from some tissues near the body and continue into the internal or external jugular vein [26]. The efferent veins are occasionally anastomosed with each other on or near the carotid body [5,26].

A similar venous plexus has been reported even in dog, cat, and rabbit carotid bodies [8]. Venous drainage of surrounding tissues into the carotid efferent veins can influence chemoreceptor activity of the carotid body [5]. Bingmann and his associates postulated in the cat that a rise in venous pressure abolished the excitatory response of the body to asphyxia [33].

4. Arterio-venous Anastomoses

Arterio-venous anastomoses in or outside the carotid body have been reported in the dog [9], cat [7,11,16], rat [5,22–24], and rabbit [29] by many authors, including some scanning electron microscopists. However, in our casts prepared by sufficient injection, such vessels were never identified [26].

5. Conclusions

In the rat, the carotid body artery arises from the external carotid artery and gives off additional branches supplying surrounding tissues. The origin of this artery shows a marked constriction, indicative of the intraarterial cushion. Within the carotid body, the artery divides into the proper carotid branches terminating within the body and pseudo-carotid branches passing through the body.

The proper carotid branches break up into thick and thin capillaries, with few anastomoses between them. The thick capillaries supply the parenchymal cells, while the thin capillaries supply the connective tissues. Both types of capillaries connect with the venous plexus in the carotid body and continue via the carotid efferent veins into the internal or external jugular vein. No arterio-venous anastomoses are observed in any areas of the carotid body.

The blood flow into the carotid body artery may be regulated by the intraarterial cushion. The pseudo-carotid branches may act as bypass routes for regulating blood flow into the parenchyma of the body.

←

Figure 15-2. A scanning electron micrograph of the vascular cast of the rat carotid body prepared by cutting off its ventral layer. Proper carotid branches (as) supplying the carotid body divide into terminal arterioles that are deep (thick arrowheads) or peripheral (thin arrowheads) in the body. Terminal arterioles break up into thick (thick arrows) and thin (thin arrows) capillaries that follow a winding course without showing anastomoses among them. CBA = carotid body artery; vp = parts of the venous plexus; V = efferent veins; ap = a broken pseudo-carotid branch passing through the body. ×470.

204

References

1. Biscoe TJ. Carotid body. Structure and function. *Physiol Rev* 51:437–495, 1971.
2. Eyzaguirre C, Fidone SJ. Transduction mechanisms in carotid body: Glomus cells, putative transmitters, and nerve endings. *Am J Physiol* 239:c135–c152, 1980.
3. Acker H, Lübbers DW. Relationship between local flow, tissue pO_2, and total flow of the cat carotid body. In: Acker S, Fidone S, Pallot D, Zyzaguirre C, Lübbers DW, Torrance RW (eds.), *Chemoreception in the Carotid Body*, Springer-Verlag, New York, pp 271–276, 1977.
4. Acker H. The meaning of tissue pO_2 and local blood flow for the chemoreceptive process of the carotid body. *Fed Proc* 39:2641–2647, 1980.
5. McDonald DM, Larue DT. The ultrastructure and connections of blood vessels supplying the rat carotid body and carotid sinus. *J Neurocytol* 12:117–153, 1983.
6. Muratori G. Ricerche anatomiche sulla vascolarizzazione sanguigna del glomo carotico. *Arch Istit Biochim Ital* 15:145–169, 1943.
7. de Castro F. Sur la structure de la synapse dans les chemorecepteurs: Leur mécanisme d'excitation et rôle dans la circulation sanguine locale. *Acta Physiol Scand* 22:14–43, 1951.
8. Chungcharoen D, Daly BM, Schweitzer A. The blood supply of the carotid body in cats, dogs and rabbits. *J Physiol* 117:347–358, 1952.
9. Serafini-Fracassini A, Volpin D. Some features of the vascularization of the carotid body in the dog. *Acta Anat* 63:571–579, 1966.
10. Al-Lami F, Murray RG. Fine structure of carotid body of *Macaca mulata* monkey. *J Ultrastruct Res* 24:465–478, 1968.
11. de Castro F, Rubio M. The anatomy and innervation of the blood vessels of the carotid body and the role of chemoreceptive reactions in the autoregulation of the blood flow. In: Torrance RW (ed.), *Arterial Chemoreceptor*, Blackwell Scientific, Oxford, pp 267–277, 1968.
12. Kobayashi S. Fine structure of the carotid body of the dog. *Arch Histol Jpn* 30:95–120, 1968.
13. Böck P, Stockinger L, Vyslonzil E. Die Finestructur des Glomus caroticum beim Menschen. *Z Zellforsch* 105:543–568, 1970.
14. Morita E, Chiocchio SR, Tramezzani JH. The carotid body of the Weddell seal (*Leptomychotes weddelli*). *Anat Rec* 167:309–328, 1970.
15. Böck P. Das glomus caroticum der Maus. *Adv Anat Embryol Cell Biol* 48:1–84, 1973.
16. Schäfer D, Seidl E, Acker H, Keller, Lübbers DW. Arteriovenous anastomoses in the cat carotid body. *Z Zellforsch* 142:515–524, 1973.
17. Seidl E. On the morphology of the vascular system of the carotid body of cat and rabbit and its relation to the glomus Type I cells. In: Purves MJ (ed.), *The Peripheral Arterial Chemoreceptors*, Cambridge University Press, New York, pp 293–299, 1975.
18. Seidl E. On the variability of form and vascularization of the cat carotid body. *Anat Embryol* 149:79–86, 1976.
19. Habeck J-O, Honig A, Pfeiffer C, Schmidt M. The carotid bodies in spontaneously hypertensive (SHR) and normotensive rats: A study concerning size, location and blood supply. *Anat Anz* 150:374–384, 1981.
20. Hesse M, Böck P. Studies on intra-arterial cushions: III. The cushions at the origins of the rat carotid body artery (CBA). *Z Mikro-Anat Forsch* 94:471–478, 1980.
21. McDonald DM. A morphometric analysis of blood vessels and perivascular nerves in the rat carotid body. *J Neurocytol* 12:155–199, 1983.
22. Habeck J-O, Honing A, Huckstorf H, Pfeiffer C. Arteriovenous anastomoses at the carotid body of rats. *Anat Anz* 156:209–215, 1984.
23. Habeck J-O, Huckstorf Ch, Honig A. Influence of age on the carotid bodies of spontaneously hypertensive (SHR) and normotensive rats. I. Arterial blood supply. *Exp Path* 26:195–203, 1984.
24. McDonald DM, Haskel A. Morphology of connection between arterioles and capillaries in the rat carotid body analyzed by reconstructing serial sections. In: Pallot DJ (ed.), *The Peripheral arterial chemoreceptors*, Oxford University Press, New York, pp 195–206, 1984.
25. Smith P, Jago R, Heath D. Glomic cells and blood vessels in the hyperplastic carotid bodies of spontaneously hypertensive rats. *Cardiovasc Res* 8:471–482, 1984.
26. Taguchi T. Blood vascular organization of the rat carotid body: A scanning electron microscopic study of corrosion casts. *Arch Histol Jpn* 49:243–254, 1986.
27. Habeck J-O. A comparison of the blood supply of the carotid body in rats and rabbits. *Anat Anz* 164:313–322, 1987.
28. Murakami T. Application of the scanning electron microscope to the study of the fine distribution of the blood vessels. *Arch Histol Jpn* 32:445–454, 1971.
29. Weigelt H, Acker H. Comparative measurements of tissue pO_2 in the carotid body, In: Acker H, Fidone S, Pallot D, Zyzaguirre C, Lübbers DW, Torrance RW (eds.), *Chemoreception in the Carotid Body*. Springer-Verlag, New York, pp 244–249, 1977.
30. Whalen WJ, Nair P. Factors affecting O_2 consumption of the cat carotid body. In: Acker H, Fidone S, Pallot D, Zyzaguirre C, Lübbers DW, Torrance RW (eds.). *Chemoreception in the Carotid Body*, Springer-Verlag, New York, pp 233–239, 1977.
31. Lahiri S. Role of arterial O_2 flow in peripheral chemoreceptor excitation. *Fed Proc* 39:2648–2652, 1980.
32. Fourman J, Moffat OB. The effect of the intra-arterial cushions on plasma-skimming in small arteries. *J Physiol* 158:374–380, 1961.
33. Bingmann D, Shultze H, Caspers H. Activity of chemoreceptors in the carotid body of the cat in relation to changes in venous pressure. In: Purves MJ (ed.), *The Peripheral Arterial Chemoreceptors*. Cambridge University Press, New York, pp 345–356, 1975.

Author's address:
Dr. Takehito Taguchi
Department of Anatomy,
Okayama University Medical School,
2-5-1 Shikata-cho,
Okayama, 700 Japan

Organization and Age-Related Changes of the Rat Hypophyseal Blood Vascular Plexuses and Portal Vessels as Observed by the Injection Replica SEM Method

TAKURO MURAKAMI

1. Introduction

Conventional light microscopy of Indiaink-injected or other tissue samples has long established in humans and various animals, including the rat, that the hypophyseal portal system originates in the median eminence and terminates in the pars distalis [1–4]. Recent scanning electron microscopy of vascular casts (injection replica SEM method) has clearly demonstrated this long portal route in the monkey, rat, and some other animals, as well as the short route connecting the infundibular process and pars distalis [5–13]. This chapter introduces our injection replica SEM findings on the hypophyseal vascular plexuses and portal vessels of newborn, pubescent, adult, and aged rats [6,12,13].

2. Hypophyseal Blood Vascular Bed of the Rat

The rat hypophyseal blood vascular bed consists of the capillary plexuses of the subependyma, median eminence, infundibular stalk, infundibular process, pars tuberalis, pars distalis, and pars intermedia, and contains the long, short, and other portal vessels (Figs. 16-1–16-6) [12,13]. These plexuses and vessels show some age-related differences or vicissitudes (Figs. 16-1–16-6)

[12,13]. The main findings are described below and are schematically diagrammed in Figs. 16-7 and 16-8.

2.1. Systemic Arteries and Veins of the Hypophysis

Throughout newborn, pubescent, adult, and aged rats, the hypophysis receives the main branches of the anterior, middle, and posterior hypophyseal arteries and emits the adenohypophyseal and neurohypophyseal veins [12,13].

The anterior and middle hypophyseal arteries arise from the internal carotid arteries and divide into the hypothalamic and hypophyseal branches (Figs. 16-1 and 16-3) [12,13]. The latter branches descend as the superior proper hypophyseal arteries, along the median eminence and infundibular stalk (Figs. 16-1 and 16-3). These arteries issue many accessory branches, the periinfundibular and infundibular ascending arteries, which run into the periventricular areas of the hypothalamus; the former ascending arteries arise at the upper margin of the eminence (Figs. 16-5B, and 16-7), while the latter ones arise on the external surface of the eminence or stalk, and penetrate the primary plexus (Figs. 16-5A, 16-5C, and 16-7) [13]. The posterior hypophyseal arteries arise from the basilar artery or its branches and

Motta, P.M., Murakami, T., and Fujita, H. (eds.), Scanning Electron Microscopy of Vascular Casts: Methods and Applications.

206

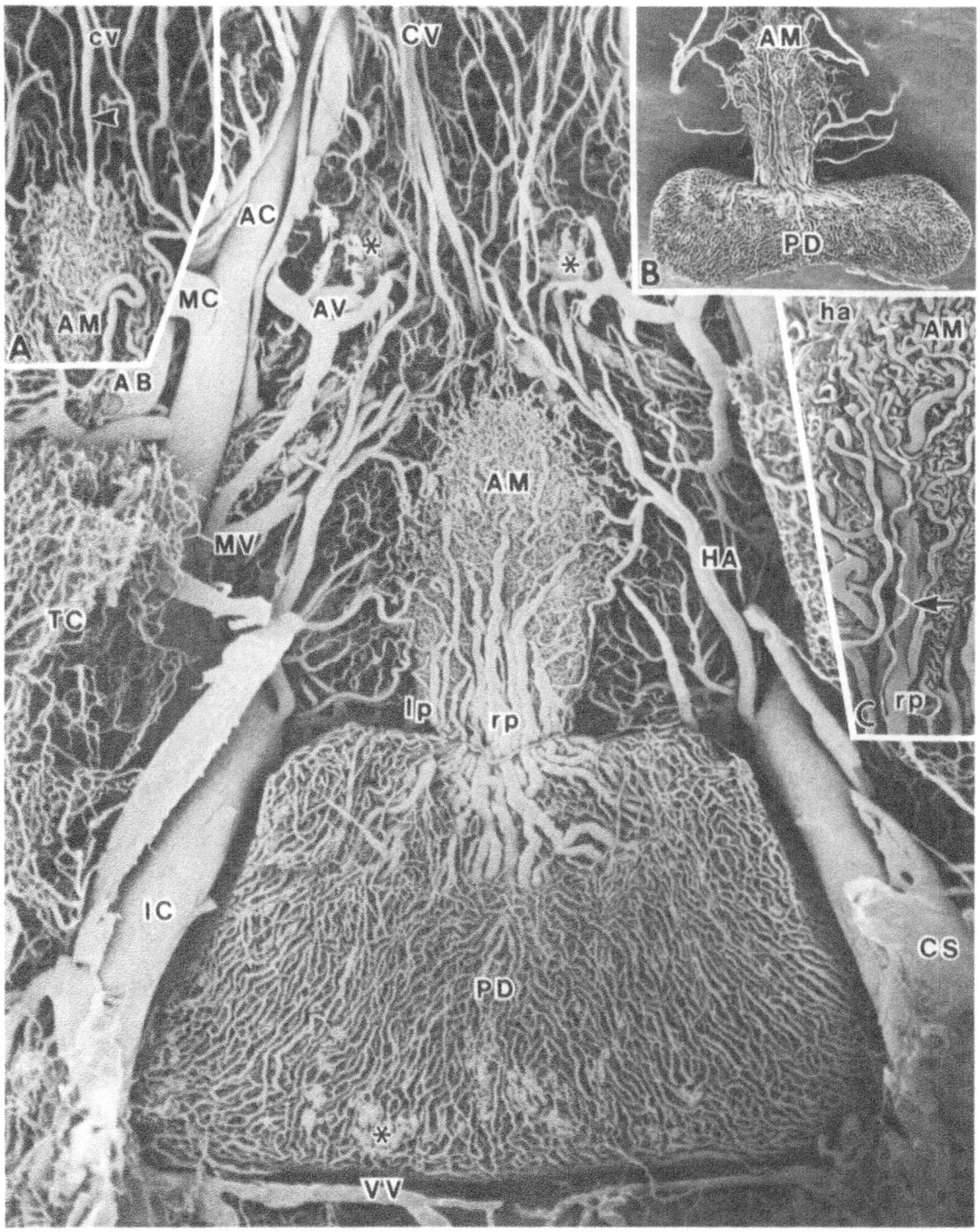

Figure 16-1. Scanning micrograph of the vascular replicas of the hypophysis and adjacent tissues (adult male rat, ventral view). Inset A shows an aberrant systemic vein of the primary plexus (pubescent female rat, ventral view). Inset B shows an isolated hypophyseal replica (male rat, 10 days after birth). Inset C shows an aberrant arteriole (arrow) continuous with a long portal vessel (adult male rat, ventral view). ×35; inset A: ×35; inset B: ×20; inset C: ×35.

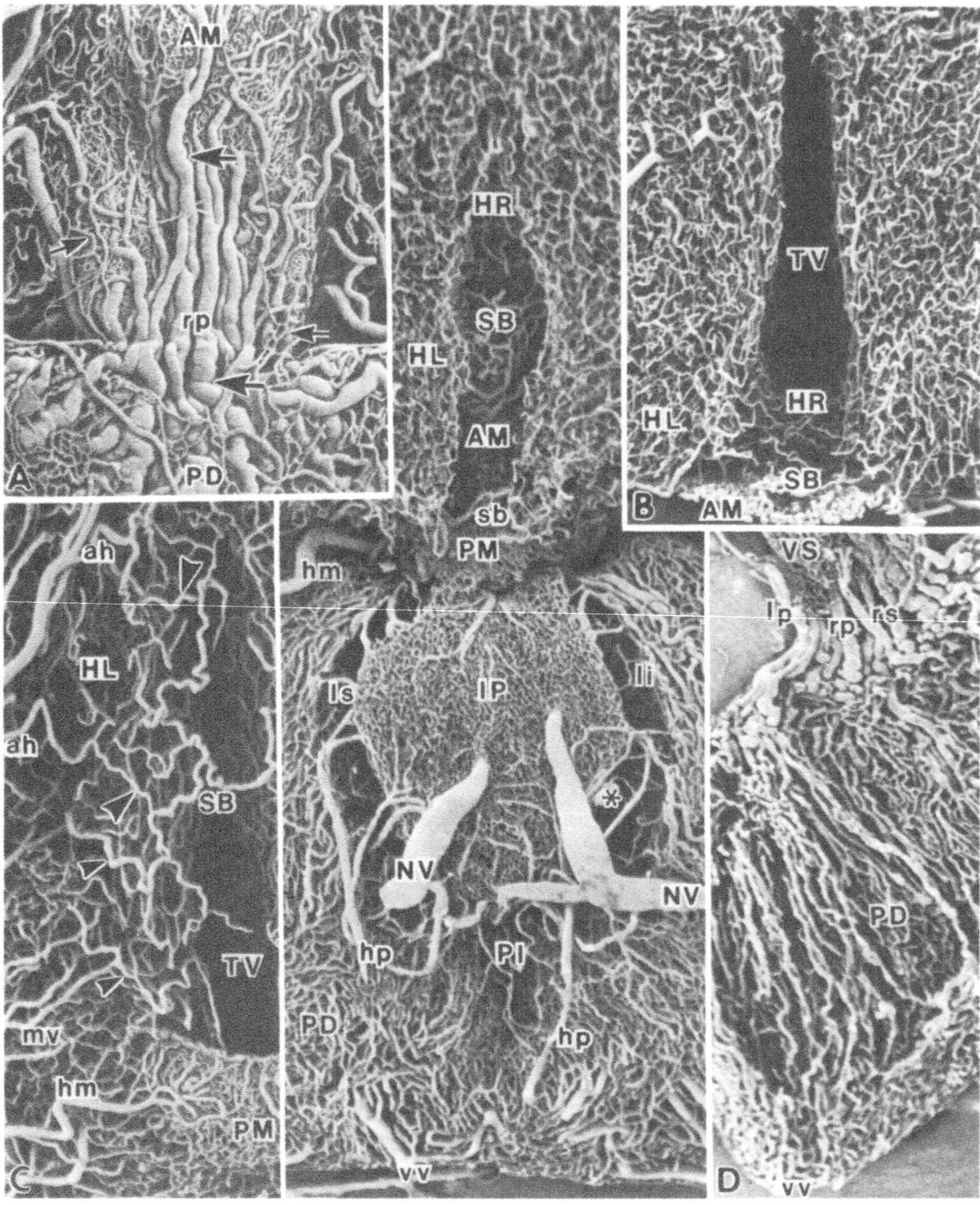

Figure 16-2. Isolated vascular replicas of the hypophysis and hypothalamus (adult male rat, dorsal view; from Takuro Murakami et al. [13], with permission). Inset A shows the plexus of the pars tuberalis (thin arrows) and the constrictions of the long portal vessels (thick arrows) (adult male rat). Inset B shows the frontally cut plexuses of the median eminence and hypothalamus (adult male rat, caudal view; from Takuro Murakami et al. [13], with permission). Inset C shows the infundibular descending arteries (thick arrowheads) and infundibular ascending veins (thin arrowheads) continuous with the subependymal network (SB) (the same specimen as shown in Fig. 16-1A, ventral view; Takuro Murakami et al. [13], with permission). Inset D shows a dissected form of the vascular bed of the pars distalis (PD) (pubescent female rat, dorsal view). ×35; inset A: ×40; inset B: ×35; inset C: ×65; inset D: ×40.

208

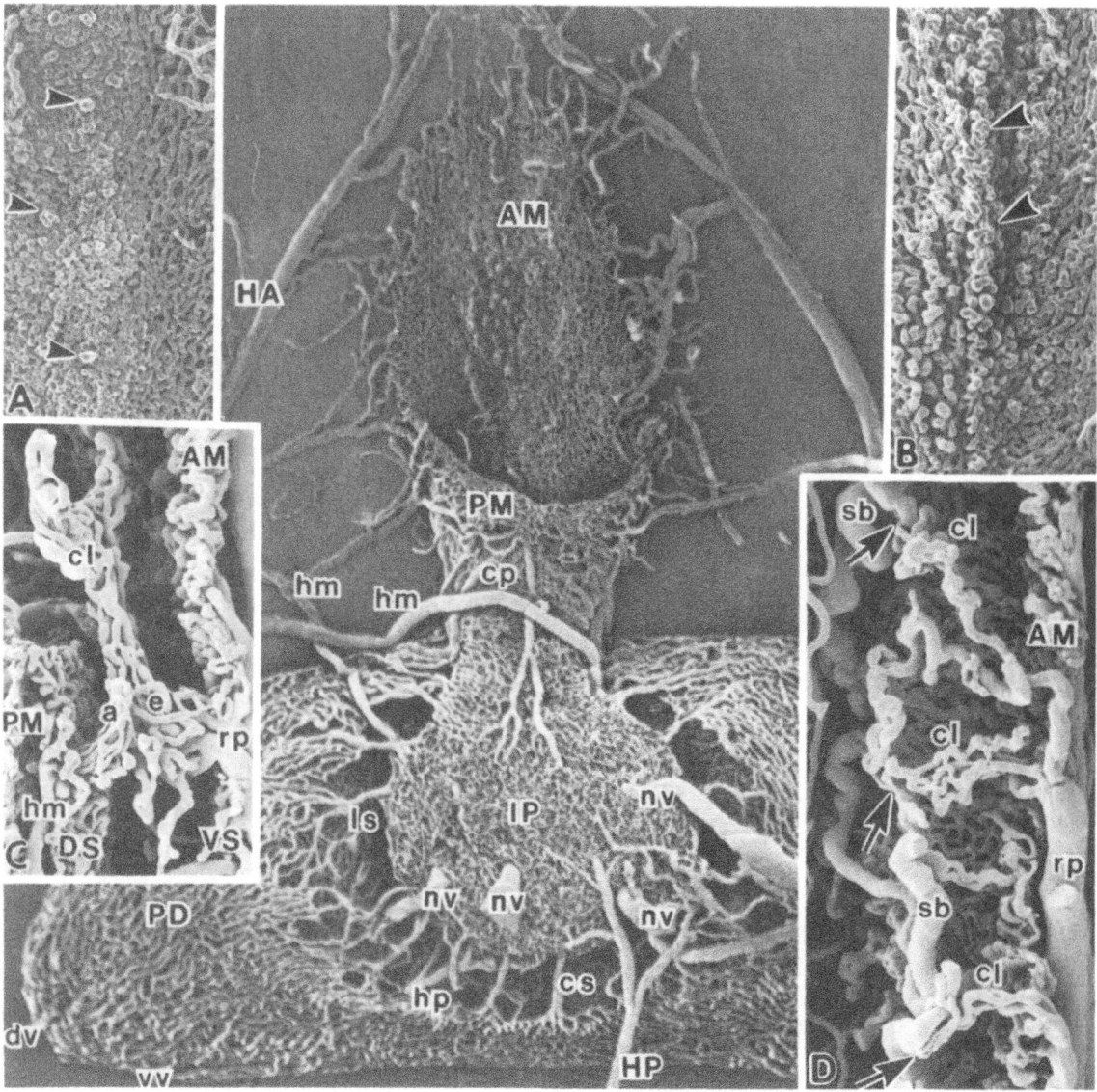

Figure 16-3. Isolated vascular bed of the hypophysis (male rat, 10 days after birth, dorsal view; from Takuro Murakami et al. [13], with permission). The subependymal network was removed. Inset A shows the initial loops (thin arrowheads) of the primary plexus (male rat, 6 days after birth). Inset B shows more developed loops (male rat, 10 days after birth). Inset C shows a fully developed loop (cl), which is provided with the proper afferent (a) and efferent (e) vessels (adult male rat). Inset D shows the subependymo-infundibular capillaries (arrows) connecting the subependymal capillaries (sb) and loops (cl) (adult male rat). ×45; inset A: ×70; inset B: ×70; inset C: ×145; inset D: ×140.

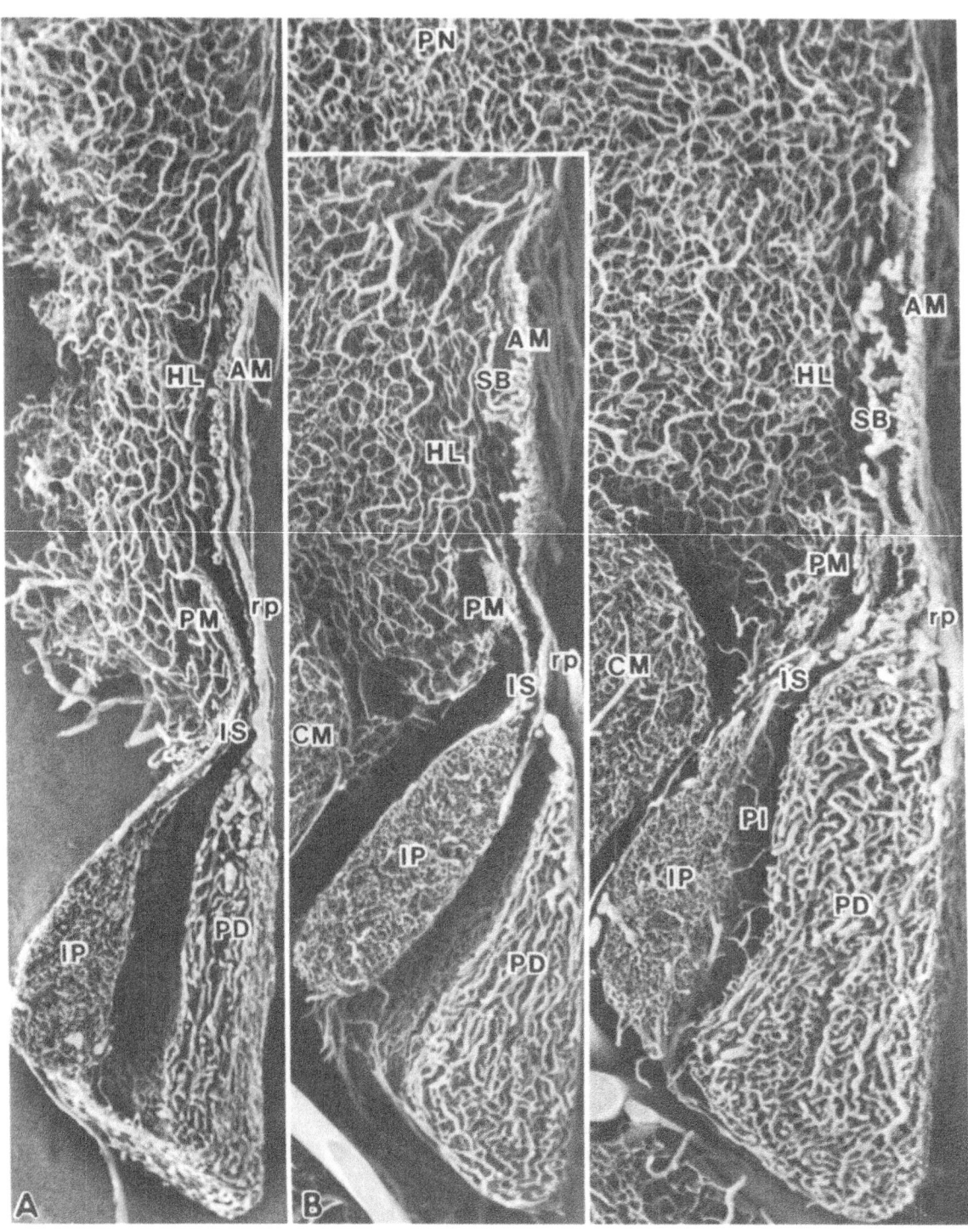

Figure 16-4. Sagittaly freeze-cut hypophyseal and hypothalamic vascular beds (adult male rat; from Takuro Murakami et al. [13], with permission). Insets A and B show the similarily cut hypophyseal and hypothalamic beds of the young rats (A: male rat, two weeks after birth; from Takuro Murakami et al. [13], with permissions, B: pubescent male rat). ×40; inset A: ×55; inset B: ×35.

210

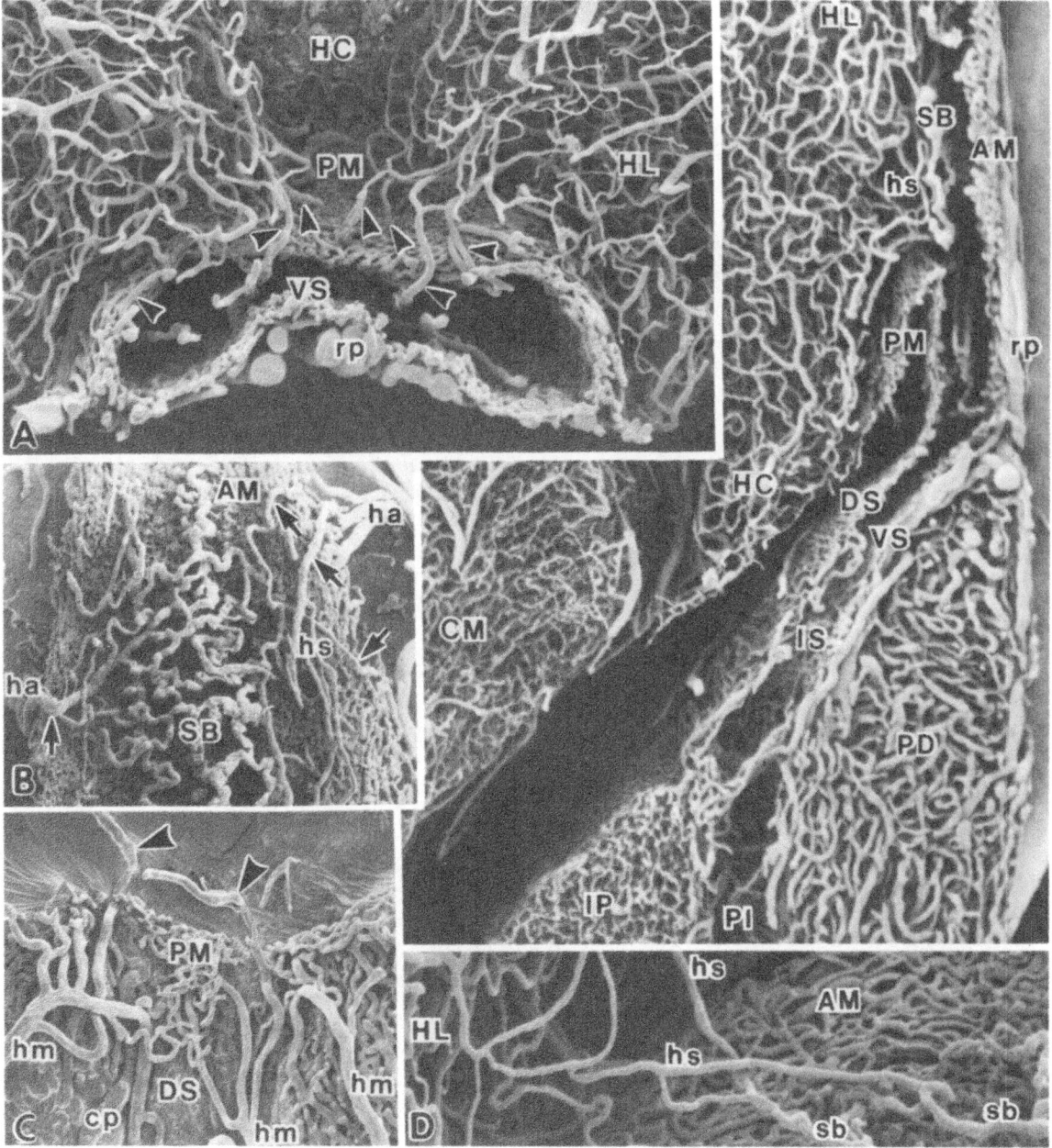

Figure 16-5. Sagitally cut hypophyseal and hypothalamic beds (adult male rat; from Takuro Murakami et al. [13], with permission). Inset A shows the infundibular ascending arteries (thin arrowheads, adult male rat). Inset B shows the peri-infundibular ascending arteries (arrows, aged male rat, dorsal view). Inset C shows the dissected infundibular ascending arteries (thick arrowheads; pubescent male rat). Inset D shows the hypothalamo-subependymal capillaries (hs) connecting the subependymal capillaries (sb) and hypothalamic vessels (HL) (adult male rat). ×55; inset A: ×100; inset B: ×35; inset C: ×100; inset D: ×120.

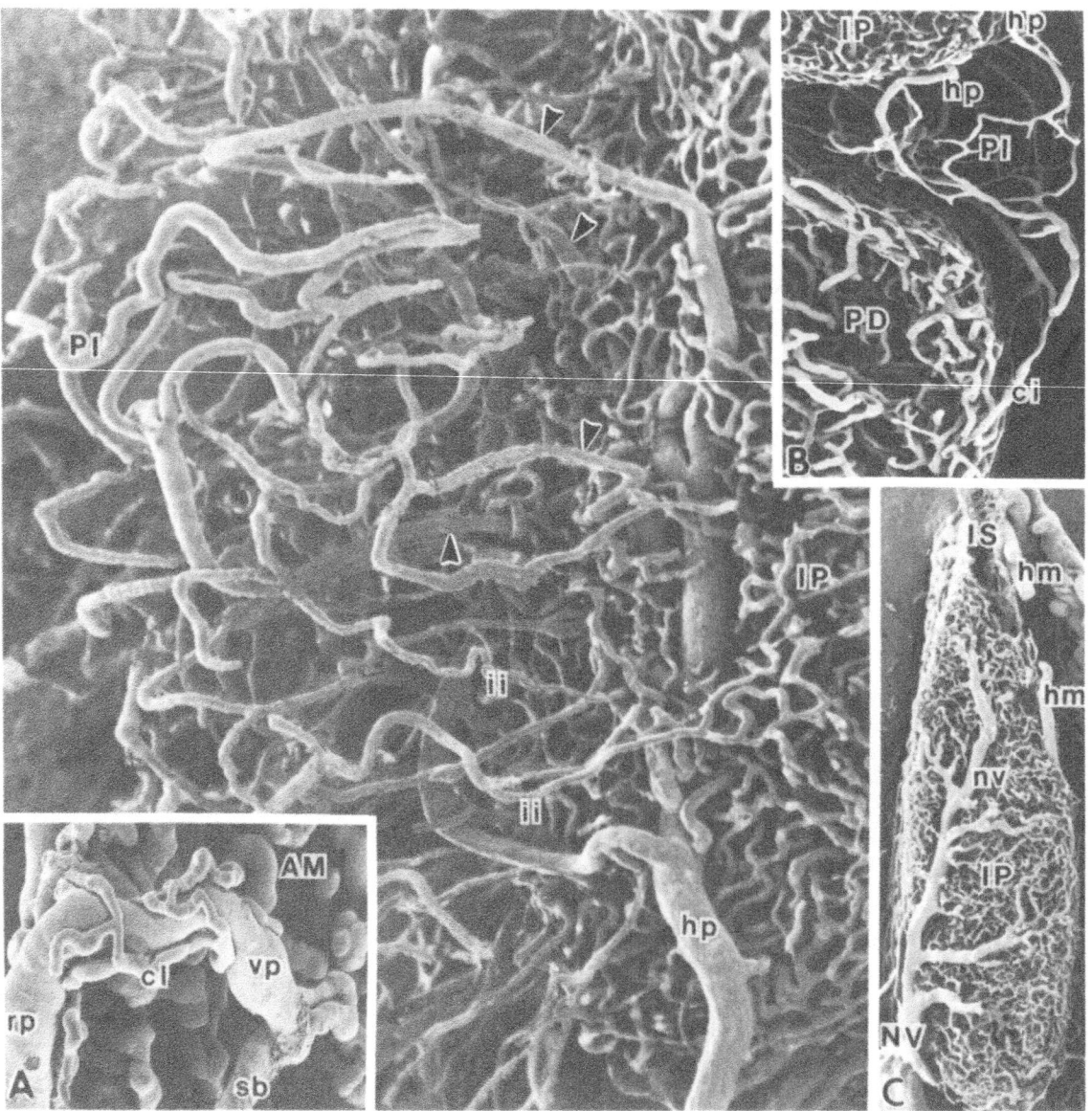

Figure 16-6. Vascular bed of the pars intermedia (adult male rat; from Takuro Murakami et al. [12], with permission). Note that the pars intermedia (PI) receives the neuro-intermedial portal vessels (ii) from the infundibular process (IP) (arrowheads; afferent arteries of the pars intermedia). Inset A shows an infundibular descending vein (vp; adult female rat). Inset B shows a posterior intermedio-distal portal vessel (ci). Inset C shows a dissected plexus of the infundibular process. ×155; inset A: ×270; inset B: ×90; inset C: ×50.

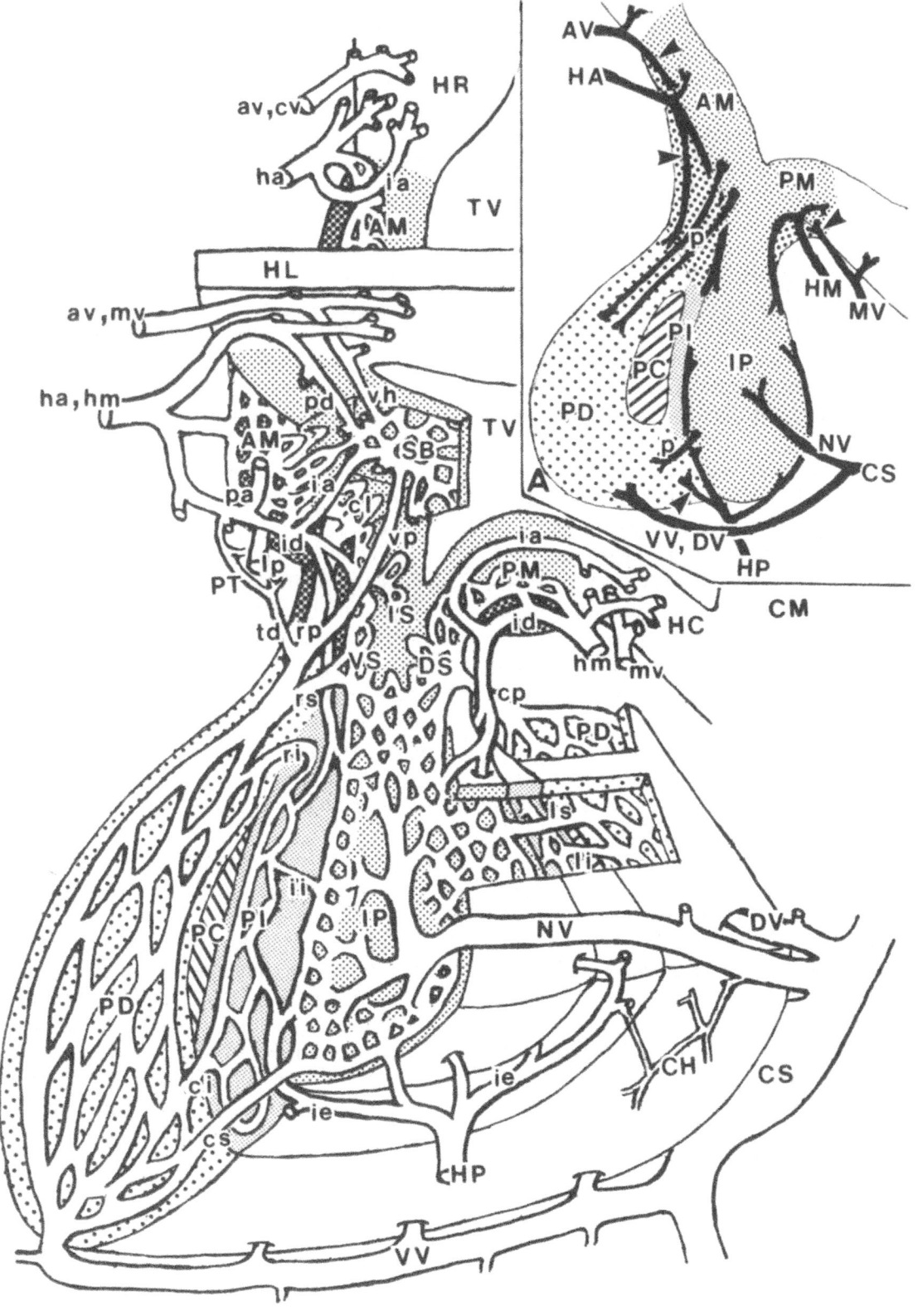

ascend, as the inferior proper hypophyseal arteries, along the infundibular process (Figs. 16-3 and 16-7) [12, 13].

The adenohypophyseal and neurohypophyseal veins originate in the pars distalis and process, respectively, and drain into the carvernous sinuses [13].

2.2. Capillary Networks and Loops of the Median Eminence and Infundibular Stalk, and their Portal Drainage into the Pars Distalis

The capillary network of the eminence has a large anterior extension and a small posterior extension, and forms the primary plexus, together with that of the stalk (Figs. 16-2B, 16-4, 16-5, and 16-7) [6,12,13].

The primary plexus receives at its upper margin or on its external surface its afferent vessels from the superior proper hypophyseal arteries, and emits from its external surface the long (infundibulo-distal) portal vessels (Figs. 16-1, 16-7, and 16-8) [13]. These vessels run into the pars distalis from the antero-ventral, antero-lateral, and antero-dorsal aspects, and form the sinusoidal or secondary plexus in the pars distalis (Figs. 16-1, 16-7, and 16-8) [13].

The secondary plexus is rather small at birth (Fig. 16-1B), though it rapidly increases its size during the newborn and pubescent periods (Figs. 16-4, 16-4A, and 16-4B) [13]. Regardless of its size, the secondary plexus emits its proper and rather thin systemic veins (ventral and dorsal adenohypophyseal veins) at the caudal margins of the pars distalis (Figs. 16-2, 16-3, and 16-7) [13].

The primary plexus additionally receives the infundibular descending arteries from the hypothalamic arteries [13] and rarely issues a few aberrant systemic veins, which drain into the hypothalamic veins (Fig. 16-1A) [13]. The secondary plexus or long portal vessels rarely receive a few aberrant arterioles from the superior proper hypophyseal arteries (Fig. 16-1C) [13].

The primary plexus projects numerous sinusoidal loops into the eminence and stalk, especially in the central area of the anterior lip of the eminence (Fig. 16-3D) [13]. The loops are faint at birth, though they rapidly develop during the newborn and pubescent stages (Figs. 16-3A, and 16-3B) [13]. The well-developed loops are provided with their proper afferent and efferent vessels (Fig. 16-3C) [13].

2.3. Capillary Plexus of the Infundibular Process and its Portal Drainages into the Pars Distalis and Intermedia

The capillary plexus of the process is supplied by the superior and inferior proper hypophyseal arteries (Figs. 16-2, 16-3, and 16-7) [13]. This plexus is continuous with the primary plexus (Figs. 16-4 and 16-5) and emits its proper and rather thick systemic veins (neurohypophyseal veins) on its postero-dorsal surface (Figs. 16-2, 16-3, 16-6C, and 16-7) [13].

In addition to the systemic veins, the process emits the short (processo-distal) portal vessels, which run into the secondary plexus from the antero-dorsal, dorso-medial, and postero-dorsal aspects (Figs. 16-2, 16-3, 16-7 and 16-8) [13]. These vessels are most conspicuous in newborn rats (Fig. 16-3); in adult and aged rats, they are regressive (Fig. 16-2) [13]. The process further emits many neuro-intermedial portal vessels, which drain into the pars intermedia (see below).

2.4. Capillary Plexus of the Pars Intermedia and its Portal Drainage into the Pars Distalis

The plexus of the pars intermedia develops after puberty and becomes denser as the animal ages (Figs. 16-4, 16-4A, 16-4B, and 16-6) [12,13].

The well-developed plexus receives from the dorsal aspect its afferent vessels from the superior and inferior proper hypophyseal arteries, and emits the intermedio-distal (intra-adenohypophyseal) portal vessels, which drain into the secondary plexus from the antero-dorsal, dorso-medial, and posterior aspects (Figs. 16-2, 16-6B, 16-7, and 16-8) [12,13].

←

*Figure 16-*7. Schematic diagram showing the vascular arrangements of the adult rat hypophysis (modified after Takuro Murakami et al. [13], with permission). Inset A shows an ideal blood supply of the fetal rat hypophysis [The vascular routes indicated by the arrowheads may be closed by the development of long and short portal vessels (p).]

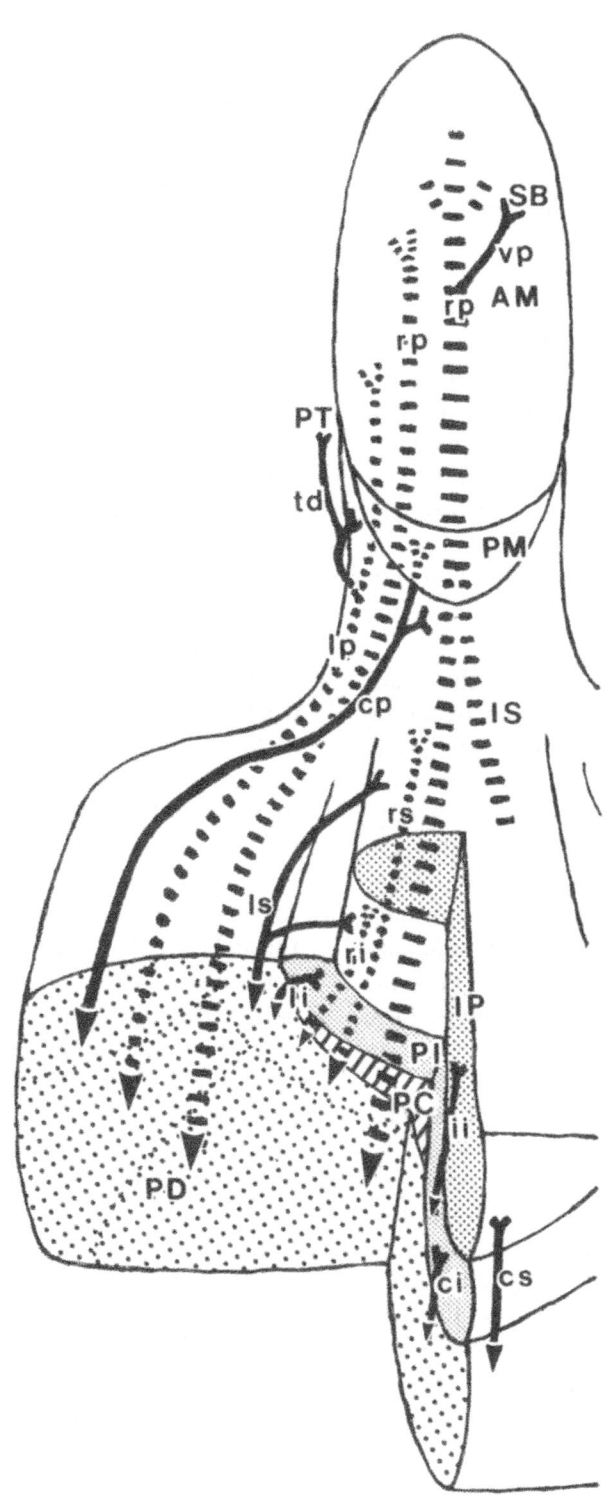

The plexus further receives the neuro-intermedial (processo-intermedial) portal vessels from the process (Figs. 16-6, 16-7, and 16-8) [12,13]. These vessels arise from the capillaries in the ventral surface of the process [13].

2.5. Subependymal Capillary Network and its Portal Drainage into the Pars Distalis

The subependymal network is coarse in newborn rats. However, it develops as the animal ages (Figs. 16-2, 16-2B, and 16-5B) [13]. This network receives some of the infundibular ascending arteries [13] and emits the infundibular descending veins (subependymo-distal or accessory long portal vessels), which penetrate the primary plexus or loops, and continue into the long portal vessels (Figs. 16-6A, and 16-7) [13]. The descending veins rarely drain into the aberrant systemic veins of the primary plexus [13].

The network further receives some of the infundibular descending or periinfundibular ascending arteries, and emits the infundibular ascending veins, which drain into the hypothalamic veins (Figs. 16-2C, 16-5B, and 16-7) [13]. In addition to these, the network always issues the subependymo-infundibular capillaries, continuous with the loops or meshwork of the primary plexus (Fig. 16-3D), and the hypothalamo-subependymal capillaries, continuous with the periventricular vessels of the hypothalamus (Figs. 16-5, and 16-5D) [13].

2.6. Capillary Plexus of the Pars Tuberalis and its Portal Drainage into the Pars Distalis

The plexus of the pars tuberalis develops after puberty [13]. This coarse plexus receives its afferent vessels from the superior proper hypophyseal arteries and emits the tuberalo-distal portal vessels draining into the long portal vessels or secondary plexus (Figs. 16-2A, 16-7, and 16-8) [13].

2.7. Constrictions of the Replicated Hypophyseal Vessels

Sharp contrictions are always imprinted at the origins of the anterior and middle hypophyseal arteries, though not as clearly in the posterior hypophyseal arteries [13]. Some shallow constrictions are consistently imprinted in the long portal vessels (Fig. 16-2A) [13].

3. Discussion

It is difficult to say who first described the hypophyseal portal system in any animal or by what method [1–4,14–16]. However, many light microscopists have confirmed that this system occurs throughout the vertebrates, though it becomes increasingly more complicated with the evolutionary order of the species [1–4,17–32]. Similar results have been obtained by modern workers, generally using the injection replica SEM method [5–13,33–36].

Our injection replica SEM findings disclose that the rat hypophyseal portal system consists of the long, short, neuro-intermedial, intermedio-distal, subependymo-distal, and tuberalo-distal routes, and that the long route is consistent throughout life, though the other routes show some age-related changes [13]. Our findings also confirm that the long route occupies the main part of the system and support the point-to-point theory [27,37,38], in which the blood in the anterior, lateral, and posterior areas of the median eminence flows into the central, lateral, and dorsal areas of the pars distalis, respectively [13]. This theory is acceptable even in the monkey and other animals, including the dog, cat, and guinea pig. These animals have a thick stalk or show numerous and thin, long portal vessels [5,6,11,13 and unpublished (u-p) data].

It is noteworthy that the plexus of the pars intermedia develops after puberty and forms the

Figure 16-8. Schematic diagram showing the origins and terminations of the rat hypophyseal portal vessels (after Takuro Murakami et al. [13], with permission).

neuro-intermedial and intermedio-distal routes [12,13]. The capillaries of this plexus show marked dilations during the mating periods (Fig. 16-6 and u-p data). This suggests that the pars intermedia becomes active after puberty and exerts some role in reproduction, along with the infundibular process and pars distalis.

It has been generally believed that the hypothalamic or parvicellular hormones are caught by the primary plexus or its loops, and are conveyed into the pars distalis via the long route [39–41]. In the perinatal or neonatal rats, with well-developed short portal vessels, the primary plexus projects few loops [13 and u-p data]. This suggests that in these rats, a considerable amount of the hormones are conveyed into the pars distalis via the short route.

In the dog and monkey, no plexus is formed in the pars intermedia and tuberalis, though a dense plexus is established in the anterior transitional zone [5,11, and u-p data]. In monkeys, dogs, cats, and guinea pigs without any marked subependymal network, the plexus of the infundibular process issues many sinusoidal loops that run into the stalk and continue into the short portal vessels or the systemic efferent rootlets in the process [5,11, and u-p data]. These indicate that the plexuses of the pars intermedia, pars tuberalis, and transitional zone, as well as the subependymal network and the loops in the process, are competitive in their occurrence.

Rats, having a marked subependymal network, never develop such loops in the process [13]. In adult and aged rats, the subependymal capillaries and infundibular ascending veins are noticeably dilated or thickened (Fig. 16-5B and u-p data). This change may leak the hypothalamic hormones into the systemic veins and may control the growth of the animal after adolescence.

Some authors described in the rat or rabbit that the hypophyseal bed is completed in the fetal stage [32,42]. However, our findings show that the loops of the primary plexus develop after birth and that the plexuses of the pars intermedia and tuberalis develop much later [13]. Our findings further demonstrate that postnatal development of the loops is accompanied by the marked enlargement of the secondary plexus and the noticeable involution of short vessels. This enhanced long portal circulation (or highly concentrated inflow of the hypothalamic hormones into the pars distalis) may activate this lobe and allow the rapid growth of the animal at the newborn and pubescent periods.

The blood flow in the long, short, intermedio-distal, subependymo-distal, and tuberalo-distal routes may be directed toward the pars distalis, since their parent plexuses are directly supplied by the arteries [12,13]. The blood flow in the neuro-intermedial route may be directed toward the pars intermedia, since this route arises near the arterial vessels in the process [12,13]. However, some immunoassay studies in the rat have detected the hypophyseal hormones, including MSH and oxytocin, in the blood in long vessels, and it is suggested that the blood in the terminal plexuses can return to the primary plexus [43–45]. Our findings also show that the systemic veins of the pars distalis are rather thin [5,13]. This limited drainage may ensure a homogeneous circulation of the portal blood in the pars distalis, so that all of the glandular cells of this lobe can constantly take up sufficient raw materials and hypothalamic hormones necessary for their own hormone synthesis.

No ascending portal vessel has been noted in our thoroughly cast specimens of the monkey, dog, cat, guinea pig, and rat [5,6,11–13, and u-p data]. Thus, it is likely that some authors [10,30, 33] mistook the infundibular ascending arteries for hypophyseal feedback vessels leading to the hypothalamus [13]. These hypothalamo-subependymal capillaries should be included in the infundibular descending arteries or ascending veins, since they usually continue into the arteries or veins in the hypothalamus [13]. Recent physiological experiments in the rat have suggested that dopamine or other nerve terminals in the median eminence regulate, as a neural ultrashort feedback system, the secretion of the respective hypothalamic or parvicellular hormones [40].

The sharp constrictions or endothelial cushions at the origins of the anterior and middle hypophyseal arteries may inhibit retrograde blood flow in these vessels, allowing a rich supply of arterial blood to the median eminence and parvicellular areas of the hypothalamus [13]. The ringlike constrictions or sphincters in the long portal vessels may minimize the retrograde blood flow from the secondary plexus to the primary plexus [13]. The plexuses of the magnocellular areas are dense, suggesting that these areas have some increased metabolic activities to produce its hormones, vasopressin and oxytocin [13].

We consider that the aberrant vessels of the primary or secondary plexus are remnants of the fetal vessels, which may be closed by the development of portal vessels. In fact, these aberrant vessels are frequently reproduced in fetal rats (u-p data). An ideal vascular pattern of the fetal hypophysis is schematically illustrated in Fig. 16-7A.

4. Conclusions

Hypophyseal blood vascular beds of newborn, pubescent, adult, and aged rats were studied by the injection replica SEM method. The primary plexus (PP) of the hypophyseal portal system consists of the capillary networks and loops in the median eminence and infundibular stalk. The loops (Ls) develop after birth. The PP issues the long portal (P) vessels, which form the secondary plexus (SP) in the pars distalis. The SP emits rather thin systemic veins. The plexus of the infundibular process (IP) is continuous with the PP and emits rather thick systemic veins; the IP issues the short P vessels, which drain into the SP, and the neuro-intermedial P vessels, which drain into the plexus of the pars intermedia (PI). The short P vessels become regressive after birth. The PI develops after puberty, and issues the intermedio-distal P vessels, which drain into the SP. The plexus of the pars tuberalis (PT) develops after puberty and drains into the SP. The sub-ependymal network (SN) is more developed in older animals; the SN issues the subependymo-distal P vessels, draining into the long P vessels, and the infundibular ascending veins, draining into the hypothalamic veins. These ascending veins markedly increase in thickness after adolescence. No ascending P vessel was replicated between the hypophysis and hypothalamus.

In neonatal rats with few Ls, the hypothalamic hormones (Hs) may be diffusely conveyed into the SP via the long and short P routes. The postnatal development of the Ls and the concomitant involution of the short P vessels may ensure a preferential and highly concentrated

inflow of the Hs into the SP via the long P route and may allow the rapid growth of the animal after birth. The SN may leak the Hs into the systemic veins and help to control the growth of the animal after adolescence. The pubescent development of the PI and its connecting P vessels may suggest that the pars intermedia becomes active after puberty and cooperates in reproduction with the pars distalis and infundibular process.

The SN, PI, and PT are inconsistent among species: In the monkey and dog, neither SN nor PI is formed; in the cat and guinea pig, no SN is formed though the PI is established. In the monkey, dog, cat, and guinea pig with no SN, the IP protrudes many long loops. In animals such as the dog with neither PI nor PT, the plexus of the anterior transitional zone is fully developed.

Acknowledgments

This work was supported in part by a grant from the Japanese Ministry of Education. The author is also grateful to Prof. T. Fujita (Editor, *Arch Histol Jpn*) for his kind permission to reproduce some of our previous micrographs in this chapter.

Abbreviations in Figures 16-1–16-8

Arteries and Veins
  AB = anterior basal vein
  AC = anterior cerebral artery
  CS = cavernous sinus
  IC = internal carotid artery
  MC = middle cerebral artery
  MB = mamillary vein.
Hypophyseal and Hypothalamic Arteries, and their Branches
  HA and *ha* = anterior hypophyseal artery and its branch
  HM and *hm* = middle hypophyseal artery and its branch
  HP and *hp* = posterior hypophyseal artery and its branch
  ah = hypothalamic branch (artery) of the anterior hypophyseal artery
  ia = infundibular ascending artery

id = superior proper hypophyseal artery

ie = inferior proper hypophyseal artery

mh = hypothalamic branch (artery) of the middle hypophyseal artery

pa = peri-infundibular ascending artery

pd = infundibular descending artery.

Hypophyseal and Hypothalamic Veins, and their Branches

AV and *av* = anterior hypothalamic vein and its branch

CV and *cv* = apical hypothalamic vein and its branch

DV and *dv* = dorsal adenohypophyseal vein and its branch

MV and *mv* = middle hypothalamic vein and its branch

NV and *nv* = neurohypophyseal vein and its branch

VV and *vv* = ventral adenohypophyseal vein and its branch

mb = branch of the mamillary vein (posterior hypothalamic vein)

vh = infundibular ascending vein

Hypophyseal Portal and Other Vessels

ci = posterior intermedio-distal (intra-adenohypophyseal) portal vessel

cp = posterior long (infundibulo-distal) portal vessel

cs = posterior short (processo-distal) portal vessel

hs = hypothalamo-subependymal capillary

ii = neuro-intermedial (processo-intermedial) portal vessel

li = lateral intermedio-distal portal vessel

lp = lateral long portal vessel

ls = lateral short portal vessel

ri = anterior intermedio-distal portal vessel

rp = anterior long portal vessel

rs = anterior short portal vessel

td = tuberalo-distal portal vessel

vp = accessory long (subependymo-distal) portal vessel (infundibular descending vein).

Tissue Components and their Capillary Plexuses or Capillaries

AM = median eminence anterior lip (extension)

CH = capsule of the hypophysis

CM = mamillary body

DS = infundibular stalk (dorsal part)

HC = hypothalamus (portero-basilar and peri-ventricular part)

HL = hypothalamus (latero-basilar and peri-ventricular part)

HR = hypothalamus (antero-basilar and peri-ventricular part)

IP = infundibular process

IS = infundibular stalk

PD = pars distalis

PI = pars intermedia

PM = median eminence posterior lip (extension)

PN = paraventricular nucleus

PT = pars tuberalis

TC = trigeminal nerve

VS = infundibular stalk (ventral part)

SB and *sb* = subependymal plexus and its capillary

cl = capillary loop or loops in the median eminence and infundibular stalk.

Others

PC = hypophyseal cleft

TV = third ventricle

* = leaked resin mass or macerated remnant of tissue element.

References

1. Daniel PM. The anatomy of the hypothalamus and pituitary gland. In: Martini L, Ganong WF (eds.), *Neuroendocrinology, Vol. 1*. Academic Press, New York, pp 15–80, 1966.
2. Christ JF. Nerve supply, blood supply and cytology of the neurohypophysis. In: Harris GW, Donovan BT (eds.), *The Pituitary Gland, Vol. 3. Pars Intermedia and Neurohypophysis*. Butterworths, London, pp 62–130, 1966.
3. Green JD. The comparative anatomy of the portal vascular system and of the innervation of the hypophysis. In: Harris GW, Donovan BT (eds.), *The Pituitary Gland, Vol. 1. Anterior Pituitary*. Butterworths, London, pp 127–146, 1966.
4. Wingstrand KG. Microscopic anatomy, nerve supply and blood supply of the pars intermedia. In: Harris GW, Donovan BT, (eds.), *The Pituitary Gland, Vol. 3. Pars Intermedia and Neurohypophysis*. Butterworths, London, pp 1–27, 1966.
5. Murakami T. Injection replica scanning electron microscope method: Use in an analysis of the dog hypophyseal portal system (in Japanese). *The Cell* (Tokyo) 7:11–18, 1975.

6. Murakami T. Pliable methacrylate casts of blood vessels: Use in a scanning electron microscope study of the microcirculation in rat hypophysis. *Arch Histol Jpn* 38:151–168, 1975.

7. Page RB, Munger BL, Bergland RM. Scanning microscopy of pituitary vascular casts. *Am J Anat* 146:273–302, 1976.

8. Lametschwandtner A, Simonsberger P, Adam H. Vascularization of the hypophysis in toad, *Bufo bufo* (L.) (Amphibia, anura). *Cell Tissue Res* 179:1–10, 1977.

9. Page RB, Bergland RM. The neurohypophyseal capillary bed: I. Anatomy and arterial supply. *Am J Anat* 148:345–358, 1977.

10. Page RB, Leure-duPree AE, Bergland RM. The neurohypophyseal capillary bed. II. Specialization with median eminence. *Am J Anat* 153:33–66, 1978.

11. Murakami T, Ohtani O., Ohtsuka A, Kikuta A. Injection replication and scanning electron microscopy of blood vessels. In: Hodges GM, Carr KE (eds.), *Biomedical Research Applications of Scanning Electron Microscopy*, Vol. 3. Academic Press, London, pp 1–30, 1983.

12. Murakami T, Ohtsuka A, Taguchi A, Kikuta A, Ohtani, O. Blood vascular bed of the rat pituitary intermediate lobe, with special reference to its development and portal drainage into the anterior lobe. A scanning electron microscope study of vascular casts. *Arch Histol Jpn* 48:69–97, 1985.

13. Murakami T, Kikuta A, Taguchi T, Ohtsuka A, Ohtani O. Blood vascular architecture of the rat cerebral hypophysis and hypothalamus. A dissection/scanning electron microscopy of vascular casts. *Arch Histol Jpn* 50:133–176, 1987.

14. Popa G., Fielding U. A portal circulation from the pituitary to the hypothalamic region. *J Anat* 65:88–91, 1930.

15. Basir MA. The vascular supply of the pituitary body in the dog. *J Anat* 66:387–398, 1932.

16. 'Espinasse PG. The development of the hypophysio-portal system in man. *J Anat* 68:11–18, 1933.

17. Wislocki GB. The vascular supply of the hypophysis cerebri of the cat. *Anat Rec* 69:361–387, 1937.

18. Wislocki GB. The vascular supply of the hypophysis cerebri of the rhesus monkey and man. *Res Publ Ass Nerv Ment Dis* 17:48–68, 1938.

19. Harris GW. The hypophysio-portal vessels of the porpoise (*Phocaena phocaena*). *Nature* 159:874–875, 1947.

20. Harris GW. The blood vessels of the rabbit's pituitary gland, and significance of the pars distalis and zona tuberalis. *J Anat* 81:343–351, 1947.

21. Green JD. The comparative anatomy of the hypophysis with special reference to its local blood supply and innervation. *Am J Anat* 88:225–312, 1951.

22. Xuereb GP, Prichard MML, Daniel PM. The hypophyseal portal system of vessels in man. *Q J Exp Physiol* 39:219–239, 1954.

23. Daniel PM, Prichard MML. The vascular arrangements of the pituitary gland of the sheep. *Q J Exp Physiol* 42:237–248, 1957.

24. Glydon RJ. The development of the blood supply in the albino rat, with special reference to the portal vessels. *J Anat* 91:237–244, 1957.

25. Landsmeer JMF. Vessels of the rat's hypophysis. *Acta Anat* 12:82–109, 1957.

26. Török B. Structure of the vascular connections of the hypothalamo-hypophyseal region. *Acta Anat* 59:84–99, 1964.

27. Adams JH, Daniel PM, Prichard MML. Distribution of hypophyseal portal blood in the anterior lobe of the pituitary gland. *Endocrinology* 75:120–126, 1964.

28. Cummings JF, Habel RE. The blood supply of the bovine hypophysis. *Am J Anat* 116:91–114, 1965.

29. Holmes RL. The vascular pattern of the median eminence of the hypophysis in macaque. *Fol Primat* 7:216–230, 1967.

30. Negm IM. The blood supply of the mouse hypophysis cerebri. *Acta Anat* 80:377–387, 1971.

31. Duvernoy H. The vascular architecture of the median eminence. In: *Brain-Endocrine Interaction, Median Eminence: Structure and Function* (Int Symp Munich, 1971). Knigge KM, Scott DE, Weindl A (eds.), S. Karger, Basel, pp 79–108, 1972.

32. Terneby UK. The development of the hypophysical vascular system in the rabbit, with particular regard to the primary plexus and portal vessels. *J Neuro-Viscer Relat* 32:311–346, 1972.

33. Bergland RM, Page RB. Can the pituitary secrete directly to the brain? Affirmative anatomical evidence. *Endocrinology* 102:1325–1338, 1978.

34. Bergland RM, Page RB. Pituitary-brain vascular relations. *Science* 204:18–24, 1979.

35. Paino G, Langella M, Caputo G. Vascular feature of the hypophysis in *Bulbalus buffelus*. *Acta Anat* 110:206–218, 1981.

36. Honma Y, Toda Y, Chiba A. Vascularization of the hypothalamo-hypophyseal complex in Japanese elasmobranchs: A scanning electron microscope study of blood vascular casts. *Arch Histol Jpn* 50:39–48, 1987.

37. Donovan BT, Harris GW. Effect of pituitary stalk section on light induced oestrus in the ferret. *Nature* 174:503–504, 1954.

38. Daniel PM, Prichard MML, Schurr PH. Extent of the infarct in the anterior lobe of the pituitary gland after stalk section. *Lancet* 1:1101–1103, 1958.

39. Halasz B, Kosaras B, Lengvari I. Ontogenesis of neurovascular link between the hypothalamus and the anterior pituitary in the rat. In: Knigge KM, Scott DE, Weindl A (eds.), *Brain-Endocrine Interaction, Median Eminence: Structure and Function* (Int Symp Munich, 1971). S. Karger, Basel, pp 27–34, 1972.

40. Fuxe K, Agnati LF, Calza L, Anderson K, Giardino L, Benfenati F. Carmurry M, Goldstein M. Quantitative chemical neuroanatomy gives new insights into the catecholamine regulation of the peptidergic neurons projecting to the median eminence. In: Usdin E, Carlsson A, Dahlstrom A, Eugel J (eds.), *Catecholamines: Neuropharmacology and Central Nervous System — Theoretical*

Aspects. Alan R. Liss, New York, pp 441–449, 1984.

41. Page RB, Dovey-Hartman BJ. Neurohemal contact in the internal zone of the rabbit median eminence. *J Comp Neurol* 226:274–288, 1984.

42. Galabov PG, Schiebler TH. Development of the capillary system in the neurohypophysis of the rat. *Cell Tissue Res* 228:685–696, 1983.

43. Oliver C, Mical RS, Porter JC. Hypothalamic-pituitary vasculature: Evidence for retrograde blood flow in the pituitary stalk. *Endocrinology* 101:598–604, 1977.

44. Gibbs DM. High concentration of oxytocin in hypophyseal portal plasma. *Endocrinology* 114:1216–1218, 1984.

45. Horn AM, Robinson CAF, Fink G. Oxytocin and vasopressin in rat hypophyseal portal blood: Experimental studies in normal and Brattleboro rats. *J Endocrinol* 104: 211–224, 1985.

Author's address:
Prof. Takuro Murakami
Department of Anatomy
Okayama University
School of Medicine
2-5-1 Shikata-cho
Okayama 700
Japan

CHAPTER 17

Blood Vascular Casts of the Thyroid Gland in Normal and Experimental Conditions

HISAO FUJITA & MASATO IMADA

1. Outline of Thyroid Structure and Function

The thyroid gland is generally covered by a capsule of connective tissue that is continuous with the interlobular and interfollicular connective tissue. However, in the small-sized thyroids of lower vertebrates and small-sized mammals, the lobular unit is not so distinct. In almost all the vertebrates, the thyroid gland consists of numerous ball-like structures called *follicles*, and of interfollicular connective tissues with blood capillaries. Each follicle is composed of numerous follicular epithelial cells arranged as a simple cuboidal epithelium to surround a lumen, and a few parafollicular cells located singly or in groups in the basal part of the follicular epithelium. Each follicular lumen surrounded by the follicular epithelium is a completely enclosed area, storing colloidal materials (thyroglobulin) secreted by the follicular epithelial cells.

Thyroid hormones released from this gland are classified into three types according to their chemical structures: iodinated amino-acid derivatives, peptides, and amines. The iodinated amino-acid derivatives (tetraiodothyronine, T_4; and triiodothyronine, T_3) are secreted from the follicular epithelial cell, while both peptide (calcitonin) and amine (serotonin) are released from the parafollicular cell.

The follicular epithelial cell goes through four steps to secrete T_4 and T_3; (1) secretion of thyroglobulin from the apical part of the cell into the follicular lumen, (2) reabsorption of thyroglobulin from the follicular lumen into the cell, (3) hydrolysis of thyroglobulin to liberate T_4 and T_3, and (4) release of T_4 and T_3 from the basal part of the follicular epithelial cell into the connective tissue space after passing through the basal lamina [1,2].

The parafollicular cell synthesizes calcitonin, as well as serotonine, and releases them into the connective tissue space. Consequently, all these thyroid hormones are released into the connective tissue space from both types of cells. The connective tissue space, which corresponds to the interfollicular loose connective tissue, contains tissue fluids, two sheets of basal lamina belonging to the follicle epithelium and to the capillary endothelium, connective tissue elements, nerve elements, and blood vessels.

The hormones released from the secretory cells into the connective tissue space, after passing through the basal lamina, enters the blood capillary by passing through the endothelial basal lamina and endothelial fenestrations [1,3].

Motta, P.M., Murakami, T., and Fujita, H. (eds.), Scanning Electron Microscopy of Vascular Casts: Methods and Applications.

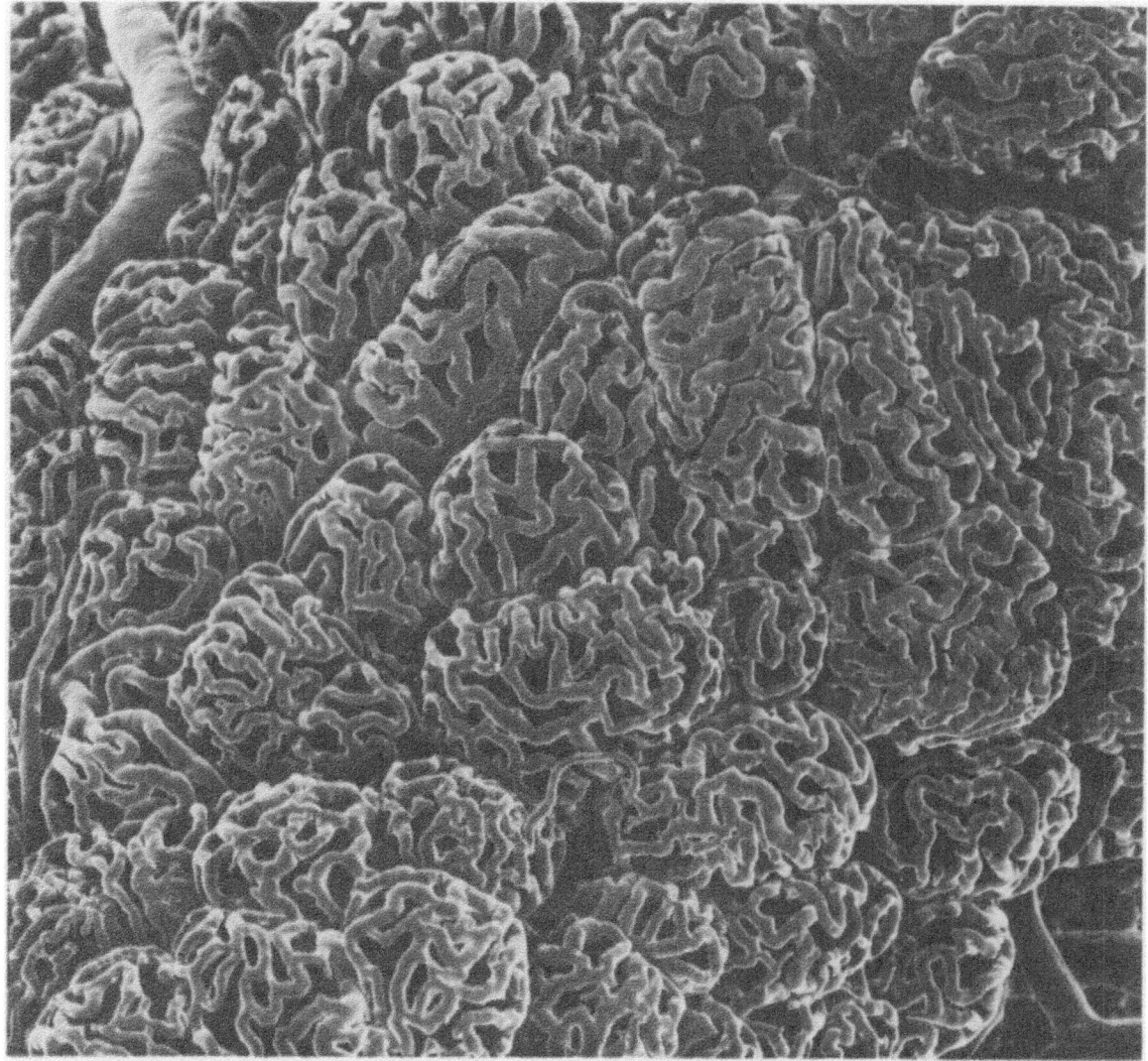

Figure 17-1. Scanning electron micrograph of corrosion cast of normal rat (12-week-old) thyroid. ×360.

2. Three-Dimensional Aspects of Blood Vessels in the Thyroid Gland

Though it is easily understood by routine histological and electron microscopical observations that the thyroid gland is supplied with rich vascularization located in the interfollicular connective tissue, three-dimensional aspects of the distribution of the blood vessels were obscure until the corrosion cast method was established and applied to thyroid study. Scanning electron microscopic study has answered many questions regarding the vascularization of the thyroid gland, which is necessary to understand the functional morphology of this organ.

2. Methods

A low-viscosity solution of methacrylate resin (Mercox, Dainihon Ink and Chemical Co. Ltd.) was used to make the corrosion casts of the thyroid gland, according to Murakami [4] and Ohtani and Murakami [5]. Normal rhesus monkeys, dogs,

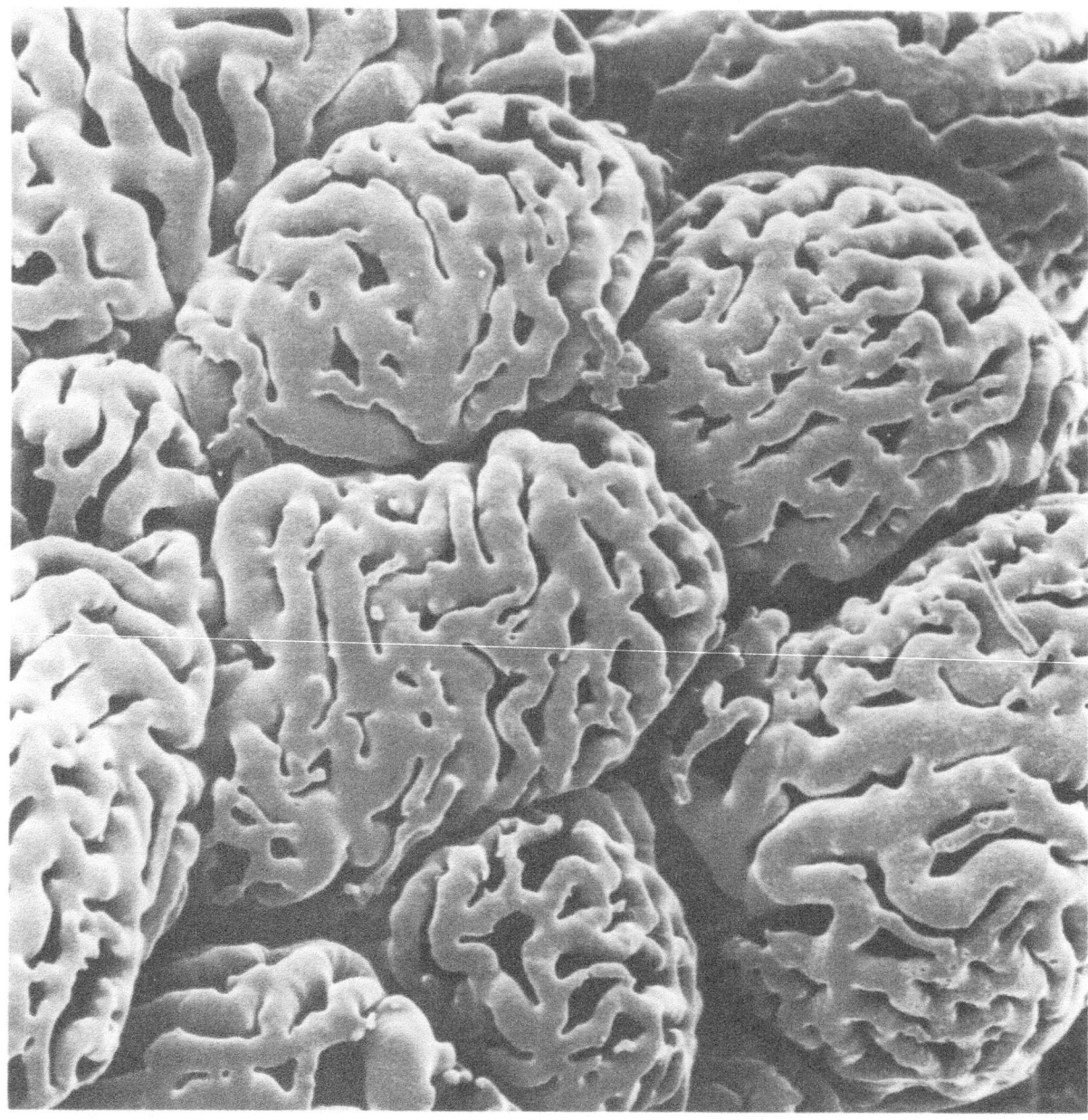

Figure 17-2. Thyroid of low-iodine-diet-treated (for 4 weeks) rat. Note thick blood capillaries, and their sproutings and fusions. ×900.

rats, chickens, tortoise, and experimental rats were used for our study. The resin was injected at a rate of 5 ml/min into the brachiocephalic arteries of the monkey and dog, and into the left ventricles of rats, chickens, and tortoise. After the injection, these animals were immersed in hot water (60°C) for 30 minutes and the thyroids were removed. The excised glands were again immersed in hot water (60°C) for 3 hours, macerated in a 20% NaOH solution for 1–2 days, and washed in H_2O. The specimens were then immersed in a sodium hypochlorite solution (60°C, household bleach) for 8 hours, washed in H_2O, and air dried. Each corrosion cast was mounted on a metal stub, coated with platinum by the use of an Eiko IB-5 ion coater, and observed in a JSM-U3 type or a

Hitachi S-800 scanning electron microscope (SEM). In order to observe sectional profiles, some casts were frozen in a 30% sucrose solution and sectioned with a cryomicrotome. After washing in H_2O, air drying, and ion coating, they were observed in the SEM.

3. Three-Dimensional Aspects of Blood Vessel in Normal Thyroid of Various Animals

Three-dimensional aspects of the thyroid gland determined by scanning electron microscopy of corrosion casts have been published by several authors [6–14].

By using scanning electron microscopy of vascular corrosion casts of normal thyroid, a small number of deep, wide, and long fissures, corresponding to the interlobular connective tissue, and many narrow grooves, corresponding to the interfollicular connective tissue, are noticed. The thyroid artery ramifies and enters the interlobular connective tissue as interlobular arteries. The interlobular artery sends its branches into the interfollicular connective tissue as interfollicular arteries and blood capillaries. Murakami et al. [14] reported, in the rat, the occurrence of anastomoses between interlobular arteries, between lobular (interfollicular) arteries, between interlobular veins, and between lobular (interfollicular) veins. Each follicle is surrounded by a network of very frequently branched blood capillaries (Fig. 17-1). The capillary network looks like a basket encapsulating the follicle. In the monkey thyroid, each capillary-network basket is almost entirely independent of adjacent baskets [6]. In thyroids of the dog and rat also, each follicle is densely enclosed by a clearly defined basketlike capillary network, though anastomoses or common capillaries between the basketlike networks of many follicles are rather more frequently seen than in the monkey thyroid [6,14].

The relationship between the capillary networks of many follicles is easily understood by the observation of sectioning profiles [11,12,14]. About 30–40% of the basketlike capillary networks of rat thyroids are not entirely independent of each other [11]. A few transfollicular capillary anastomoses or common capillaries of two adjacent follicles are sometimes seen in the rat [11,14].

However, each network always receives proper afferent vessels from the interfollicular arteries and issues proper efferent vessels to the interfollicular vein. In the thyroid of Japanese quail, Mikami and Miyasaka [15] reported that the basketlike unit is difficult to recognize, though the interfollicular capillaries are extremely rich in distribution and form a capillary network around each follicle. As we [11,13] and Murakami et al. [14] reported, each follicular capillary network is considered to be an independent functional unit in the microcirculation. Murakami et al. [14] found the occurrence of the sphincterlike constrictions only in the efferent vessel vein and considered that the constriction may limit the outflow of the blood from the follicular capillary network into the efferent vessel. In this way follicular epithelial cells can easily pick up the raw materials sufficient for hormone synthesis. With regard to the actual structure of this microvascular apparatus, more detailed observations using the transmission electron microscope are needed.

4. Rat Thyroids in Experimental Conditions

Interesting and important observations are related to the three-dimensional changes in the thyroidal vascular vessel distribution in hyperfunctional or hypofunctional conditions.

4.1. Hyperfunctional State

The data dealing with this subject have already been reported by our group [11]. For this purpose, we used TSH-, low-, iodine-, or propyl thiouracil (PTU)-treated rats. Male Wistar rats, 8 weeks of age, weighing 180–200 g, were fed standard pellets (Oriental Yeast Co. Ltd.) or a low-iodine diet (Oriental Yeast Co. Ltd.) and tap water for 4 weeks. Some animals fed standard pellets were subcutaneously injected twice a day with 5 U/day of thyroid stimulating hormone (TSH, Sigma) for 7 days. Other animals were fed the standard pellets and tap water containing 0.02% propyl thiouracil (PTU, Sigma) for 4 weeks.

The thyroid gland secretes thyroxine by stimulation of TSH. In long-term low-iodine-diet-treated, or long-term PTU-treated animals, secre-

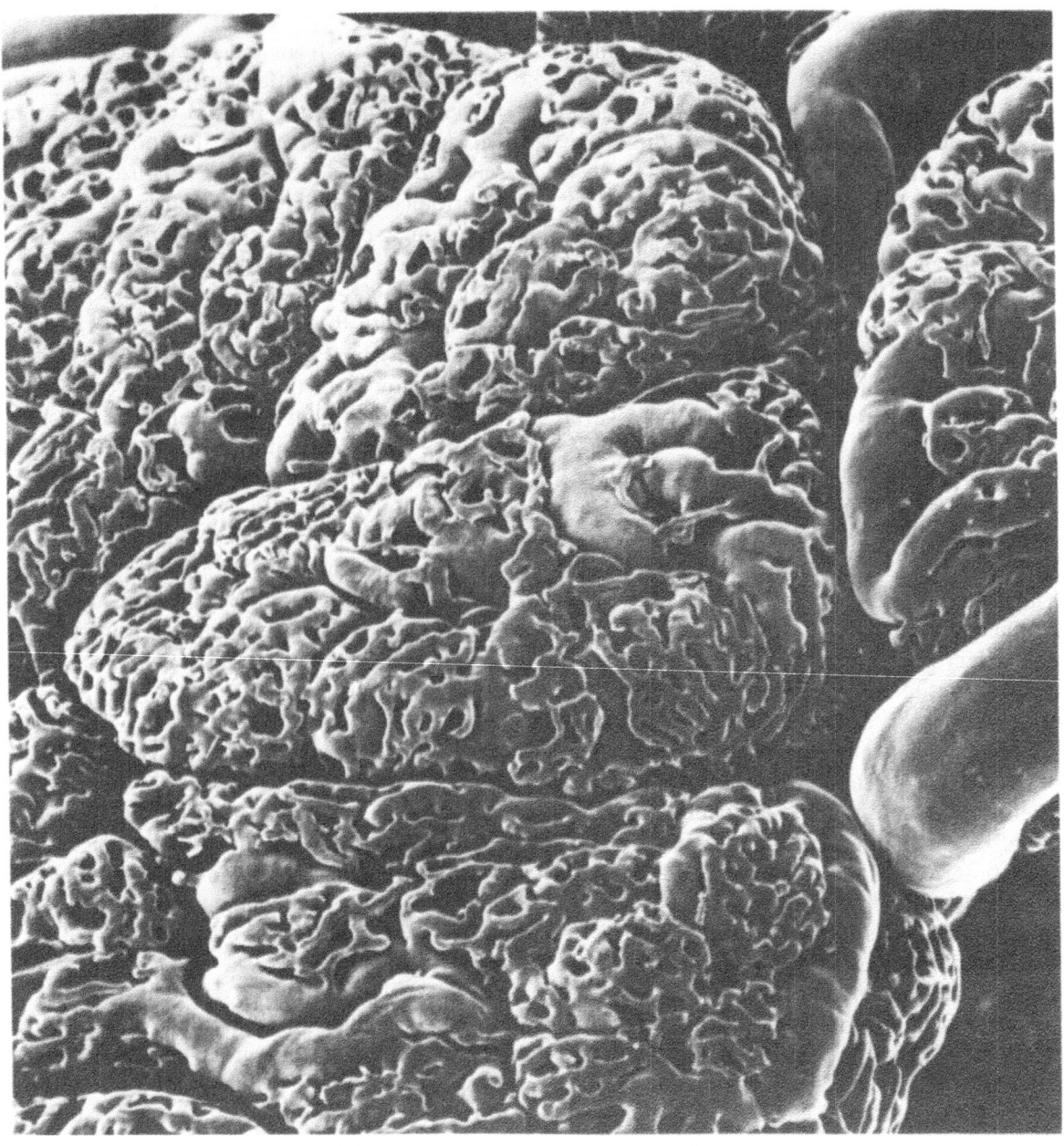

Figure 17-3. Thyroid of PTU-treated (for 4 weeks) rat. The basketlike unit of capillaries of each follicle becomes irregular in shape and indistinct, while the lobules are clear. Blood capillaries are irregular and variable in diameter. ×560.

tion of thyroxine is markedly reduced, and the low concentration of thyroxine in the blood stimulates the hypothalamus-pituitary system to secrete an excess dose of TSH by a negative feedback mechanism. Then the thyroid follicle epithelial cell in chronically hyperstimulated by an excess dose of TSH in the blood.

4.1.1. TSH- or low iodine-treated animals. In TSH-, or low iodine-treated rats, the thyroid follicle epithelial cells appear taller and the follicle lumina are reduced in size by light microscopy (Fig. 17-3). By scanning electron microscopy of corrosion casts, the grooves corresponding to the

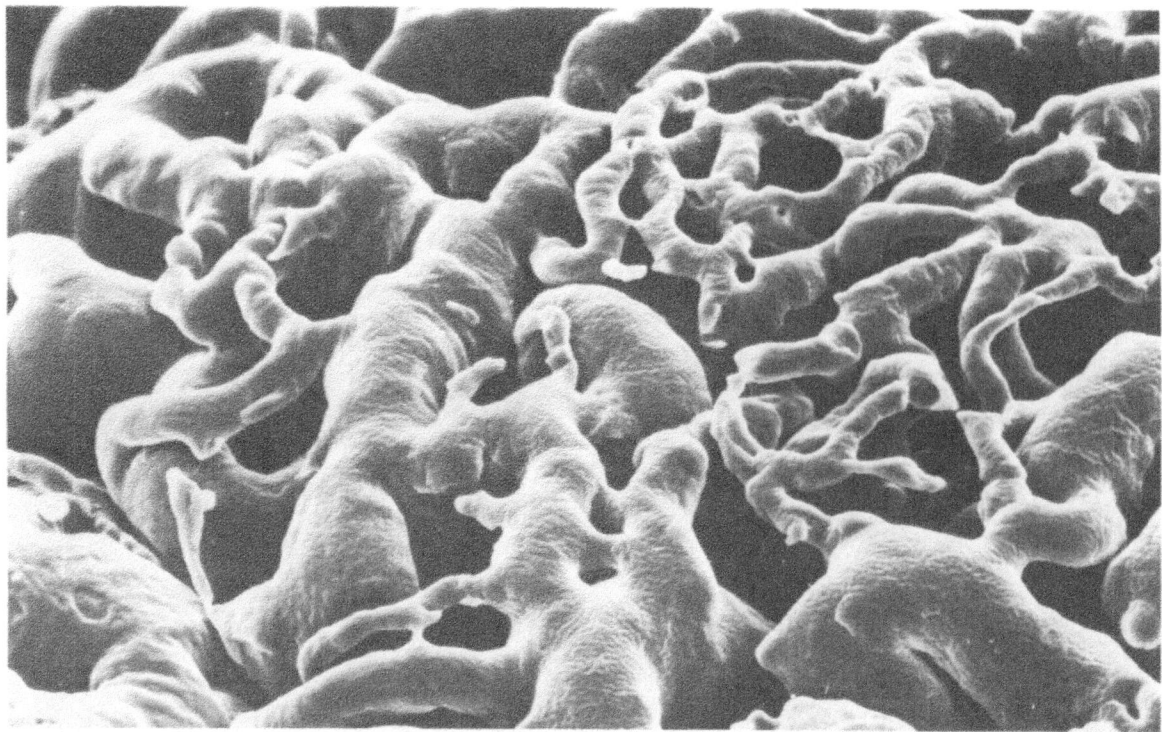

Figure 17-4. Thyroid of PTU-treated (for 4 weeks) rat. Note sprouting and fusion from irregularly shaped blood capillaries. ×1100.

interfollicular connective tissue and the basket-like capillary networks are distinctly visible. The capillaries in each basket are markedly dilated. The thickest part is about 25 μm (normal 5–15 μm) in diameter. Among these enlarged segments are occasionally narrower parts, the thinnest one being about 5 μm in diameter [11]. The capillary bed covers about 80% (normal 50%) of the follicular surface area [11]. Using the freeze-fracture method, our group [16] reported an increase of endothelial fenestrations in number and population density in long-term TSH-treated mice [16].

4.1.2. PTU-treated animals. In PTU-treated rats, all thyroid follicles become smaller in size and irregular in shape by light microscopy, while the follicle epithelial cells are considerably taller than those of both low-iodine-diet-treated and TSH-treated animals. The follicle lumina are very small in size and are distorted. The interfollicular connective tissue is strikingly increased in volume, and the blood capillaries surrounding the follicle become greatly dilated and irregular in shape [11].

By scanning electron microscopy (Figs. 17-3 and 17-4), the basketlike network of the blood capillaries is found to be rather well preserved in some parts, though each basket is extremely irregular is shape and variable in size [11]. Most baskets are distorted. The capillaries around the follicle are usually heterogeneous in size and shape. In the capillary baskets, some parts are markedly thicker than other parts. The thickest part is 60 μm in diameter, and the thinnest part is only 2 μm [11]. As a whole, the diameter of the capillaries in the PTU-treated animals is much larger than that of the other experimental animals. The capillary bed covers about 70% of the follicular surface area.

Small protrusions (sproutings) of the blood capillaries, suggesting the neogenesis of the capillary; fusions; and anastomoses of the capillaries are often recognized in each basket by scanning as well as the transmission electron microscope in all these hyperstimulated thyroid glands [11,13] (Fig.

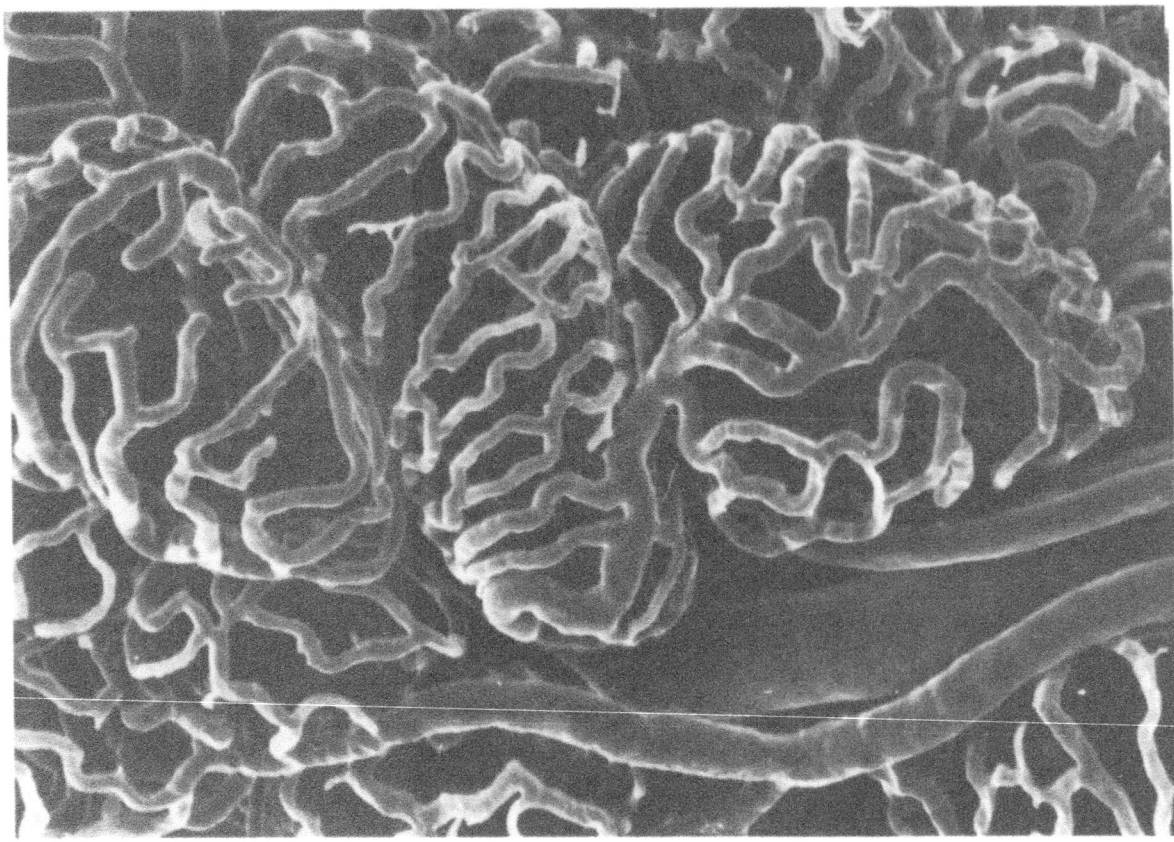

Figure 17-5. Thyroid of levothyroxine sodium-treated (for 4 weeks) rat. Each follicle become larger in size, and the capillaries in each basketlike network are very thin in diameter and poor in distribution. ×630.

17-4). The capillary endothelial cells of the thyroid show protein-synthesizing signs, such as well-developed rough endoplasmic reticulum and Golgi apparatus, and an increase of mitosis in low-iodine-diet-treated or PTU-treated animals.

4.2. Hypofunctional State

The following data in the hypofunctional state of the thyroid have reported earlier by our group [12]. Male sistar rats, 8 weeks of age, were fed standard pellets (Oriental Yeast Co., Ltd.) and tap water, while the another group of rats was fed standard pellets and tap water containing 0.5% levothyroxine sodium (Thyradin-S, Teikoku-zoki Co., Ltd.) for 4 weeks. In long-term levo-thyroxine-treated animals, the thyroxine level in the blood is usually high, and TSH secretion from the anterior pituitary is suppressed by a negative feedback mechanism. The thyroid goes into a hypofunctional state during thyroxine secretion. The thyroid gland after this treatment is much smaller in size than in normal rats. The follicle epithelial cells are excessively shorter, and the follicle lumina, round or oval in shape, are intensively enlarged as compared with those of the normal rat. The rough endoplasmic reticulum and Golgi apparatus in the follicular epithelial cells are markedly reduced in size. Interfollicular connective tissue elements are very thin and the blood capillaries are scant in distribution (Fig. 17-5).

Scanning electron microscopic findings in the hypofunctional state of the thyroid were published earlier by our group [12,13]. The basketlike capillary network surrounding each follicle in the levothyroxine-treated animals is as well preserved as that of normal animals. The basket of capillaries is generally independent of that surrounding the adjacent follicle, though the transfollicular inter-

228

connections are sometimes seen, as in normal cases. The capillaries covering each follicle in these experimental animals become very narrow in diameter, the thickest enlarged part measures about 8 μm in diameter, while the thinnest part measures about 3 μm. The capillaries are more poorly distributed as compared with those in normal and hyperstimulated animals, and anastomoses of the capillaries within each basketlike network are also strikingly decreased in number. The capillary bed covers about 25% (normal 50%) of the area of the follicular surface.

In our observations [16] using freeze-replica images, endothelial fenestrations are markedly reduced in number in long-term thyroxine-treated mice.

These findings indicate that the number and population density of endothelial fenestrations are changeable and flexible, depending on the functional state of the thyroid gland.

5. Development of Basketlike Blood Capillaries

In the newborn rat, it was reported by Murakami et al. [14] that the basketlike capillary network system corresponding to each thyroid follicle is not well differentiated and its differentiation begins after birth. They speculated that transfollicular capillaries are undifferentiated remnants of fetal thyroid capillaries. Further studies on the differentiation and development of the vascularization pattern of the thyroid are needed to better clarify these ontogenetical aspects.

6. Conclusions

Throughout our experimental studies [11,12,17], we wish to emphasize that the microvascular system in the thyroid gland is flexible and changeable, and shows a plasticity in three-dimensional images, distribution patterns, and endothelial morphologies, which reflects the functional state of the gland.

References

1. Fujita H. Fine structure of the thyroid. *Int Rev Cytol* 40:197–280, 1975.

2. Fujita H. Functional morphology of the thyroid. *Int Rev Cytol* 113:145–185, 1988.

3. Ekholm R. The ultrastructure of the blood capillaries in the mouse thyroid gland. *Z Zellforsch* 46:139–146, 1957.

4. Murakami T. Application of the scanning electron microscope to the study of the fine distribution of the blood vessels. *Arch Histol Jpn* 32:445–454, 1971.

5. Ohtani O, Murakami T. Peribiliary portal system in the rat liver as studied by the injection replica scanning electron microscope method. *Scan Electron Microsc* II: 241–244, 1978.

6. Fujita H, Murakami T. Scanning electron microscopy on the distribution of the minute blood vessels in the thyroid glands of the dog, rat and rhesus monkey. *Arch Histol Jpn* 36:181–188, 1974.

7. Raj S, Meserve LA. Thyroid vascular casts from mice with chemically altered thyroid status. *Micron* 13:455–456, 1982.

8. Ohtani O, Kikuta A, Ohtsuka A, Taguchi T, Murakami T. Microvasculature as studied by microvascular corrosion casting/scanning electron microscope method. I. Endocrine and digestive system. *Arch Histol Jpn* 46:1–42, 1983.

9. Meserve LA, Klak AT. Vascular casts of thyroids from euthyroid and thiouracil fed young rats: Quantitative comparison. *Micron Microsc Acta* 15:115–116, 1984.

10. Kikuta A, Ohtsuka A, Ohtani O, Murakami T. Microvascularization of endocrine glands as studied by injection-replica scanning electron microscope method. In: Motta PM (ed.), *Ultrastructure of Endocrine Cells and Tissues*, Martinus Nijhoff, Norwell, MA pp 313–320, 1984.

11. Imada M, Kurosumi M, Fujita H. Three dimensional aspects of blood vessels in thyroids from normal, low iodine diet-treated, TSH-treated, and PTU-treated rats. *Cell Tissue Res* 245:291–296, 1986.

12. Imada M, Kurosumi M, Fujita H. Three-dimensional imaging of blood vessels in thyroids from normal and levothyroxine sodium-treated rats.

13. Fujita H, Imada M. Three dimensional aspect on the functional morphology of the thyroid gland. In: Motta PM (ed.), *Cells and Tissues. A Three-Dimensional Approach by Modern Techniques in Microscopy*. Alan P. Liss, New York, pp 227–233, 1989.

14. Murakami T, Miyake T, Uno Y, Ohtsuka A, Taguchi T, Sano T. The blood vascular architecture of the rat thyroid gland. A scanning electron microscope study. *Arch Histol Jpn* 52:15–30, 1989.

15. Mikami S, Miyasaka S. Vascularization and its fine structure of the thyroid gland of Japanese quail, *Coturnix coturnix japonica, Iwate Daigaku Nogakubu Hokoku* 16: 81–91, 1982.

16. Ishimura K, Okamoto H, Fujita H. Freeze-etching studies on ultrastructural changes of endothelial cells in the thyroid of normal, TSH-treated and thyradin-treated mice. *Cell Tissue Res* 175:313–317, 1976.

17. Kux E. Uber muskulare Drosselvorrichtung ("Zellknospen", "Porster") in den Arterien der Schilddruse. *Virchows Arch* 294:358–364, 1935.

18. Wollman SH, Herveg JP, Zeligs JD, Ericson LE. Blood capillary enlargement during the development of thyroid

hyperplasia in the rat. *Endocrinology* 103:2306–2314, 1978.

19. Ericson LE, Wollman SH. Ultrastructural aspects of capillary fusion during development of thyroid hyperplasia. *J Ultrastr Res* 72:300–315, 1980.

Author's address:
Prof. Hisao Fujita
Department of Anatomy,
Osaka University Medical School
2-2 Yamadaoka
Suita-city
Osaka-fu, 565, Japan

Blood Vascular Beds of Rabbit Uterus and Oviduct: A Study of Corrosion Casts by Scanning Electron Microscopy

WOLFGANG KÜHNEL & LÜDER C. BUSCH

1. Introduction

During the past 20 years scanning electron microscopy (SEM) has emerged as a powerful tool for studying the three-dimensional surface morphology of cells and organs, and other biological structures of complex multicellular systems. This is due to improvements both in SEM instrumentation and in the preparation of biological specimens for SEM. Additionally, intracellular structures could be observed by SEM because various kinds of cracking methods have been developed, such as cryofracture [46], the frozen resin cracking method [44,45], high-resolution SEM [47], and others [17].

The concept of using corrosion casts for anatomical studies is many hundred years old. It has had periods of intensive use [12]. The technique to fill cavities with liquids becoming rigid after application dates back to Leonardo da Vinci (1452–1591). He made wax casts of the cerebral ventricles and heart chambers of humans. Hyrtl [18] used gelatin as an injection medium and introduced celloidin for preparation of corrosion casts in 1860. This approach was modified by other investigators [41,48]. Schummer [43], whose corrosion casts are well known up to today, used a polymerizing plastic (Plastoid). Grosser [15] initiated the use of India ink to fill blood capillaries. Serial sections of an organ injected with the pigment medium had to be studied by light microscopy in order to reconstruct the microvascular pattern. The studies provided interesting information on the general microcirculatory pattern of an organ, but the findings failed to show three dimensionality, and the resolution of microvascular branches was poor. Thus, complex three-dimensional relationships were often misinterpreted.

In order to circumvent these problems, microvessel corrosion casts were studied by SEM procedure, as reported by numerous studies. A casting medium such as Mercox®, a polymethylmetacrylate, is injected into the vascular bed and, after polymerization of the resin and digestion of the tissue by HCl or NaOH, a replica of the microvascular tree is obtained. When the casts are viewed in SEM, the architecture can be studied at low and high magnifications [31,37,39].

We reported by histological, histochemical, and biochemical findings, 20 years ago, how hormones regulate preimplantation stages in pregnant rabbits, as well as in rabbits with hormonally induced pseudo-pregnancy [1–5,7,21,24]. We described the morphological transformation of the endometrium during the preimplantation stage (6 days p.c. and 7 days p.c.) and correlated the results with electrophoretic and immunological studies on the protein pattern of uterine secretions [10,20,22,23,25]. There are proteins specific

Motta, P.M., Murakami, T., and Fujita, H. (eds.), Scanning Electron Microscopy of Vascular Casts: Methods and Applications.

for uterine secretions, e.g., uteroglobin and some glycoproteins. The chronological sequence of protein pattern secretion and the histochemical changes during preimplantation were assumed to be caused by complex hormonal regulations. Changes in the microcirculatory organization were not considered. Subsequently, we examined surface profiles of the rabbit endometrium and the rabbit oviduct during estrus and pregnancy by the use of SEM [9,26–29]. The endometrium forms interconnecting circumferentially arranged pillows, short mounds, and ridges. Straight, continuous, longitudinal folds do not occur. The oviduct mucosa of the ampullary region appears as high, unbranched, longitudinally aligned folds. An intricated system of low mucosal ledges lying slanted occurs between the high longitudinal folds [9,10,26–28]. We also studied alterations of tight junctions of uterine epithelial cells, as well as cell membrane events involved in the processes of the preimplantation stage [11,53–57].

Precise knowledge of the tubal and uterine microcirculation is indispensable to understand the microvascular adaptations of the oviduct and the uterus in pregnancy. Such adaptations possibly mediate the striking increase of secretion during the preimplantation stages. For this reason, we investigated the microvessel patterns in the oviduct and uteri of rabbits in estrus and during the early phase of pregnancy, i.e., up to the sixth day, in order to clarify the relationship of microvascular architecture to surface structures [30,31,40]. Information on the main vascular supply of the genital tract of the female rabbit is available [13, 14,16,19,38]. This knowledge is based on studies carried out with India ink injected intravascularly followed by lightening of preparations. As mentioned above, the technique does not allow to depict microvascular patterns. In parallel, other authors examined microvasular corrosion casts by SEM [30,31,35–37,50–52], extending our understanding of the microvascular architecture of rabbit oviduct and uterus, respectively.

2. The Gross Vascular Supply

The gross vascular supply of the rabbit female genital tract is well known from the investigations of Parry [38] and Del Campo and Ginther [13].

Blood comes to the uterus by way of the ovaric arteries, as well as by the uterine arteries. The ovaric arteries are long, slender vessels that originate from the ventral portion of the abdominal aorta, just below the origins of the renal arteries. The ovaric arteries then spread laterally towards the hilum of the ovary. A caudal branch occurs that divides into branches for the uterine portion of the oviduct and into a small branch for the tip of the uterine horn. The uterine branch anastomoses with the proximal branch of the uterine artery. The uterine arteries deriving from the internal iliac arteries, as seen in other species, reach the cranial portion of the cervix and the adjacent vagina. Each artery runs parallel to the uterine horn and divides into a medial branch and a lateral branch. The medial branch for the caudal portion of the uterine horn anastomoses with the corresponding uterine artery of the other side. The lateral branch runs to the cranial portion of the uterine horn, contributes to the arterial supply of the uterine portion of the oviduct, and anastomoses with the ovaric artery [13]. The anatomy of the uterine and ovaric veins is similar to that of the corresponding arteries.

In the rabbit, arteries and veins of the uterus are found within the mesometrium of the horn (mesometrial artery and mesometrial vein), where vessels branch and reach the antimesometrial side of the uterine horn as circumferentially arranged arteries. The circumferentially arranged arteries give rise to many ventrally located uterine arteries, which are connected with arteries on the other side. The circumferential, as well as the ventral, uterine arteries run in the myometrium and are the source of arterioles that spread radially in the endometrium towards the uterine surface epithelium. These arterioles terminate as capillaries, which form the subepithelial capillary plexus. This plexus is connected with venules that return to the myometrium in order to reach the circular veins juxtaposed to the circumferential arteries. At the basalis of the endometrium, venules are organized in another microvascular plexus [49].

3. Morphology of the Oviduct and of the Uterus

The rabbit oviduct is divided into the preampulla, with the infundibulum, ampulla, isthmus, and the utero-tubal-junction. In cross section, each area reveals a characteristic mucosa when viewed in SEM and TEM [9,20,21,23,26,27,29]. In the ampulla, we see extremely long, thin villi that are often bent or folded. In the isthmus, there are broad-based, leaf-shaped protrusions, which are rich in connective tissue and are rarely branched. Longitudinal mucosal ridges emerge from the leaf-shaped protrusions. Between ridges another system of fine corrugations can be detected. They are intricately branched. Some of the branches anastomose with one another. The surface architecture is, in general, not oriented along the axis of the oviduct. Shallow or deep depressions, as well as narrow channels, are found between the branched ridges. In the isthmus, longitudinal folds are more developed. They are thicker and more extended. At the bottom of the channels, narrow bridgelike structures appear that connect longitudinal folds at right angles. With respect to the highly prismatic epithelial cell layer of the ampulla, we find secretory and ciliated cells arranged in a quite regular pattern [6,8,20,23,42]. In the isthmus, secretory cells predominate.

Cross sections of rabbit uteri show that the endometrium bulges into the lumen and forms broadly based mounds and narrow folds [9,10,27]. According to Hafez and Tsutsumi [16], the mucosal mounds or hills appear in the entire uterine horn and are oriented longitudinally. The longitudinal folds run in pairs, symmetrical to the mesometrium, the mesometrial plane acting as the plane of symmetry. However, when observed by SEM, the endometrium of estrus rabbits lacks folds or mounds separated by valleys and channels running longitudinal to the axis of the uterine horn. A transversely oriented array of folds is seen instead, which displays various patterns. The mesometrial side of the uterus mainly demonstrates wide, cushionlike protrusions, which become more narrow towards the periplacentar direction, where they look like thin, elevated areas. The mesometrial cushionlike protrusions are separated by deep valleys, sometimes by short, shallow ditches or by mere slits. The bottom of the valleys of different depth is — as light microscopy and transmission electron microscopy have shown — lined by single columnar epithelium having cells comparable to the columnar cells of mucosal mounds [6,39]. The appearance of the

antimesometrial mucosal surface architecture reminds one of tire tracks. Short mucosal folds spread from the middle of the antimesometrial area as clumpy mounds. The pattern of folds and mounds arranged diagonally to the longitudinal axis of the uterine horn is repetitive, thus creating a regular system of rows. The antimesometrial mucosal ridges are directed towards the periplacentar region as coarse, irregularly formed mucosal bulges, which are interconnected by attenuated, often-bended, mucosal folds. Wide or narrow slits may appear between them.

The surface epithelial cells are polygonal in appearance, with the exception of some round epithelial cells. The surface is densely covered with short, often club-shaped microvilli. Some ciliated cells are randomly located.

4. Microvessel Corrosion Casts

When Mercox casting medium is applied by way of the abdominal aorta, satisfactory microvessel corrosion casts of the oviduct and uterus are obtained, even though casts could show incomplete resin filling.

Uterine arteries approach uterine horns from the mesometrium and along the course of the arteries, arterioles originate that are continuous with the capillary bed of the mesometrial connective tissue and of the serosa (Fig. 18-1). After reaching the mesometrium, uterine arteries diverge and come to lie between the longitudinal and the circular muscle cell layers, supplying numerous small branches to both layers. In cross sections of uterine horns, in common with many organs, larger arteries and veins run in parallel (Fig. 18-1). Arteries are smaller in diameter than the accompanying veins. Some small myometrial arteries attain the basalis of the endometrium, where the vessels are oriented in a longitudinal manner. The longitudinal and circumferential arteries give rise to arterioles. Some of them spread radially towards the surface epithelium, transforming into capillaries of the subepithelial capillary plexus (Fig. 18-1). The appearance of the lumen surface is characterized by a honeycomblike network of capillaries (Fig. 18-2). The meshes of the network presumably reflect locations of the outlets of uterine glands. Veins

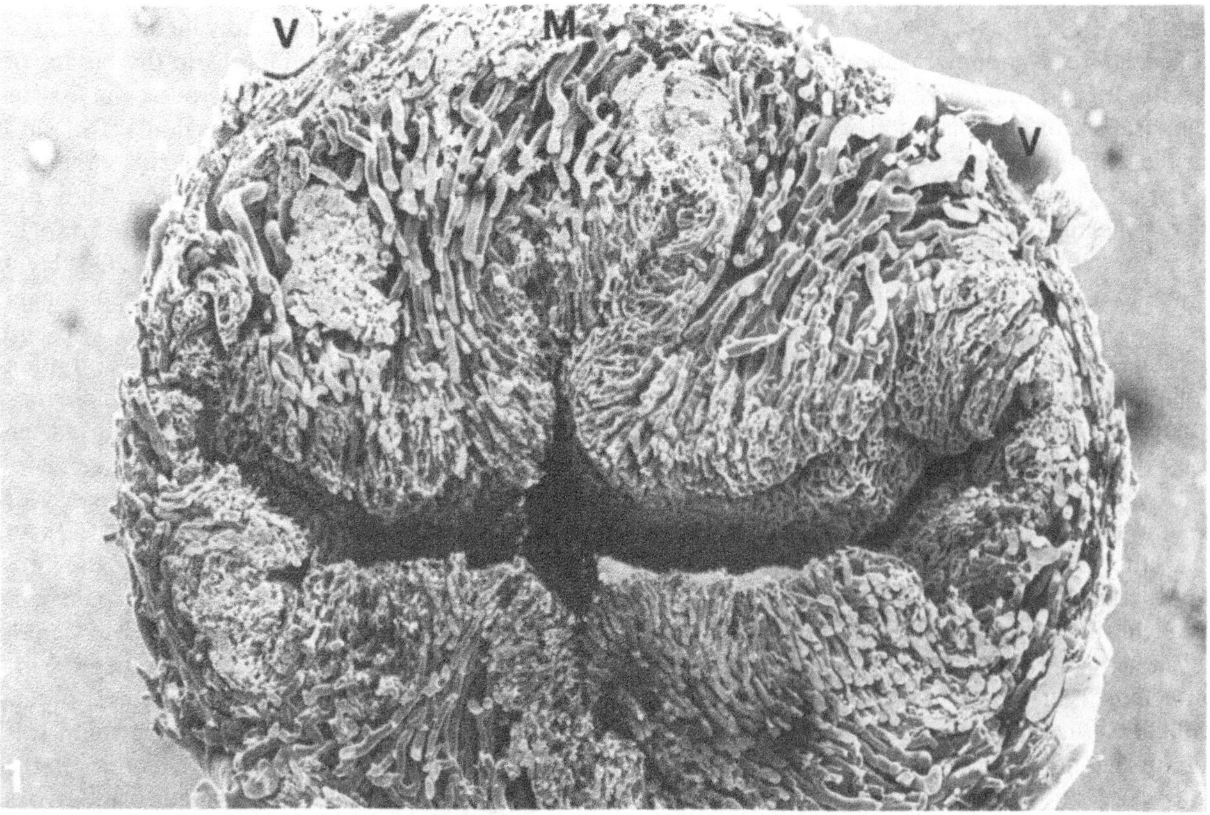
V
M
V
1

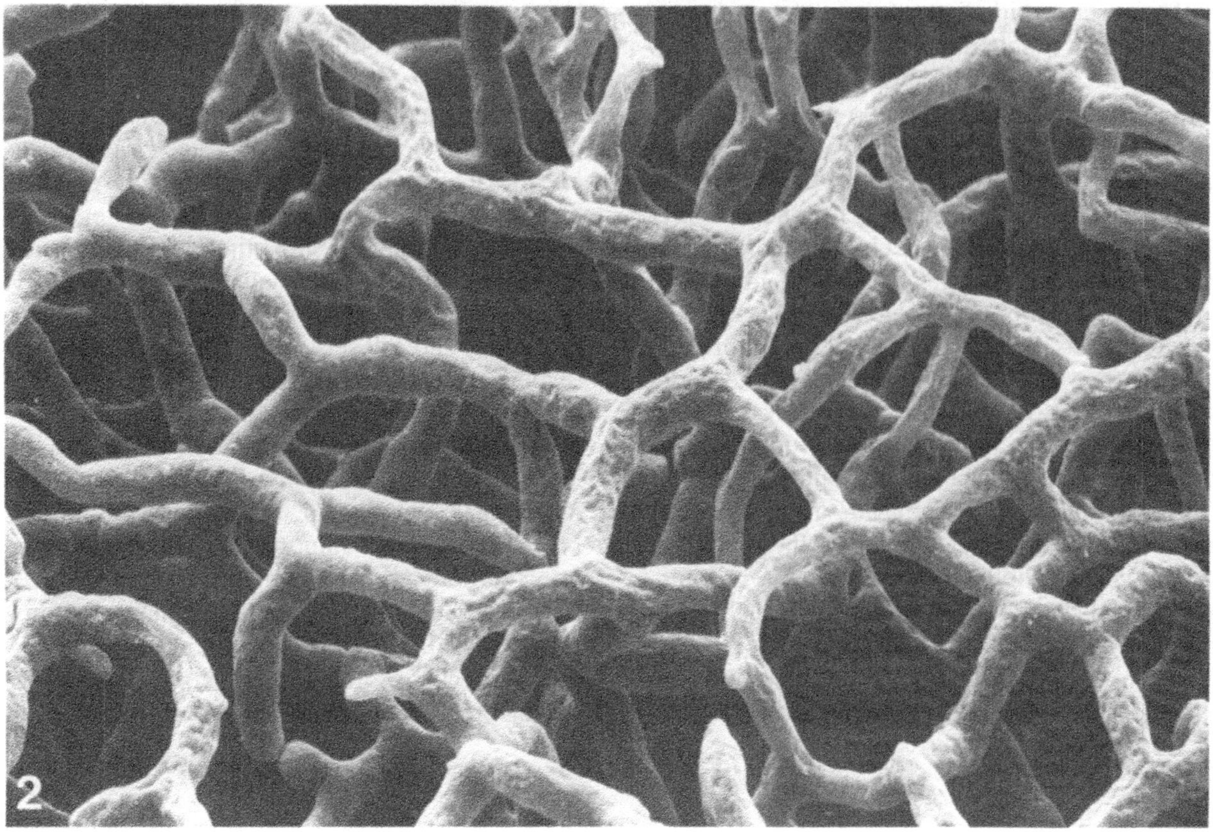
2

originate from the capillary plexuses and run alongside corresponding arteries. The described features of the microvascular arrangement are comparable for the placental, obplacental, and periplacental folds.

Large, circumferentially arranged, and ventrally collecting veins branch at the area basalis of the endometrial stroma and constitute a network of medium-sized veins. They are always found to use arteries as a scaffold (Fig. 18-3). Close connections can be seen in semithin sections. Occasionally bulging extensions of the vein wall are apparent in close proximity to the arterial wall (Fig. 18-3). Whether direct contact exists between veins and arteries remains to be clarified. Our assumption that these contacts may be vasa vasorum can neither be confirmed nor repudiated. Apart from the contacts, sphincters, and sphincterlike constrictions, as well as arterial cushions, can be found at intersections of vessel branching (Fig. 18-4). An irregular network of capillaries running preferentially longitudinally to the uterine horn can be found in the serosa (Fig. 18-5).

The surface epithelium of the endometrium of nonpregnant rabbit uterus is relatively smooth. During the preimplantaion stage of pregnancy or during an hCG-induced pseudo-pregnancy (days 0–6) the surface morphology of the endometrium changes strikingly. The filigreelike transformation of the endometrium is typical for the late preimplantation stage of day 6. On the second day of pregnancy, the surfaces of the endometrial mounds and of mucosal cushions are covered with branching slits. This is the onset of the subdivision and splitting of cushionlike bulges of the endometrium. Cell proliferation and cell transformation of the endometrium become pronounced on days 5 and 6 of pregnancy or pseudo-pregnancy. Mucosal ridges are arranged in a gyruslike manner, and the cellular lining of the mucosa is reminiscent of cobblestone pavement. The endometrium is characterized by conspicuous branching of folds with narrow holes.

From the day of mating or of hCG injection, up to day 6 of pregnancy, the microvascular bed of the endometrium is markedly increased. Large vessels become more numerous in the basalis of the endometrium, and the microvascular bed is more apparent in superficial regions. The subepithelial capillary plexus extends rapidly. Capillaries have dilated and elongated. They run back and forth following the gyruslike arrangement of the endometrial convolutions. On day 4 of pregnancy, we find a close and dense arrangement of capillary curls and loops (Fig. 18-6 and 18-7). Marginal capillaries are tortuous and often form an arabesque network. The capillaries merge into venules, which drain into veins in the center of endometrial folds.

Thus, during the first 6 days of pregnancy or pseudo-pregnancy, the microvascular pattern of the endometrium develops parallel to the proliferation of the endometrium. These conditions are considered to be adequate for increased endometrial metabolism and secretory production. It remains to be clarified, however, whether sprouting of capillaries occurs in the endometrium. It is also possible that the capillary network of the nonpregnant endometrium simply undergoes hypertrophy.

Before we describe the vascular system of the oviduct, we want to remind the reader of a few morphological and physiological details. Fertilization, development of cleavage stages, and blastocyst formation take place within the oviduct. The maternal organism provides the essential extrinsic conditions for the preimplantation development of the mammalian embryo by means of tubal and uterine cell transformation, cell proliferation, and cell secretion. For this reason, the oviduct cannot be regarded as a transport organ. It must be taken into account that the secretory products of the tubal epithelium contribute to the survival of the developing embryo. The macromolecular composition of these secretions is probably of major importance. During the preimplantation stage of pregnancy (days 1–6 p.c.), the serum-identical proteins albumin and transferrin prevails in oviductal secretions, whereas uterus-specific proteins, such as prealbumin and uteroglobin, predominate in the uterine fluid in

Figure 18-1. Rabbit uterus, 4 days p.i. hCG. Cross section showing the microvascular pattern. ×17. V = veins; M = mesometrium.
Figure 18-2. Rabbit uterus, estrus stage (0 days). Note the honeycomblike network of the subepithelial capillary plexus. ×560.

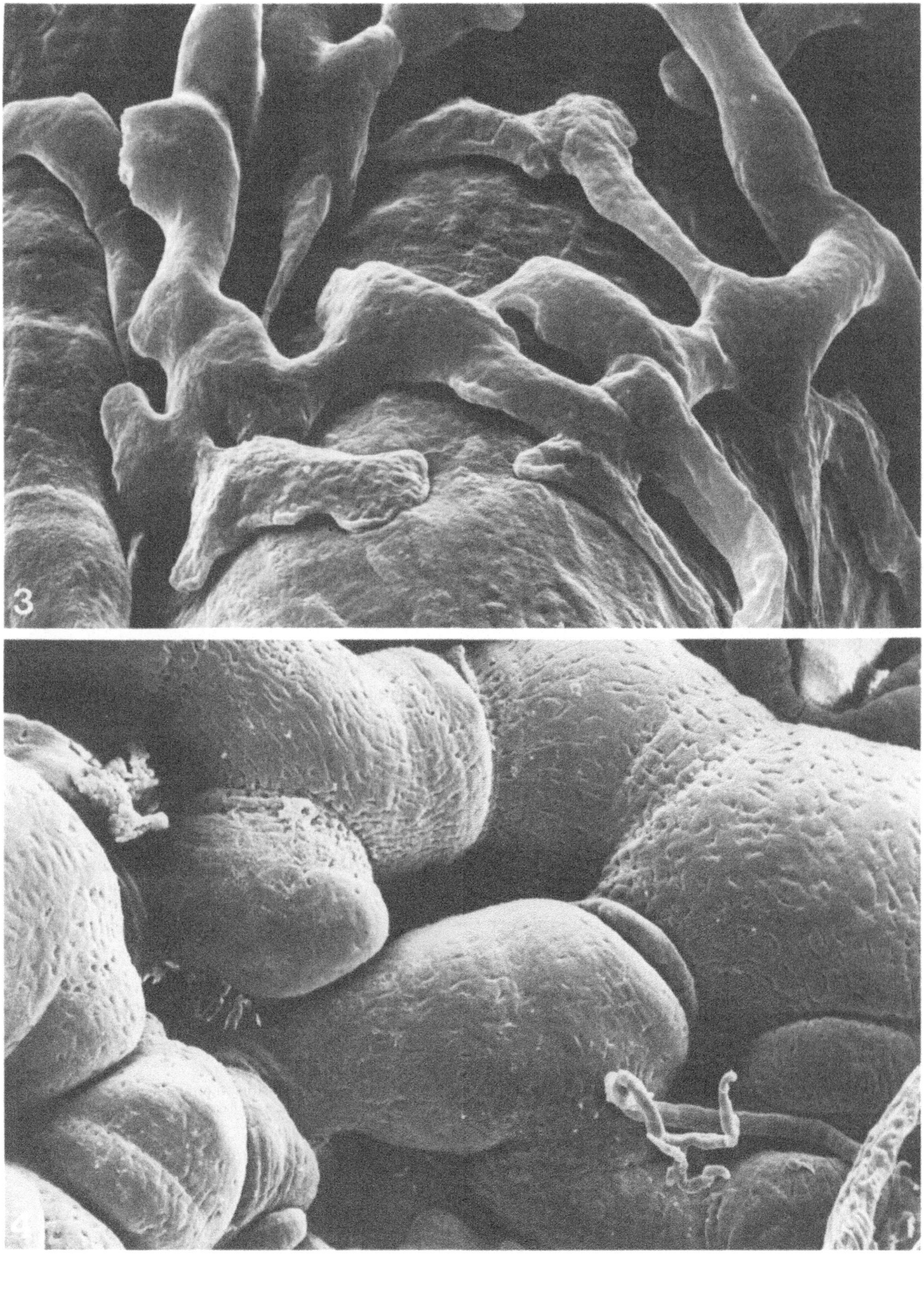
3
4

time-specific patterns. An acid mucoprotein fraction can also be detected in the oviductal fluid.

The rabbit oviduct is supplied by branches of uterine and ovaric arteries that anastomose and run parallel to the oviduct in the broad ligament (Fig. 18-8). SEM examination of transversely or longitudinally cut segments of microvessel casts of the oviduct reveals several surface veins below the serosa. The veins are continuous with veins of the hilus region of the ovary. Veins located in the serosa are connected to capillaries of the subserosal plexus. All veins are derived from capillaries of mucosa folds of the oviduct.

Arteries of the oviduct ramify adjacent to the subserosal plexus. The main vessels of these arteries reach the submucosa. We find longitudinally oriented submucosal arteries from where branches originate before terminal ramification occurs in high mucosal folds (Fig. 18-9). Arteries extending into mucosal folds form a narrow capillary network, which supplies mucosal folds near the apex. A dense capillary network (Figs. 18-10–18-12) is noticed with the fimbriae and the infundibulum region. In contrast to the endometrium and compared to estrous oviducts, SEM reveals only few changes in the tubal microvasculature during the first 6 days of pregnancy.

5. Concluding Remarks

A properly timed preimplantation phase is crucial to the normal development of mammalian embryos. Optimal conditions for their nutriment, migration, and development must be obtained in both the oviduct and uterus. Under the influence of ovarian hormones, the epithelium of oviduct and uterus produces a secretion that is absolutely necessary to the developing blastocyst and its implantation.

Subsequent to earlier investigations about the functional morphology of the oviductal mucosa and the endometrium, this paper reviewed various morphological aspects of uterine and tubal microvasculature of rabbits during the estrus stage and during the preimplantation phase up to day 6 of pregnancy or of an hCG-induced pseudopregnancy as they appear in vascular casts. During the first 6 days of pregnancy or pseudopregnancy, the microvascular pattern of the endometrium developed parallel to the proliferation of the endometrium, and this proliferation is considered to be adequate for increased endometrial metabolism and secretory production. As shown, it is possible to determine a great deal of information related to the functional morphological features of the microvasculature using the combined technique of vascular cast preparation and scanning electron microscopy. This method does provide an opportunity to demonstrate the need of a closely meshed network of capillaries and sinusoids for the implantation of the blastocyst, and from the functional standpoint the most important features of the endometrium during pregnancy are these increased circulatory systems.

Our SEM findings of the microvascular pattern are necessary and helpful for further physiological studies of intrauterine oxygen tension and for investigations to assess the interrelationship between uterine and tubal blood flow and intraluminal oxygen tension. No doubt much of the data collected in such a physiological study can be correlated with these observations concerning the uterine and tubal vasculature.

Acknowledgment

We gratefully thank Mrs. R. Jönsson and Mrs. R. Münzinger for typing the manuscript.

References

1. Beier HM, Petry G, Kühnel W. Sekretion des Endometriums und frühe Keimesentwicklung. *Hoppe-Seyler's Z Physiol Chem* 351:423, 1970.
2. Beier HM, Kühnel W, Petry G. Uterine secretion proteins as extrinsic factors in preimplantation development. *Adv Biosci* 6:165–189, 1971.
3. Beier HM, Kühnel W, Petry G. Morphologische und biochemische Befunde am pseudograviden Kaninchen-

Figure 18-3. Rabbit uterus, estrus stage (0 days). Flat and tonguelike extensions of the veins in close proximity to the arteries. ×1680.

Figure 18-4. Rabbit uterus, estrus stage (0 days). Sphincters and sphincterlike constrictions. ×150.

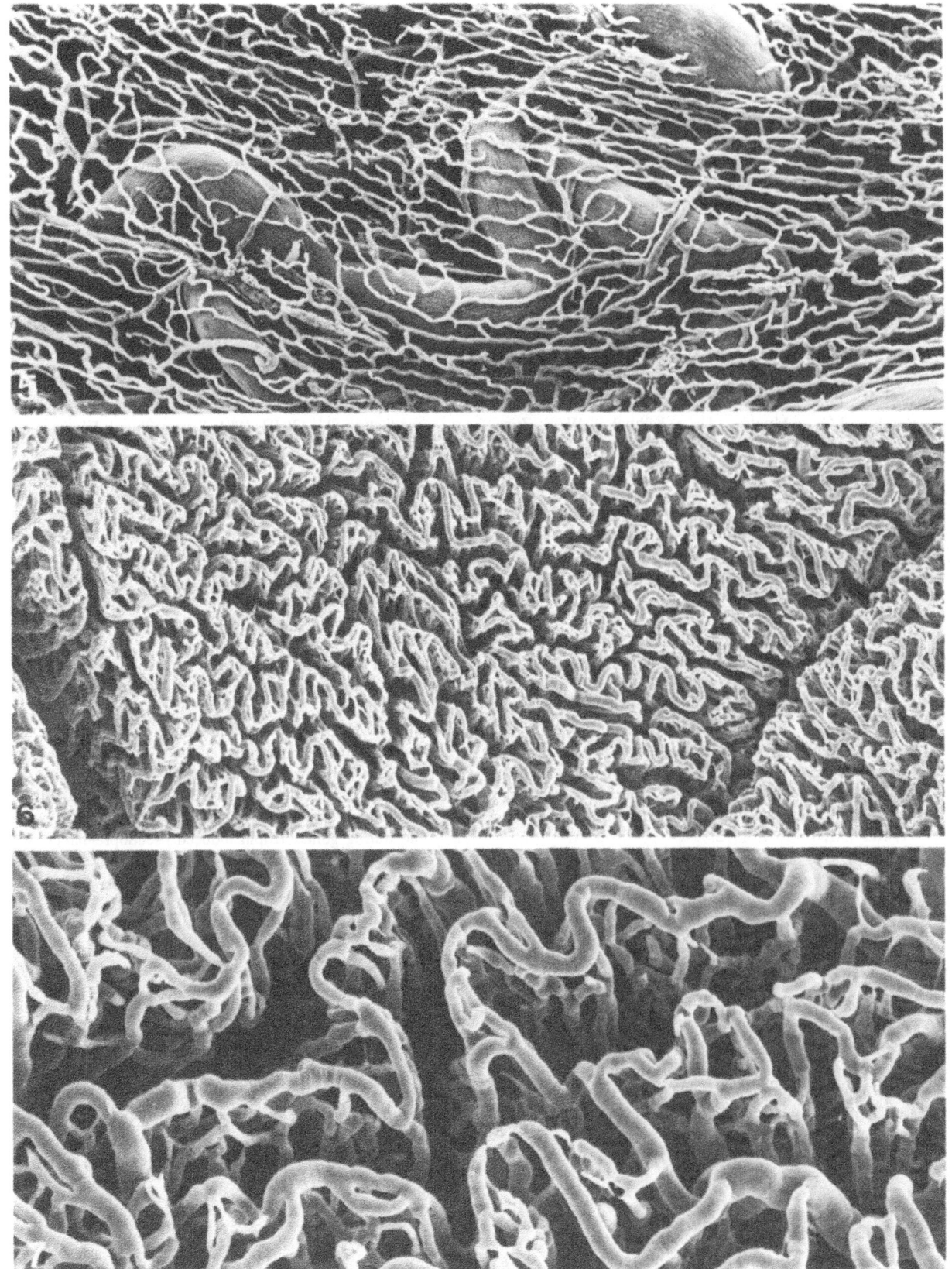

endometrium nach gonadotroper Stimulierung. *Verh Anat Ges* 66:445–457, 1972.

4. Beier HM, Kühnel W. Die verzögerte Uterussekretion nach Oestrogen-injektionen beim graviden Kaninchen. *Verh Anat Ges* 67:567–575, 1973.

5. Beier HM, Kühnel W. Pseudopregnancy in the rabbit after stimulation by human chorionic gonadotropin. *Hormone Res* 4:1–27, 1973.

6. Beier HM, Kühnel W. Untersuchungen zur funktionellen Morphologie des Epithels der Endosalpinx und des Endometriums. *Verh Anat Ges* 70:831–838, 1976.

7. Beier HM, Mootz U, Kühnel W. Endokrinologische Studien an der östrogeninduzierten verzögerten Transformation und Sekretion des Kaninchenendometriums. *Acta Anat* 99:250, 1977.

8. Borisch B, Kühnel W. Lektinmarkierung des Tuben- und Uterusepithels von Kaninchen. *Verh Anat Ges* 80: 707–708, 1986.

9. Busch LC, Mootz U, Kühnel W. Zur Oberflächenbeschaffenheit der Schleimhaut von Tube und Uterus des Kaninchens im Oestrus. *Verh Anat Ges* 71:525–530, 1977.

10. Busch, LC, Kühnel W, Mootz U. Scanning electron microscopical studies of the rabbit endometrium during estrus and preimplantation. In: DiDio LJA, Motta PM, Allen DJ (eds.), *Three-Dimensional Microanatomy of Cells and Tissue Surfaces*, Elsevier-North Holland, pp 267–278, 1981.

11. Busch LC, Winterhager E, Kühnel W. Symplasmatische Umwandlung des Cavumepithels im Uterus pseudogravider Kaninchen. *Acta Anat* 111:22, 1981.

12. Cole FJ. The history of anatomical junctions. In: Singer C (ed.), *Studies in the History and Method of Science*, Vol. 2, Clarendon Press, Oxford, pp 285–343, 1921.

13. Del Campo CH, Ginther OJ. Vascular anatomy of the uterus and ovaries and the unilateral luteolytic effect of the uterus: Guinea pigs, rats, hamsters, and rabbits. *Am J Vet Res* 33:2561–2578, 1972.

14. Duval M. Le placenta du lapin. *J Anat* (Paris) 26: 273–344, 1890.

15. Grosser O. *Frühentwicklung, Eihautbildung und Placentation des Menschen und der Säugetiere* J.F. Bergmann, München, 1927.

16. Hafez ESE, Tsutsumi Y. Changes in endometrial vascularity during implantation and pregnancy in the rabbit. *Am J Anat* 118:249–282, 1966.

17. Humphreys WJ, Spurlock BO, Johnson JS. Critical point drying of ethanol-infiltrated cryofractured biological specimens for scanning electron microscopy. In: *Scanning Electron Microsc* pp 275–281, 1974.

18. Hyrtl J. Die Korrosionsanatomie und ihre Ergebnisse. Braumüller, Wien, 1873.

19. Krichesky B. Vascular changes in the rabbit uterus and in intraocular endometrial transplants during pregnancy. *Anat Rec* 87:221–234, 1943.

20. Kühnel W. The mucosae of the oviduct and of the uterus. In: *Satellite Symposium: Proteins and Steroids in Early Mammalian Development, Aachen, 15.–17.07. 1976.* Monographs on Endocrinology. Springer, Berlin 1977.

21. Kühnel W. Morphological studies of structural changes in the tubal mucosa of the rabbit at estrus and during hCG-induced pseudo-pregnancy. In: Dellenbach-Hellwig (ed.), *Functional Morphologic Changes in Female Sex Organs by Exogenous Hormones*, Springer, Berlin, pp 146–167, 1980.

22. Kühnel W. Transformation und Sekretion des Endometriums in der frühen Graviditätsphase. *Folia Anatom Jugoslav* 11:141, 1981.

23. Kühnel W. Morphological aspects of the fallopian tube and of the uterus. III. World Congress of Human Reproduction, 22.–26.03. 1981, Kongreßband, Berlin, p 69, 1981.

24. Kühnel W, Beier HM, Petry G. Untersuchungen zur hormonellen Regulation der Praeimplantationsphase der Gravidität. II. Histologische, topochemische und biochemische Analysen am hormonbehandelten Kaninchenuterus. *Cytobiologie* 4:9–40, 1971.

25. Kühnel W, Beier HM, Busch LC, Mootz U, Scheele G. Oberflächenveränderungen am Kaninchenendometrium zwischen Östrus und Implantation. *Acta Anat* 99:286, 1977.

26. Kühnel W, Busch LC. Surface morphology of the rabbit uterus and oviduct during estrus. *Anat Embryol* 156: 189–195, 1979.

27. Kühnel W, Busch LC. Functional morphology of the oviductal mucosa and the endometrium as viewed by SEM. *Biomed Res* 2:341–353, 1981.

28. Kühnel W, Busch LC, Beier HM. SEM-studies of structural changes in the tubal mucosa of the rabbit at estrus and during gonadotropin induced pseudopregnancy. In: DiDio LJA, Motta PM, Allen DJ (eds.), *Three-Dimensional Microanatomy of Cells and Tissue Surfaces*, Elsevier/North Holland, pp 279–289, 1981.

29. Kühnel W, Busch LC. REM-Studien am Eileiterepithel des Kaninchens. *Anat Anz* 149:93, 1981.

30. Kühnel W, Busch LC. Zur Vascularisation des Kaninchenuterus im Oestrus und in der Praeimplantationsphase. *Gegenbaurs Morph Jahrb* (Leipzig) 134:433–434, 1988.

31. Kühnel W, Busch LC. Microcirculation of the genital tract of the female rabbit during estrus and pregnancy. In: Motta PM (ed.), *Developments in Ultrastructure of Reproduction*, Alan R. Liss, New York, pp 383–397, 1989.

Figure 18-5. Rabbit uterus, 8 days p.i. hCG. Irregular network of capillaries of the serosa. ×90. A = arteries.

Figure 18-6. Rabbit uterus, 4 days p.i. hCG. The subepithelial capillary plexus has rapidly extended in amount. ×70.

Figure 18-7. Rabbit uterus, 4 days p.i. hCG. Capillaries of the subepithelial plexus are elongated, running tortuously around the outlets of the uterine glands. ×190.

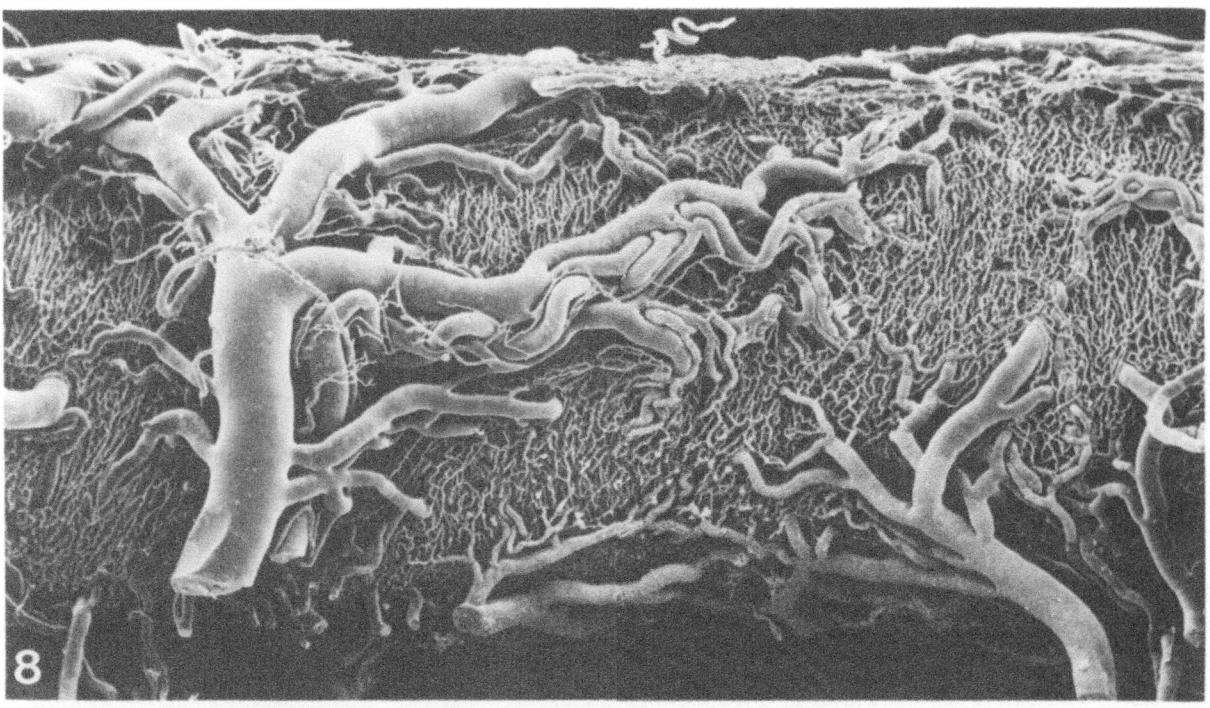

Figure 18-8. Rabbit oviduct, estrus stage (0 days). Longitudinal segment of a vascular cast of the ampulla, demonstrating the subserosal ramification of vessels. ×30.

32. Matsusaka T, Fujibashi T. Ultrastructral specialization appearing on the plastic cast of ocular blood vessels. *J Clin Electron Microsc* 7:227–228, 1974.
33. Matsusaka T, Fujibashi T. Angio-architecture of the choroid. *Jpn J Ophthalmol* 20:330–346, 1976.
34. Murakami T. Application of the SEM to the study of the fine distribution of the blood vessels. *Arch histol Jpn* 32:445–454, 1971.
35. Nakamura T, Ninomiya H. Vascular architecture at the implantation site of the rabbit. *Proc Jpn Acad* 55:441–444, 1979.
36. Ninomiya H, Nakamura T. Vascular architecture of the endometrium in early pseudopregnancy of the rabbit. *Proc Jpn Acad* 54:455–458, 1978.
37. Ninomiya H, Nakamura T. A scanning electron microscopic study on vascular changes in the endometrium at the pre-implantation stage in the rabbit. *Proc Jpn Acad* 56:317–321, 1980.
38. Parry HJ. The vascular structure of the extraplacental uterine mucosa of the rabbit. *J Endocrinol* 7:86–99, 1950.
39. Petry G, Kühnel W, Beier HM. Untersuchungen zur hormonellen Regulation der Praeimplantationsphase der Gravidität. I. Histologische, topochemische und biochemische Analysen am normalen Kaninchen-uterus. *Cytobiologie* 2:1–32, 1970.
40. Sahle E, Kühl P, Busch LC, Kühnel W. Über das Gefäßmuster des Kaninchenuterus in der frühen Pseudogravidität. *Verh Anat Ges* 79:541–542, 1985.
41. Spanner R. Neue Befunde über die Blutwege der Darmwand und ihre funktionelle Bedeutung. *Morphol Jahrb* 69:394–454, 1932.
42. Schramm U, Kühnel W. Glycogen particles associated with ER-convolutions and tubular structures in the ciliated cells of the rabbit oviduct. *Biomed Res* 2:4–10, 1981.
43. Schummer A. Ein neues Mittel (Plastoid) und Verfahren zur Herstellung korrosionsanatomischer Präparate. *Anat Anz* 81:177–224, 1935.
44. Tanaka K. Frozen resin cracking method for scanning electron microscopy of biological materials. *Naturwiss.* 59:77, 1972.
45. Tanaka K. Frozen resin cracking method and its role in cytology. In: Hayat MA (ed.), *Principles and Techniques of Scanning Electron Microscopy*, Vol. 1, Van Nostrand Reinhold, New York, pp 125–134, 1973.

Figure 18-9. Rabbit oviduct, estrus stage (0 days). Cross section of the ampulla with mucosal folds. ×60.
Figure 18-10. Rabbit oviduct, estrus stage (0 days). Border of a fimbria. ×310. A= artery.

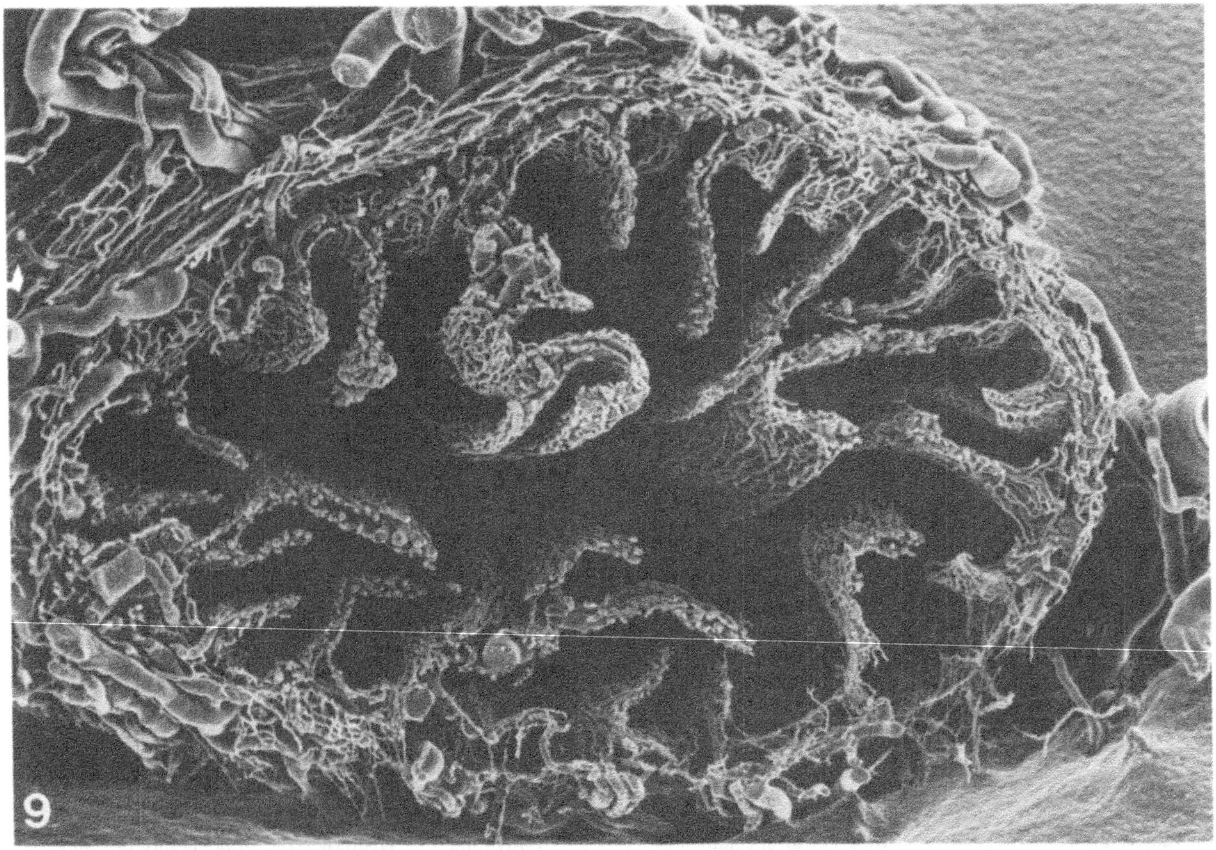

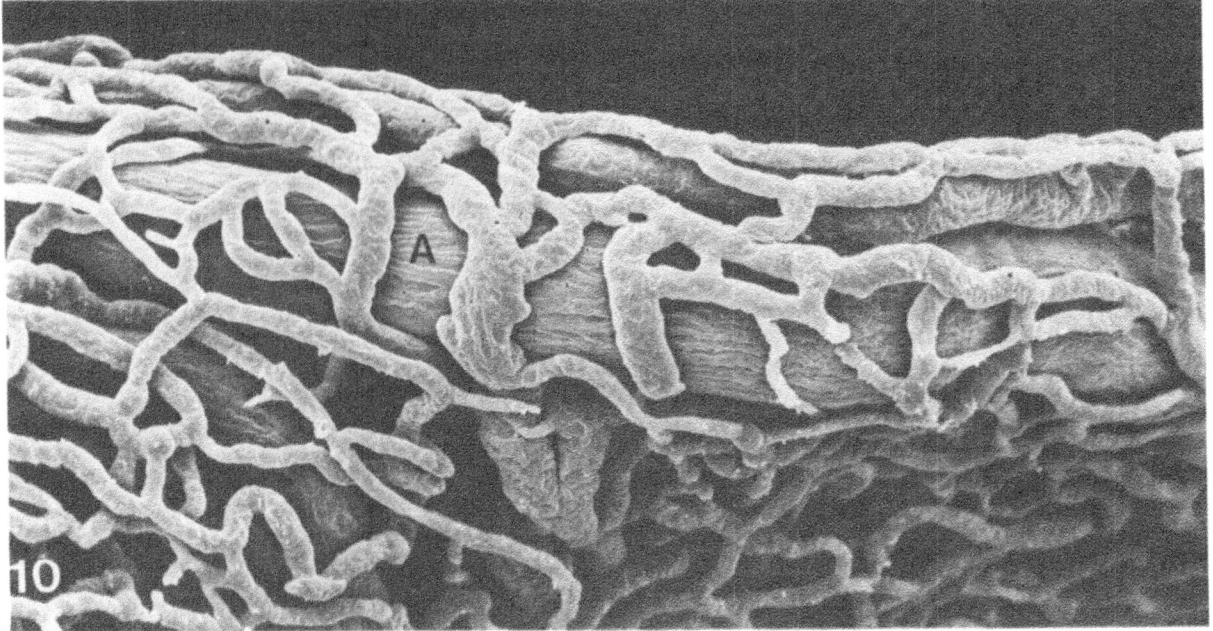

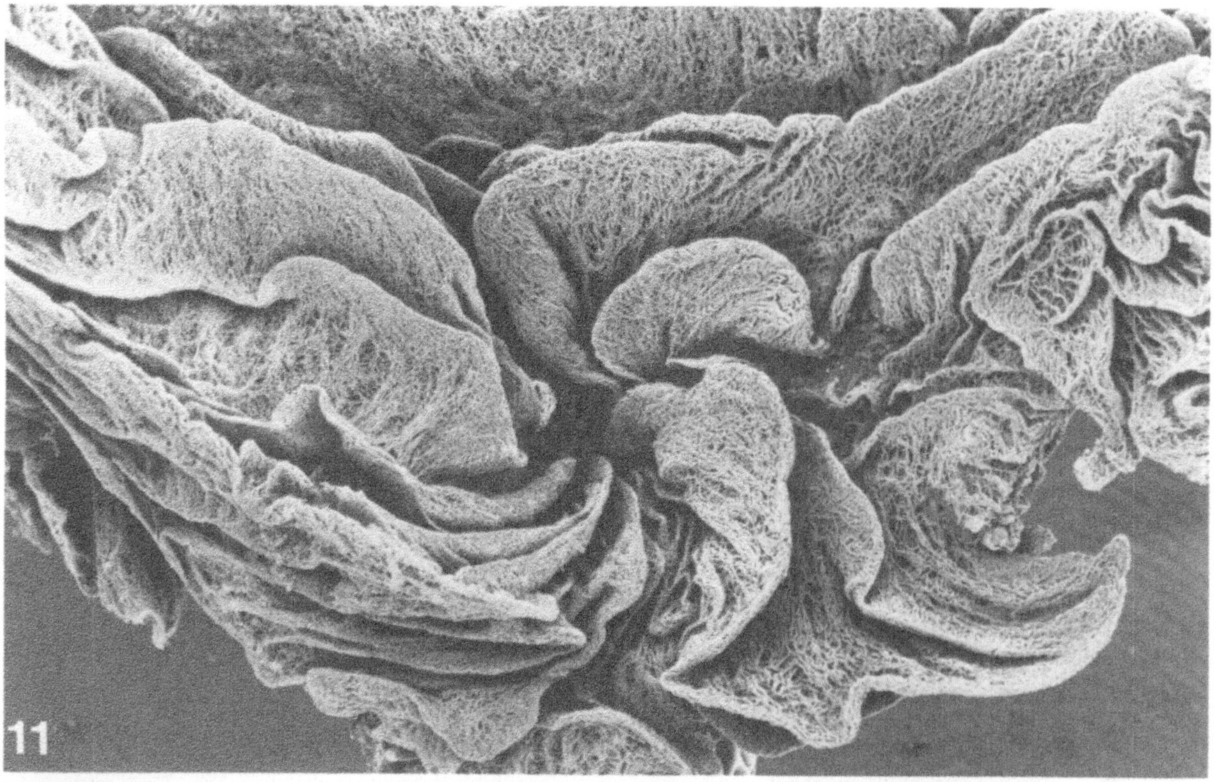
11

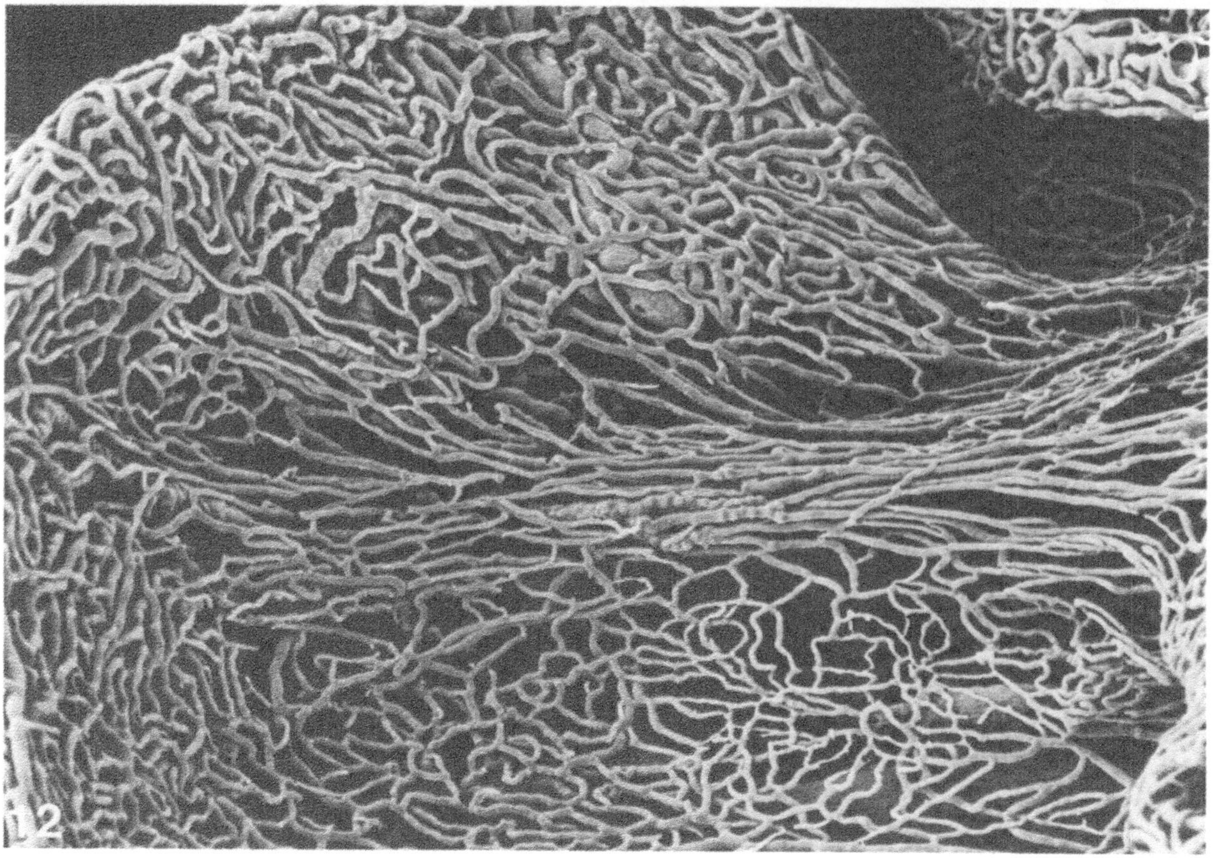
12

46. Tanaka K, Iino A, Naguro T. Scanning electron microscopic observation on intracellular structures of ion-etched materials. *Arch Histol Jpn* 39:165–175, 1976.

47. Tanaka K, Naguro T. High resolution scanning electron microscopy of cell organelles by a new specimen preparation method. *Biomed Res* 2 (Suppl.):63–70, 1981.

48. Taniguchi Y, Ohta Y, Tajiri S. New improved method for injection of acrylic resin. *Okajimas Folia Anat Jpn* 24: 259–267, 1952.

49. Tsutsumi Y, Hafez ESE. Endometrial vascularity during pseudo-pregnancy in the rabbit. *J Morphol* 118:43–56, 1966.

50. Verco CJ, Gannon BJ, Jones WR. Microvascular architecture of the mechanically induced hydrosalpinx in the rabbit. *Clin Reprod Fertil* 1:311–321, 1982.

51. Verco CJ, Gannon BJ, Jones WR. Uterine tube microvascular architecture after oophorectomy in the rabbit. *Acta Anat* 117:193–200, 1983.

52. Verco CJ, Gannon BJ, Jones WR. Fallopian tube microvasculature in the rabbit. *Aust J Exp Biol Med Sci* 61: 127–138, 1983.

53. Winterhager E, Kühnel W. Über den Strukturwandel der Zellkontakte im Uterusepithel pseudogravider Kaninchen. *Anat Anz* 149:108, 1981.

54. Winterhager E, Kühnel W. Alterations of intercellular junctions of the uterine epithelium during the preimplantation phase of the rabbit. *Cell Tissue Res* 224:517–526, 1982.

55. Winterhager E, Busch LC, Kühnel W. Membranveränderungen während der Zellfusion im Uterusepithel pseudogravider Kaninchen. *Verh Anat Ges* 77:431–433, 1983.

56. Winterhager E, Busch LC, Kühnel W. Membrane events involved in fusion of uterine epithelial cells in pseudopregnant rabbits. *Cell Tissue Res* 235:357–363, 1984.

57. Winterhager E, Kühnel W, Denker HW. Proliferation of tight junctions and fusion in uterine epithelial cells of the rabbit. *Eur J Cell Biol* (Suppl. 5) 33:38, 1984.

Author's address:
Dr. W. Kühnel
Institut für Anatomie
Medizinische Universität zu Lübeck
Ratzeburger Allee 160
D-2400 Lübeck 1
Germany

←

Figure 18-11. Rabbit oviduct, estrus stage (0 days). Microvascular cast of the fimbriated infundibulum. ×18.
Figure 18-12. Rabbit oviduct, estrus stage (0 days). Note the dense subepithelial capillary plexus of fimbria. ×115.

CHAPTER 19

The Ovary: Three-Dimensional Morphodynamics of the Luteo-Follicular Complex by SEM of Vascular Corrosion Casts and Other EM Techniques

GUIDO MACCHIARELLI, STEFANIA A. NOTTOLA, AKIO KIKUTA,
OSAMU OHTANI, TAKURO MURAKAMI, & PIETRO M. MOTTA

1. Introduction

The main ovarian activities are morphofunctionally related to the dynamic structure called the *luteo-follicular complex* (LFC). The LFC, in fact, is the morphodynamic expression of the continuous and cyclical modifications of the parenchymal ovarian components: the developing follicles and corpora lutea.

In mammals, during the reproductive period developing follicles, in order to ovulate, undergo rapid growth, characterized by prominent structural and ultrastructural changes of oocyte and follicle wall components (granulosa and theca layers). After ovulation, the follicle remnants are again involved in structural remodelling and morphological changes, with transformations first in the growing corpus luteum and then in the degenerating corpus luteum. Naturally, all of these well-known phenomena are accompanied by intense changes in the LFC microvasculature.

As recently reviewed, numerous studies have described the ultrastructure of the developing follicles and corpora lutea [1,2]. Furthermore, correlated scanning and transmission electron microscopy (SEM and TEM) [3–5] studies have been particularly focused on the three-dimensional morphological organization of LFC [6–10]. The ovarian microvasculature has been studied in different mammals, including humans, using dye injections and casting methods [11–15]. Furthermore, only recently, due to studies performed by means of SEM of vascular corrosion casts, has it been possible to detail the complex morphodynamic changes of the ovarian vascular bed of some mammals[16–23].

In the present chapter, the most salient information on the three-dimensional ultrastructural organization of the LFC and accompanying microvascular changes will be reviewed.

2. The Follicle

In mammals, cyclically a selection of primordial follicles (i.e., the follicles that are structurally characterized by an oocyte blocked in the prophase of the first meiotic divison, surrounded by a single layer of flattened granulosa cells) undergoes "recruitment" [24]. The "recruited" follicles develop and change their morphological features as a prerequisite to ovulation. Developing follicles include preantral, antral, and ovulatory follicles, and are destined for the ovulation and the consequent formation of the corpora lutea. However, a number of follicles do not take part in LFC development because they halt their normal growth and undergo atresia [1,5,25].

Motta, P.M., Murakami, T., and Fujita, H. (eds.), Scanning Electron Microscopy of Vascular Casts: Methods and Applications.

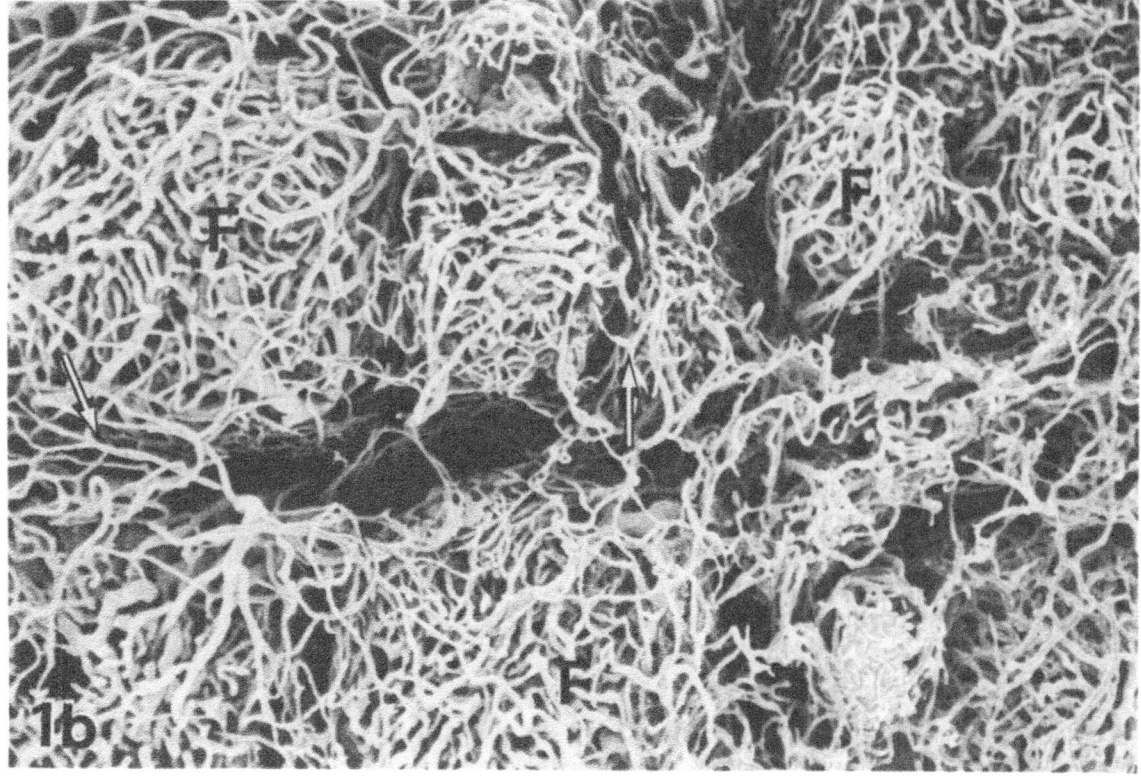

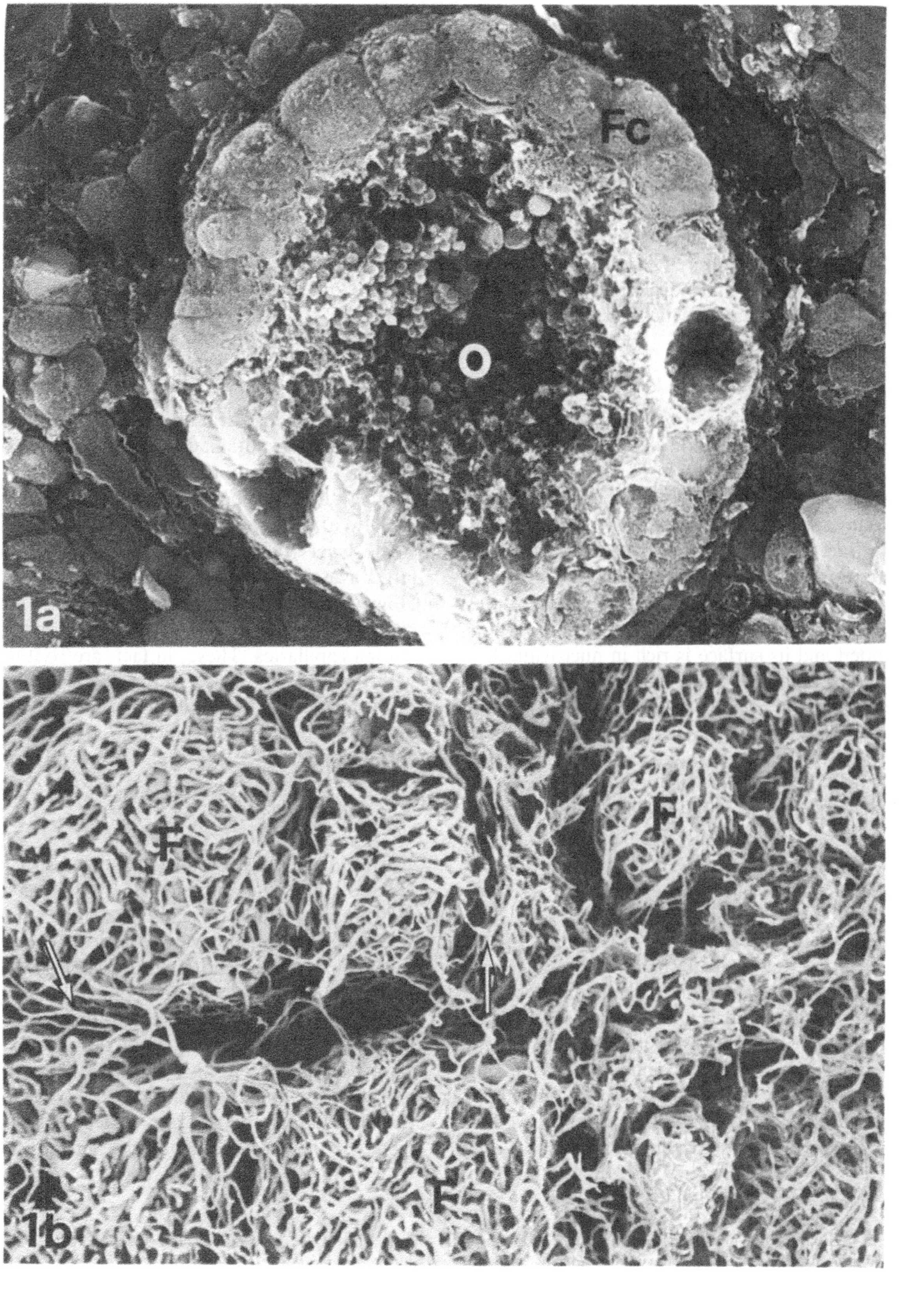

Follicular development is characterized by two phases. In the first phase (preantral) the simultaneous growth of the oocyte and follicular cells predominates. In the next phase, which starts as soon the oocyte reaches the appropriate size for ovulation, is characterized by the continuation of follicular cell development and the formation of the antrum [1–3]. Therefore, during follicular growth the theca-layer microvasculature changes its morphological features in order to supply adequate nutrients and hormonal stimuli to the growing granulosa layer, which is devoid of vessels [26].

2.1. Preantral Follicle

As seen by SEM, preantral follicles appear as smooth, rounded structures of about 130 μm in diameter, mainly localized in the periphery of the ovarian cortex [1,2].

In fractured samples, the oocyte is seen at the center of one (Fig. 19-1a) or a few layers of follicular (granulosa) cells, surrounded by a thin vascularized theca cell layer [1,2]. The oocyte is rounded and its surface is rich in microvilli. The oocyte cytoplasm has a rounded central nucleus and a variety of organelles distributed throughout the cytoplasm [1–3,27]. This is even more evident in SEM of O-D-O macerated samples observed at high resolution [2,28]. The oocyte is covered by patches of a dense, homogeneous material, corresponding to the forming zona pellucida. In larger follicles, the patches of the zona pellucida tend to gradually coalesce to form a continuous layer [1,29].

The granulosa cells have a smooth regular surface, are cuboidal in shape, and are closely packed to form an uninterrupted envelope for the oocyte [2,10,30] (Fig. 19-1a). In fact, these cells present numerous desmosomes and tight junctions [31]. However, communications among the granulosa cells and between granulosa cells and the oocyte compartment are allowed by the presence of homocellular (granulosa cell-granulosa cell) and heterocellular (granulosa cells-oocyte) gap junctions [1,31]. Further, granulosa cells lie on a continuous basal lamina, which separates them from the theca layer [1].

As seen by SEM, theca cells are grouped in a few layers of elongated elements, resembling fibroblasts, and are intermingled with a few collagen fiber bundles [26]. The inner layers (theca interna) are more highly vascularized than the outer layers (theca externa), which are formed mostly by patched connective tissue elements. By SEM of vascular corrosion casts, preantral follicles are recognized as capillary baskets of 80–200 μm in diameter [32] (Fig. 19-1b). These baskets are supplied by one or two branches of the cortical arteries [17,21,32]. Follicular baskets are drained by one or two efferents vessels converging in the cortical veins. Preantral follicles show a proper vascular bed, characterized by a poor capillary wreath, organized in a single layer (theca interna), and surrounding a large avascular space (granulosa-oocyte areas). However, smaller follicle vascularization is not completely independent from the interstitial stroma capillaries. These, in fact, are continuous with those of the theca externa [32] (Fig. 19-1b). Follicular vessels are thin and present a polygonal-mesh network, being similar to the interstitial stromal capillaries.

2.2. Antral Follicle

Antral follicles are recognized as large structures located in the ovarian cortex (Fig. 19-2a). During development, their position becomes progressively deeper in the ovarian cortex. They vary greatly in number and size, the largest follicles being less numerous. When adequately exposed, antral follicles appear rounded or, more frequently, ovoid in shape (Fig. 19-2a). In fractured samples, smaller follicles present, among a compact and thick granulosa layer, a few small cavities. Larger follicles are, instead, characterized by a unique large cavity (the antrum), which

←

Figure 19-1. Preantral follicle. a: Human. SEM view of a primary follicle. The oocyte (0) is at the center of a monolayer of smooth, cuboidal follicular cells (Fc). The oocyte nucleus was removed during preparation. 1564×. (From Makabe et al. [2], with permission.) b: Rabbit. Vascular corrosion cast of preantral follicles. The capillaries forming the follicular vascular baskets (F) are thin. Note the anastomoses with the interstitial-stromal capillaries (arrows). 114×.

O
G
TL
2a
ff
zp
a
O
co
P
Gc
2b
a
ff
GL
V
V
TI
V
TE
2c

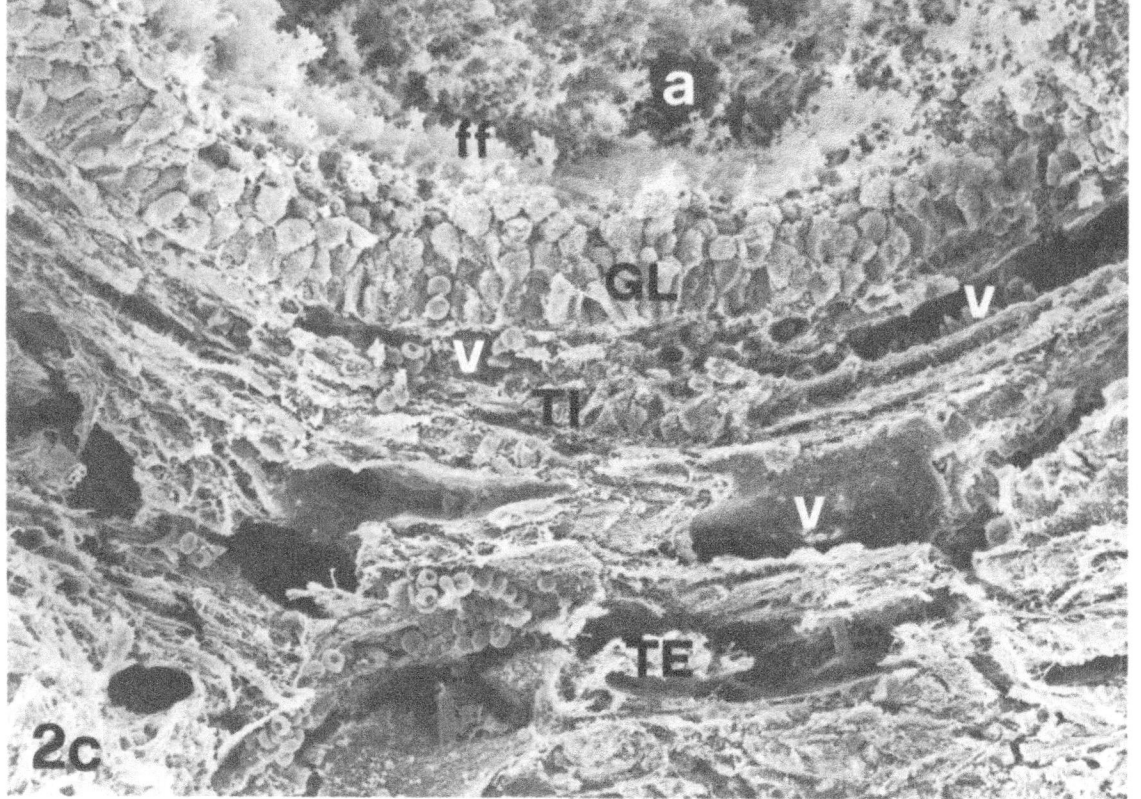

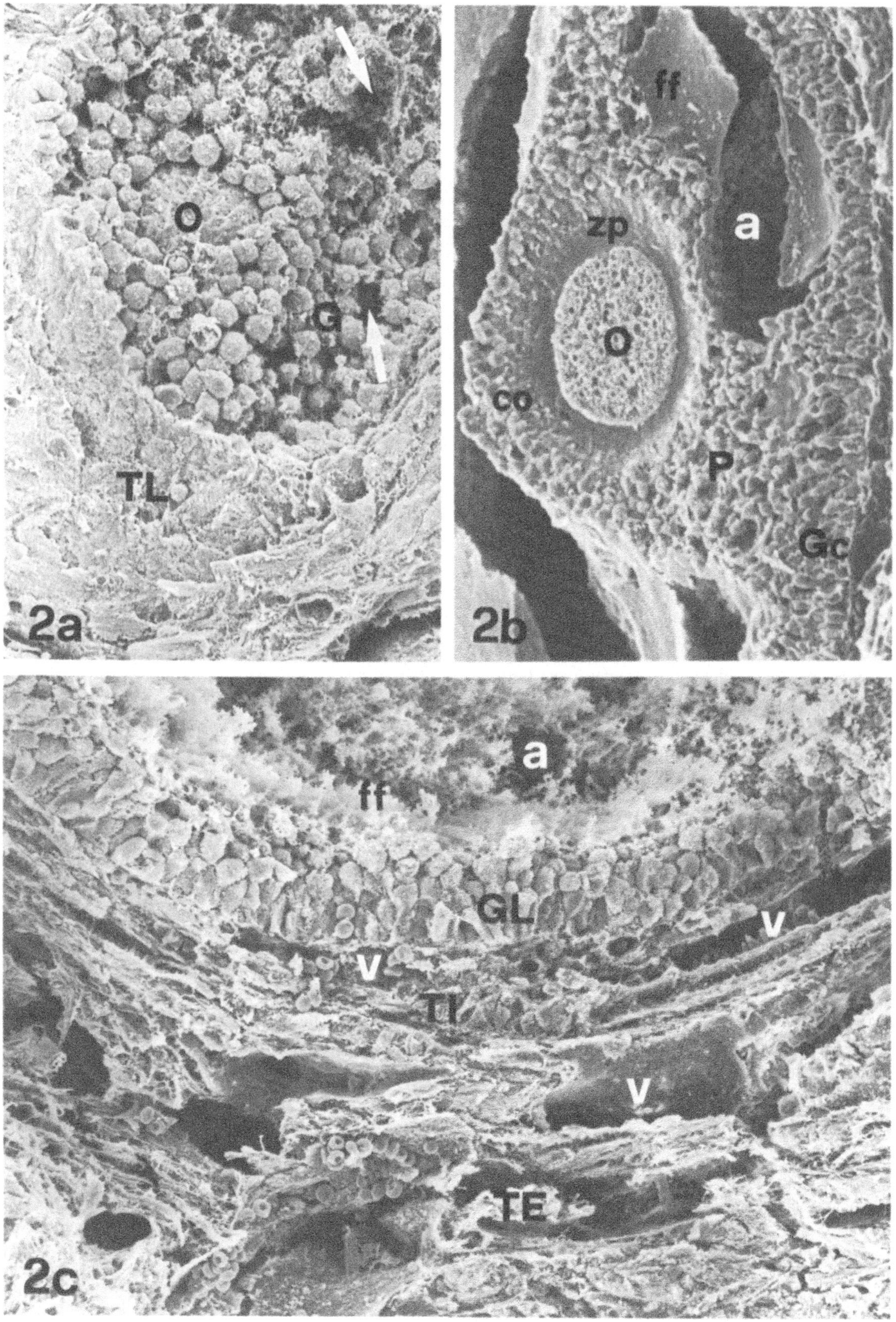

present an eccentric cluster of small cells inside (the cumulus oophorus), connected to the parietal granulosa layer (Fig. 19-2b). Thick theca layers surround the granulosa cells of these follicles [1,2, 10,30] (Fig. 19-2c).

The oocyte may be recognized when the fracture line passes through the cumulus oophorus. By SEM, the oocyte cytoplasm of developing antral follicles has been recently studied in rat and human oocytes, after removing of the cytoplasmic matrix by means of the O-D-O method. Cytoplasmic organelles have been adequately visualized. The Golgi apparatus distribution in the cortical areas and the topographical arrangement of mitochondria in terms of their perinuclear localization has particularly been elucidated [2, 28]. The oocyte appears characteristically surrounded by an inner layer of cylindric cells (corona radiata) and by several outer layers of cells belonging to the cumulus [6,10,33]. The oocyte is larger in antral than in preantral follicles. In fact, at this stage the oocyte reaches its final size. When cumulus cells are occasionally removed, wide areas of the oocyte surface can be seen. In such cases, the oocyte is covered by a dense-amorphous material corresponding to the zona pellucida. The surface of the zona pellucida is quite irregular. This is due to numerous infoldings or small crypts (zona pellucida fenestrations) and to fine granulosity, which is better visualized at high magnification [6].

The cumulus cells are mainly arranged in longitudinal groups, radially placed around the oocyte (Fig. 19-2b). Cumulus cells present a heterogenous morphology. They may be irregularly flattened, polygonal, or star shaped. Their surface is often covered by precipitated antral fluid. The cells directly surrounding the oocyte (corona radiata) possess long cytoplasmic evaginations, which are often seen to penetrate the zona pellucida [6].

The antrum is delimited by granulosa cells possessing long cytoplasmic protrusions. The surface of the cells looking towards the antral cavity is smooth and is often covered by a thin layer of a granular/filamentous substance, probably corresponding to precipitated follicular fluid [6] (Fig. 19-2c). The outer follicle cells have a variable morphology. In smaller follicles they present homogeneous surface characteristics. Usually they are polygonal, with smooth surfaces, and lie on an outer basal lamina. Gradually, with follicular growth, these cells may assume different shapes [1,6,33] (Fig. 19-2c). Further, within the membrana granulosa, small cavities, delimited by a thin membrane and surrounded by radiating follicle cells, may be seen. These vesicles are likely to correspond to the Call-Exner bodies [1,6]. The granulosa cells lying peripheral to these follicles are externally delimited by a basal lamina to which many thick bundles of collagen fibers are attached.

In antral-growing follicles, the theca layer becomes pluristratified (Fig. 19-2c), and three types of theca cells may be recognized: steroidogenic, transitional, and fibroblastlike cells. By SEM, the theca interna consists of elongated cells, which are polyhedral or flattened in shape (Fig. 19-2c). Flattened cells are the fibroblastlike cells, whereas polyhedric cells are the steroidogenic cells. Intermediate forms probably correspond to a transitional type. Steroid-secreting cells also present numerous microvilli and blebs, which often project into small intercellular lacunae. Theca cells are intermingled with a dense connective amorpous substance. Furthermore, numerous vascular lacunae, mainly arranged around the basal lamina, or dispersed among the connective tissue of the outer theca layers, are observed [1,26] (Fig. 19-2c).

As seen by means of SEM of vascular corrosion casts, the antral follicles may be recognized

Figure 19-2. Antral follicle. a: Mouse. SEM view of the surface of a forming antral follicle. Note the presence of small cavities (arrows) among the granulosa cells (G). O = oocyte; TL = outer aspect of the theca layer. 920×. (Courtesy of Prof. G. Familiari.) b: Dog. SEM view of a fractured antral follicle. The oocyte (O) is surrounded by the zona pellucida (zp) and the cumulus oophorus (co). a = antrum; ff = follicular fluid; Gc = parietal granulosa cells, connected to the cumulus oophorus by a cellular pedunculus (P). 644×. (From Motta PM, Van Blerkom J [10], with permission. c: Mouse. SEM view of a fractured large antral follicle. The granulosa layer (GL) is formed by several layers of cells showing different shapes. The inner granulosa cell layer is covered by precipitated follicular fluid (ff). The theca layer is richly vascularized. a = antrum; TI = theca interna; TE = theca externa; V = vascular lacunae. 920×. (Courtesy of Prof. G. Familiari.)

250

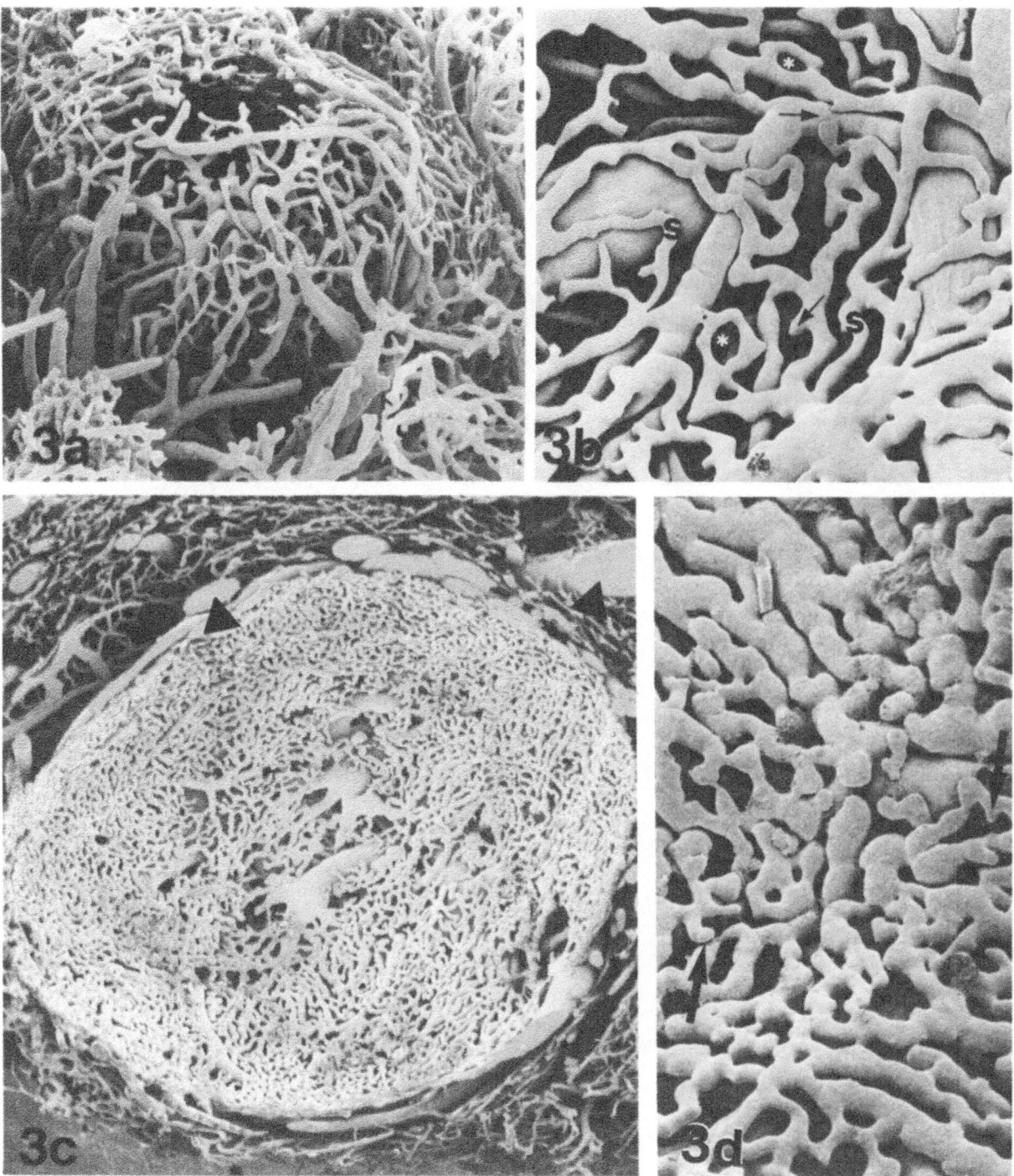

Figure 19-3. Vascular corrosion casts of the antral follicle. a: Rabbit. Small antral follicle. The follicular vascular basket is multilayered and is formed by an inner capillary plexus surrounded by an outer layer of capillaries, arterioles, and venules. 90×. b: Rabbit. Inner aspect of the capillary plexus of an antral follicle. Capillaries run sinusoidally (s), forming typical round meshes (*). Blind ends, related to angiogenetic sprouts, are present (arrows). 275×. c: Rabbit. Inner aspect of the capillary plexus of a large antral follicle. The outer vascular layer, formed by larger vessels (arrowheads), is also seen. 40×.d: Rabbit ovary. Inner aspect of the capillary plexus of a large antral follicle. Note capillary thickening and the capillary sprouts (arrows). 345×.

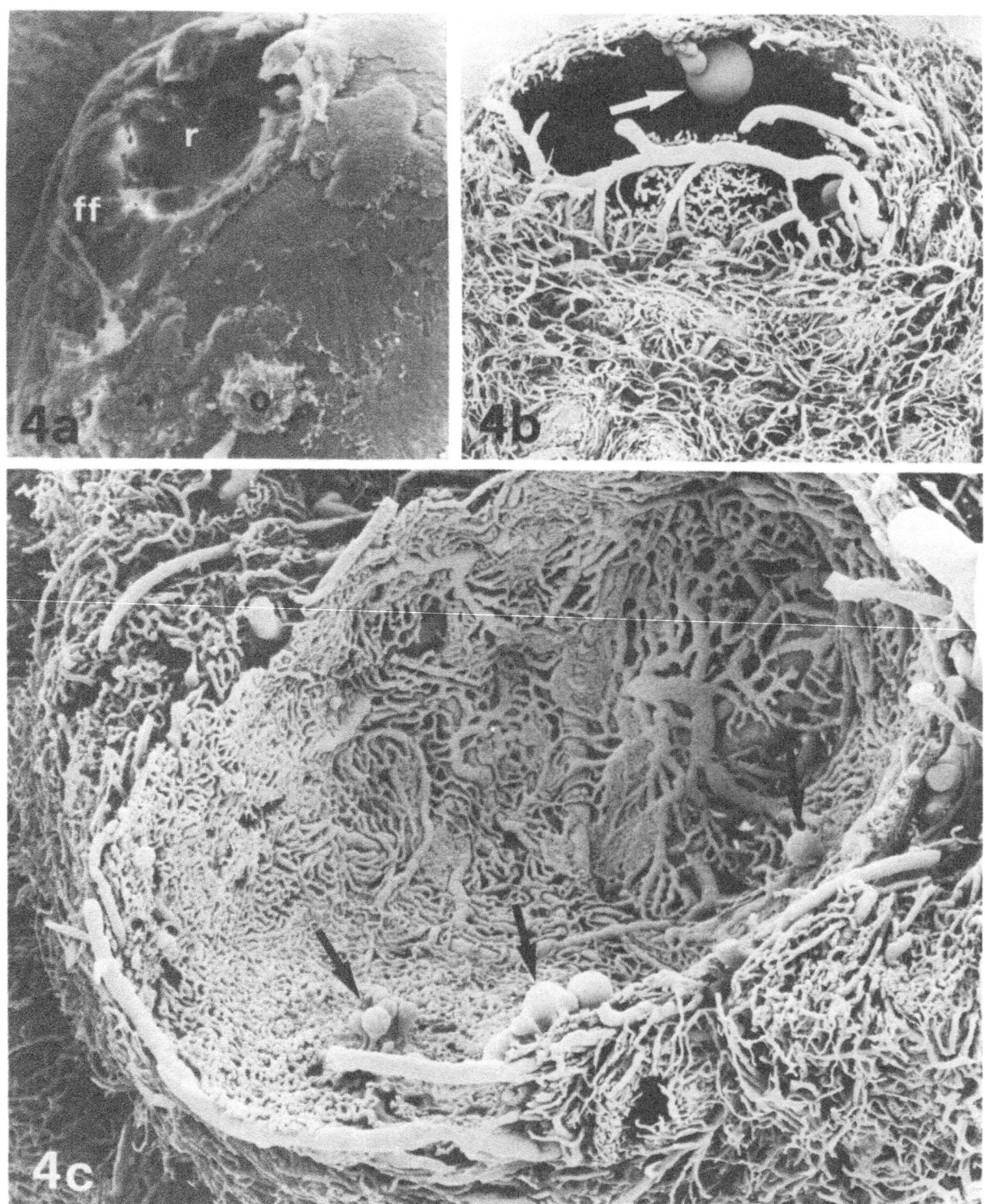

Figure 19-4. Ovulation. a: Rabbit. Ovulatory follicle 9 hours after coitus. Note the presence of a ruptured area (r) on the surface of the ovary corresponding to the site of oocyte extrusion at the time of ovulation. Follicle cells immersed in the follicular fluid (ff) can be seen in the areas adjacent to the follicular rupture. O = ovulated oocyte surrounded by corona cells and spermatozoa. 95×. (From Motta PM, Van Blerkom J, [8], with permission.) b: Rabbit. Ovulatory follicle 12 hours after hCG stimulation. Note the large apical avascular (ischemic) area, which is likely to correspond to a just ovulated follicle. In the follicular basket cavity, note resin leakages (arrow) due to inter-endothelial cell gaps. 45×. c: Rabbit. Inner aspect of the capillary plexus of an ovulatory follicle 12 hours after hCG stimulation. Resin leakages (arrows) are also seen. 65×.

by their size, when compared to light or electron microscopy observations (Fig. 19-3a). The follicular baskets of smaller antral follicles (about 200–300 µm in diameter) have a layer of capillaries, which tend to assume a sinusoidal arrangement, drained by thicker outer vessels (Figs. 19-3a and 19-3b). As the follicles mature, both inner and outer vascular layers show angiogenetic phenomena and enhancement of vessel tortuosity. Follicular baskets may reach 1 mm in diameter. The outer surface of the largest follicular baskets has numerous small and tortuous vessels of capillary, venular, and arteriolar nature. These vessels form a multilayered large-mesh vascular network, surrounding a inner sinusoidal capillary layer. In fractured samples, the inner capillary layer is very rich [32] (Fig. 19-3c). These capillaries, running sinusoidally, present a round-mesh network and show numerous blind ends. These aspects are enhanced in larger follicles (Figs. 14-3b and 19-3d). These structures resemble vascular sprouts related to the enhanced angiogenesis present during the follicular growth [34].

2.3. Ovulatory Follicle

Mature preovulatory and ovulatory follicles may be easily recognized by means of SEM, because they are very large. They tend to protrude from the ovarian surface and exert pressure on the flattened surface epithelium. Furthermore, as a result of events preceding follicular rupture, in mature preovulatory follicles, a smooth "apical" rounded area, in which cells of surface epithelium tend to become necrotic and slough off [1], is seen (stigma). Otherwise, in ovulated follicles, the apical area is always interrupted by a large and irregularly shaped hole, corresponding to follicle rupture for oocyte extrusion [23] (Fig. 19-4a). Ovulatory follicles are characteristically formed by thick outer theca layers, which are richly vascularized, and by an inner large cavity, which is limited by several layers of granulosa cells. Granulosa cells are connected with the eccentrically placed oocyte-cumulus cell complex by means of a cellular pedunculus, which is a remnant of the cumulus oophorus. Generally, this connection is lost just before ovulation begins [1,8–10].

The oocyte at the time of ovulation characteristically shows dissolution of the nucleolemma (breakdown of the germinal vesicle) with meiotic fusion and polarization of chromosomes. In ovulated oocytes, this is followed by the extrusion of the first polar body [1]. This event has been also described in naked oocytes (without a zona pellucida) by means of SEM [35,36]. The oocyte cytoplasm shows abundant cortical granules having a typical subplasmalemmal localization. Furthermore, numerous rounded mitochondria, with typical curved cristae, are seen in the proximity of smooth endoplasmic reticulum membranes or cytoplasmic large vesicles [1]. The oocyte surface is richly covered by short microvilli. These are uniformly distributed, except in the smooth area, where the extrusion of the first polar body occurs [35,36]. At this stage, the oocyte is still surrounded by the zona pellucida. As seen by SEM, the outer surface of the zona pellucida of periovulatory oocytes presents numerous fenestrations (spongelike structure). This structure has been recently studied in in-vitro cultured human oocytes. These results revealed that the zona pellucida surface possesses a more compact structure in immature and atretic oocytes than in mature oocytes [37].

Corona radiata cells surround the oocyte and send numerous thin cytoplasmic extensions, which pass through the zona pellucida and reach the oocyte surface. At the time of ovulation, corona radiata cell prolongations retract, with consequent metabolic uncoupling between oocyte and cumulus cells. As seen by SEM, cumulus cells show polygonal, irregularly flattened, or star shapes. Many of these cells are covered with a granular filamentous material corresponding to precipitated follicular fluid or mucous material, which is secreted by these cells [1]. Furthermore, numerous cumulus cells are arranged in longitudinal groups that are placed radially around the oocyte [6,8]. These cells, remaining attached to the oocyte, remain with the oocyte after ovulation occurs. Their three-dimensional ultrastructural characteristics have been studied in the human and in other mammals. In humans (in oocytes taken from hormonally stimulated cycles for an in-vitro fertilization project), cumulus cells present a cytoplasm that gradually transforms into that of steroidogenic cells. Their surface is characterized by numerous blebs of varying size, a few short microvilli, and occasional ruffles [38].

In the granulosa layer of periovulatory follicles, the outer cells in contact with the follicle basal lamina undergo some surface changes. They have a polyhedric shape, as in earlier stages, but their size is moderately increased. Further, voluminous pseudopodia and cytoplasmic foldings may be seen on their surface, and the number of blebs and microvilli appear to increase in number when compared with earlier stages [1]. Theca cells increase in number in the periovulatory periods; furthermore, they increase in steroidogenic activity, as evidenced by clear cellular ultrastructural evidence for this function. Therefore, their surface characteristics are a remarkable increase of microvilli and blebs, accompanied by enlargement in size. Follicular vascularization shows significant changes. Vascular lacunae are clearly increased and enlarged, and are often filled with blood cells. Blood cells may be also seen in the interstitial perivascular spaces (periovulatory follicular hemorrage) [26].

By SEM of vascular corrosion casts, several features related to the follicular hyperemia, edema, and ischemia occurring at the moment of ovulation characterize vascular baskets of periovulatory follicles. These are very large and have a large avascular area located on the apical pole (Fig. 19-4b). The basal pole presents an enriched vascularization. The wall of the baskets is plurilayered. Inner capillaries are clearly enlarged and show (in the rabbit, but not in the rat) many resin blebs (leakages), which have been related to the passage of resin among newly formed intercellular spaces [17,18,21,32] (Figs. 19-4b and 19-4c). In fact, in TEM studies with tracers, sinusoidal endothelium showed increased permeabilization, with formation of numerous large intracellular fenestrations and of intercellular gaps, the latter allowing the passage of larger moleculae [39].

3. The Corpus Luteum

The follicles undergo repair within a few moments after ovulation [7]. These events mainly consist in a coagulation of the follicular fluid with blood cells, debris, granulosa, and connective cells (fibrin clot), which seal the follicle rupture. Within a few days, the apex of the follicle is completely covered with connective tissue.

Furthermore, the superficial epithelium tends to proliferate above the repairing follicle to rehestablish the ovarian surface continuity. This is followed by connective tissue organization within the follicular cavity, accompanied by connective cell migration, vascular growth, and rapid functional activation of the granulosa and theca cells. Two types of corpora lutea may be considered: (1) growing corpus luteum, including menstrual and pregnancy corpus luteum, which are characterized by their gradual morphofunctional transformation into a gland, which mainly secretes progesterone and (2) regressing corpus luteum, in which the gradual loss of the functional endocrine activity is a prelude to the fibrotic (formation of corpus fibrosus) or cystic degeneration of the corpus luteum. In addition, the corpus fibrosus, when it undergoes collagen hyalinization, transforms into the corpus albicans or into the corpus nigricans (when residual hematic pigments are seen) [40].

3.1. Growing Corpus Luteum

Usually, in mammals the growth of corpus luteum takes a long time — a few days. However, a great variability exists in different species. The main morphofunctional transformation within the repairing follicle consists of the enlargement of the granulosa cells, followed by their migration in the follicular cavity and accompanied by prominent changes in theca cells. By TEM, both granulosa and theca cells present ultrastructural features of luteinization. These mainly consist of an increase in lipid droplets, formation of mitochondria with tubular cristae, and enlargement of smooth endoplasmic reticulum [41–43]. The latter structure, by SEM, shows a labyrinthine arrangement, seeming to form a three-dimensional network in continuity with other cytoplasmic membranes [7]. During corpus luteum formation, the aspects of luteinization are mainly evident in the granulosa cells. In fact, theca cells, which are fewer in number and smaller in size, show only little ultrastructural changes [9,41–43].

As seen by SEM, corpora lutea are recognized in the cortical area as spheroidal structures delimited by a thick connective tissue layer [7,9,44] (Fig. 19-5a). In fractured samples, the follicular

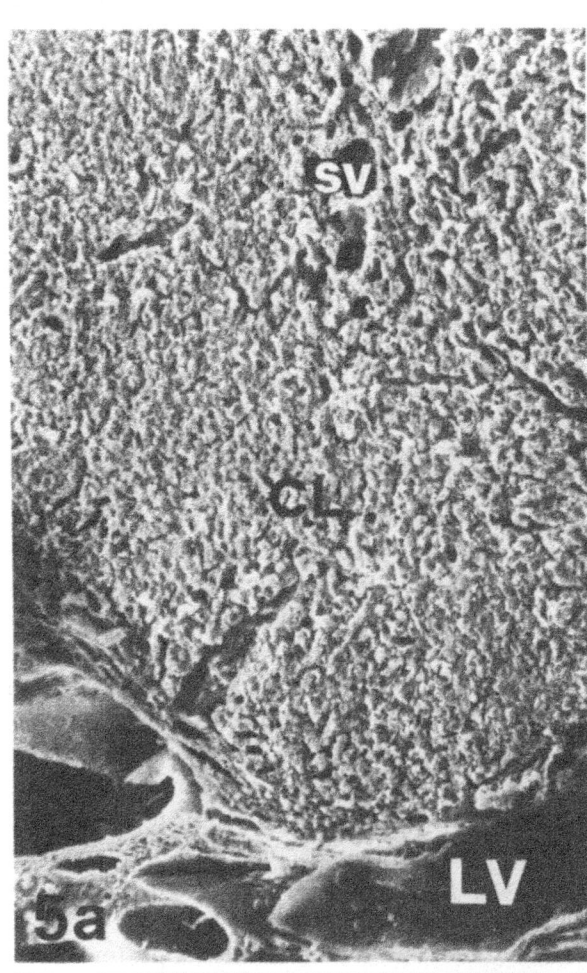

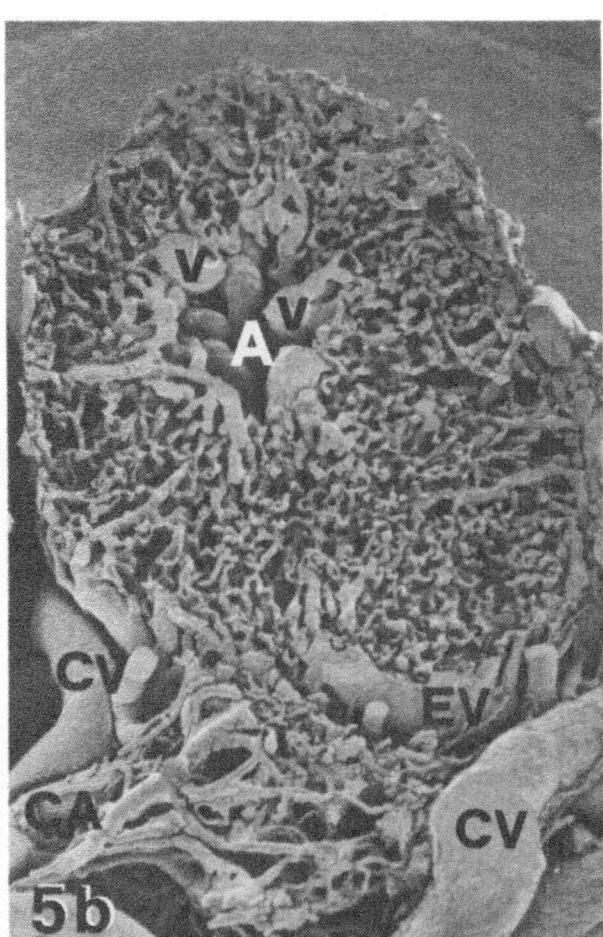

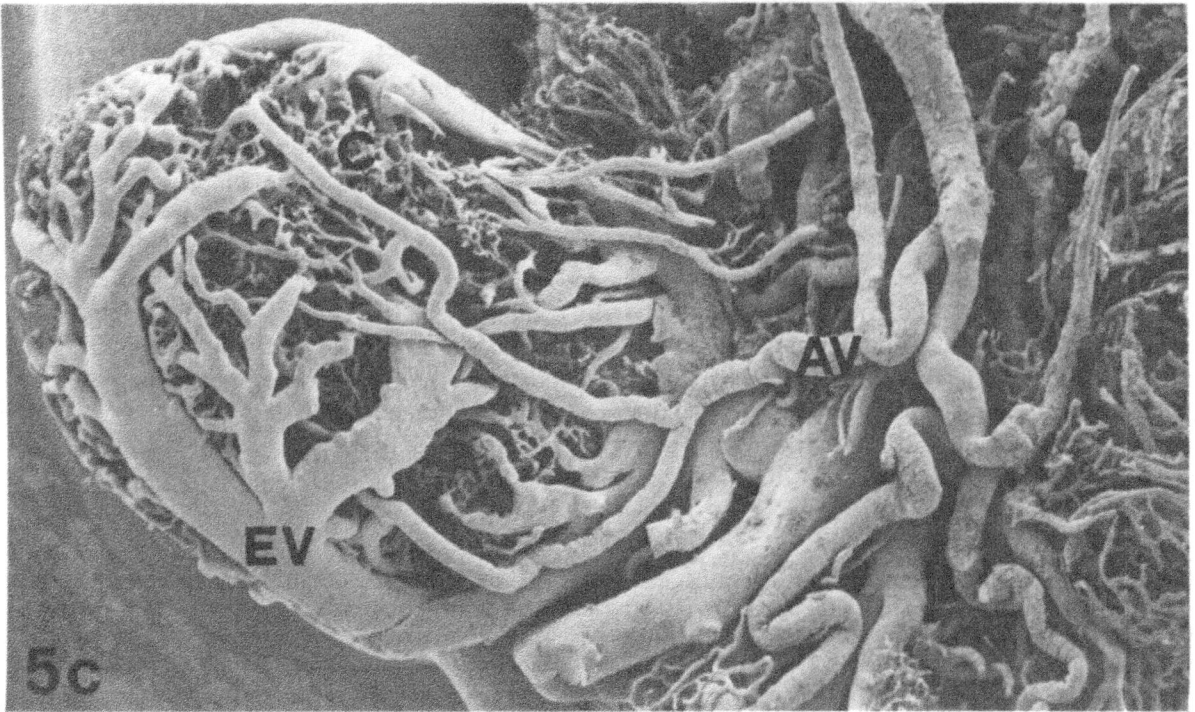

cavity is seen to be gradually filled by connective-vascular elements, intermingled with cords of radiating luteal-transforming cells, delimiting vascular lacunae [7,9] (Fig. 19-5a). Granulosa-luteal cells enlarge, assume a polyhedral shape and are seen to be intermingled with connective fibers and blood cells. The latter are typically found in clusters surrounded by a fibrin clot. The luteal cell surface is rich in blebs and microvillous proliferation. In addition, cytoplasmic projections and deep invaginations, probably related to the presence of previous reciprocal contacts between adjacent cells, are frequently seen [7]. In the periphery of the corpus luteum, numerous flat, elongated cells, resembling fibroblasts, proliferate and completely surround the corpus luteum [9,42,43].

During corpus luteum formation, an exuberant neovascularization is seen. This event is mainly related to capillary angiogenesis. By SEM, numerous capillaries, originating from peripheral twigs, follow the proliferating endocrine cell cords, which radiate towards the center of the repairing follicle [7]. These vascular lacunae resemble dilated sinusoids, are often filled with blood cells, and may present discontinuity and fenestrations [45–47]. In addition, large perivascular lacunae, forming a labyrinthine structure of interconnected spaces, are seen. In these spaces, probably acting as a reservoir for interstitial fluids, lymph perivascular cells, may be also present.

By SEM of vascular corrosion casts, the postovulatory follicles show important microvascular changes. In rats and rabbits [17,21], large openings (site of oocyte dehiscence) characterize the vascular baskets. Numerous collapsed baskets are often observed surrounding the opened larger basket (ovulated follicle). In the inner capillary layer, a large amount of leakage (related to endothelial permeabilization, blood extravasation, and following the formation of the central clot) is always observed. The capillaries appear enlarged and show enhanced sinusoid formation. Furthermore, numerous thin protrusions arising from these capillaries invade the granulosa cell area and radiate toward the follicular cavity centrum. These sprouts are probably related to angiogenetic phenomena, as numerous new capillary anastomoses are formed. As a result of these changes, the capillary layer become thicker, richly anastomosed, and plurilayered. The thickness of the capillary plexus of the corpus luteum increases parallel to gland development [32].

In growing corpus luteum, the apical opening is rapidly invaded by new capillaries. The tortuosity and permeability (resin leakages) of capillaries is also increased. The follicular cavity is invaded by many new vessels, except for a small avascular central area, which disappears, in fully developed corpora lutea (Fig. 19-5b). In the latter, however, a further change is observed in the capillaries. In the fully developed corpus luteum of the rat, although the capillaries remain tortuous and richly branched, they become thinner and similar in appearance to those of neighboring stroma or primary follicles [21,32].

Arteriolar and venular vessels of postovulatory follicles do not show great changes in comparison to their earlier stages. In developing corpus luteum, the peripheral arterioles originate by cortical branches, run along the corpus luteum surface, and supply twigs for the capillary plexus. In later stages, the twigs duplicate and give off short branches to the outer layer of the corpus luteum and longer vessels to the deeper stratum. Developing corpus luteum presents thick venules located within the capillary plexus, which in fully developed corpus luteum may assume a rosette-like configuration. The venules are drained by radiating branches that empty into the peripheral vessels [29] (Fig. 19-5b).

←

Figure 19-5. Corpus luteum. a: Rat. SEM view of the fractured surface of the corpus luteum (CL) during midpregnancy. The periphery of the luteal mass contains large blood vessels (LV). Smaller vessels (SV) are seen in the interior portions. 140×. (From Van Blerkom J, Motta PM [7], with permission.) b: Rat. Capillary plexus of a fully developed corpus luteum. Note that the original segments (v) of the deep twigs of the efferent vessels surround, a centrally located avascular area (A), like a rosette. CA, CV = cortical artery and veins; EV = efferent vessel of the corpus luteum. 85×. c: Rat. Vascular corrosion cast of a degenerative corpus luteum. EV = efferent vessels; AV = afferent vessels; C = capillary plexus. 75×. Figs. 5b and 5c are from Murakami et al. [21], with permission.)

256

3.2. Regressing Corpus Luteum

Decreasing steroid function results in rapid degeneration of corpus luteum. Such an involutive process has similar morphological aspects, in the corpus luteum in both menstruation and pregnancy [40]. The luteal mass is soon invaded by hematic cells. This is accompanied by connective tissue proliferation, mainly characterized by fibrosis. The luteal cells show nuclear alterations (picnosis), cytoplasmic vacuolization (related to lipid extraction), numerous autophagic vacuoles, residual bodies, and myelinic figures [42,48]. Degenerating luteal cells decrease in size and undergo cellular lysis. At the same time, fibroblasts producing collagen [40] and activated macrophages, phagocytizing degenerating luteal cells [49], proliferate within the luteal mass.

As seen by SEM, the degenerating corpus luteum is characterized by a honeycomb appearance [7], due to large intercellular spaces containing vacuolized luteal cells. Connective fibers are abundant and are organized to form a thick peripheral capsula, which has numerous projections running within the luteal mass and surrounding the degenerating luteal cells.

In vascular corrosion casts, the microvascularization of degenerating corpus luteum shows great changes (Fig. 19-5c). The luteal mass loses its vascular independence and has numerous connections with the vascularization of the surrounding stroma. The degenerating corpus luteum capillaries appear to be generally reduced in thickness. Their size is no longer homogeneous. They are usually thready, but in some areas keep their sinusoidal arrangement. These capillaries may show focal interruptions. In addition, large avascular areas are seen. In highly degenerated corpus luteum, the capillary plexus is rather scarce and is similar to that of the stromal-interstitial tissue. Residual corpus albicans also has a poor capillary supply [21,32]. Larger vessels are arranged in a fashion similar to that of fully developed corpus luteum. However, venules may often appear flattened, mainly in the later stages. Furthermore, some direct connections between arteries and veins have been demonstrated in the degenerating corpus luteum of rats [21].

4. Conclusions

The three-dimensional aspects of LFC have been briefly reviewed in this chapter. SEM techniques allowed a better understanding af all dynamic changes of this unique structure. In fact, SEM studies performed in the last two decades [6–10] have clearly demonstrated that LFC should be considered as different morphofunctional aspects (growing and ovulatory follicle, growing and regressing corpus luteum) of the same physiological event (the ovarian cycle). In these studies, a realistic three-dimensional picture of follicular and luteal cell microtopography has been produced. More recently, new techniques of tissue exposure for SEM viewing have been introduced. These techniques, such as the O-D-O method [50] for the three-dimensional study of intracellular structures, have already provided further details on the three-dimensional microstructure of the ovary [2,28]. Their combined use, in the near future, will surely provide useful information for better visualization and understanding of the oocyte and luteofollicular cell subcellular changes during LFC evolution [2,28].

In this chapter it has been attempted to correlate en-face SEM micrographs with SEM views of vascular corrosion casts, which could provide a better definition of our morphofunctional understanding of the ovarian cycle. In fact, as shown by the recent literature, a close morphofunctional connection between steroid-producing cell changes and microvascularization does exist. Vascular corrosion cast SEM studies have demonstrated that LFC microvascularization presents significant modifications during the ovarian cycle [21,32]. Such changes mainly consist of progressive growth of the follicular theca vascularization, which first becomes independent of the stromal vascularization and then shows peculiar changes in the capillary plexus. The capillary plexus gradually adapts its morphology to ovulation, which follows: formation of sinusoids, neo-angiogenesis, and finally, vascular dilatation and permeabilization. These aspects appear to be enhanced during corpus luteum formation. This reflects the transformation of a capillary net supplying a simple epithelium (primary follicle) into a typical sinusoidal network supplying an endocrine gland.

However, if from a pure morphological point of view this transformation is rather expected, nevertheless, it is not yet clear what the role of these changes is during LFC revolution. Do these changes reflect only an epiphenomenon of LFC cellular changes? Or do microvascular changes play a primary role in inducing or controlling certain events, such as ovulation, corpus luteum formation, or corpus luteum regression? The following considerations may be of some significance in this matter.

Many data suggest that microvessels show structures, such as microsphincters, that are able to regulate blood flow within the follicle [17,21, 32]. Therefore, blood transport of hormones may be regulated at the follicular or corpus luteum level.

The dramatic vascular changes seen in pre-ovulatory follicles have been understood as typical phenomena of an inflammatory process (ovulation as an inflammatory event) [51]. In fact, capillary dilatation and endothelial permeabilization are rapid vascular changes that are a prelude to blood extravasation and interstitial edema, which are epiphenomena of acute inflammatory processes. However, it is rather interesting to see how vascular changes, such as angiogenesis and the formation of sinusoids, start to occur at an early time (antral follicles) and are established rather slowly prior to when blood levels of inflammation-inducing substances reach their highest peaks. In addition, the follicles that are not ovulated (atretic follicles) have a vascularization that is quite similar to that of the interstial-stromal tissue, without the formation of sinusoids [32]. This suggests that ovulation should occur only in those follicles that have a particular microvascular pattern.

Finally, there is evidence that a vascular mechanism, i.e., redistribution of blood flow within the ovary through the opening of artero-venous shunts, may be involved, possibly causing luteolysis [32,52]. These data suggest that microvascular changes play a significant role in LFC morphofunctional evolution. Certainly, further studies using combined SEM techniques will be essential for obtaining a complete three-dimensional picture of LFC cell microtopography and microvascularization.

Acknowledgments

Funds for this work were provided by M.U.R.S.T.-C.N.R, Italy and the Ministry of Education, Japan.

References

1. Familiari G, Makabe S, Motta PM. The ovary and ovulation: A three-dimensional ultrastructural study. In: Van Blerkom J, Motta PM (eds.), *Electron Microscopy in Biology and Medicine, Vol. 5: Ultrastructure of Human Gametogenesis and Early Embryogenesis* Kluwer Academic Publishers, Boston 85–124, 1989.
2. Makabe S, Naguro T, Nottola SA, Pereda J, Motta PM. Migration of germ cells, development of the ovary, and folliculogenesis. In: Familiari G, Makabe S, Motta PM (eds.), *Electron Microscopy in Biology and Medicine, Vol. 9: Ultrastructure of the Ovary* Kluwer Academic Publishers, Boston, pp 1–27, 1991.
3. Dvorak M, Tesarik J. Ultrastructure of human ovarian follicles. In: Motta, PM Hafez ESE (eds.), *Biology of the Ovary*, Martinus Nijhoff Publishers, The Hague, pp 121–137, 1980.
4. Gulyas BJ. Fine structure of the luteal tissue. In: Motta PM (ed.), *Electron Microscopy in Biology and Medicine, Vol. 1: Ultrastructure of Endocrine Cells and Tissues* Martinus Nijhoff Publishers, Boston, pp 238–254, 1984.
5. Greenwald GS, Terranova PF. Follicular selection and its control. In: Knobil E, Neill J (eds.), *The Physiology of Reproduction*, Raven Press, New York, pp 387–445, 1988.
6. Motta PM, Van Blerkom J. A scanning electron microscopy study of the luteo-follicular complex. I. Follicle and oocyte. *J Submicrisc Cytol* 6:297–310, 1974.
7. Van Blerkom J, Motta PM. A scanning electron microscopy study of the luteo-follicular complex. Formation of the corpus luteum and repair of the ovulated follicle. *Cell Tissue Res* 189:131–153, 1978.
8. Motta PM, Van Blerkom J. A scanning electron microscopy study of the luteo-follicular complex. II. Events leading to ovulation. *Am J Anat* 143:241–264, 1975.
9. Van Blerkom J, Motta PM. *The Cellular Basis of Mammalian Reproduction*. Urban & Schwarzenberg, Baltimore, 1979.
10. Motta PM, Van Blerkom J. Scanning electron microscopy of the mammalian ovary. In: Motta PM, Hafez, ESE (eds.), *Biology of the Ovary*, Martinus Nijhoff Publishers, The Hague, pp 162–175, 1980.
11. Clark JC. The origin, development and degeneration of the blood-vessels of the human ovary. *Bull Johns Hopkins Hosp* 9:593–676, 1900.
12. Delson B, Lubin S, Reynolds SRM. Vascular patterns in human ovaries. *Am J Obstet Gynec* 57:842–853, 1949.
13. Gillet JY. Etudes anatomo-physiologiques. La microvascularisation de l'ovaire. *Gynecol Obstet* 70:251–272, 1971.

258

14. Basset DL. The changes in the vascular pattern of the ovary of the albino rat during the estrous cycle *Am J Anat* 73: 251–291, 1943.

15. Burr JH, Davies JI. The vascular system of the rabbit ovary and its relationship to ovulation. *Anat Rec* 111: 273–297, 1951.

16. Kardon RH, Kessel RG. SEM studies on vascular casts of the rat ovary. *Scann Electron Microsc* III:743–750, 1979.

17. Kanzaki H, Okamura H, Okuda Y, Takenaka A, Morimoto K, Nishimura T. Scanning electron microscopic study of rabbit ovarian follicle microvasculature using resin injection-corrosion casts. *J Anat* 134:697–704, 1982.

18. Kitai H, Yoshimura Y, Write KH, Santulli R, Wallach EE. Microvasculature of preovulatory follicles: Comparison of in situ and in vitro perfused rabbit ovaries following stimulation of ovulation. *Am J Obstet Gynecol* 152:889–895, 1985.

19. Takada S, Shimada T, Nakamura M, Mori H, Kigawa T. Vascular pattern of the mammalian ovary with special reference to the three-dimensional architecture of the spiral artery. *Arch histol Jpn* 50:407–418, 1987.

20. Hees H, Koenig HE, Hees I. Recherches sur la structure vasculaire du systeme de vaisseaux sanguins des structures elaborees au cours du cycle ovarien chez la jument. Une recherche au microscope optique et au microscope a balayage electronique. *Contracept Fertil sex* 16:521–526, 1988.

21. Murakami T, Ikebuchi Y, Ohtsuka A, Kikuta A, Taguchi T, Ohtani O. The blood vascular wreath of rat ovarian follicle, with special reference to its changes in ovulation and luteinization: A scanning electron microscopic study of corrosion casts. *Arch Histol Cytol* 51:299–313, 1988.

22. Konig HE, Amselgruber W, Russe I. La microcirculation dans les follicles et les corps jaunes d'ovaires de bovins. Une etude anatomique par corrosion. *Contracept Fertil Sex* 17:179–186, 1989.

23. Murdoch WJ, Cavender JL. Effect of indomethacin on the vascular architecture of preovulatory ovine follicles: Possible implication in the luteinized unruptured follicle syndrome. *Fertility Sterility* 51:153–155, 1989.

24. Hodgen GD, Kenigsburg D, Collins RL, Schenken RS. Selection of the dominant ovarian follicle and hormonal enhancement of the natural cycle. *Ann NY Acad Sci* 422:23–37, 1985.

25. Byskov AG. Follicular atresia. In: Jones RE (ed.), *The Vertebrate Ovary*, Plenum Press, New York, pp 533–562, 1978.

26. Familiari G, Vizza E, Miani A, Motta PM. Ultrastructural and functional development of the theca interna. In: Familiari G, Makabe S, Motta PM (eds.), *Electron Microscopy in Biology and Medicine, Vol. 9: Ultrastructure of the Ovary*, Kluwer Academic Publishers, Boston, pp 113–128, 1991.

27. Baca N, Zamboni L. The fine structure of human follicular oocytes. *J Ultrastruct Res* 19:354–381, 1967.

28. Makabe S, Naguro T, Motta PM. Three-dimensional fine structure of ovarian follicles revealed by the ODO method (abstract). In: Abstract book of the IX International Symposium on Morphological Sciences, Nancy, 9–13 September, p 96, 1990.

29. Kang Y. Development of the zona pellucida in the rat oocyte. *Am J Anat* 139:535–566, 1974.

30. Okamura T. Morphological observations on ovulation. *Prog Clin Biol Res* 296:79–90, 1989.

31. Motta PM, Takeva S, Nesci E. Etude ultrastructurale et histochimique des rapports entre les cellules folliculaires et l'oocyte pendant le developpement du follicule ovarien chez les mammiferes. *Acta Anat* 80:537–562, 1971.

32. Kikuta A, Macchiarelli G, Murakami T. Microvasculature of the ovary. In: Familiari G, Makabe S, Motta PM (eds.), *Electron Microscopy in Biology and Medicine Vol. 9: Ultrastructure of the Ovary* Kluwer Academic Publishers, Boston, pp 239–254, 1991,in press.

33. Motta PM, Makabe S. Morphodynamic changes of the mammalian ovary in normal and some pathological conditions. A scanning electron microscopic study. *Biomed Res* 2 (Suppl.):325–339, 1981.

34. Spanel-Borowsky K, Amselgruber W, Sinowatz F. Capillary sprouts in ovaries of immature superstimulated golden hamster: A SEM study of microcorrosion casts. *Anat Embryol* 176:387–391, 1987.

35. Longo FJ. Fine structure of the mammalian egg cortex. *Am J Anat* 174:303–315, 1985.

36. Ebensperger C, Barros C. Changes at the hamster oocytes surface from the germinal vesicle stage to ovulation. *Gamete Res* 9:387–397, 1984.

37. Familiari G, Nottola SA, Micara G, Aragona C, Motta PM. Is the sperm-binding capability of the zona pellucida linked to its surface structure? A scanning electron microscopy study of human in vitro fertilization. *J In Vitro Fert Embryo Transfer* 5:134–143, 1988.

38. Nottola SA, Familiari G, Micara G, Aragona C, Motta PM. The role of the cumulus-corona cells surrounding in vitro human oocytes and polypronuclear ova: An ultrastructural study. *Prog Clin Biol Res* 296:345–354, 1989.

39. Okuda J, Okamura H, Kanzaki H, Takenaka A. Capillary permeability of rabbit ovarian follicles prior to ovulation. *J Anat* 137:263–269, 1983.

40. Balboni GC. Structural changes: Ovulation and luteal phase. In: Serra GB (ed.), *The Ovary*, Raven Press, New York, pp 123–141, 1983.

41. Motta PM. Electron microscopy study of the human lutein cell with special reference to its secretory activity. *Z Zellforsch Mikrosk Anat* 98:233–245, 1969.

42. Adams EC, Hertig AT. Studies on the human corpus luteum I. Observations on the ultrastructure of development and regression of the luteal cells during the menstrual cycle. *J Cell Biol* 41:696–715, 1969.

43. Meyer GT, Bruce NW. Quantitative cell changes and vascularization in the early corpus luteum of the pregnant rat. *Anat Rec* 197:369–374, 1980.

44. Pendergrass PB, Principato R, Reber M. Scanning electron microscopy of corpus luteum in the golden hamster. *J Submicrosc Cytol* 13:527–536, 1981.

45. Morris B, Sass MB. The formation of lymph in the ovary. *Proc R Soc London* 164:577–591, 1966.

46. Bruce NW, Meyer GT, Dharmarajan AM. Rate of blood flow and growth of the corpora lutea of pregnancy and of previous cycles throughout pregnancy in the rat. *J Reprod Fert* 71:445–452, 1984.

259

47. Dharmarajan AM, Bruce NW, Meyer GT. Quantitative ultrastructural characteristics relating to transport between luteal cell cytoplasm and blood in the corpus luteum of the pregnant rat. *Am J Anat*, 172:87–99, 1985.

48. Lennep EW, van, Madden LP. Electron microscopic observation on the involution of the corpus luteum of menstruation. *Z Zellforsch* 66:365–380, 1965.

49. Paavola LG, Boyd CO. Surface morphology of macrophages in the regressing corpus luteum as revealed by scanning electron microscopy. *Anat Rec* 195:659–682, 1979.

50. Tanaka K. Eukaryotes: Scanning electron microscopy of intracellular structures. *Int Rev Cytol* 17 (Suppl.):89–120, 1987.

51. Espey LL. Ovulation as an inflammatory reaction. A hypothesis. *Biol Reprod* 22:73–106, 1980.

52. Goding JR, Baird DT, Cumming IA, McCracken JA. Functional assessment of autotransplanted Endocrine glands. *Acta Endocrinol* 158 (Suppl.):169–191, 1972.

Author's address:
Prof. Guido Macchiarelli
Department of Anatomy
Faculty of Medicine
University "La Sapienza"
Via A. Borelli 50
00161
Rome
Italy

Structural and Functional Aspects of Placental Microvasculature Studied from Corrosion Casts

RUDOLF LEISER & BÄRBEL KOOB

1. Introduction

There are two blood circulations in the placenta, one originating from the maternal/uterine and one from the fetal vessel system. They are strictly separated from each other by the placental barrier or interhemal membrane and carry substances to and from the interhemal membrane. The capacity of transplacental substance exchange across this interhemal membrane is vital for growth of the fetus and, depends on the arrangement of the capillary bed between the two vasculatures. Therefore, the materno-fetal interrelationship of the vascular network is very important for the general effective function of a given type of placenta [review: 1,2].

Three different patterns of placental vessel arrangement have been morphologically detected thus far, and they have been classified by physiologists in order of increasing effectiveness: multivillous flow, cross-current flow and counter-current flow [1,2]. Mixed types of vessel arrangements can be observed as well. A concurrent type was also proposed by physiologists in a mathematical model; however, morphologically, it has not been found and, therefore, it may not exist as a functioning principle in nature [3,4]. In this comparative placental study, emphasis is laid on the characterization of the three-dimensional morphology of vessel systems as a prerequisite for their functional interpretation. The placentas of the human, guinea pig, cat, goat, and pig are described here in order to give a series of examples representing the increase in interhemal

membrane drawn by Grosser's [5] and Enders' [6] scheme, namely, from hemomonochorial (human and guinea pig) and endotheliochorial (cat) to epitheliochorial (goat and pig). Special attention is referred to the capillary system, as it represents the main site for materno-fetal exchange. The precapillary and postcapillary vessel systems have been thoroughly described by Ramsey and Donner [7] and by Kaufmann et al. [8] in humans, by Kaufmann [9] in the guinea pig, by Leiser and Kohler [10, 11] in the cat, by Leiser [12] in the goat, and by Leiser and Dantzer [13] in the pig.

With regard to the method of microcorrosion casting and a three-dimensional demonstration of placental vasculature by scanning electron microscopy the reader is referred to other literature [4,10,14–16].

2. Human

The discoidal, near-term human placenta is divided into 10–38 *maternal cotyledons* (lobes) [17], each overlapping several *fetal cotyledons* or *placentones* [18,19], 60–70 of which, in turn, exist [17,20]. The maternal cotyledon, viewed from the basal surface of the placenta, is a slightly elevated area; at its periphery it consists of variably shaped maternal septa projecting from the basal plate without a connection to the chorionic plate. The fetal cotyledon or placentone includes a fetal villous tree (Fig. 20-1). Its stem rests on the chorionic plate and its final branchings partly fuse

Motta, P.M., Murakami, T., and Fujita, H. (eds.), Scanning Electron Microscopy of Vascular Casts: Methods and Applications.

262

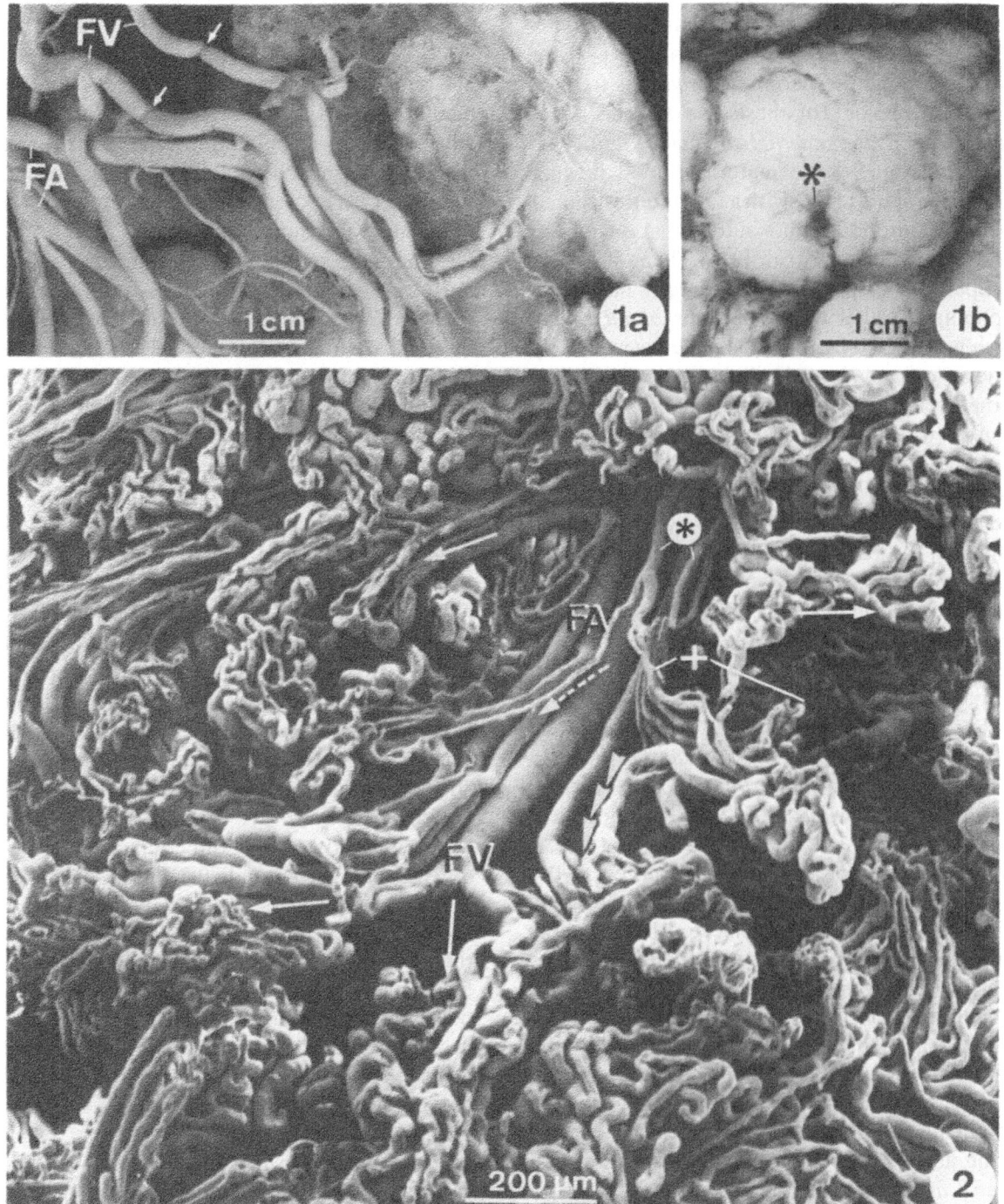

Figure 20-1. Fetal vascular cast of human hemomonochorial placenta with fetal cotyledons of different forms and sizes, at term. a: Side of chorionic plate with chorionic (extracotyledonar) arteries (FA) and veins (FV), the latter marked by valvular constrictions (arrows). b: Side of basal plate of placenta, demonstrating a cotyledon with a central cavity (asterisk).

Figure 20-2. Magnification of human central cavity corresponding to Fig. 20-1b, situated beneath the maternal arterial inlet. Arterial (FA) and venous (FV) rami-chorii substantiating stem vessels of about the 10th order of the fetal villous tree ramify into arterioles and venules of mature intermediate villus. They continue into numerous capillary convolutes of terminal villi (arrows). A paravascular capillary network (asterisk) extends directly into left-sided terminal villi (dashed arrow). On the right, it feeds into the weblike vessels of an immature intermediate villus (cross), which is followed by a terminal villus (arrowheads).

with the basal plate [8]. On the inner surface of the basal plate, about 100 spiral arteries [17] have an inlet located close to the centers of villous trees [18], whereas 50–200 venous outlets [17] are arranged around the periphery of villous trees.

The morphology of these two systems functionally effects the *maternal circulatory unit*. This unit is initiated by the maternal blood flowing through the spiral artery with a rather unrestrained blood pressure, since no autoregulation of the spiral artery lumen exists [21]. Then, it enters the open blood space as a jet directly facing the trophoblast, called the *intervillous blood space*, which is bordered by the basal plate, as well as by the chorionic plate and villous tree. The intervillous blood space around the inlet of the spiral artery is widened to the central cavity of villous tree (Figs. 20-1b and 20-2), from which the blood streams radially to the periphery of the villous tree and slowly flows back into several venous outlets on the basal plate [7]. From the periphery of the villous tree, however, the blood can also intermix with maternal blood of neighboring placentones, or even maternal cotyledons, since the villous trees and maternal septa do not substantiate in strict anatomical borders (see Fig. 20-18). Therefore, the *human placenta has an open system of maternal blood space* [19].

The *fetal blood flows in a closed vascular system* [for review 8,17,22–24]. Two umbilical arteries (φ 3 mm) and one umbilical vein (φ 5.5–6 mm) run in a spiral course and represent the vessels of the umbilical cord. Near the chorionic plate, these arteries are connected by the short HYRTL-anastomose or may even fuse [17] before they dichotomously branch up to four times in chorionic arteries of the chorionic plate (Fig. 20-1a). The collateral veins run almost in the same manner, with regard to their branching pattern as well as their peripherally oriented radial course. A total of 60–70 vertical branches of the above chorionic arteries and veins, called *truncal arteries* (φ 1.5 mm) and *truncal veins* (φ 2 mm), enter the stem villi. The stem villi, as the base and central part of villous trees, determine the number of fetal cotyledons or placentones (see above). The truncal vessels, then as stem vessels, start ramifying dichotomously in parallel to the branching patterns of the villous tree, i.e., rami chorii of the first to fourth order and con-

secutive ramuli chorii of the first to about the tenth order [compare 25]. In the last generations of stem villi, the arteries and veins transform into arterioles and venules (less than two layers of muscle cells in the media) and enter the intermediate villi (Fig. 20-2) [20]. The intermediate villi consist of an immature type, which is relatively thick and has growth capacity and is infrequent (5%) at term, and a mature type, which is slender and numerous (95%) at term. The terminal villi, as the ending segments of the villous tree, are thick, bulbous, and rather few, consecutive to immature intermediate villi, and they are slender and numerous when continuing in mature intermediate villi. Terminal villi exclusively bear capillaries (see below), and only half of their volume is occupied by stroma.

On casts, *fetal arteries and veins* can be easily distinguished by the different impression patterns made by endothelial-cell borders and endothelial-cell nuclei (Fig. 20-3). On arteries, they are deep, slender, and regularly arranged, parallel to the long axis of the vessel, whereas on veins, they are rounder, shallow, and rather randomly oriented [22]. In contrast, *arterioles and venules* only form a few impressions without any orientation. Collateral venules are characterized by their smaller diameter (15–20 μm, φ arterioles 20–40 μm), circumferential constrictions, and flattened parts; however, both vessel types can be well identified by their three-dimensional course on casts (Figs. 20-2 and 20-3) [24].

The *capillaries of terminal villi* (φ 5–20 μm) spread out directly from terminal arterioles and then serially connect capillary loops of three to five neighboring terminal villi before entering a postcapillary venule (Fig. 20-3). Thereby, such a capillary system ranges in length from 3000 to 5000 μm; shortcuts, however, are the exceptional case [24]. Many sinusoidal dilations (φ up to 50 μm), preferably seen on the top of capillary loops of near-term placentas (Fig. 20-3) [8] may create a reduction in blood flow resistance over a long distance, according to the theory of Boyd and Hamilton [17]. This is further supported by the fact that the extent of sinusoids multiplies with the increased length of the capillary loops [3,8]. This capillary system may also result in retardation of blood flow, which fosters maternofetal exchange across the hemomonochorial inter-

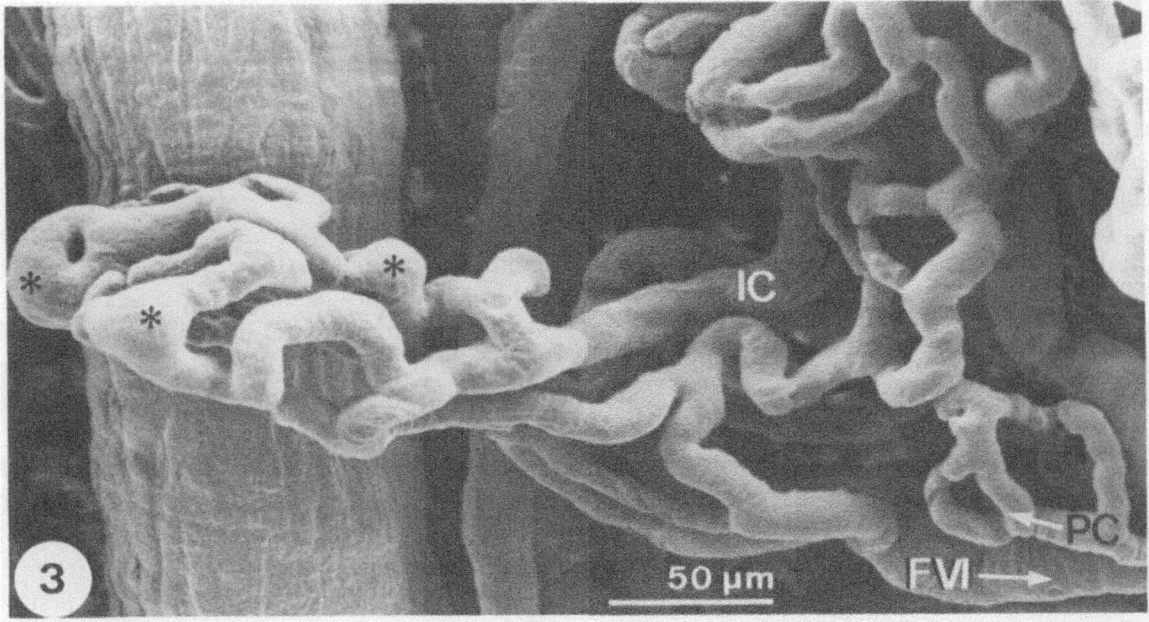

Figure 20-3. Human ramified capillary convolutes with sinusoids (asterisk) of two terminal villi. Note a paravascular capillary (PC ←) entering and an intervillous capillary (IC) bridging the two villi. A postcapillary venule (FVI →) leaves the capillary system. In the left background are an artery and an arteriole of the stem villus.

hemal membrane or placental barrier [26]. The barrier on the sinusoidal dilations, for example, measures less than 2 μm [23,27].

Subtrophoblastic capillaries of the peripheral types of stem villi and intermediate villi consist of the so-called paravascular net [26,28]. These capillaries follow a rather straight course, running mostly parallel to the longitudinal villous axis, but also form hairpinlike loops [8,22]. As arterio-arterial or arterio-venous shunts measuring up to 2000 μm [8], they are connected to the arteries and veins located centrally in the villus by rather short arteriolar and venular segments. Near the terminal villi, they rarely continue directly in the capillary system of terminal villi (Figs. 20-2 and 20-3). According to Leiser et al. [22] and Kaufmann et al. [8], the paravascular network may represent the remainder of the immature placenta, where terminal villi are absent and the paravascular network has to sustain materno-fetal exchange. The development of better vascularized terminal villi renders it an unfunctional relic.

As morphologically shown and described in this study, the human placenta has an open

system of intervillous blood space on the maternal side and a closed intravillous blood space on the fetal side. The *human placenta*, therefore, *represents the multivillous flow type* (see Fig. 20-18), as proposed by physiologists [overview: 1,2,29] and morphologists [overview: 3]. With regard to the amount of diffusible substance exchange across the interhemal membrane, this is the least effective among the different system types [2]. This may be supported, to some extent, by the fact that, with respect to other species, the weight ratio of neonate to placenta, 1:6, is low in the human [3]. Whether this ratio in the human relates to problems concerning pathological growth of the fetus and placental malfunction, which in the human are more severe than in the hitherto placentologically investigated mammals [8,30,31], remains an enigma.

3. Guinea Pig

The near-term placenta of guinea pig, as visible on arterial filling of a maternal cast in Fig. 20-4, is discoid shaped and is subdivided into 60–80 cylindrical lobules (φ 3–4 mm), which are supplied with maternal lacunae and fetal capillaries, and have a materno-fetally oriented axis.

The *maternal placental vasculature*, as shown by cast, starts with two to four coiled intramural arteries (φ 1 mm, Fig. 20-4), which enter the placenta at its endometrial pole. After they have branched and lost their vessel wall, they turn into four to six ascending main lacunae (φ 0.3–1.0 mm) [light and electron microscopic overview: 3,9,32]. Those main lacunae, ascending inside the so-called interlobium (sponge of small maternal lacunae), traverse approximately two thirds of the thickness of the placenta in a fetal direction and parallel to the lobular surface. Then, as the radial main lacunas, they bend at right angles, taking a horizontal direction into the center of lobule, where two branches, centrolobular lacunae, follow the axis of the cylindrical lobule in the opposite direction [9]. From these centrolobular arterial lacunae, many capillary lacunae ramify, which cross the lobule radially — forming a labyrinth with the fetal capillaries (see next paragraph) [4] (Figs. 20-5 and 20-6) — and, on the lobular surface, meet to form venous lacunae. As marginal venous lacunae, they first run into the interlobium parallel to the surface of lobes, and then around the margin or in the center of the placenta. Finally, these vessels run out of the organ and are collected in a basal venous lacunal ring (φ 0.5–1.0 mm), located at the maternal subplacental (endometrial) entrance of organ [32].

The *fetal placental vasculature* originates in two umbilical arteries, which ramify in a superficial, dichotomously branching system of chorionic arteries on the abendometrial or fetal pole of the placental disc [9,32]. From these arteries, interlobular arteries ramify, giving rise to another dichotomously branching arterial system between the maternal arterial ascending main lacunae and the lobular surface in the interlobium mentioned above (Fig. 20-5). From this departure short arterioles bend at right angles into the lobular labyrinth, proliferating in numerous capillaries. These capillaries converge centripetally to venules and collecting veins in the center of lobules (Fig. 20-5). The collecting veins, in an opposite direction from that of the centrolobular maternal arterial lacunae (see above), follow the per-

pendicular lobular axis and meet, converging into larger veins, which are oriented at a level that crosses this axis and is some distance away from the fetal end of the lobules [3,32]. The umbilical vein arises from the center of this venous system, leading out of the organ and through the umbilicus.

The *placental vessel labyrinth* of guinea pig is shown in detail in Figs. 20-5 and 20-6. The maternal capillary lacunae have a general radial direction from the center of lobule to the periphery. However, they are more coiled and ramified in a networklike system near the center (Fig. 20-6) and are more parallel near the periphery. The diameter of lacunae averages 15 μm, the cross section is roundish to irregular, and, on cast, a superficial roughness is recognizable. In contrast, the fetal capillaries converge from the periphery to the center of the lobule; the first two thirds along this way, the capillaries are very tortuous and are anastomosed, forming elongated bundles (Fig. 20-5); whereas in the last third, the capillaries show a more straight and parallel course (Figs. 20-5 and 20-6). The fetal labyrinth capillaries approximately 10 μm in diameter, are roundish in cross section and have a smooth superficial structure (Fig. 20-6) [4,31].

The materno-fetal blood flow interrelationship in the guinea pig — not only from physiological calculations [overview: 29,33], but also from histological [overview: 3] and three-dimensional studies [4], as also shown here — represents a nearly ideal *countercurrent blood flow system* (see Fig. 20-18) [2]. With respect to physiological data [1,2,29,33], this system has the greatest capacity for diffusible substance exchange across the interhemal membrane. The fact that the arrangement of labyrinthine vessels is very tight (Fig. 20-6), the nonvascular tissue amounts to only 30% [32], and the interhemal membrane measures only 1 μm [9,32], as well as other factors, e.g., hormones, carrier-influenced materno-fetal substance transfer, self-regulatory mechanisms of vessel lumen, etc. [3], may contribute to the capacity of this system. This may partly explain why the neonate/placental weight ratio in the guinea pig, 1:20 — in addition to the Chinchilla, which has a very similar placental structure — is the best among investigated species in this respect [3].

266

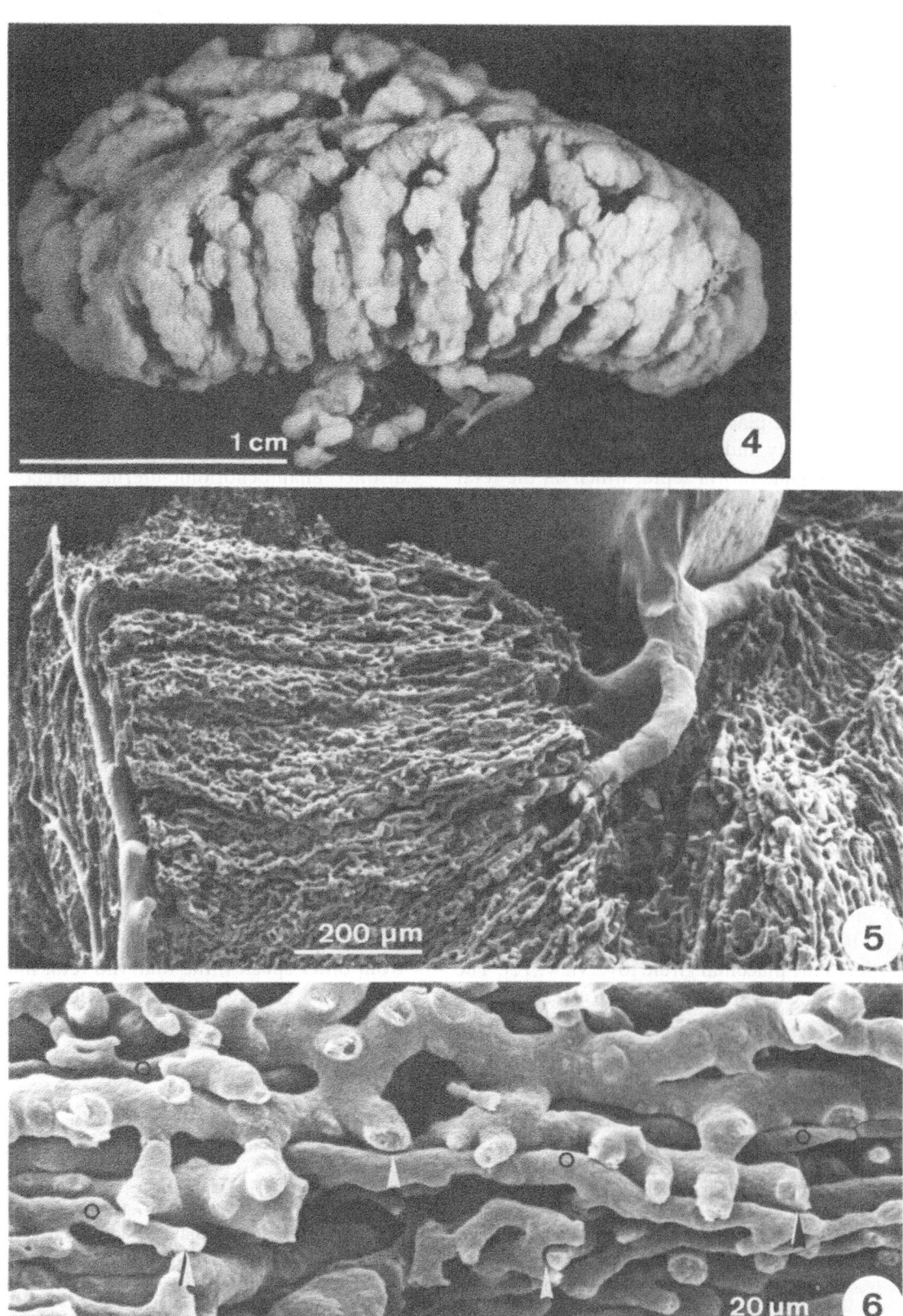

4. Cat

The cat placenta, centered between two paraplacental cupules, is characterized by a zonary girdle, which, however, at the early stage of the 22nd day of pregnancy, as shown in Fig. 20-7, is rather "spheroid." The girdle consists of maternal and fetal capillaries, which are complexly folded into lamellas and alternate with each other.

The *maternal placental vasculature* is supplied by stem arteries, each originating from arcadic arteries along the mesometrial line (Fig. 20-7). Then, they course straight pass the myometrial and placental layers, and branch several times before forming a funnel-shaped area on the fetal side of the placenta [10]. From this system, arterioles ramify and enter the capillary network of maternal lamelliform septa, which on the fetal side are basket shaped and on the maternal side of the placenta become narrowed (Figs. 20-8 and 20-9). Near term, these lamellae become complexely wrinkled and reduced in thickness (22nd day = $25\,\mu m$; 58th day = $17\,\mu m$). The maternal capillaries, in cast detail, form a rather coarse network; the capillary cross section is oval to irregular and the superficial structure is rough [4,10]. The capillary lamellae converge to form venules, which continue in stem veins. These veins, leading out of the organ, connect with the deep endometrial layers, the venous plexus in the myometrium, and finally join the superficial network of uterine veins (Fig. 7) [3,10].

The *fetal placental vasculature* is initiated by two umbilical arteries and one umbilical vein, which both ramify to branches following a course parallel to the placental girdle [overview: 11]. The arterial branches and consecutive arterioles join a capillary network, which on the fetal side of placenta covers the maternal arteries while forming double-layered, parallelely oriented lamellae with peduncular or tuftlike endings towards the maternal side of the placenta. These lamellas, alternating with single-layered maternal lamellae (see above), together form a very tightly folded placental system (Fig. 20-9). The fetal venules arise from different levels inside this system and leave the organ between the two capillary networks of a fetal lamella on its fetal side (Fig. 20-9) [11]. The fetal capillaries, in cast detail, form an irregular, rather fine network (ratio of vessel lumen to mesh size = 40:60%), and they are smooth and round, with a diameter averaging $15\,\mu m$ [3].

The maternal blood pathway in the septal capillary network of the feline placenta has a distinct allantochorionic-uterine direction. The fetal blood flow in the capillary network of the chorionic lamellae can be described as a more or less horizontally flow, crossing the maternal blood flow (see Fig. 20-18). Therefore, the cat placenta can be typified according to the predominant blood flow interrelationship as *one-way cross-current*. With respect to substance transfer across the endotheliochorial placental barrier, the mainly cross-current blood flow system in the cat is not as efficient as a countercurrent system, e.g., guinea pig and rabbit (see Section 3) [2,3,29]. The neonatal/placental weight ratio, 8:1, also refers to the rather moderate effectiveness of placental substance transfer [3]. However, the cat placenta may compensate this towards the end of pregnancy, when the lamellar system becomes extremely densely packed [11] and, different from other species, the materno-fetal interhemal distance is reduced to a conspicuously constant width of only $2\,\mu m$ [4,34].

←

Figure 20-4. Maternal vascular cast of guinea-pig hemomonochorial placenta, near term; overview. Four coiled intramural maternal arteries supply the cylindrically lobuled capillary system of placenta from the endometrial pole. The cavities between the lobules partly correspond to the fetal placental vasculature.

Figure 20-5. The fetal placental vasculature shown on a cracked specimen of guinea pig. On the outer surface of the lobule, fetal arteries and arterioles (left) ramify at right angles to a capillary labyrinth, which is arranged in elongated bundles leading to venules and veins in the center of lobule (middle right). The capillaries, on their way through the labyrinth, are first tortuous, and are then rather straight. The cavities between the venules mainly correspond with maternal arterial lacunas.

Figure 20-6. Capillary lacunar labyrinth shown on a fractured materno-fetally combined cast from the more central part of the lobule. The smaller fetal capillaries show a rather straight course, with little ramification (circles; see also Fig. 20-5), whereas the larger maternal lacunas are highly ramified and tortuous. Note the narrow cleft between the two vessel systems, comprising the materno-fetal interhemal distance (arrowheads).

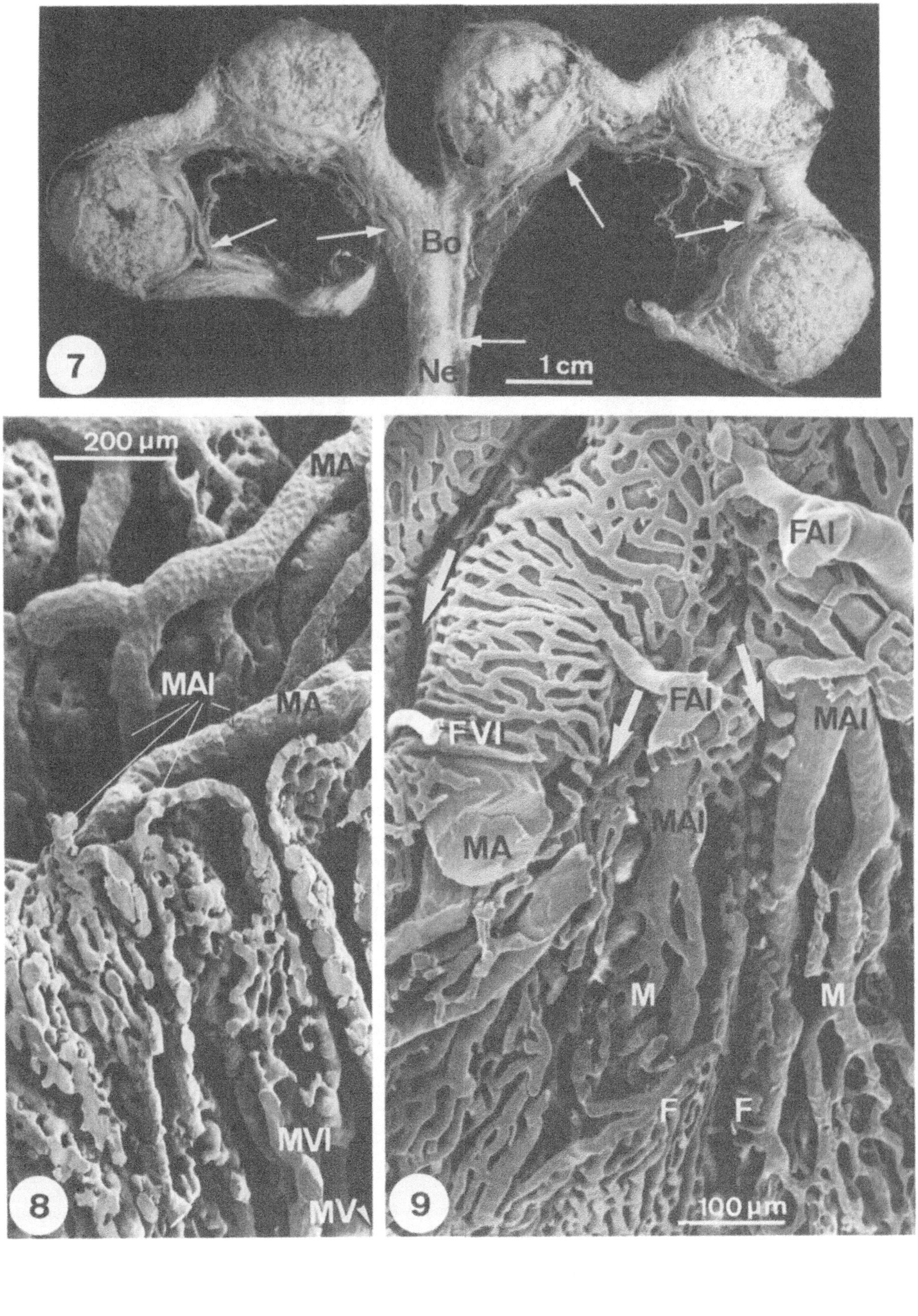

7
Bo
Ne
1 cm
8
200 µm
MA
MAI
MA
MVI
MV
9
FAI
FAI
FVI
MA
MAI
MAI
M
M
F
F
100 µm

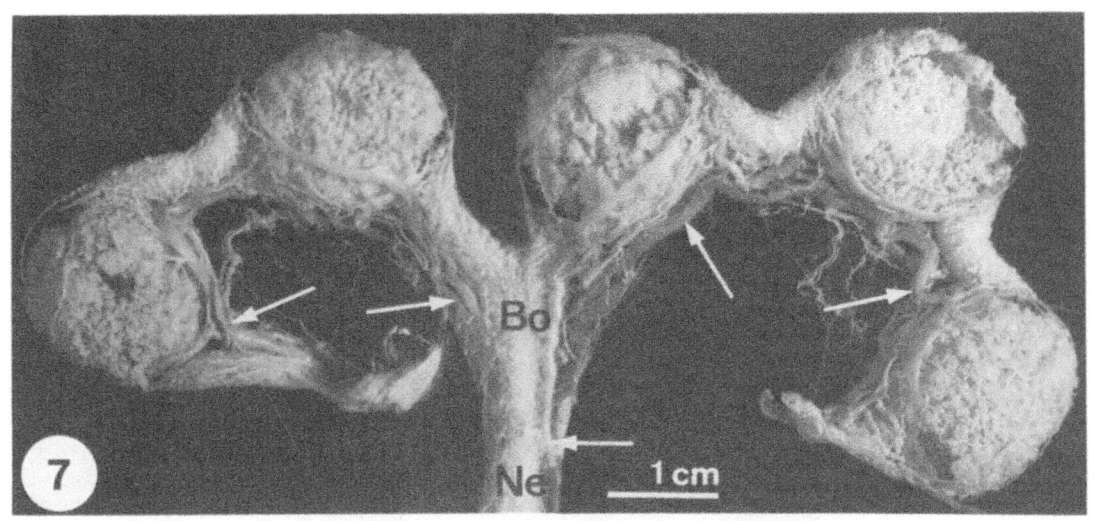

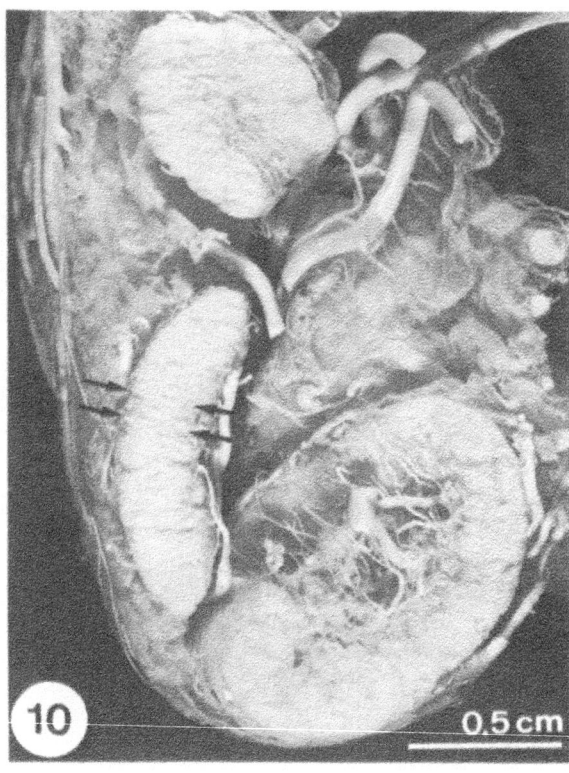

Figure 20-10. Cracked materno-fetally filled vessel cast of goat placenta, 120th day of pregnancy. The irregularly shaped placentones are supplied by maternal vessels on their convex side; whereas on their concave side, there are fetal vessels. The arrows point to the main direction of intraplacentonal vessels. (From Leiser [12], with permission.)

5. Goat

The goat placenta is cotyledonary or multiplex, with maternal caruncles and fetal cotyledons as part of 80–100 placentones (Fig. 20-10) [3]. The caruncles, as shown by blood vessel casts, are built of capillaries, forming maternal septa with crypts, which are complementary to the opposing fetal cotyledonary villous capillaries.

The *maternal placental vasculature* begins with stem arteries as branches originating from uterine arteries and arcadic arteries [12]. These stem arteries arise from the caruncular convex side of the placentone (Figs. 20-10 and 20-11) and follow a fetal direction through the septa towards the top of septa. At the fetal side of the caruncle, the stem arteries branch in a tuft of arterioles, which bend in a feto-maternal direction and join to form

←

Figure 20-7. Overview of maternal vascular cast from a cat uterus and five placentae at the 22th day of pregnancy. Smaller arcadic arteries and larger veins (arrows) build a "mesometrial line" along the neck (Ne), body (Bo), and two horns of the uterus. The five "spheroid" zonary girdles, predominantly formed by endometrial and placental vessels, are visible through the wide-meshed network of superficially located perimetrial and myometrial vessels. This network is particularly dense in the narrowed interplacental zones.

Figure 20-8. Cat fractured maternal corrosion cast corresponding to Fig. 20-7, viewed from the fetal (top) and lateral (cracked) side. Arterioles (MAI) link arteries (MA) to the septal capillary networks. These networks form basketlike lamellas on the fetal side, which become narrow towards the myometrial side (bottom) and converge into venules (MVI) and veins (MV). Note the reticular endothelial impressions on the arterial vessels.

Figure 20-9. Materno-fetally combined cast from the cat placenta on day 62 of pregnancy, viewed from the fetal (top) and lateral (cracked) side. The fetal side of placenta comprises arterioles (FAI), a venule (FVI), and (double) capillary networks (F), forming lamellas (arrows); whereas the maternal side is represented by an artery (MA), arterioles (MAI), and the capillary network (M), alternating with the fetal network (F). (From Leiser et al. [4], with permission.)

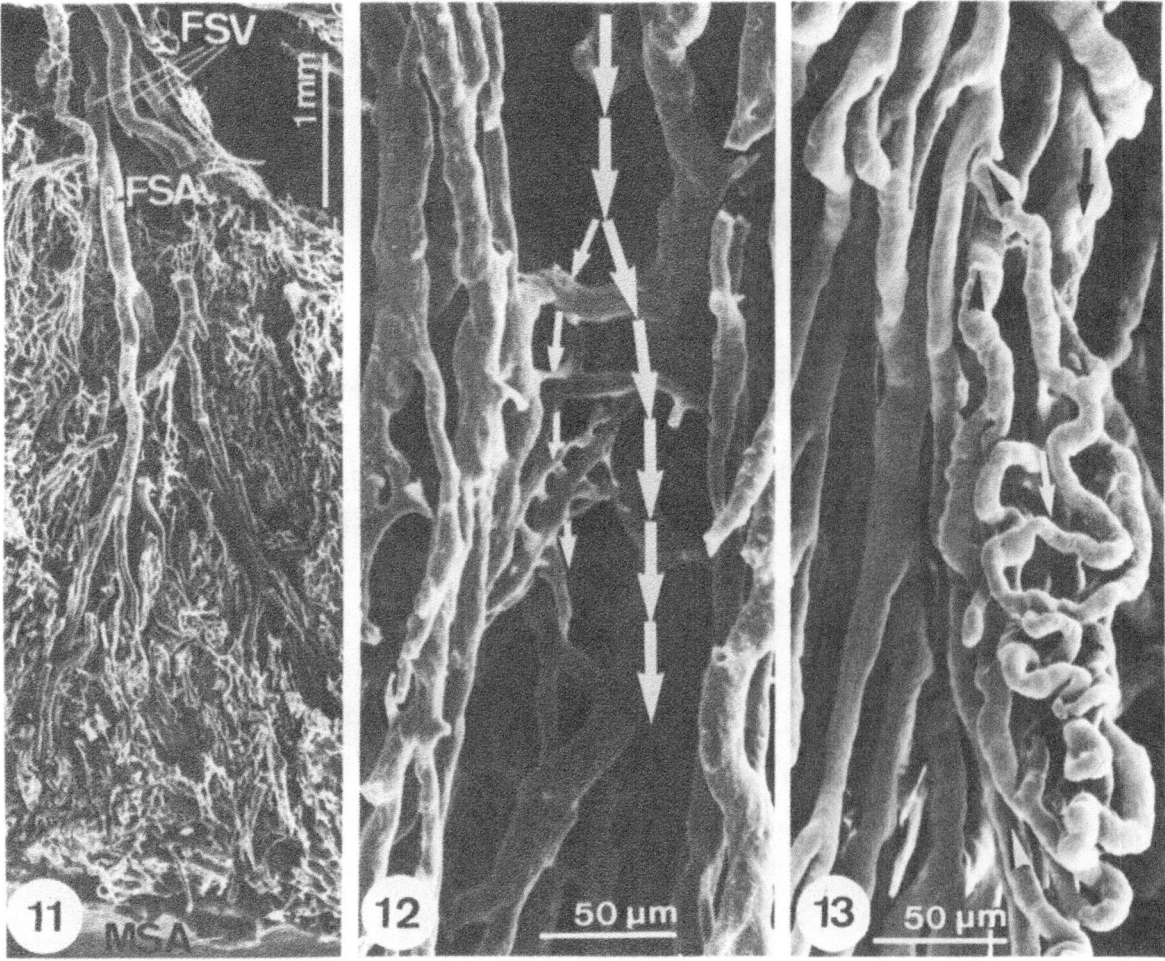

Figure 20-11. Goat combined materno-fetal corrosion cast corresponding to Fig. 20-10. A fetal stem artery (FSA) and fetal stem veins (FSV) enter the capillary villous system from the top, whereas on the bottom maternal stem arteries (MSA) can be observed. (From Leiser [12], with permission.)

Figure 20-12. Maternal cast of septal capillary meshwork on the 130th day of pregnancy in goat. Main crypts (big arrows) are followed by rather large feto-maternally oriented capillaries (top to bottom), whereas smaller or ending crypts (small arrows) are characterized by smaller and irregularly meshed capillaries. (From Leiser [12], with permission.)

Figure 20-13. Capillary convolute of fetal terminal villus, 133th day of pregnancy in goat. This convolute is supplied by a barely visible central arterial capillary limb (arrows) and is drained by several venous limbs, which connect (arrowheads) to an anastomosed bundle of materno-fetally oriented venules (bottom to top).

a capillary network. This network follows the irregular contours of the branched maternal crypts to the maternal side (Fig. 20-12), where the network converges with the maternal venules and the veins of the caruncular stalk. The maternal capillaries, in detail, form an irregular-coarse network. Cross sections are very irregular, with regard to their form and size, and their superficial structure on casts is rough (Figs. 20-12 and 20-14) [4,12].

The *fetal placental vasculature* enters the placenta from its concave or fetal side (Fig. 20-10). Several fetal arteries and veins that originally ramified from two umbilical arteries or one umbilical vein, respectively, often penetrate the stem of fetal villous tree as a bundle and ramify as arterioles or venules, according to the branching pattern of villous tree (Fig. 20-11) [compare also 3]. Near the villous tip, an arteriole continues in capillary coils and loops, capillary convolutes,

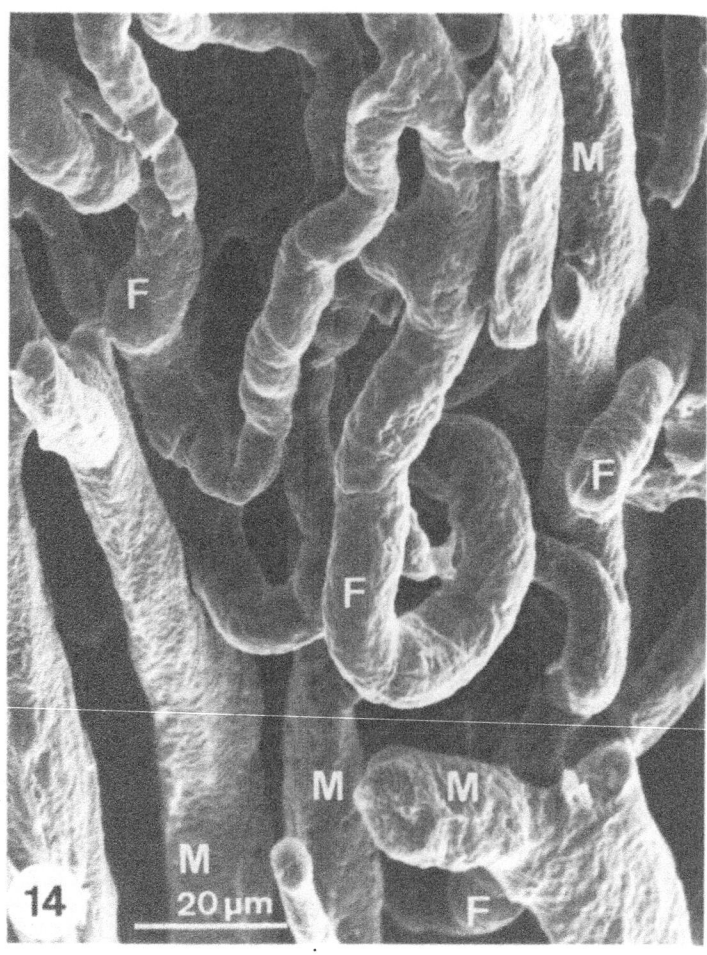

Figure 20-14. Combined materno-fetal corrosion cast, 120th day of pregnancy in the goat. Capillary loops of fetal villus tip (F) reach from top to a coarse capillary network of maternal septa (M). Note the superficial roughness of capillary cast, which is more distinct on the maternal part.

with rather few anastomosing interbranchings. These convolutes are asymmetrically arranged, with a shorter, more centrally positioned arterial capillary limb and a larger, superficial venous capillary limb. Neighboring capillary convolutes often are serially linked. At the base of the villous tip, the capillary convolutes, which are inter-connected by anastomoses, continue in fetal venules (Fig. 20-13) [12]. As is clearly visible in Fig. 20-14, the fetal capillaries, in detail, are roundish in cross section (diameter about 10 μm) and show a rather smooth superficial structure in casts [4].

As studied here by maternal, fetal, and materno-fetally combined blood vessel casts [compare also 12], the cryptal capillary system

meets the fetal capillary convolutes of villous tips partly in a *multivillous to countercurrent* type, partly in a more or less cross-current blood-flow system (see Fig. 20-18). The cross-current system also has been confirmed by Makowski [35]. The results of physiological experiments taken from sheep — which also may be valid for the goat [36] — would largely fit a concurrent vascular arrangement [2,37]. Such an anatomical situation, however, has not been observed in the placentae investigated. Therefore, the results of anatomical and physiological blood flow relationships in the goat are controversial and do not refer to the efficiency of diffusional transplacental sub-stance transfer. This function does not seem to be facilitated by the materno-fetal interhemal

272

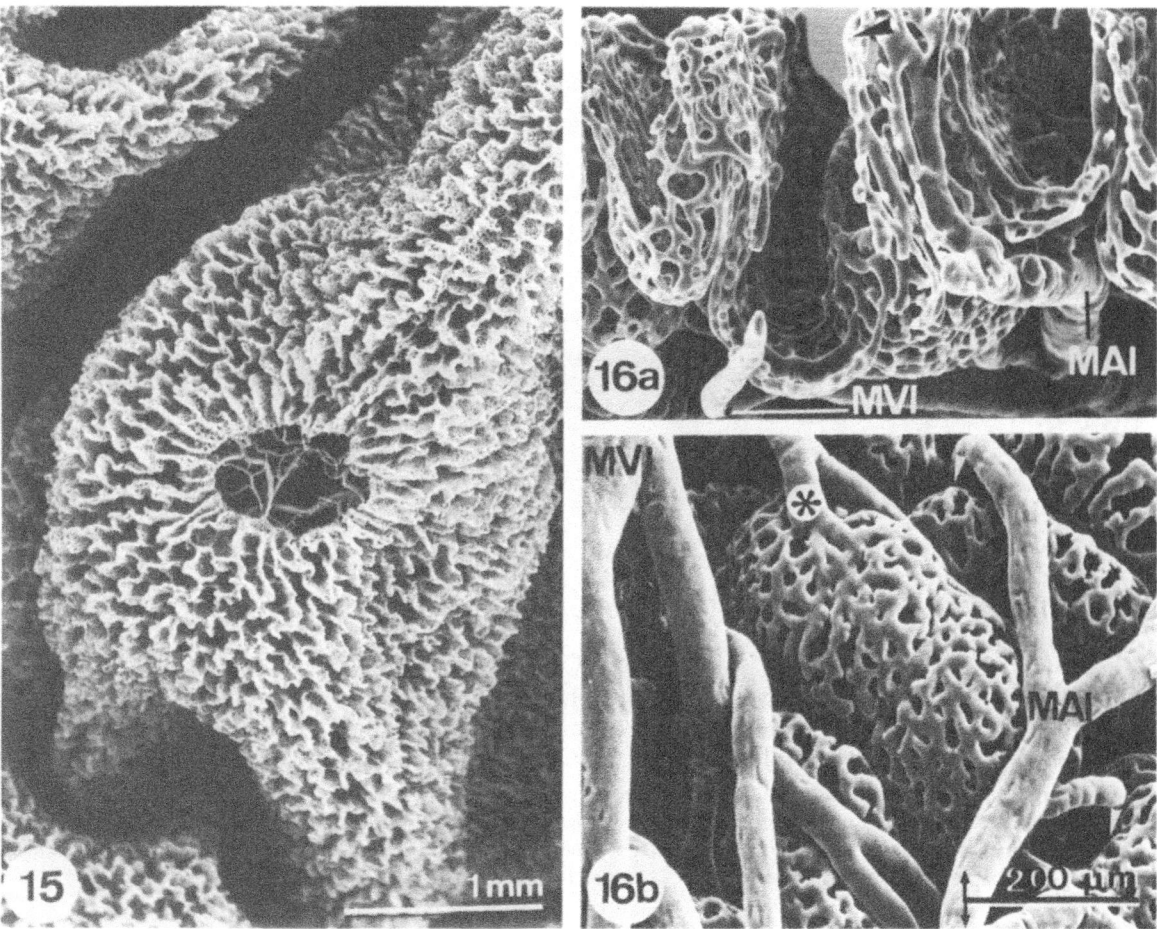

*Figure 20-15.*Overview of maternal cast of folded pig placenta, seen from the fetal side, on the 99th day of pregnancy. A capillary pattern of troughs (fossae) between ridges (rugae) is clearly visible. On the top of the fold, where this pattern is lacking, a glandular mouth or maternal part of the so-called areola gives sight to the endometrial vessels beneath.

Figure 20-16. Maternal vessel cast corresponding to Fig. 20-15, viewed from the lateral (cracked) side (Fig. 20-16a) and from the maternal side (Fig. 20-16b). Maternal arterioles (MAI) reach the capillary network on the top of ridges (arrowheads), whereas venules (MVI) leave the capillary system on the base of basketlike troughs (asterisk).

distance of the goat epitheliochorial placenta. Its width is highly variable as a consequence of the coarse and irregular network of maternal capillaries, which cannot follow the fetal capillary curling (Fig. 20-14) [4]. This discrepancy, in addition to unknown factors, may be compensated by the neonatal/placental weight ratio, 10:1, which is rather effective in the goat [3].

The materno-fetal blood flow interrelationship in the goat placenta is comparable to the vascular arrangement on the distal part of human villous tree, which represents the multivillous flow type (see Section 2). However, as opposed to the human, the maternal blood flow of goat is of a closed form (see Fig. 20-18).

6. Pig

The near-term pig placenta is a diffuse type, characterized by interlocking circular folds covering almost the entire placental surface, which arise from the endometrium and from the fetal chorion. The folds, mainly consisting of capillary networks on casts, are subdivided on the maternal side into microscopic folds or troughs (fossae)

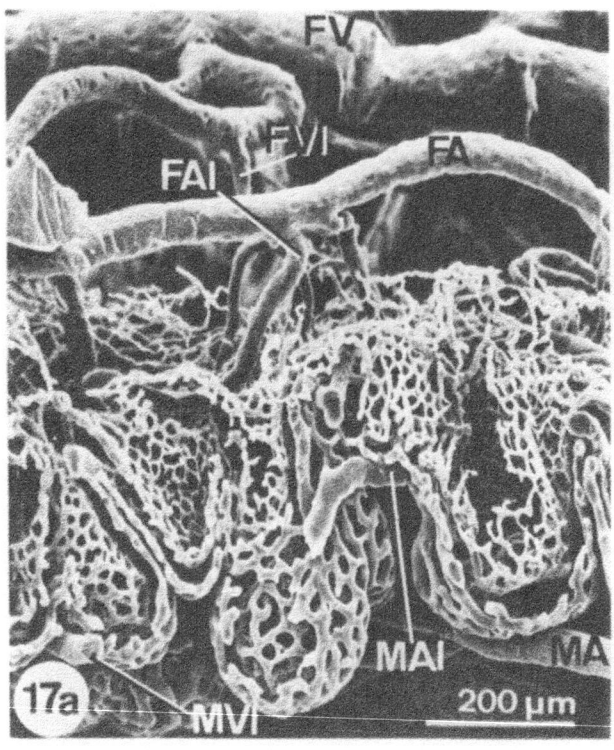
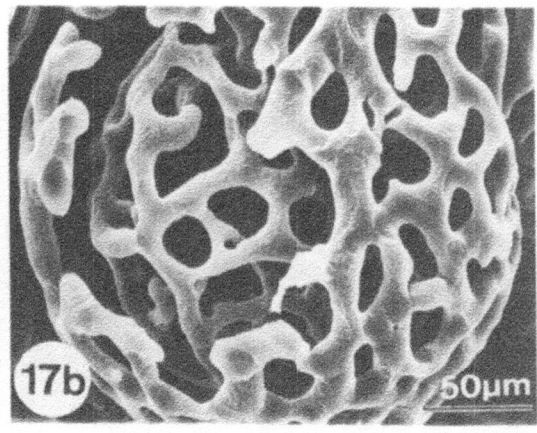

Figure 20-17. Materno-fetally combined placental vessel cast of pig at day 99 of pregnancy. a: The maternal capillary network (below) — forming convex troughs with rather narrow openings oriented to the fetal side — interlocks with the fetal capillary network of bulbous protrusions (from the top; see also area magnified in Fig. 20-17b). MA = maternal artery; MAI = maternal arteriole; MVI = maternal venule; FA = fetal artery; FAI = fetal arteriole; FV = fetal vein; FVI = fetal venule. b: The maternal (peripheral) capillary network is less meshed than the fetal (central) network.

between ridges (rugae) (Fig. 20-15) and on the fetal side show numerous bulbous protrusions [3,13,38–40].

On the maternal side of *maternal placental vascular casts*, the branching of arteries into arterioles and venules into veins can be followed (Fig. 20-16b). The arterioles run into the space between the troughs and can only be observed in a cracked area (Fig. 20-16a), where they are seen to ascend to the top of the maternal ridges and to continue in a capillary network. The network on the ridges is very often fused to form larger capillaries, but it continues at the slopes of the ridges into the troughs, which are subdivided by secondary ridges, thereby dissecting the troughs into rows of basketlike structures (Figs. 20-16a and 20-16b) [13]. At the bottom of the troughs, the blood is collected by maternal venules (Figs. 20-16b and 20-17a). In detail, the capillaries on the top of ridges generally are roundish in cross

section, with a diameter averaging 15 μm. On slopes and at the bottom of the troughs, the capillaries have a more flattened, irregular or oval form, with a capillary volume/mesh-size ratio of 3:2, which results in a rather coarse network (Figs. 20-16b and 20-17b) [compare 13].

The *fetal placental vasculature* of porcine placenta arises from two umbilical arteries, branching several times until arterioles enter fetal ridges or bulbous protrusions on the fetal side of the placenta at a right angle (Fig. 20-17a). These arterioles from the center of protrusions give off capillary branchings to the side and the top of protrusions, which continue in a capillary network on the surface of the protrusions. The capillaries converge to one or two fetal prevenous capillaries per protrusion at the sides of the fetal troughs. This arrangement evolves, since the fetal troughs are lined by columnar trophoblast, which is free of a so-called intraepithelial capillary

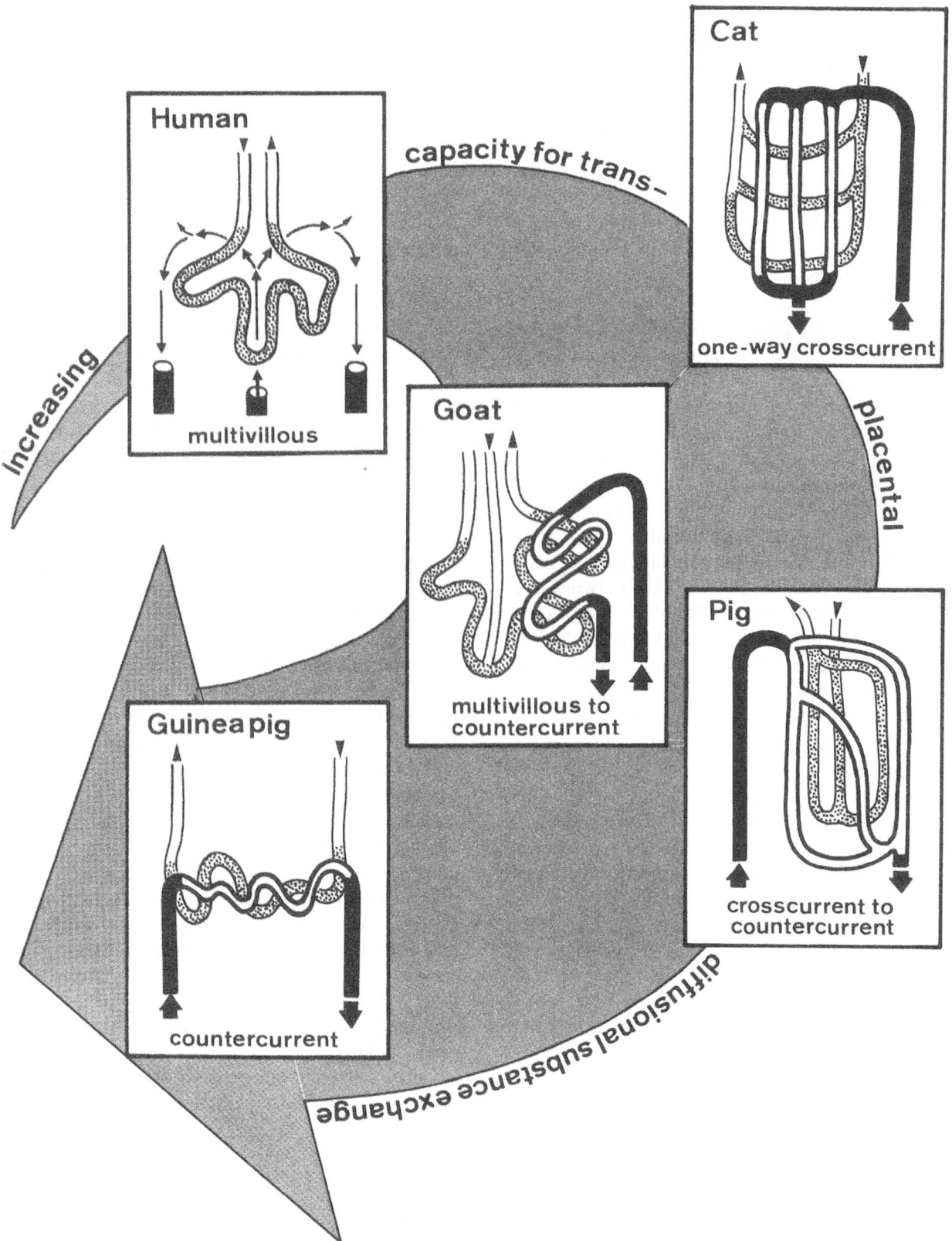

Figure 20-18. Five different types of placentas, schematically illustrated by maternal (arrow) and fetal (arrowhead) vessel arrangements. Capillaries, which are involved in the most intensive materno-fetal substance exchange, are drawn by tubes (maternal) and by dots (fetal). Other vessels represent precapillary arteries and arterioles, as well as postcapillary venules and veins.

network, as is the case at the sides and the top of the fetal protrusions. Prevenous capillaries may often function as connecting capillaries, which link several protrusions serially. As venules and veins, they lead out of the organ (Fig. 20-17a) [13]. The fetal capillaries, in detail, are round, about 10 μm in diameter, show a smooth surface on casts [4], and form networks with a capillary/ mesh-size ratio of 1:1. This fetal capillary network, in contrast to the maternal one, looks more meshed (Fig. 20-17b).

As demonstrated in materno-fetally combined casts, the bulbous fetal protrusions complementarily interlock with the cavenuous, basketlike maternal troughs between ridges, with a slightly broadened top, thus creating an anchoring effect (Figs. 20-17a and 17b). The maternal capillary blood, as shown from the casts described above, mainly flows in the chorionic-uterine direction. The reverse is valid for the fetal capillary blood flow. However, due to the capillary side branches of arterioles reaching the capillary network during their course to the top of fetal protrusions (see above), maternal and fetal blood flow partly cross each other. Therefore, the materno-fetal blood-flow interrelationship in the pig is *mainly one-way cross-current with a countercurrent component* (see Fig. 20-18) [13]. In relation to other species of this research the pig blood-flow type refers to a favorable transplacental exchange for diffusible substances. The physiological data on oxygen tension, however, indicate that the pig placenta has a rather poor exchange capacity, although it seems to increase during the last third of gestation [41]. During this time, the epitheliochorial porcine interhemal membrane is progressively thinned at the site of the so-called intraepithelial capillaries of uterine epithelium and trophoblast [4,42]. The neonatal/placental weight ratio, 9:1, reflects the effectiveness of placental transfer and takes a middle position between those of other species [3].

7. Concluding Remarks

The five chosen placental types of very different macroscopical aspects studied by vessel casts also reveal a very variable placental vasculature morphology.

The histological classification of the placental barrier by Grosser [5] — hemochorial, endotheliochorial, and epitheliochorial — does not show any correlation with feto-maternal blood flow interrelationships, multivillous flow, crosscurrent, and countercurrent (Fig. 20-18). However, according to physiologists [2,29,37], the transfer of readily diffusible inert substances through the placenta increases from multivillous flow to crosscurrent and to countercurrent, thus classifying the placental types by its corresponding efficiency. The multivillous flow system exists in the discoidal hemochorial human placenta, as well as partly in the cotyledonary epitheliochorial goat placenta. Crosscurrent conditions are found in the zonary endotheliochorial cat placenta, partly in the diffuse epitheliochorial porcine placenta, and to some extent also in the goat placenta. The typical blood flow type in the discoidal hemochorial placenta of guinea pig is countercurrent and it participates with some component in the pig placenta (see Fig. 20-18).

The weight ratios of neonate and placenta, reflecting, to some extent, the efficiency of transplacental exchange, also show a good correlation with the materno-fetal blood flow relationship, i.e., in the guinea pig, cat, and human. In addition, a positive neonate/placental weight ratio and a favorable blood-flow condition also may be associated with the complexity of materno-fetal interdigitation and with the decreasing thickness of the placental interhemal barrier, as in the guinea pig and pig [3,4]. For other species, these criteria seem to be less decisive factors. Other criteria, such as the stage of gestation, umbilical and uterine blood flow rates, active transplacental transport, growth hormones, sexual hormones, fetal growth rate, etc. [1–3,41] may also play a role in guaranteeing the appropriate fetal development. Many of these factors are partly subject to future research, but the placental vasculature is somehow linked with them as a carrier medium in all species.

References

1. Carter AM. Placental circulation. In: Steven DH (ed.), *Comparative Placentation. Essays in Structure and Function.* Academic Press, New York, pp 108–160, 1975.

2. Faber JJ, Thornburg KL. Placental physiology. In: *Structure and Function of Fetomaternal Exchange*. Raven Press, New York, 1983.

3. Dantzer V, Leiser R, Kaufmann P, Luckhardt M. Comparative morphological aspects of placental vascularization. *Troph Res* 3:235–260, 1988.

4. Leiser R, Dantzer V, Kaufmann P. Combined microcorrosion casts of maternal and fetal placental vasculature. A new method of characterizing different placental types. In: Motta PM (ed.), *Developments in Ultrastructure of Reproduction*, Progr Clin Biol Res 296, Alan R. Liss, New York, pp 421–433, 1989.

5. Grosser O. Vergleichende Anatomie und Entwicklungsgeschichte der Eihäute und der Plazenta. Braumüller, Wien, 1909.

6. Enders AC. A comparative study of the fine structure of the trophoblast in several hemochorial placentas. *Am J Anat* 116:29–68.

7. Ramsey EM, Donner MW. Placental Vasculature and circulation. Georg Thieme, Stuttgart, 1980.

8. Kaufmann P, Luckhardt M, Leiser R. Three-dimensional representation of the fetal vessel system in the human placenta. *Troph Res* 3:113–137, 1988.

9. Kaufmann P. Electron microscopy of the guinea-pig placental membranes. *Placenta* 2 (Suppl.):3–10, 1981.

10. Leiser R, Kohler T. The blood vessels of the cat girdle placenta. Observations on corrosion casts, scanning electron microscopical and histological studies. I. Maternal vasculature. *Anat Embryol* 167:85–93, 1983.

11. Leiser R, Kohler T. The blood vessels of the cat girdle placenta. Observations on corrosion casts, scanning electron microscopical and histological studies. II. Fetal vasculature. *Anat Embryol* 170:209–216, 1984.

12. Leiser R. Mikrovaskularisation der Ziegenplazenta, dargestellt mit rasterelektronisch untersuchten Gefäßausgüssen. *Schweiz Archiv Tierheilk* 129:59–74, 1987.

13. Leiser R, Dantzer V. Structural and functional aspects of porcine placental microvasculature. *Anat Embryol* 177:409–419, 1988.

14. Risco JM, Nopanitaya W. Ocular microcirculation. Scanning electron microscopic study. *Invest Ophthalmol Vis Sci* 19:5–12, 1980.

15. Christofferson RH, Nilsson BO. Microvascular corrosion casting with analysis in the scanning electron microscope. *Scanning* 10:43–63, 1988.

16. Leiser R. Fetal vasculature of the human placenta: Scanning electron microscopy of microvascular casts. *Contrib Gynecol Obstet* 13:27–31, 1985.

17. Boyd JD, Hamilton WJ. The human placenta. Heffer, Cambridge, 1970.

18. Schuhmann RA. Plazenton: Begriff, Entstehung, funktionelle Anatomie. In: Becker V, Schiebler TH, Kubli F. (eds.), *Die Plazenta des Menschen*. Georg Thieme Verlag, Stuttgart, pp 109–207, 1981.

19. Schuhmann RA. Placentone structure of the human placenta. *Bibl Anat* 22:46–57, 1982.

20. Kaufmann P. Development and differentiation of the human placental villous tree. *Bibl Anat* 22:29–39, 1982.

21. Schneider H, Luckhardt M. Entwicklung der Plazenta und des utero-plazentaren Kreislaufes aus morphologischer und funktioneller Sicht. *Geburtsh u Frauenheilk* 49:843–851, 1989.

22. Leiser R, Luckhardt M, Kaufmann P, Winterhager E, Bruns U. The fetal vascularization of term human placental villi. I. Peripheral stem villi. *Anat Embryol* 173:71–80, 1985.

23. Schiebler TH, Kaufmann P. Reife Plazenta. In: Becker V, Schiebler TH, Kubli F (eds.), Die Plazenta des Menschen, Georg Thieme Verlag, Stuttgart, pp 51–100, 1981.

24. Kaufmann P, Bruns U, Leiser R, Luckhardt M, Winterhager E. The fetal vascularization of term human placental villi. II. Intermediate and terminal villi. *Anat Embryol* 173:203–214, 1985.

25. O'Neill JEG. Vascularizacao da Placenta Humana. Dissertacao Universidade Nova de Lisboa, Portugal, 1983.

26. Arts NFT. Investigations on the vascular system of the placenta. Parts 1 and 2. *Am J Obstet Gynec* 82:147–158, 159–166, 1961.

27. Sen DK, Kaufmann P, Schweikhart G. Classification of human placental villi. II. Morphometry. *Cell Tissue Res* 200:425–434, 1979.

28. Boe F. Studies on the vascularization of the human placenta. *Acta Obstet Gynecol Scand* 32 (Suppl. 5):1–92, 1953.

29. Martin CB. Jr. Models of placental flow. *Placenta 1* (Suppl.):65–80, 1981.

30. Villee CA, Page EW, Villee DB. Human reproduction. Essentials of Reproductive and Perinatal Medicine 3rd ed., W.B. Saunders, Philadelphia, Chapt. 9, 1981.

31. Scheffen I, Philippens L, Kaufmann P, Leiser R, Mironov V. Maternal oxygen supply as a regulator of fetal placental capillarisation. *Placenta* 10:501, 1989.

32. Kaufmann P, Davidoff M. The guinea-pig placenta. *Adv Anat Embryol* 53, Fasc. 2, Springer, Berlin, 1977.

33. Moll W, Kastendieck E. Transfer of N_2O, CO, and H_2O in the artificially perfused guinea pig placenta. *Resp Physiol* 29:283–302, 1977.

34. Leiser R. Development of the trophoblast in the early carnivore placenta of the cat. *Bibl Anat* 22:93–107, 1982.

35. Makowski EL. Maternal and fetal vascular nets in placentas of sheep and goat. *Am J Obstet Gynecol* 100: 283–288, 1968.

36. Silver M. An assessment of the chronically catheterized fetal preparation in sheep and other species. *Placenta* 2 (Suppl.):89–108, 1981.

37. Longo LD. The interrelations of maternal-fetal transfer and placental blood flow. *Placenta* 2 (Suppl):45–64, 1981.

38. Tsutsumi Y. The vascular pattern of the placenta in farm animals. *J Facul Agr Hokkaido Univ Sapporo* 52: 372–482, 1962.

39. Macdonald AA. Uterine vasculature of the pregnant pig: A scanning electron microscope study. *Anat Rec* 18:689–698, 1976.

40. Macdonald AA. The vascular anatomy of the pig placenta: A scanning electron microscope study. *Acta Morphol Neerl Scand* 19:171–172, 1981.

41. Caton D, Bazer FW. Respiratory gases in uterine circulation of pregnant domestic swine as sampled by indwelling catheters. *Am J Physiol* 234:R25–R28, 1978.
42. Dantzer V. Cell Biological Aspects of Porcine Placentation. A Scanning and Transmission Electron Microscopic Study. Ph.D. Thesis Roy. Vet. Agr. Univ. Copenhagen. A/S Carl Fr. Mortensen, Copenhagen, 1986.

Author's address:
Prof. Rudolf Leiser
Institut fur Veterinar-Anatomie
Histologie and Embryologie der
  Justus-Liebig-Universität
Frankfurter Strasse 98 D6300 Giessen
FRG

Three-Dimensional Microvascular Architecture of the Testis and Excurrent Duct System Studied by Corrosion Casting-Scanning Electron Microscopy

AKIO KIKUTA, AIJI OHTSUKA, & TAKURO MURAKAMI

1. Introduction

The mammalian testis and the excurrent duct system are composed of an extraordinarily long duct system, seminiferous tubules, epididymal ductules, and vas deferens, in which spermatogenesis, hormone secretion, and protein secretion and absorption continuously occur. In particular, the seminiferous tubules are compactly packed within the oval space by the tunica albuginea. In humans and many mammals, such as the rat and mouse, in which spermatogenesis is labile to high temperature, the testes are localized within the scrotum and spermatogenesis proceeds safely at a temperature 2–3° lower than that of the intraperitoneal cavity [1–3]. The vascular system that supplies and drains the scrotal testis has been considered to be organized in special vascular arrangements for cooling arterial blood before entering the testis, as well as enabling continuous and homogeneous blood flow throughout the testis.

The gross blood vascular system of the mammalian testis, and the excurrent system, including humans, has been extensively studied [4,5]. The microvascular arrangement of the testis and the excurrent duct system has been studied by light microscopy using various methods: India-ink injection [6], silicone rubber injection and clearing method [7], or microangiography [8–13]. However, these observations could not provide sufficient three-dimensional visualization of the microvascular system, especially of the capillary networks. Vascular corrosion casting-scanning

electron microscopy (SEM) is one of the most efficient methods to study that system, providing three-dimensional and wide viewing of microvascular beds with a large depth of focus [14]. Recently, using this vascular corrosion casting-SEM method, the microvascular system of the testis and the excurrent duct system has been studied in the human [15,16] and several other mammals, rat [15,17–20], mouse [21], bull [22], and pig [23].

In this chapter, we describe the three-dimensional organization of the microvascular system of the testis, epididymis, and vas deferens in the rat, especially showing the regional transition in subepithelial capillary network patterns along the male reproductive duct system, the seminiferous tubules, epididymal ductules, and vas deferens. We also briefly review the microvascular system of the testis and the excurrent duct system in other mammals having scrotal testies, including humans.

2. Corrosion Casting — Scanning Electron Microscopy

Wistar rats to be examined are anesthetized and the vascular beds are thoroughly perfused with Ringer's solution through the thoracic aorta to flush out the blood. Subsequently, laboratory-prepared [14] or commercially available (Mercox, Oken Shoji Co. Tokyo, Japan) casting media are injected under manual pressure. The perfused animals or organs are kept at room temperature

Motta, P.M., Murakami, T., and Fujita, H. (eds.), Scanning Electron Microscopy of Vascular Casts: Methods and Applications.

280

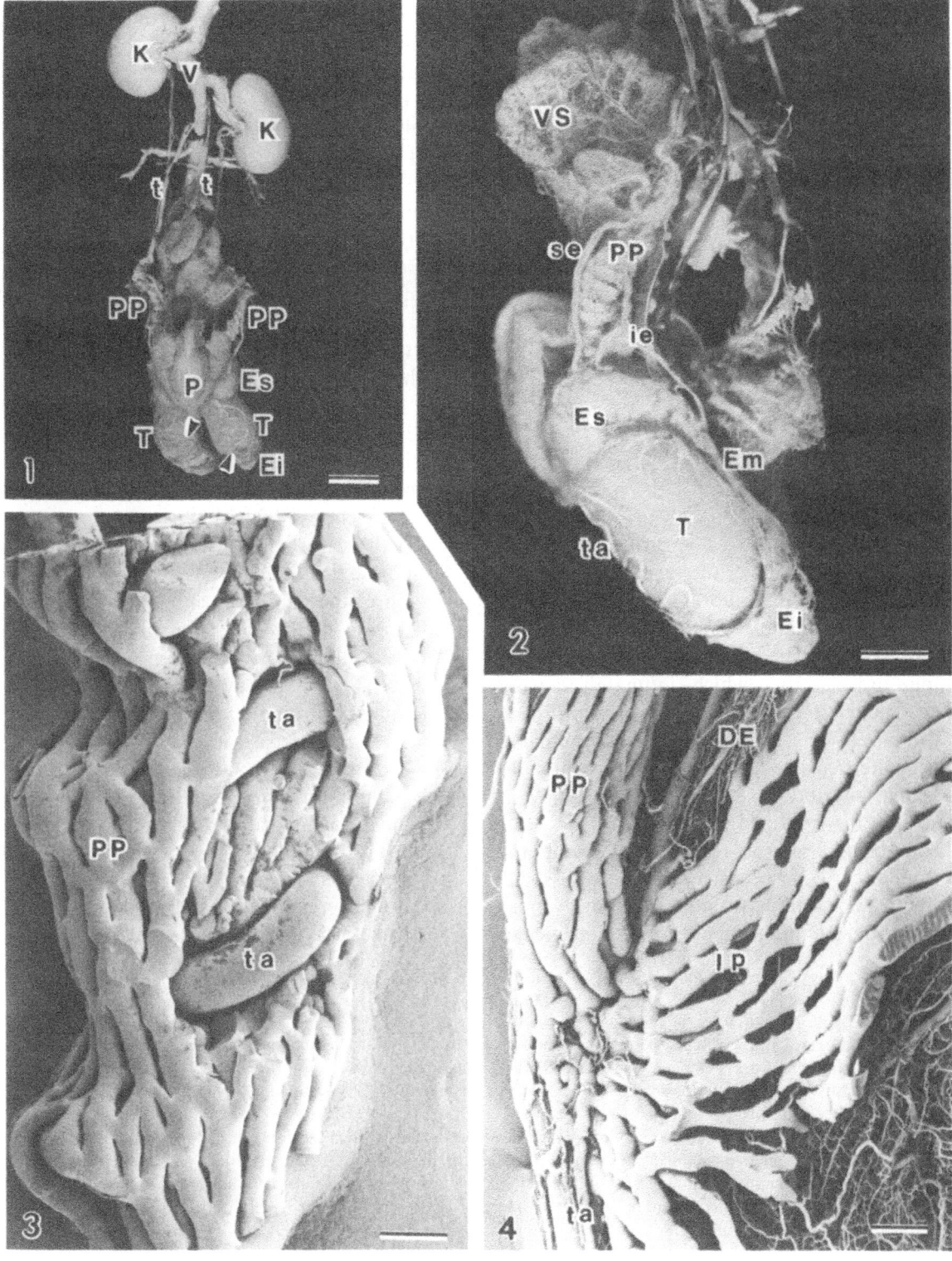

for about 10 minutes until the resin hardens, then they are immersed for 2 hours in a hot-water bath (60°C) to completely polymerize the injected resin. The specimens are then corroded in a hot 10% NaOH solution (60°C) overnight and washed in running tap water to completely remove the tissue elements and air dried. The blood vascular casts thus prepared are dissected with sharpened forceps or needles, or freeze cut with razor blades in order to expose the required parts. The dissected or freeze-cut specimens are sputter coated with gold and observed with a scanning electron microscope.

3. Microvasculature of the Testis

3.1. Testicular Artery and Pampiniform Plexus

The gross vascular supplies of the human testis and other scrotal mammals are principally similar. The testicular artery arises from the abdominal aorta below the renal artery and descends along the vertebral column and then runs in the spermatic cord to supply the testis. Venous vessels issued from the testis form the pampiniform plexus, which intimately covers the testicular artery and then converges into the testicular vein (Figs. 21-1 and 21-2). In humans, the right testicular vein directly drains into the vena cava inferior, while the left testicular vein drains into the left renal vein.

The segment of the testicular artery in the spermatic cord descends in intimate association with the countercurrently ascending venous vessels of the pampiniform plexus or the testicular vein (Figs. 21-2 and 21-3). In many species with scrotal testes, this segment of the testicular artery, interwoven with the pampiniform plexus, shows peculiar coiling or a convoluted configuration. In the rat, the testicular artery, after giving off the superior and inferior epididymal arteries, begins to coil and convolute (Fig. 21-3). Coiling and convolution are repeated 20–30 times as it runs to the rostral end of the testis [4,17,24,25]. The convolutions of the testicular artery become more closely packed at the distal end, where it is surrounded by thick venous anastomoses of the pampiniform plexus [17]. Well-developed coiling and convolutions of the testicular artery in this portion have been reported in many mammals with scrotal testes, such as the mouse [4], monkey [24,26], ram [27], bull [22], camel [28], dog [4, 8,24], and cat [24]. In humans, although the testicular artery is intimately associated with the pampiniform plexus, arterial convolutions are not so conspicuous [4,24,25,29].

In the camel, the testicular artery is small in caliber near its origin and increases in size in the region of the pampiniform plexus and thereafter [28]. Also in the rat, an increase in caliber of the testicular artery has been noted as it nears the testis [19]. Measurements taken on light and scanning electron micrographs indicate that there is a 1.6-fold increase in the luminal diameter of the testicular artery during the passage through the pampiniform plexus [19].

The testicular artery reaches the testis from the superior aspect and courses down on the surface of the testis beneath the scrotal skin. Striking species differences are noted in the courses of the testicular artery on the surface of the testis. In the

←

Figure 21-1. Photograph of the vascular cast of the testis and the excurrent ducts of an adult rat, viewed from the ventral aspect. Note the dense venous network, pampiniform plexus (PP), and remarkable convolutions (arrowheads) of the testicular artery on the surface of the testis (T). V = inferior vena cava; Es = caput epididymidis; Ei = cauda epididymidis; K = kidneys, P = penis; t = testicular artery and vein. Bar = 10 mm.

Figure 21-2. Photograph of the vascular cast of the pampiniform plexus, testis, and excurrent ducts of the adult rat, viewed from the lateral aspect. A well-developed pampiniform plexus (PP) ensheathes the coiling testicular artery. Note marked convolutions of the testicular artery (ta) on the surface of the testis (T). Ei = cauda epididymidis; Em = corpus epididymidis; Es = caput epididymidis; VS = vesicula seminalis; ie = inferior epididymal artery and vein; se = superior epididymal artery and vein. Bar = 5 mm.

Figure 21-3. Scanning electron micrograph of the cast of the pampiniform plexus (PP). A sheath of the pampiniform plexus cast was removed to expose coiling and a convoluted segment of the testicular artery (ta) within the plexus. Bar = 0.5 mm.

Figure 21-4. Frozen-cut vascular cast of the rat testis, showing cranial pole region. Note the well-developed intraalbugineal venous plexus (ip), which is continuous with the pampiniform plexus (PP). DE = loose capillary plexus of the ductuli efferentes; ta = testicular artery. Bar = 0.5 mm.

human, monkey, dog, rat, mouse, camel, and buffalo, the testicular artery encircles the testis one time or less. In the rabbit, the testicular artery completely circles the organ two or three times before entering the parenchyma [7]. The rat testicular artery shows five to seven marked convolutions (Figs. 21-1 and 21-2) [4–6,8,19]. Such convolutions are also noted in the monkey [26] and camel [28]. In the mouse [7,21], rabbit [7], and dog [4,25], convolutions of the testicular artery running on the surface of the testis are weak or absent.

Spermatogenesis is labile to high temperature and spermatogenetic cells are easily damaged if the testis is suspended in the abdominal cavity where the temperature is 3–5°C higher than that in the scrotum [1–3,25]. Thus, the testes of mammals with high body temperature are situated in the scrotum, outside the abdominal cavity. The scrotum and specific vascular system serve as cooling systems of mammalian scrotal testis. The scrotal skin is the primary portion where the heat of the testis is dissipated to environmental air, while the vascular system of the testis is arranged in various ways to precool the arterial blood before entering the testis.

First of all, the testicular artery courses in the spermatic cord in intimate association with the countercurrently coursing testicular vein and the pampiniform plexus. Secondly, the venous vessels composing the pampiniform plexus extensively repeat dividing and anastomosing to form a highly dense venous network, which closely surrounds the testicular artery. In this countercurrent heat-exchange system, increasing the number of tubes in proximity to each other will increase the surface across which heat can be conducted and, consequently, may facilitate the exchange of heat. Thirdly, the testicular artery repeats coiling and convolution in the spermatic cord. Thus the areas of the intimate association between the testicular artery and the venous vessels are extraordinarily widened. Furthermore, the coiling and convolution of the testicular artery may cause a reduced blood flow velocity. In the rat, the testicular artery within the testis are relatively pulseless [27]. An increase in the caliber of the testicular artery in the rat and camel is also considered to cause a reduction in the blood velocity. Re-

duced blood velocity in the testicular artery may enhance the efficiency of the countercurrent heat exchange.

We should note that such a countercurrent system works simultaneously as a warming system of the blood recurring from the testis and flowing into the abdominal cavity, thus efficiently keeping the abdominal temperature to prevent body temperature loss.

A considerable long run of the testicular artery on the surface of the testis, which is beneath the scrotal skin, may also play an important role in cooling the arterial blood. The convolutions of the testicular artery on the surface of the testis, which are especially conspicuous in the monkey [26] and rat [4–6,8,19], may also contribute to this function. Surface coursing of the major veins of the testis may be important to further cool the venous blood. In the rat, the venous plexus is highly developed to form the intraalbugineal venous ramification, which cover the rostral portion of the testis and are continuous with the pampiniform plexus [17,19,20] (Fig. 21-4).

On the other hand, it is also well known that high concentrations of the male sex hormone, testosterone, are essential for the spermatogenesis within the testis and epididymis. Measurements of hormone concentrations in the blood of the testicular artery showed a significant high level of the hormone in the testicular artery, suggesting the presence of hormone transfer from the pampininform plexus to the testicular artery [25,29,30]. Thus, since the 1970s this countercurrent exchange system has been postulated by several investigators to play another physiological role, i.e., maintaining a high concentration of testosterone in the testis and epididymis through local transfer of hormones from veins to arteries [26,30,31].

3.2. Intratesticular Arterial System

In the rat, after encircling about two thirds of the testis, the testicular artery enters into the testicular blood vascular bed from the antero-superior aspect at the level of the caudal border of the caput epididymis and between the original segment of the subalbugineal veins, and then again descends postero-caudally in the superficial layer of the testis, giving off a dozen or more intratesticular arteries [20]. These intratesticular arteries repeatedly divide and give off smaller branches, radiate arteries (Fig. 21-5). Although the initial segments of the radiate arteries run in an indefinite relation to the direction of the surrounding seminiferous tubules, their smaller, terminal segments course through the inter-tubular space. The radiate arteries are distributed evenly throughout the testis (Fig. 21-5). In the rat, the intratesticular arteries are not end arteries but usually anastomose with each other at the level of the radiate arteries or their branches, forming arterio-arterial anastomoses [7,19, 20]. Furthermore, the radiate arteries or their branches are occasionally seen to flow into the radiate veins, forming arterio-venous anas-tomoses [19,20]. These spatial relationship between minute vessels were first unequivocally demonstrated by microdissection of vascular corrosion casts and scanning electron microscopic observations [19,20].

In the human testis, the main branches of the testicular artery divide into branches towards the rete testis and are called *centripetal arteries*. Many of the major branches of the centripetal arteries run in an opposite direction and are therefore called the *centrifugal arteries* [11]. The human intratesticular arterial vessels frequently show peculiar coilings, independent of the age of the subject [11]. In the mouse, small arteries after the initial branching of the intratesticular arteries present unique tight coils [7]. In the buffalo, the intratesticular arteries run centripetally, pursuing spirally coiled courses [13]. These coilings may also be related to the hemodynamics of the organ, although their physiological significance in intra-testicular arteries has not been proved.

There are considerably large differences in the configuration of the testicular arterial system; the coiling and convolution of the testicular artery in the testicular vascular pedicle in most mammals with scrotal testes so far investigated, encircling the run of the testicular artery around the testis; convolutions of the testicular artery on the testic-ular surface; and the coils or kinks in the branches of the intratesticular arteries, as observed in the human, buffalo, and mouse. When we com-

pare the testicular arterial systems among the mammalian species reported so far, we recognize that each species possesses one or more of these special vascular structures. Only a few excep-tions have been reported in several species with abdominal and inguinal testes, such as elephants and moles [5]. These suggest that each mam-malian species has developed many mechanisms to decrease arterial blood-fluctuations and blood-flow velocity within the testis.

3.3. Microvascular Architecture of the Testicular Capillary Plexus

Kormano and his collaborators investigated the microvascular organization of the testes of various mammals by microangiography and light microscopy, and stated that the basic organization in the human testicular parenchyma is essentially similar to that in the rat testis [8,11]. However, recent vascular corrosion casting-scanning elec-tron microscopic studies showed that there are considerable wide differences in the microvascular organization of the testicular parenchyma be-tween rodents [15,17–21] and humans [15,16].

In the rat, the radiate arteries repeat, further dividing and finally give rise to the intertubular arterioles that course in the intertubular connec-tive tissue columns (Figs. 21-5–21-7). Each of the intertubular arterioles sends off many capillaries of various sizes (Fig. 21-7). Müller classified them into two groups, intertubular and peritubular capillaries [6]. The intertubular capillaries course within the intertubular columns, which are the triangular columnar interstices among the three to four adjacent seminiferous tubules containing many interstitial tissues. The peritubular capil-laries are localized within the thin interstitial spaces between the two adjacent seminiferous tubules. In the rat and mouse, the peritubular capillaries predominantly run parallel to each other and course at right angles to the inter-tubular arterioles and capillaries, forming a rope-ladder-like pattern around the individual seminiferous tubules (Figs. 21-5 and 21-7). This rope-ladder-like capillary network is single layered and is shared by two adjacent tubules. The intertubular and peritubular capillaries anastomose with each other and gather into the

284

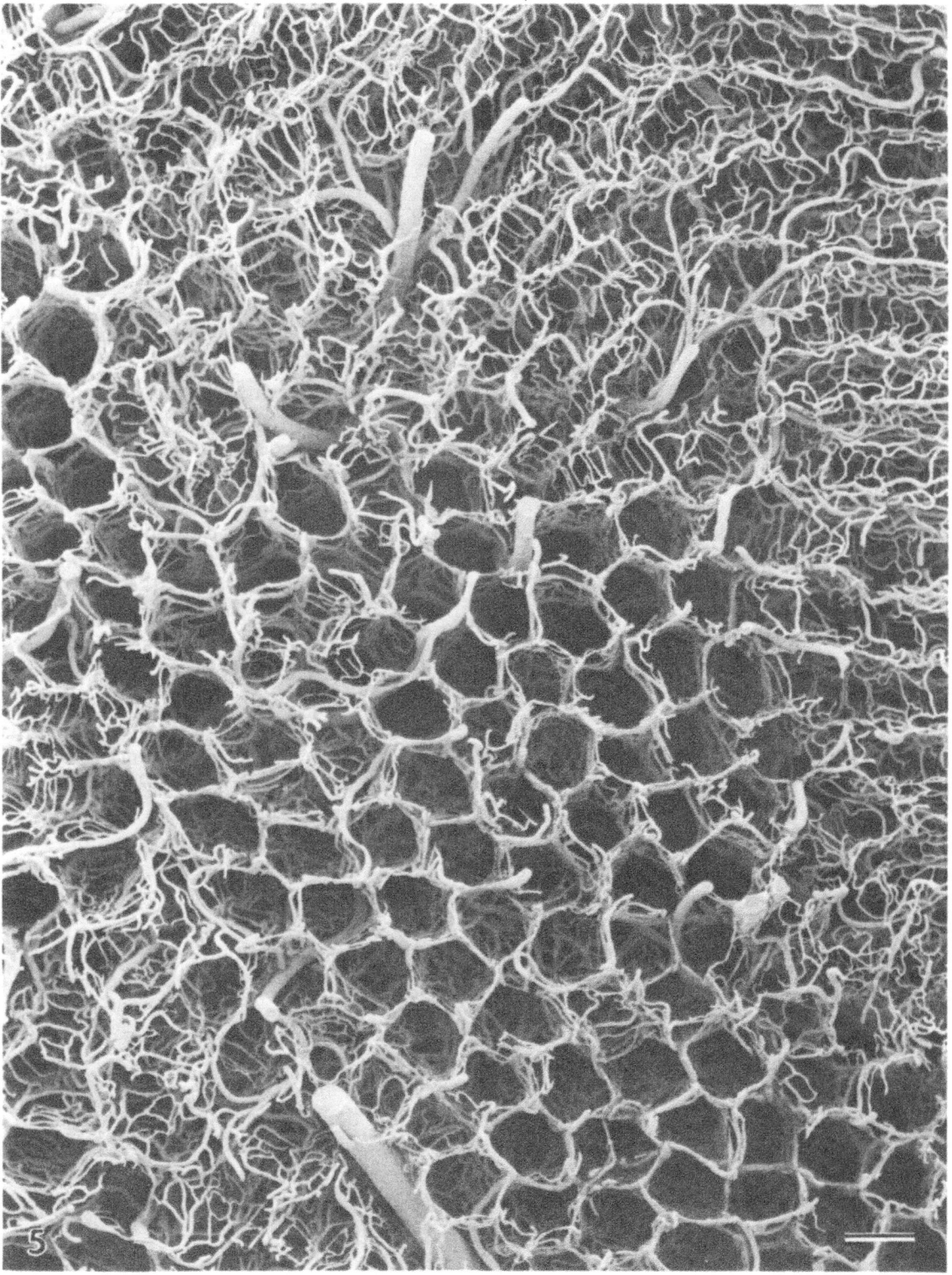

Figure 21-5. Frozen-cut vascular cast of the rat testis. Note honey-comb-like appearance of the cross-sectioned portions of the testicular capillary plexuses and rope-ladder-like appearances of the longitudinally sectioned portions. This photograph is composed of both the hexagonal or pentagonal columnar capillary networks. Bar = 300 µm.

intertubular venules in the intertubular connective tissue columns. In the rat and mouse, the seminiferous tubules are compactly packed and tend to course straight and parallel with each other [5,6], and one seminiferous tubule tends to be surrounded by five or six seminiferous tubules running in the same direction. Thus, the cross-sectioned portions of microvascular casts of the testicular parenchyma of the rat and mouse present honey-comb-like appearances (Figs. 21-5 and 21-6).

Most testosterone-releasing Leydig cells are localized within the intertubular regions, in the vicinity of the intertubular capillaries and venules. Throughout the testicular capillary vascular beds, there are no regional specialities in this microvascular architecture of the rat testis, except for the region of the rete testis. This suggests that within the rat testis a homogeneous microvascular environment, that is, a hormonal microenvironment, as in long seminiferous tubules, is provided by this honey-comb-like microvascular system composed of the two capillary systems.

The microvascular beds of the human testis has been described using microangiography and light microscopy as having a basic organization of capillary architecture consisting of intertubular capillaries within Leydig tissue and rope-ladder-like peritubular capillaries connecting these networks [11]. However, recent studies on human testicular microvascular organization by the vascular corrosion casting-scanning electron microscopy method present other views [15,16]. In the human testis, seminiferous tubules convolute and do not show parallel arrangements, and the peritubular interstitial tissues are considerably thicker than those of the rat and mouse [20]. Thus the microvascular beds of the human testis does not present a highly organized appearance, as do those of the rodents [20,21]. There are no clear differences between the intertubular and peritubular capillaries. Some capillaries run spirally around the tubules and others run among the tubules and form a considerably fine capillary plexus that is coarser than that of rodents. The rope-ladder-like pattern was not found in the capillary plexus of humans [15,16].

The peritubular and intertubular capillaries of the rat testicular parenchyma drain into the inter-tubular venules running in the intertubular connective tissue columns. The thick intertubular venules or their parent vein are consistently accompanied by a few intertubular capillaries that arise from the peritubular capillary network and flow into these thick intertubular venules or parent veins (Fig. 21-7) [20]. The perivenular intertubular capillaries are sometimes continuous with the periarteriolar intertubular capillaries.

The intertubular arterioles and venules never course along with each other within the same intertubular connective tissue space [19,20]. Usually intertubular arterioles and venules are alternatively distributed in the adjoining intertubular connective tissue columns. Furthermore, in many cases, each neighboring intertubular arteriole and venule are arranged parallel to each other and run in the same direction. These peculiar arrangements of the intertubular arterioles and venules may also contribute to forming a homogeneous vascular environment without any gradient along the seminiferous tubules.

The intertubular venules collecting the intertubular and peritubular capillaries drain into the radiate veins. These radiate veins further join into much larger collecting veins, running centrifugally toward the surface of the testis, and drain into the subalbugineal veins or intra-albugineal venous plexus [19,20]. The intertubular venules arising in the superficial layers directly drain into the branches of the sub-albugineal veins [20].

The rete testis of the human testis possesses a microvascular supply that is distinctly different from that of the seminiferous tubules [11]. The microvasculature of the rete testis consists of a poorly organized plexus [16]. The rete testis of the rat, adhering to the surrounding connective tissues and the tunica albuginea, is provided with few proper blood capillaries [20]. The rete testis gives off the efferent ductuli, which continue into ductus epididymis in the caput epididymis. The cystic protrusion and efferent ductuli of the rete testis are provided with few blood capillaries (Fig. 21-4) [20]. The blood vascular bed of this portion consists of randomly distributed capillaries and is continuous with the capillary bed of the caput epididymis. However, no capillary vascular connection is present between the capillary beds of these structures and that of the testis [20].

286

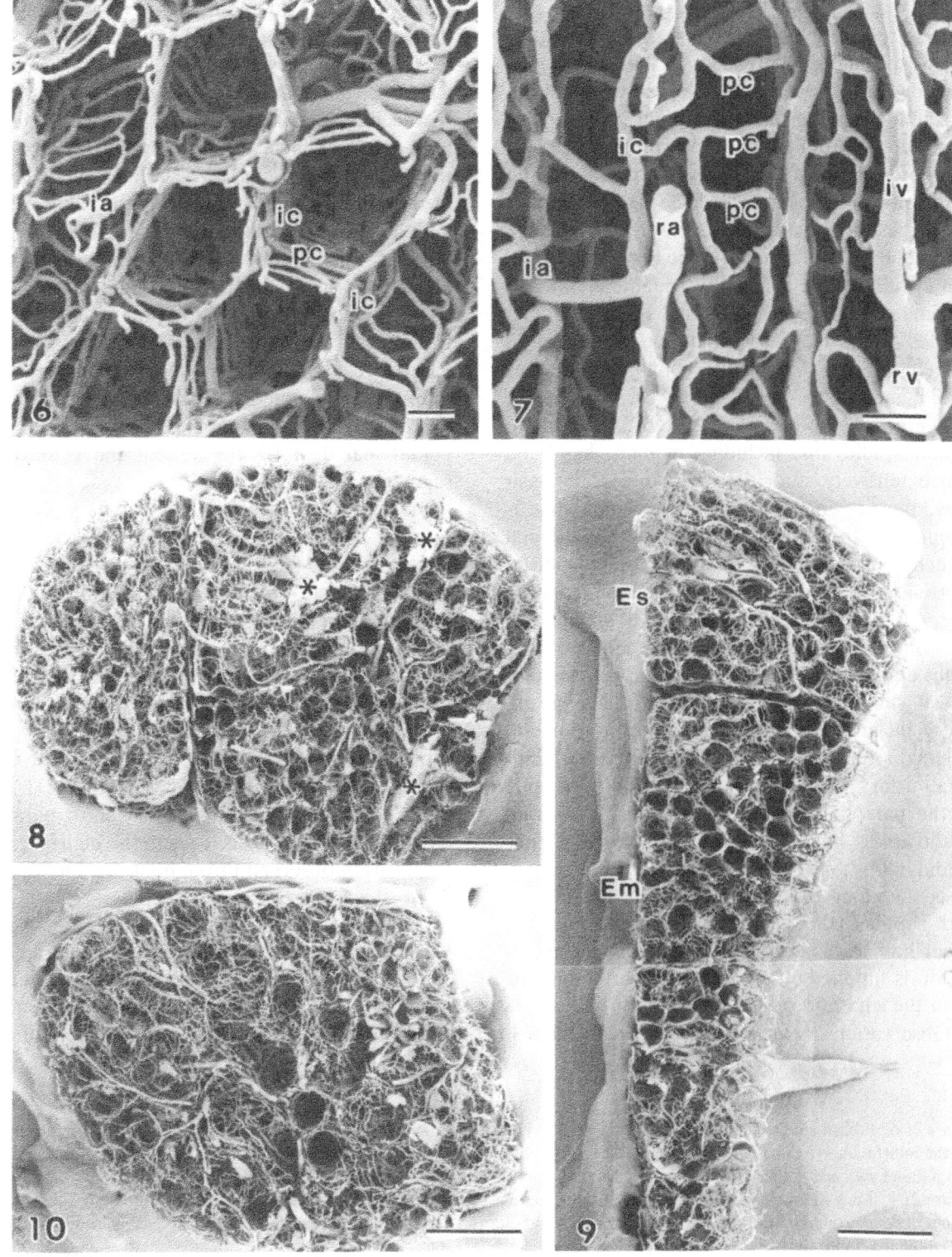

4. Microvasculature of the Epididymis

Further remarkable transformation and maturation of spermatozoa proceeds in the epididymis. It is now well known that the epididymal epithelium influences the luminal content by means of its secretory products and absorptive processes, and both of these functions contribute to the maturation of spermatozoa [32,33]. The epididymis is customarily subdivided into three regions, caput, corpus, and cauda. At the distal end of the cauda, the ductule gradually straightens out and continues as the ductus deferens. The epididymal tubules, which are the continuation of the ductuli efferentes, form a highly convoluted ductus epididymidis.

The gross blood vascular system of the mammalian epididymis, including in man, has been also extensively studied [7,9,12,16,21,23]. Basically, the proximal portion of the epididymis is supplied by the superior and inferior epididymal branches of the testicular artery, and the distal portion of the epididymis receives blood from the artery of the vas deferens, branches of which form anastomosing connections with arteries of the caput epididymis [4,17]. In the rat, the testicular artery, shortly after crossing over the iliac artery and vein, gives off the superior and inferior epididymal arteries (Fig. 21-2). The superior epididymal artery descends along the anterior side of the pampiniform plexus to supply the caput epididymidis and the upper half of the corpus epididymidis. The inferior epididymal artery descends along the posterior side of the plexus, and supplies the cauda epididymidis and the lower half of the corpus [17]. The cauda epididymidis is supplied by the deferential artery arising from the superior vesical artery [36].

Observations by angiography and light microscopy reported that the microvascular arrangement of the human epididymis is basically similar throughout the epididymis, although there are marked differences in the histological structure of various segments of the epididymal body and tail. In the human epididymis, the peritubular capillary network encircles each tubule cylindrically throughout the organ [16]. The microvasculature of the initial segment was not remarkably dense, and the regional variation of the microvasculature may be more limited, both in density and type. However, the present and other observations using a vascular corrosion casting-scanning electron microscopy method in mammalian species — mouse [21] and boar [23] — show distinct regional differences in microvascular network patterns of the subepithelial capillary networks of the epididymal ductules throughout the epididymal segments (Figs. 21-8–21-13).

In the mouse, two forms of capillary architecture are described in the testis and excurrent duct system [21]. The microvascular system of the middle segment of the epididymis consists of the testicular type, in which longitudinal, intertubular vessels interconnected with peritubular capillaries in a rope-ladder-like fashion are present. In the initial and terminal segment of the epididymis, the other deferential type, in which the peritubular capillaries form a network encircling each tubule in the subepithelial layer, is seen [21]. The initial segment is endowed with almost parallel, circular capillaries. This leads to the highest capillary density observed throughout the epididymis.

In the rat, epididymal capillaries follow a tortuous course and form a relatively dense network surrounding epididymal ductules. In the caput and corpus epididymidis, peritubular capillaries course tortuously and repeat branching and anastomosing to form a cylindrical, subepithelial

←

Figure 21-6. Frozen-cut vascular cast of the rat testis. The radiate arteries give off the intertubular arterioles (ia), which further give off the intertubular (ic) and peritubular capillaries (pc). Many peritubular capillaries run perpendicular to the intertubular vessels and parallel with each other, thus presenting a rope-ladder-like appearance. Bar = 100 µm.

Figure 21-7. Longitudinally frozen-cut aspect of the columnar capillary network, in which a seminiferous tubule resides in intact tissues. The intertubular arterioles (ia) give off the intertubuler (ic) and peritubular (pc) capillaries. These capillaries flow into the intertubular venules (iv). ra = radiate arteries; rv = radiate veins. Bar = 50 µm.

Figure 21-8. Frozen-cut sectional view of a vascular cast of the rat caput epididymidis. * = leaked resin. Bar = 1 mm.

Figure 21-9. Frozen-cut sectional view of a vascular cast of the last segment of the caput epididymis (Es) and corpus epididymis (Em) of the rat. Note the tube-like capillary plexuses compactly packed in the corpus epididymis. Bar = 1 mm.

Figure 21-10. Frozen-cut sectional view of the vascular cast of the rat cauda epididymidis. Bar = 1 mm.

288

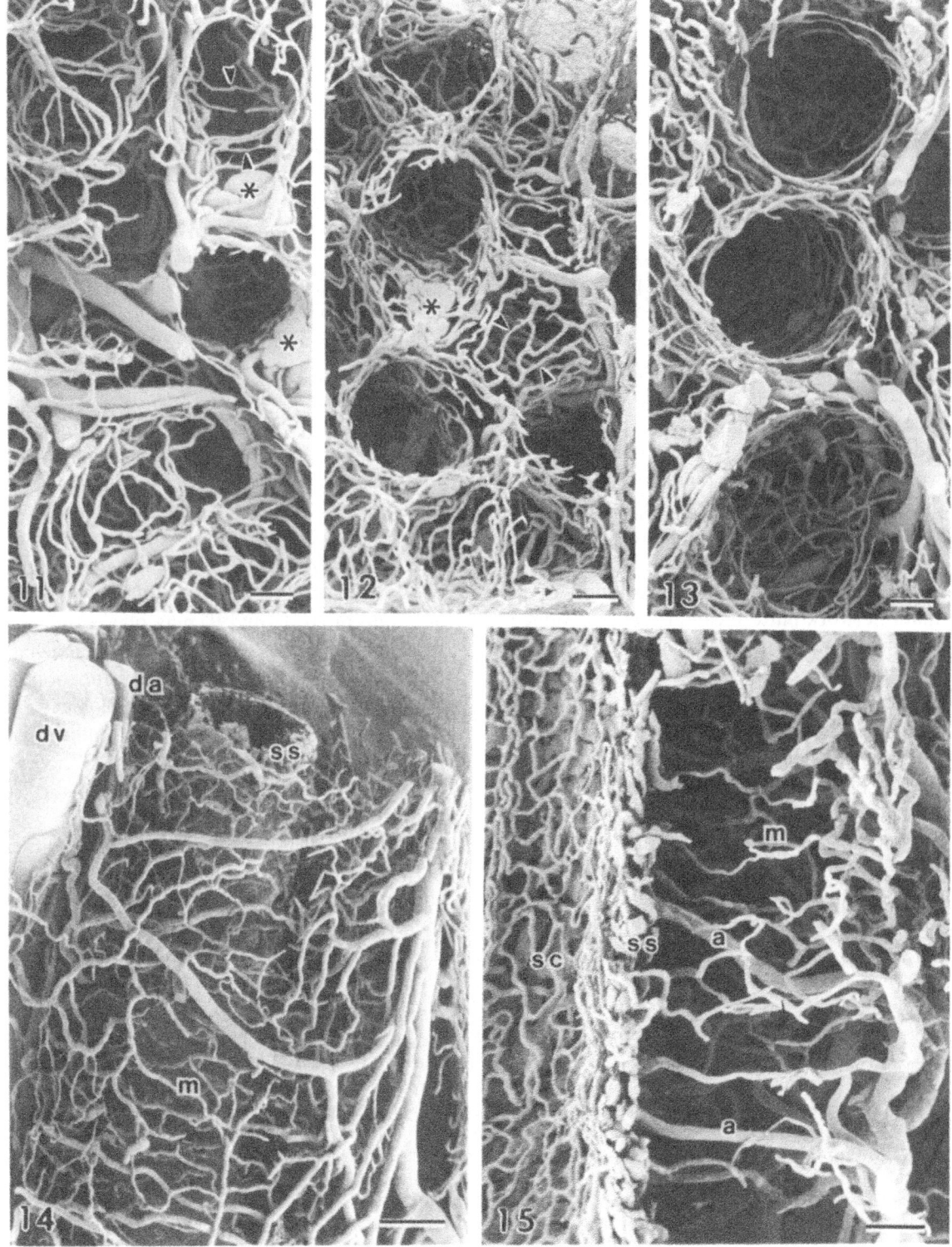

capillary network encircling each epididymal tubule. The slight abundance of circularly running segments gives these capillary networks with an incomplete rope-ladder-like appearance (Figs. 21-8, 21-9, 21-11, and 21-12). In the caudal portion, as the muscular coat of the epididymal ductules thickens, the capillaries tend to form two separate microvascular networks around the ductules. One network is located in the interductular connective tissue. The other is the subepithelial capillary network, which is circular in shape instead of the polygonal pattern usually found in the testis (Figs. 21-10 and 21-13). The subepithelial capillaries of the cauda epididymidis run somewhat straight to form a capillary network that is freely branching and anastomosing. Each ductule has a proper capillary network around it and does not share this subepithelial capillary network with the adjacent ductules. In the cauda epididymidis, circularly running segments are not conspicuous, thus giving an appearance similar to that of the vas deferens (Fig. 21-13).

In the boar a clear regional variation of the capillary network pattern can also be reproduced by the corrosion casting-SEM method [23]. The initial segment is endowed with almost parallel, circular capillaries. This leads to the highest capillary density observed throughout the epididymis. As the density decreases, this annulate feature is progressively lost and the cauda possesses a polygonal capillary network [23].

In general, the capillary network pattern and density of capillaries of a given region are considered to reflect the metabolical activity of the region. Such striking changes observed in the epididymal peritubular, subepithelial capillary network pattern may be related to the functional changes in the epididymal segments. Recent biochemical and histochemical studies have revealed these regional differences in secretory activity [33,34].

In the rat, the epididymal capillaries are drained into venules that run centrifugally in the epididymis. These venules finally appear on the surface of the epididymis and gather into the superior and inferior epididymal veins. Vascular corrosion casting-SEM observations have revealed that the superior and inferior epididymal veins ascend along their respective arteries and enter into the pampiniform plexus [17]. Over the course of the epididymal arteries, each of them forms a thin venous plexus (epididymal venous plexus) that envelops the epididymal arteries. Interestingly, the inferior epididymal venous plexus is so well developed that it entirely surrounds the proximal half of the inferior epididymal artery. This epididymal venous plexus, which has not been demonstrated by angiography and light microscopy [9], may play a role in cooling and the androgen-concentrating mechanism, as discussed in the section on the function of the pampiniform plexus. In a physiological experiment on the rat in which the epididymis was surgically moved to the abdominal cavity, the maturation of the spermatozoa in the cauda epididymis was inhibited, although it was not affected in the caput and corpus epididymidis [35]. Temperature measurements of the unstrained conscious rat showed significantly lower temperatures in the epididymis than those in the

←

Figure 21-11. Frozen-cut sectional view of the vascular cast of the caput epididymidis of the rat. Circularly running segments (arrowheads) of the peritubular capillaries show an incomplete rope-ladder-like appearance. * = leaked resin. Bar = 100 μm.

Figure 21-12. Frozen-cut sectional view of the vascular cast of the corpus epididymidis of the rat. Peritubular capillaries course tortuously to form tube-like subepithelial capillary networks. Circularly running segments (arrowheads) of the peritubular capillaries are relatively more conspicuous. * = leaked resin. Bar = 100 μm.

Figure 21-13. Frozen-cut sectional view of the vascular cast of the cauda epididymidis of the rat. Subepithelial capillaries form a freely branching and anastomosing network with an isotropic meshwork pattern. A dense capillary plexus is developed in the interductular connective spaces. Bar = 100 μm.

Figure 21-14. Scanning electron micrograph of the vascular cast of the rat vas deferens. Note the well-developed sinusoidal network in the lamina propria (ss) and the surrounding loose capillary plexus in the smooth muscle layer (m). dv = vas deferens vein; da = vas deferens artery. Bar = 200 μm.

Figure 21-15. Longitudinally frozen-cut view of the vascular cast of the rat vas deferens (longitudinal section). Arterial vessels (a) penetrate through the loose capillary plexus of the muscular layer (m) and break into the sinusoidal network (ss) in the lamina propria. Note the fine-meshed subepithelial capillary network (sc). Bar = 100 μm.

abdominal cavity [2]. These may suggest that the inferior epididymal venous plexus, by which the upper segments of the inferior epididymal artery is closely covered, may play a role in the thermal control of the cauda epididymis.

5. Microvasculature of the Vas Deferens

The microvascular architecture of the rat vas deferens shows clear regional changes. The proximal third of the vas deferens is composed of an adventitial layer, a vascular network in the muscle layer, and the subepithelial capillary network [9,36,37]. In the distal two thirds of the duct, a well-developed sinusoidal network is formed in the lamina propria, which is localized between the muscle layer and the luminal epithelium, thus showing a peculiar vascular patten (Figs. 21-14 and 21-15) [36,37]. In the mouse, in this layer a prominent venous plexus is formed. The vascular architecture of the mouse vas deferens is fundamentally similar to that in the rat [21].

In the human vas deferens, Kormano and Reijonen [12], using microangiography, reported two laryers of microvascular networks, the outer microvascular network surrounding the smooth muscle layer and the inner network located immediately beneath the duct epithelium. A sinusoidal plexus in the subepithelial lamina propria, as observed in the rat, has not been reproduced [12].

The functions of the well-developed subepithelial capillary network may be related to the secretory/absorptive functions of the epithelial cells of the vas deferens [37].

The significance of the well-developed sinusoidal plexus has not been revealed. Some investigators suggest that it may function in the diffusion of materials out of the duct [21,36]. However, at the beginning of the venous vessels draining the sinusoidal network, constricted imprints, suggesting the presence of sphincterlike structures, have been observed in casts [37]. Constrictions caused by these sphincterlike structures, or rapid and tremendous influx of blood in the sinusoidal network, may induce engorgement, and the sinusoidal layer may work as a tough and elastic wall that maintains ductal structure

[37]. The sinusoidal plexus also might be resistant against the pressure caused by the contraction of ejaculation, which would make it possible to transport the luminal fluid.

6. Concluding Remarks

The vascular corrosion casting-scanning electron microscopy method has provided us a new technique for observing the microvascular organization of the testis and its excurrent duct system. We can microdissect vascular casts and selectively demonstrate the arterial supply and venous drainage of interesting portions, and observe their characteristic microvascular network pattern three-dimensionally and extensively with scanning electron microscopy.

In mammals with scrotal testes, including man and rat, venous vessels draining the testis form a tight plexus, pampiniform plexus, and entirely envelop the testicular artery, which repeats convolution and coiling in this segment. In the testicular vascular system, many other special structures with wide species variations have been found, such as coiling and convolution of the testicular artery, and the intraalbugineal venous plexus in the rat. These specialized vascular structures provide morphological bases for close contact of the artery and venous vessels, slowing down the blood velocity, and make it possible to function as an effective countercurrent heat exchange apparatus; it might be a condenser of testosterone that is secreted in the testis and is required to be kept at a high concentration level in the testis and epididymis.

The microvascular system within the testis shows species differences. In the rat and mouse, seminiferous tubules are vascularized by the hexagonal prismatic capillary plexuses, consisting of rope-ladder-like peritubular capillaries and intertubular capillaries, while in the human testes, also, considerably fine capillary plexuses are formed, but no highly organized network pattern is found. This difference in vascular pattern may be due to the differences in the method of packing of the seminiferous tubule and in the amount of interstitial tissues.

Regional differences in the microvascular architecture of the testis and excurrent duct system can be well demonstrated by the vascular corrosion casting-scanning electron microscopy method. There are also species differences in this regional transition of the pericapillary network pattern. The regional differences in the microvascular architecture may reflect the regional differences in their functions. However, despite considerable advances in our knowledge of epididymal secreting and absorbing functions, the precise relationship between them remains to be revealed by further investigation.

References

1. Collins P, Lacy D. Studies on the structure and function of the mammalian testis. II. Cytological and histochemical observations on the testis of the rat after a single exposure to heat applied for different lengths of time. *Proc R Soc Lond* [Biol] 172:17–38, 1969.

2. Brooks DE. Epididymal and testicular temperature in the unrestrained conscious rat. *J Reprod Fertil* 35:157–160, 1973.

3. Jones TM, Anderson W, Fang VS, Landau RL, Rosenfield RL. Experimental cryptorchidism in adult male rats: Histological and hormonal sequelae. *Anat Rec* 189:1–28, 1977.

4. Harrison RG. The distribution of the vasal and cremasteric arteries to the testis and their functional importance. *J Anat* 83:267–282, 1949.

5. Setchell BP. Testicular blood supply, lymphatic drainage, and secretion of fluid. In: Johnson AD, Gomes WR, Van Demark NL (eds.), *The Testis*. Academic Press, New York, pp 101–218, 1970.

6. Müller I. Kanälchen- und Capillararchitektonik des Rattenhodens. *Z Zellforsch Mikroskop Anat* 45:522–537, 1957.

7. Chubb C, Desjardins C. Vasculature of the mouse, rat and rabbit testis-epididymis. *Am J Anat* 165:357–372, 1982.

8. Kormano M. An angiographic study of the testicular vasculature in the postnatal rat. *Zeitschr Anat Entwickl* 126:138–153, 1967.

9. Kormano M. Microvascular structure of the rat epididymis. *Ann Med Exp Fenn* 46:113–118, 1968.

10. Sasano N, Ichijo S. Vascular patterns of the human testis with special reference to its senile changes. *Tohoku J Exp Med* 99:269–280, 1969.

11. Kormano M, Suoranta H. Microvascular organization of the adult human testis. *Anat Rec* 170:31–40, 1971.

12. Kormano M, Reijonen K. Microvascular structure of the human epididymis. *Am J Anat* 145:23–32, 1976.

13. Dhingra LD. Angioarchitecture of the buffalo testis. *Anat Anz* 146:60–68, 1979.

14. Murakami T. Application of the scanning electron microscope to the study of the fine distribution of the blood vessels. *Arch Histol Jpn* 32:445–454, 1971.

15. Takayama H, Tomoyoshi T. Microvascular architecture of rat and human testis. *Invest Urol* 18:341–344, 1981.

16. Suzuki F, Nagano T. Microvasculature of the human testis and excurrent duct system. Resin-casting and scanning electron-microscopic studies. *Cell Tissue Res* 243:79–89, 1986.

17. Ohtsuka A. Microvascular architecture of the pampiniform plexus-testicular artery system in the rat: A scanning electron microscope study of corrosion casts. *Am J Anat* 169:285–293, 1984.

18. Kikuta A, Ohtsuka A, Ohtani O, Murakami T. Microvascularization of endocrine glands as studied by injection-replica SEM method. In: Motta M (ed.), *Ultrastructure of Endocrine Cells and Tissues*, Martinus Nijhoff Publishers, Boston, pp 313–320, 1984.

19. Weerasooriya TR, Yamamoto T. Three-dimensional organization of the vasculature of the rat spermatic cord and testis. A scanning electron-microscopic study of vascular corrosion casts. *Cell Tissue Res* 241:317–323, 1985.

20. Murakami T, Uno Y, Ohtsuka A, Taguchi T. The blood vascular architecture of the rat testis: A scanning electron microscopic study of corrosion casts followed by light microscopy of tissue sections. *Arch Histol Cytol* 52:151–172, 1989.

21. Suzuki F. Microvasculature of the mouse testis and excurrent duct system. *Am J Anat* 163:309–325, 1982.

22. Hees H, Leiser R, Kohler T, Wrobel KH. Vascular morphology of the bovine spermatic cord and testis. I. Light- and scanning electron-microscopic studies on the testicular artery and pampiniform plexus. *Cell Tissue Res* 237:31–38, 1984.

23. Stoffel M, Kohler T, Friess AE, Zimmermann W. Microvasculature of the epididymis in the boar. *Cell Tissue Res* 259:495–501, 1990.

24. Harrison RG, Weiner JS. Vascular patterns of the mammalian testis and their functional significance. *J Exp Biol* 26:304–316, 1949.

25. Dahl EV, Herrick JF. A vascular mechanism for maintaining testicular temperature by counter-current exchange. *Surg Gynecol Obstet* 108:697–705, 1959.

26. Dierschke DJ, Walsh SW, Mapletoft RJ, Robinson JA, Ginther OJ. Functional anatomy of the testicular vascular pedicle in the rhesus monkey: Evidence for a local testosterone concentrating mechanism. *Proc Soc Exp Biol Med* 148:236–242, 1975.

27. Waites GMH, Moule GR. Blood pressure in the internal spermatic artery of the ram. *J Reprod Fertil* 1:223–229, 1960.

28. Osman DI, Tingari MD, Moniem KA. Vascular supply of the testis of the camel (*Camelus dromedarius*). *Acta Anat* (Basel) 104:16–22, 1979.

29. Hundciker M, Keller L. Die Gefässarchitektur des menschlichen Hodens. *Gegenbaurs Morph Jahrbuch* 105:26–73, 1963.

30. Free MJ, Jaffe RA, Jain SK, Gomes WR. Testosterone concentrating mechanism in the reproductive organs of the male rat. *Nature New Biol* 244:24–26, 1973.

31. Ginther OJ, Mapletoft RJ, Zimmerman N, Meckley PE, Nuti L. Local increase in testosterone concentration in the

testicular artery in rams. *J Anim Sci* 38:835–837, 1974.

32. Amann RP. Function of the epididymis in bulls and rams. *J Reprod Fertil* [Suppl.] 34:115–131, 1987.

33. Cooper TG, Yeung CH, Bergmann M. Transcytosis in the epididymis studied by local arterial perfusion. *Cell Tissue Res* 253:631–637, 1988.

34. Dacheux F, Dacheux JL. Androgenic control of antagglutinin secretion in the boar epididymal epithelium: An immunocytochemical study. *Cell Tissue Res* 255: 371–378, 1989.

35. Bedford JM. Influence of abdominal temperature on epididymal function in the rat and rabbit. *Am J Anat* 152:509–522, 1978.

36. Hamilton DW, Cooper TG. Gross and histological varia-

tions along the length of the rat vas deferens. *Anat Rec* 190:795–810, 1978.

37. Ohtani O, Gannon BJ. The microvasculature of the rat vas deferens: A scanning electron and light microscopic study. *J Anat* 135:521–529, 1982.

Author's address:
Dr. Akio Kikuta
Department of Anatomy
Okayama University Medical School
2-5-1 Shikata-cho
Okayama 700
Japan

CHAPTER 22

Blood Vessels of the Eye and their Changes in Diabetes

ANDRZEJ W. FRYCZKOWSKI

1. Introduction

The ocular vasculature has been the subject of detailed study in the past [2,6,7,13,16,21–24, 27–29,31,34,37–39,41,43,44,46,47], however, in several areas controversies among researchers still exist. For example, does the choroidal vasculature extend into the lamina cribrosa (LC)? What is the exact blood supply to the lamina cribrosa, the region crucial for optic disc changes, where many disorders occur? Further controversies include the choriocapillaris structure, the location of the feeding arteriole(s) and collecting venules (central position of arteriole and peripheral venules) [22,24,37–39,46], or vice versa [27,43], or the mixed distribution of these vessels changing in different areas [13], and the submacular choriocapillaris (with regard to the presence [6,39] or absence [13,23,37,38,46] of the submacular artery). Some authors have described the presence of regular vascular dilations in the equatorial choroid [2,13], but they differ in their interpretation of the role of these vascular structures, which Ashton [2] considered to be a simple flattening of the veins by the underlying artery. On the other hand, these vascular dilatations undergo significant changes in long-standing insulin-dependent diabetes mellitus (IDDM) [14,20], and their role could be far more important than implicated by Ashton [2,3]. Some authors have attempted to simplify the choroidal structure and to present it in schematic form [22,24,38,46]. The choroid does not fit such schematics, which differ from the real angio-architecture. In addition, data from animal studies, especially monkey, was in the past directly applied to data from human studies [24,34,37]. The human choroid is a highly complicated and variable structure, and classifying it as a "lobular" structure is true with regard to only part of the choroid. Some of these controversies were partially solved by dynamic angiographic vascular study, in which fluorescein was used for retinal or choroidal vascular study [1,12,32, 36] and indocyanine green high-speed digital cineangiography (ICG) was used for choroidal vasculature study [8,9,26]. However, there are some limitations for in vivo and postmortem study [11,14,36,45], and the results of postmortem study do not always support these first findings. This creates a forum for more future correlative study, in which using better, more reliable techniques could provide outstanding data and solve some controversies. One of the best presently known techniques for microvascular study is the vascular cast and scanning electron microscopy (SEM) technique [13–20, 31,33,34,37,40,43,46]. Although this technique has some limitations, as do the others, the SEM three-dimensional images of the ocular vasculature are far superior to others, including some of the injected techniques used in the past [2,6, 21,39,44]. In addition, this vascular cast and SEM technique allows one to distinguish between the arterial and venous sides of the choroidal circulation. In this way, the most controversial points regarding which part is venous and which part is arterial in such a complicated vascular structure

Motta, P.M., Murakami, T., and Fujita, H. (eds.), Scanning Electron Microscopy of Vascular Casts: Methods and Applications.

294

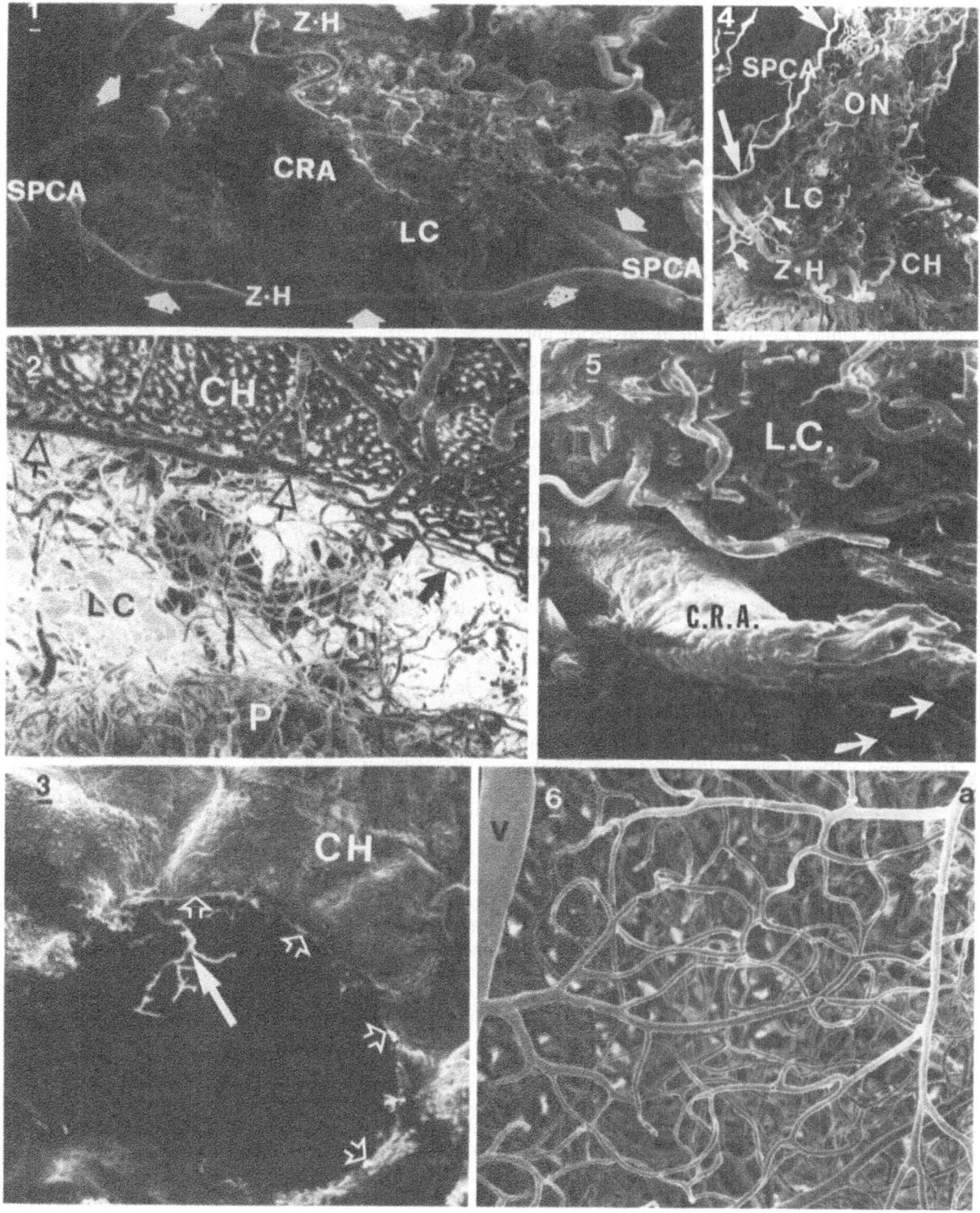

as the choroid can be solved. In this chapter some of the SEM images of normal as well as the abnormal ocular vasculature are presented, and the interpretation of their significance is discussed.

2. Normal Ocular Vasculature

2.1. Lamina Cribrosa and Peripapillary Area

The lamina cribrosa (LC) is supplied by: (1) the short posterior ciliary artery branches (SPCA), direct to the lamina cribrosa (in its posterior and medial parts), or from the so-called Zinn-Haller scleral arterial circle (Fig. 22-1); (2) the medial and anterior parts of the lamina cribrosa receive their blood supply from branches from the Zinn-Haller circle, as well as from the third division of the short posterior ciliary arteries (SPCA), which participate in forming the peripapillary choroid (Fig. 22-2); (3) the posterior part of the lamina cribrosa receives its blood supply from the centripetal branches of the pial arterioles and the longitudinal intraneural capillary network. A detailed description of the LC vasculature was been published elsewhere [16].

The peripapillary choroid anatomically consists of a dense capillary network with a honeycomb pattern and without clear differentiation of the lobuli (Fig. 22-3). The Juxtapapillary margin of the choroid ends abruptly on the border of the optic nerve head. The feeding arterioles on the posterior side of the choriocapillaris have arteriolo-arteriolar anastomoses. Several of the branches from the larger choroidal branches pass the juxtapapillary margin from the posterior to the anterior part of the optic disc, approach the prelaminar area, and contribute to the blood supply to this region. There are recurrent branches from the larger choroidal vessels (third or fourth division of the SPCA) to the pia of the optic nerve (Fig. 22-4). All these relationships to the lamina cribrosa area are presented in Diagram 22-1.

2.1.1. The retinal vasculature.
The central retinal artery (CRA) and vein (CRV) are the major vessels participating in the retinal and anterior optic nerve blood supply. The CRA originates from the ophthalmic artery and more often from branches of the ophthalmic artery and, after an interorbital course, approaches the optic nerve, where it is located in the axial part, going forward toward the optic disc. The CRA has interorbital, intervaginal, and intraneural branches that participate in the optic nerve capillary network. The CRA has no branches in the lamina cribrosa region, however, branches from the CRA are observed directly behind the lamina cribrosa (Fig. 22-5). Usually at the level of the optic disc (OD), the CRA branches into four main branches and supplies the inner and medial part of the retina up

Figure 22-1. The arterial circle of Zinn-Haller (Z-H; arrows) is formed by branches of the short posterior ciliary arteries (SPCA) surrounding the optic nerve capillary plexus in the lamina cribrosa region (LC). CRA = central retinal artery. SEM photomicrograph of the vascular cast. ×62. (From Fryczkowski et al. [16], with permission from Dr. Junk Publishers.)

Figure 22-2. Juxtapapillary margin of the choroid — open arrows (posterior view). Some branches from the choroidal vessels extend to the prelaminar area (arrows). P = pial vessels; LC = lamina cribrosa; CH = choroid. ×65.

Figure 22-3. Peripapillary choriocapillaris (anterior view). The optic nerve has been removed and open arrows indicate the juxtapapillary choroidal margin with arteriolo-arteriolar anastomoses. The solid arrow points to a larger choroidal venule that supplies the lamina cribrosa. CH = choroid. ×54.

Figure 22-4. Vascular cast of the distal optic nerve (ON), the lamina cribrosa region (LC), and peripapillary choroid (CH) viewed from above. A branch of the short posterior ciliary artery (SPCA) circles around the optic nerve, forming the scleral arterial circle of Zinn-Haller (Z-H). Recurrent pial branches from the scleral portion of the SPCA (large arrows) coursing within the previously digested pial sheath can be traced to their distal supply, the retrobulbar optic nerve capillary plexus. Recurrent choroidal branches (small arrows) can be traced to the lamina cribrosa region. ×30. (From Fryczkowski et al. [16], with permission from Dr. Junk Publishers.)

Figure 22-5. The central retinal artery (C.R.A.) in its course through the lamina cribrosa (L.C.), posterior view. The optic nerve was cut off. Intraneural branches from the C.R.A. to the optic nerve capillary network originate just behind the L.C. (arrows). ×170.

Figure 22-6. SEM photomicrograph from human vascular cast: retinal capillaries, anterior view. The area approximately 1 mm from the optic disc closes with the temporal-superior arcade. a = feeding arteriole; v = venule. A three- to four-layer capillary network is present. ×138.

296

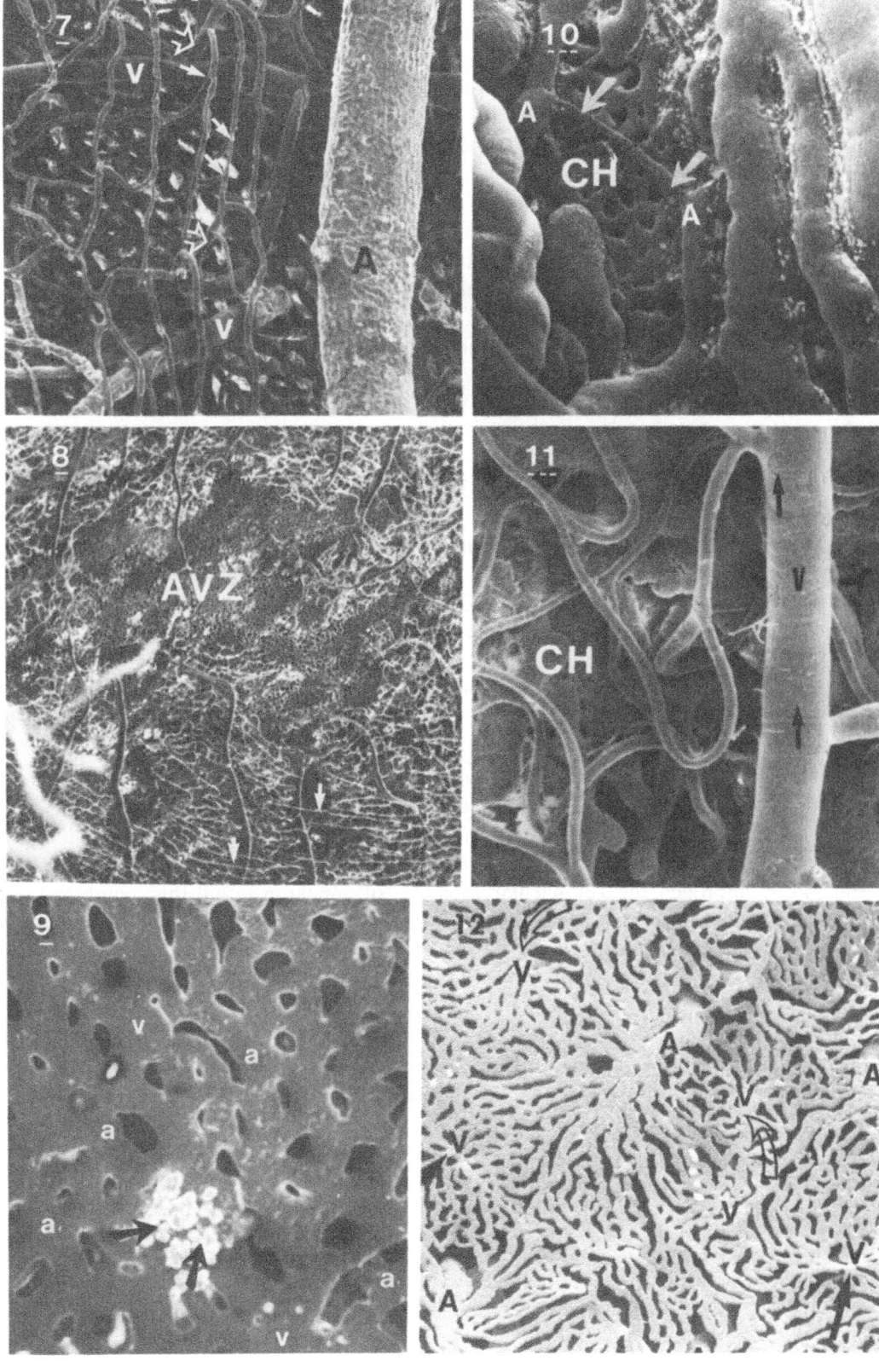

to the nuclear layer. Close to the optic disc there are three or four layers of capillary networks (Fig. 22-6). A special system, called the *radial peripapillary capillaris* (RPC), can be observed along the temporal and nasal arcades at some distance from the optic disc. This RPC has a special arrangement, with both sides interconnected with the major vein trunks, while the feeding arterioles reach these capillaries in their center (Fig. 22-7). The kind of anatomical arrangement of the capillary network could be more susceptible to increased intraocular pressure, which can be an important cofactor in understanding the pathophysiology of Bejrum's scotoma in glaucoma.

2.2. Submacular Area

The retinal arterioles from temporal arcades give rise to the two-layered arrangement of the perimacular capillaries, which form an avascular zone (AVZ) ranging from 350 to 650 μm in diameter. This zone, which in humans is mostly round, is elongated in the horizontal direction in primates (Fig. 22-8). As observed in the monkey, the RPC are located as close as 100 μm away from the AVZ.

The choriocapillaris in the submacular area is similar to the peripapillary area in that it has the same dense honeycomb pattern and lacks differentiation of the lobuli (Fig. 22-9). From the posterior scleral side, the feeding arterioles from the SPCA approach the submacular area centripetally. The arteriolo-arteriolar anastomoses in this area are frequent (Fig. 22-10).

In addition to the main openings, the feeding arterioles in this choriocapillaris, have additional openings. Thus, the arteriolar openings outnumber the venular ones in this area.

2.3. Posterior Pole

The posterior pole retinal vascular arrangement is interconnected with the temporal and nasal arcades, and forms mostly two layers of capillaries (Fig. 22-11).

The choroidal angioarchitecture in this region has several important features, such as the lobular arrangement of the choriocapillaris (Fig. 22-12), the segmental vascular supply of these lobuli (Fig. 22-13), additional vascular channels on venules and large veins levels, and several different anastomoses — venulo-venular, vein-vein, arteriolo-arteriolar, and arteriolo-venular (Figs. 22-14 and 22-15; also see Fig. 22-22). The presence of direct arterio-venous anastomoses allows arterial blood shunts to the veins. Because of this, it is easier to understand the higher oxygenation of the venous side of the choroid compared to other places of the body.

The number of choriocapillaris interconnections markedly decreases outside of the submacular zone. This allows some of the feeding arterioles and collecting venules to be found in the plane of the choriocapillaris. The balance of the feeding arterioles and collecting venules in place of the choriocapillaris have a branching type of appearance and this, combined with the focal type of opening perpendicular to the choriocapillaris, gives this area its lobular structure.

Figure 22-7. Radial peripapillary capillaries (RPC) in a characteristic parallel arrangement of RPC between two venous trunks (open arrows). The feeding arterioles join these RPC at a 90° angle (arrows). A = arteriole; V = venule. ×165.

Figure 22-8. SEM photomicrograph of the monkey perimacular retinal vasculature and submacular choriocapillaris. Anterior view. Radial peripapillary capillaries (arrows) are seen within 100 μm from the avascular zone (AVZ). The AVZ has a more elongated and horizontally oriented oval shape. ×26.

Figure 22-9. Human submacular choriocapillaris, anterior view, with a dense, honeycomb appearance in the capillaries. a = arteriolar openings; v = venular openings. Artifacts (arrows). ×210.

Figure 22-10. Human SEM photomicrograph from a vascular cast. Submacular choriocapillaris (CH), posterior view, from 22-year-old male. arrows = feeding arteriolo-arteriolar anastomoses; A = arteriole. ×510. (With permission from S. Karger AG, *Acta Anatomica* 132:265–269, 1988.)

Figure 22-11. SEM photomicrograph from the human ocular vascular cast: two layers of the retinal capillaries, anterior view. V = vein; arrows = endothelial cells; CH = background of the choriocapillaris. ×360.

Figure 22-12. Fifty-five-year-old male: equatorial area, anterior view, showing the lobular pattern of the choriocapillaris. The two main venular (V) opening are "round", 90° degrees to the choriocapillaris (solid arrow), and "in plane" with the choriocapillaris (open arrow). A = arteriolar openings. ×64. (From Fryczkowski [14], with permission from Dr. Junk Publishers.)

RADIAL ARTERIOLAR VASCULAR SUPPLY TO THE LAMINA CRIBROSA

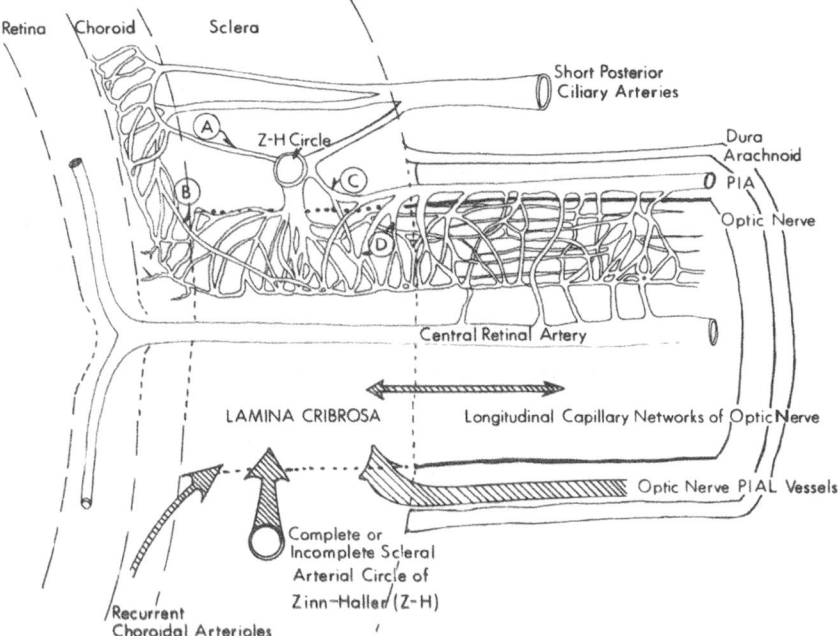

(A) Choroidal Branches from Z-H Circle   (B) Recurrent Choroidal Branches to the Lamina Cribrosa   (C) Recurrent Branches to PIA from Z-H Circle   (D) Centripetal PIAL Arterioles

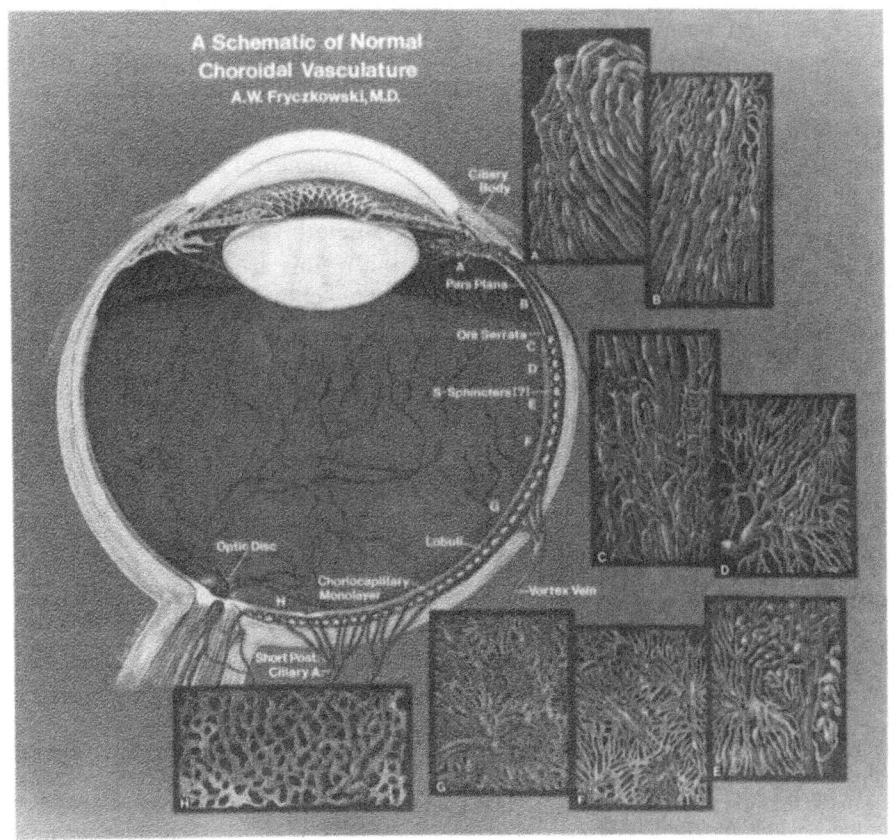

The number of venular openings outnumber the arteriolar openings in this area (Fig. 22-16).

The position of the feeding arterioles and collecting venules within the lobuli changes. In the posterior pole sometimes the central portion of the lobule is occupied by the feeding arteriole. However, this is not a rule; many times in the center of lobuli it is possible to find one or more collecting venules, while at the same time the feeding arteriole(s) are located more peripherally, supplying a few lobuli which form the "choroidal unit." One larger vein obtains blood from several collecting venules (Fig. 22-17).

In commenting upon the above results, it appears that the anatomical observations are somewhat in conflict with the present interpretation of the fluorescein angiography results. However, the explanation for the more central occurrence of the fluorescein in early-stage angiography in the posterior pole is the location of the feeding arterioles and collecting venules in the same plane as the choriocapillaris. In addition, many arteriolar openings are perpendicular to the posterior surface of the choriocapillaris. Thus, the blood that approaches the lobuli has both a more direct and elongated path to the choriocapillaris. One can explain that the fluorescein that approaches the lobuli from the peripherally located arterioles fill the collecting venule, and the concentration of the fluorescein is seen flowing in a retrograde direction from the center to the periphery. At the same time, in the other lobuli the fluorescein starts to fill the lobuli from the centrally located feeding arteriole. Results from anatomical observations may be used toward explanation of functional blood circulation in the choroid. According to our findings, one "main" arteriolar opening supplies the "choroidal unit", which consist from a few lobuli. On the FA images, this will look like a centrally filled "lobuli", which actually is the choroidal unit. Thus, on FA photographs, the size of these "lobuli" looks much larger than a true anatomical

lobuli. This is probably a source of mistake done previously by other author in the study on this topic [24].

Additional or supportive smaller arteriolar openings are noted across lobular part of the choroid. Their existence may explain faster resolving ischemia in some choroidal lobuli. This study did not support Hayreh's theory [24] in which the central feeding arteriole and peripheral collecting venules are observed in the *entire* choroid. Rather, SEM vascular cast study presents a changeable position for the feeding arteriole and collecting venules in the lobuli. From the posterior (scleral) side, it is possible to recognize many additional vascular channels that interconnect this same vessel along its course (see Fig. 22-14). The smaller vascular channels interconnect the collecting venule with major veins many times. As a constant feature, we observed very small venular channels originating from the choriocapillaries, which crossed over several lobuli and joined larger veins (Fig. 22-18; also see Fig. 22-16). Their role could be as a safety valve in the choroidal blood circulation. Perhaps they could play some role in the mass control of the choroidal circulation. This should be explored in the future.

2.4. Equatorial Area

A single layer of retinal capillaries is most common in this area.

The choriocapillaris in this area has a more elongated, lobular structure, with mostly collecting venules present at the center of the lobuli and feeding arterioles located more peripherally (Fig. 22-19). Between the lobuli in this region are dilated vascular structures, with some evidence of constrictions on their borders (Figs. 22-20 and 22-21). Intervascular venous channels, similar to those observed in the posterior pole, are noted also in the equatorial area. In many cases, espec-

Diagram 22-1. Schematic drawing of the radial arteriolar vascular supply to the lamina cribrosa (A.W. Fryczkowski, M.D., Ph.D.; B.S. Grimson M.D., R.L. Peiffer, D.V.M., Ph.D.). A: choroidal branches from Zinn-Haller circle; B: recurrent choroidal branches to the lamina cribrosa; C: recurrent branches to pia from Z-H circle and short posterior ciliary arteries; D: centripetal pial arterioles. (From Fryczkowski et al. [16], with permission of Dr. Junk Publishers.)

Diagram 22-2. Schematic drawing of normal choroidal vasculature drawn based on SEM micrographs. The distribution of the arterial and venous blood within the choriocapillaris was recognized based on a higher magnification of these SEM images.

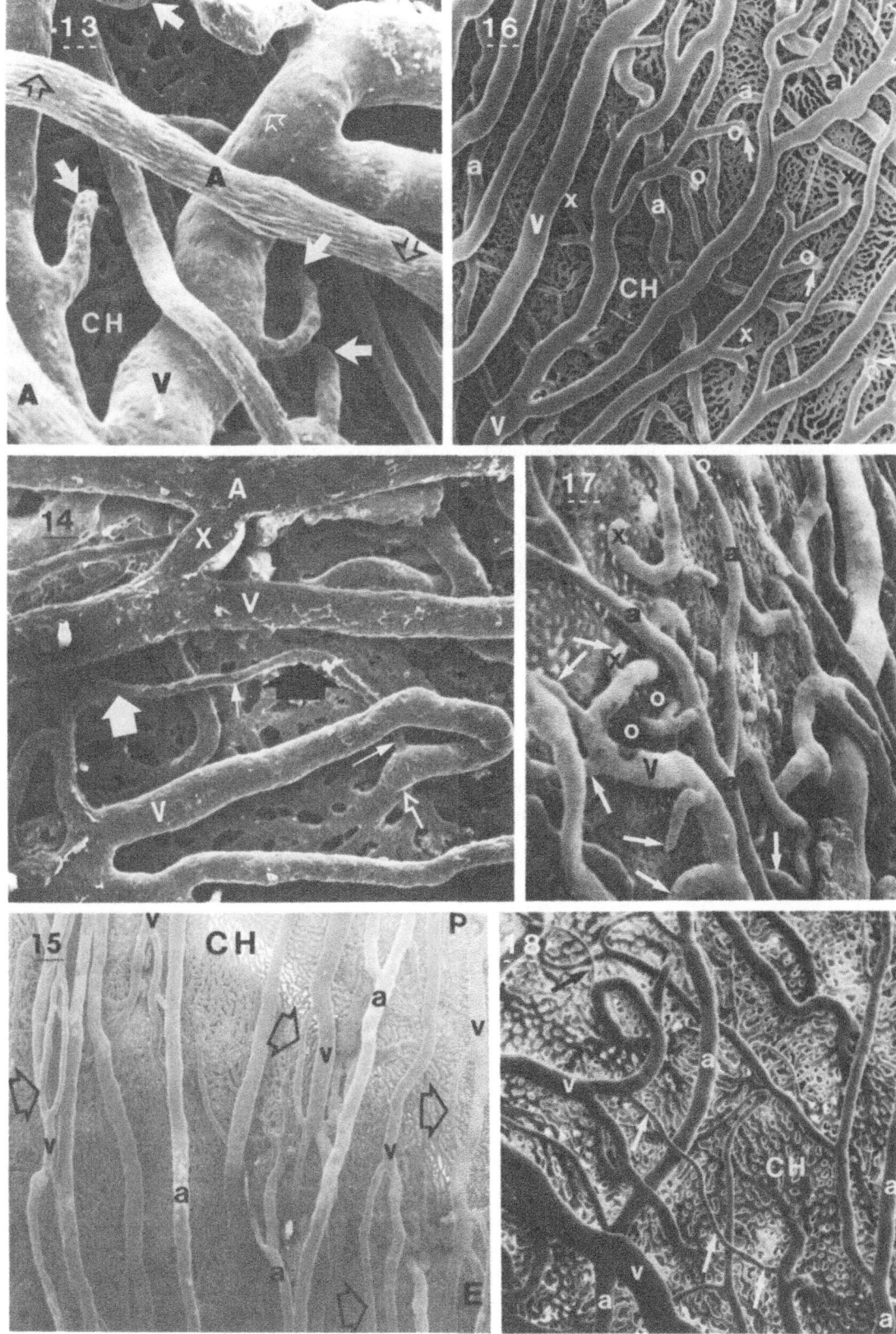

ially from elder subjects with a known history of hypertension, it is possible to observe venous splashing while crossing the underlying artery in the choroidal vessels (Fig. 22-22). This could be the same arterio-venous phenomenon observed in the retinal vessels in the subject who had hypertension (Fig. 22-23).

2.5. Pars Plana, Ciliary Processus, and Iris

The choriocapillaris forms a palmlike arrangement and ends at the ora serrata, forming arcades. A monolayer of mostly veins (up to 95% of total number of vessels in this area) with multiple vein-to-vein anastomoses is seen in the pars plana (Fig. 22-24). Two different kinds of ciliary processes are present in the ciliary body — narrowed and broadly angulated. The iris vasculature presents two layers of interconnected capillaries, with perpendicularly directed feeding arterioles and collecting venules. The capillaries form arcades that are interconnected in a circular arrangement parallel to the iris pupillary margin (Fig. 22-25, Diagram 22-2).

3. Diabetic Ocular Vascular Changes

The retinal vascular changes in diabetic retinopathy, observed in vascular cast under SEM, provide significant insight into the potential mechanism of their formation. The microaneurysms observed in our cases could be divided into four categories: (1) end-capillaries microaneurysms, (2) microaneurysms along the capillary wall, (3) microaneurysms on the capillary division (branching), (4) microaneurysms forming from capillary loops (Figs. 22-26–22-29).

One of the most interesting mechanisms of microaneurysm formation from the capillary loop is that described by Ashton [3] and Wolter [42]. Both of these authors believe that the bridgelike inner mesodermic connections between the retinal capillaries are responsible for contraction and kinking of the retinal capillaries. Such intervascular mesodermal strands, which perhaps are related to the damage of pericytes many which release an actinlike substance, could contract and create tension on the already weakened capillary wall [15]. The disappearance of retinal capillary pericytes is a well-known pathological phenomenon in diabetic retinopathy [4,5]; thus such a dislocation of actinlike substance from pericytes could be one of the cofactors in this scenario. A potential mechanism of microaneurysm formation is discussed in a separate paper [15].

Briefly, damage to the capillary basement membrane, selective pericyte loss, endothelial cell proliferation, hyaline degeneration, and changes in intraluminal rheology have been

←

Figure 22-13. Vascular cast of the choroid from 20-year-old male: posterior pole, posterior view. The distribution of the collecting venules (solid arrows) within the lobuli is shown. CH = choriocapillaris; V = vein; A = arteriole. Endothelial cell nuclear indentations spindle shaped in the arteriole and round-oval in the venule (open arrows). ×360. (With permission from S. Karger AG, *Acta Anatomica* 132:265–269, 1988.)

Figure 22-14. Human choroid: posterior pole, posterior view. The venular collateral channel is shown (solid arrows). The open arrow indicates a venular opening in the choriocapillaris. A = arteriole; V = venule; X = large arteriolar-venular anastomosis. ×435.

Figure 22-15. The human choroid — posterior pole, posterior view — of "splitting" veins that form additional vascular channels (open arrows) along the course of the veins. a = arteriole; v = veins; CH = choriocapillaris; E = equator area; P = peripheral area. ×26.

Figure 22-16. The monkey choroid — posterior pole, posterior view — of venular openings "in plane" (x) and at 90° (o) to the choriocapillaris (CH). v = vein; a = arteriole. The central position in the lobuli of the collecting venule is indicated by arrows. The course of the arteriolar branches with main openings in the center of the choroidal units could be traced. ×30.

Figure 22-17. Posterior pole of the human choroid — posterior (scleral) view — showing the changing position of the feeding arterioles (a) and collecting venules in neighboring lobuli. In one lobule the arterial openings are located in the center and are surrounded by venular openings; in a neighboring lobule there is a more central (x) venular position and a peripheral arteriole (o). The larger vein (V) drains the choriocapillaris by multiple collecting venules (arrows). ×22. (From Fryczkowski et al. [20], with permission from Altier & Maynard Comm. Inc.)

Figure 22-18. The choroid, posterior pole, posterior view. a = arteriole; v = venule; CH = choriocapillaris. Small venular communication channels (arrows) between collecting venules from different lobuli. ×45.

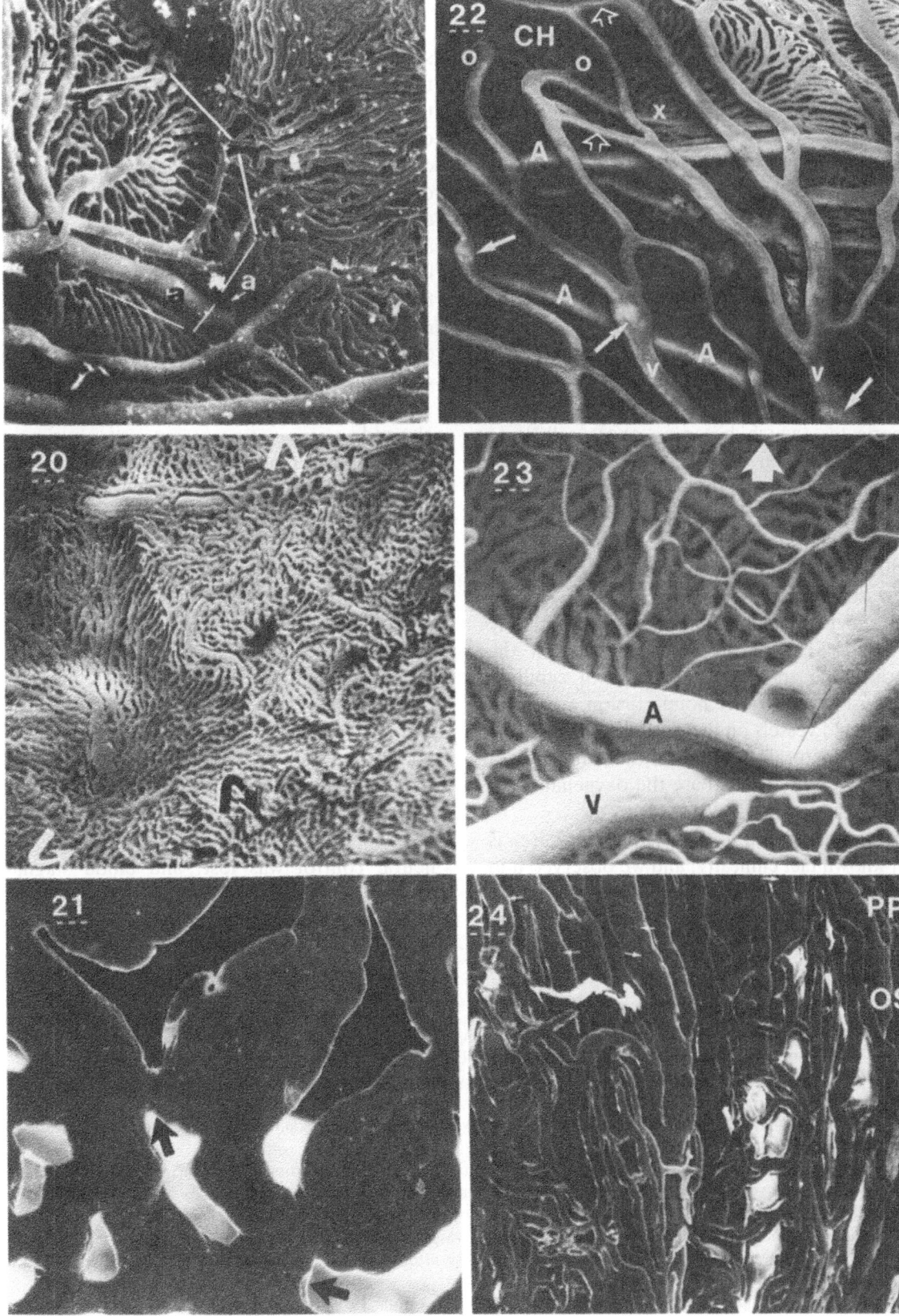

implicated as major contributing factors in microaneurysm formation. Clot formation on retinal capillary bifurcations, with weakened walls due to pericyte loss, endothelial cell stretching and proliferation with increased vascular resistance, and normal or increased blood pressure could be responsible for end-capillary microaneurysm formation. These same factors without clot formation could lead to microaneurysm formation on capillary branchings. The acute loop theory suggests the formation of microaneurysms by the contraction of intravascular fibrous strands, in addition to all known vascular and rheological factors. This theory could be modified by the implication that the actin migrates from the pericytes (which become "ghost cells") to the fibrous vascular tissue. The fact that actin, rather than collagen, is the substance responsible for membrane contraction (intervascular fibrous strands), will support this speculation as to the potential microaneurysm formation from the acute vascular loop.

The microaneurysms observed in randomly distributed areas of the retina, mostly in the posterior pole and on the borders of capillary nonperfusion areas. The nonperfusion areas had larger dilated vessels on their borders, which may indicate a natural autoregulation tendency of the retinal vessels to increase the oxygenation of ischemic tissue (Fig. 22-30).

Neovascularization in the peripapillary area is a well-known fact clinically, and we observed it in a few cases. Peripapillary retinal neovascularization is somewhat related to the capillaries that originate from prelaminar branches from larger choroidal branches.

The choroid from insulin-dependant diabetic subjects showed changes in all cases. The duration of disease was an important factor that contributed to its intensity as widespread localization of the vascular changes in the choroid. Early changes in mid-peripheral choriocapillaries were observed in subjects with 8–9 years of disease duration. These changes included dilatations and narrowings in the vascular lumen, hypercellularity, increased tortuosity, and vascular loop formation (Fig. 22-31). The deformation of normally smooth and regular interlobular vascular dilatations (see Figs. 22-20 and 22-21) was also observed. These changes were more extensively observed in the temporal aspect of the choroid.

With an increased duration of the disease, the vascular changes in the choroid showed additional progression. An increased number of endothelial cell indentations on the venous side of the choroidal circulation was observed on the vascular casts under SEM. The frequent formation of vascular loops and microaneurysms in the midperipheral choriocapillaris could be somewhat related to the hypercellularity of these endothelial cells (Fig. 22-32). Areas of capillary nonperfusion and, as result, capillary focal dropout, were noted.

Significant changes were observed in the interlobular vascular dilatations. The sinuslike structures that were formed in this area suggest that neural control was lost, and blood stagnation could have created some focal zones of relative ischemia. Some of these irregular structures were $90 \times 150\,\mu m$ (Fig. 22-33).

Marked dilatation of one vessel adjacent to the

←

Figure 22-19. The human choroid — equatorial area, posterior view — showing the central position of the venule (v) and the peripheral position of the feeding arteriole (a) in the lobuli. The lobuli borders are marked by the broken lines. ×52.

Figure 22-20. A 20-year-old male choriocapillaris in anterior view. Note the regular, smooth vascular dilations between the lobuli (arrows). ×52.

Figure 22-21. Magnified view of the vascular dilations seen between the lobuli, with some filling defects on their interconnections with the regular capillaries (arrows). ×625.

Figure 22-22. Choroidal equatorial area, posterior view showing an arteriovenous crossing (solid arrows). A = arteriole; v = venule. Two kinds of venular openings, (1) forming a 90° angle to choriocapillaris (o) and (2) in the same plane of the choriocapillaris (x). Venulo-venular anastomoses are indicated by open arrows. CH = choriocapillaris. ×304.

Figure 22-23. SEM photomicrograph from the human ocular vascular cast in a patient with long-standing insulin-dependant diabetes mellitus (IDDM) with hypertension: anterior view. The arterio-venous crossing phenomenon in the retina is shown. A = arteriole; V = venule; microaneurysm (arrows). ×190.

Figure 22-24. Human peripheral choroid: anterior view. OS = ora serrata; PP = pars plana. The palmlike arrangement of the choriocapillaris that forms arcades and ends in the ora serrata is shown. Multiple venous endothelial cell nuclei indentations are indicated by arrows. Most vessels in the pars plana are veins. ×40.

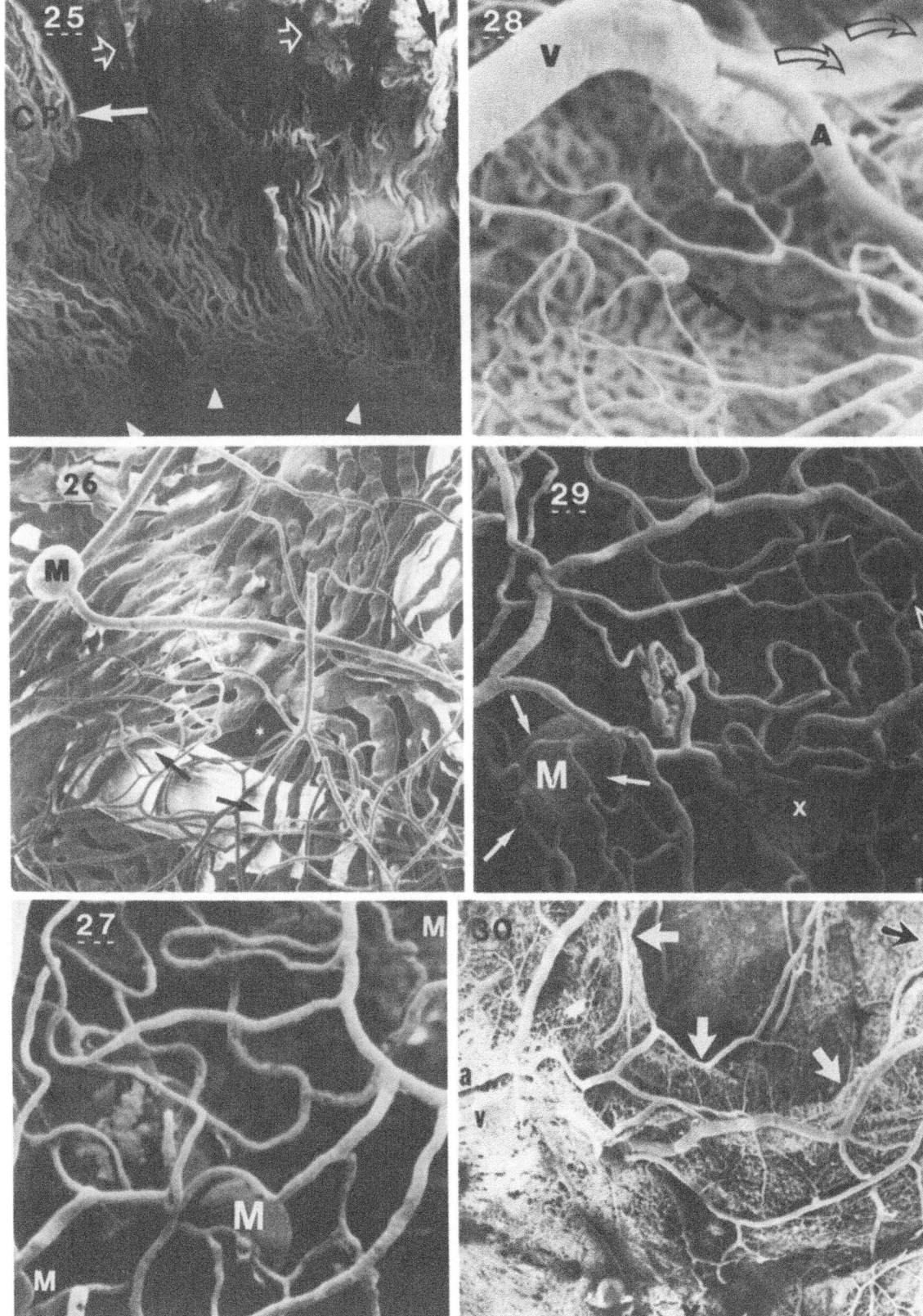

narrowing of others, increased tortuosity, out-pouchings, and hypercellularity were observed in the larger vessels in the choroid when viewed from the scleral side.

The arteriovenous anastomoses occasionally observed in different choroidal areas in diabetics were somewhat enlarged, and their presence was noted in the areas that had some dropout of capillaris. The severity of the vascular changes in the choroid increased significantly with the duration of the disease (Fig. 22-34).

All observed vascular changes were randomly distributed, however, it is believed that they occur focally rather than generally. The extensive presentation of vascular changes was noted in the midperipheral and peripheral areas in the choriocapillaris and in the perimacular area in larger choroidal vessels. In these cases, evidence of arteriosclerosis was also observed.

In the choroidal peripapillary area, the tortuosity of the vessels, dilations, and areas of capillary nonperfusion (dropout of capillaris) were observed. However, in the present study, we did not observe increased vascular changes in the choriocapillaris in this area; even larger choroidal vessels showed the same changes as noted in other choroidal regions. In the iris, changes included significant focal dilation and narrowing. In a few cases that had documented rubeosis, iris neovascularization was observed.

In one case of proliferative diabetic retinopathy, significant neovascularization from the retinal vessels was present in the vascular cast (Fig. 22-35). This florid type of neovascularization, growing toward the vitreous-retinal border, was different from the extensive loop neovascularization growing into the vitreous in form of the loops (Fig. 22-36). This latter kind of neovascularization clinically causes traction retinal detachment.

In one case of diabetic retinopathy in vivo, laser treatment was performed. Despite an 11-year lapse between this treatment and death, the vascular cast showed a significant delay in the choriocapillaris repair (Fig. 22-37). This indicates that the choriocapillaris in diabetic patients can repair after laser treatment in a different fashion than has been recognized based upon the experimental study.

In one case of preproliferative diabetic retinopathy, significant thrombosis of the branches of the central retinal vein was recorded in vivo. This same image of thrombi inside the retinal veins, observed in vascular cast by SEM, allowed us to better understand the mechanism of its attachment to the vascular wall and the fact that many times the blood flow is sustained on fluorescein angiography, even if a significant portion of the vein is literally packed with plaques and thrombi (Fig. 22-38). In such cases, full occlusion of the vein lumen can occur at any place.

In a comparative study on diabetic monkeys,

Figure 22-25. SEM of the ciliary processes (CP) and iris, posterior view, of a 46-year-old male. Narrowed (open arrows) and broadly angulated (solid arrows) ciliary processes, and the iris margin (arrowheads) are shown. ×24. (From Fryczkowski et al. [18], with permission from Dr. Junk Publishers.)

Figure 22-26. Posterior pole anterior view in a 45-year-old male with IDDM. SEM micrographs show end-capillary retinal microaneurysm (M) on the long stipule. Some drop-off in the retinal capillaries (arrows) as well as drop-off in the choriocapillaries (asterisk) are shown. ×385.

Figure 22-27. This same case as in Fig. 22-26, showing the retinal microaneurysms (M) along the retinal capillaries. Loss of pericytes, increased vascular resistance, endothelial cells stress, and proliferation could be contributing factors to the development of such microaneurysms. ×385.

Figure 22-28. Posterior pole, anterior view, of a 56-year-old female with IDDM and hypertension showing a retinal microaneurysm (solid arrow) located on the retinal capillary branching. A = artery; V = vein. An arterio-venous crossing is present. Filling defects (open arrows) probably represent replicas of the plaques attached to the veins on the endothelial side. A significant dropout of the retinal capillaris is seen. ×336. (From Fryczkowski [14], with permission from *Scanning Microscopy International.*)

Figure 22-29. Same case as in Fig. 28. A retinal saccular vascular microaneurysm (M) (arrows) and spherical microaneurysm (open arrow) are shown. This spherical microaneurysm could be formed from a capillary loop that fused the inner walls. X = artifact. ×385.

Figure 22-30. IDDM, low-power SEM photomicrograph, anterior view, showing an overall view of the retina with a significant area of dropout of the capillaris (arrows). Enlarged veins on the border of this area indicate a tendency to compensate for this decrease of oxygenation. a = artery; v = vein; X = macular area. ×28. (From Fryczkowski et al. [17], with permission from the American Medical Association.)

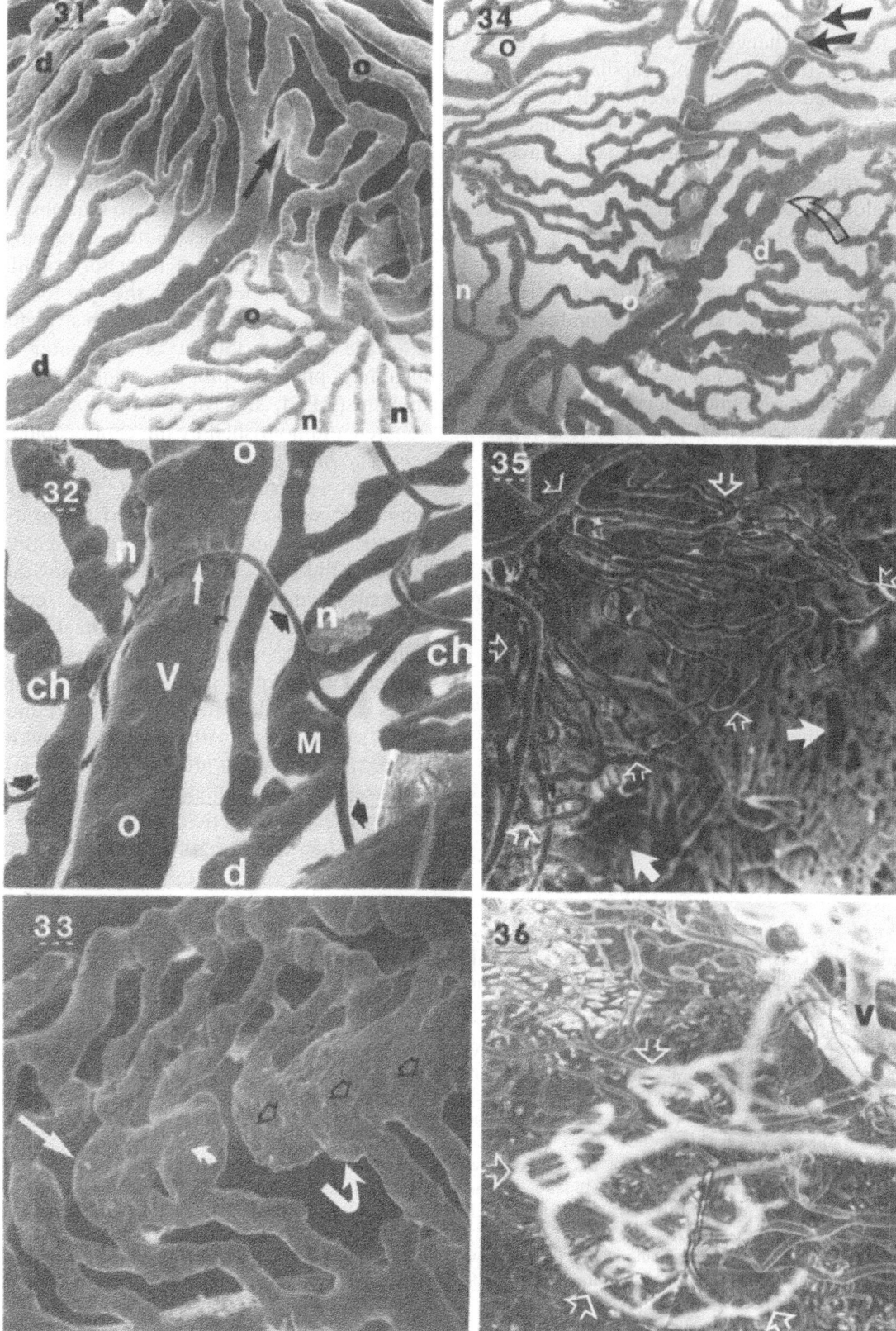

changes in the choroidal and retinal vasculature similar to human changes were also noted. Micro-aneurysms and dropout of capillaries, (Fig. 22-39), as well as retinal neovascularizations, were particularly present (Fig. 22-40). This also indicates that the insulin-dependant diabetic monkey with a long duration of disease is the closest model to the human in terms of diabetic retinopathy and should be the subject of future study.

Vascular changes in the diabetic eye leading to serious complications and potential blindness has drawn the attention of clinicians and researchers. Vascular stress oxygenation-vasodilatation [3,10], rheological and biochemical changes [10,30], and angiogenic hypotheses [3,10] have all been impli-cated in the pathogenesis of diabetic retinopathy [18]. Retinal vascular pathology; vein dilata-tions and arterial narrowing, loops, and micro-aneurysm formation; intraretinal microvascular abnormalities (IRMA); increased vascular per-meability and leakage from the vessels; hemor-rhages; vascular shunts; and vessel proliferation, which could lead to retinal traction detachment, are all clinically easy to observe and are well known to physicians. Studies of the choroidal changes in diabetes, however, have been few in number thus far [12,14,19,20,25,35,45].

From our study, it is evident that the changes in the choroid in long-standing insulin-dependent diabetes mellitus are similar to changes that have been documented in the retinal vasculature. Dilatations, narrowing, and increased tortuosity of the choroidal vessels; outpouching; hyper-cellularity and microaneurysm formation; vas-cular shunts; neovascularization; and finally dropout of the capillaries were noted in our material. We believe that the mechanism of their formation is somewhat similar to that observed in the diabetic retinal vasculature. Knowing that the proper functioning of the choriocapillaris is crucial for metabolism of the photoreceptor/ retinal pigment epithelium/Bruch's membrane interface (up to 130 μm of the outer retina, including photoreceptors, retinal pigment epi-thelium (RPE), and Bruch's membrane, is oxy-genated and nourished by the choriocapillaris), it seems possible that any deterioration of choroidal function will adversely affect this important anatomic complex. The detailed localization of the focal changes in the choroidal circulation (which is hidden behind the RPE), using the ICG high-speed digital cineangiography technique, could be important for better understanding the failure of laser treatment (a presently well-accepted technique) in a significant number of cases with proliferative diabetic retinopathy (PDR). Our case with obvious delay of the choriocapillaris repair process, even 11 years after the laser treatment, supports a recently published study [40]. The diabetic choroidal cellular repair process after laser treatment could be different

Figure 22-31. 67-year-old female with 15-year duration of IDDM: choriocapillaris, midperipheral area, anterior view. Increased tortuosity, dilations (d), outpouchings (o), narrowings (n) of the vessels, and acute vascular loop (arrow) are shown. ×173. (From Fryczkowski et al. [18], with permission from Kluwer Academic Publishers.)

Figure 22-32. Long-standing IDDM: peripheral choroid (ch) seen from the retinal side. Outpouchings (o), tortuosity, dilations (d), endothelial hypercellularity microaneurysm (M), and a vein (V) are seen. Note the unusual course of the retinal capillaries (arrows). ×265. (From Fryczkowski et al. [20], with permission from Altier & Maynard Comm. Inc.)

Figure 22-33. Long-standing IDDM: choriocapillaris in equatorial area as seen from the retinal side. A magnified view of the coalesced vascular sinuslike structures (solid arrows) that occurred in place of regular interlobular dilations (compare with Figs. 22-20 and 22-21) is shown. Endothelial cell indentations (open arrows) are shown. ×312. (From Fryczkowski et al. [20], with permission of Altier & Maynard Comm. Inc.)

Figure 22-34. The SEM photomicrograph of a 72-year-old man with IDDM of 26 years' duration. Peripheral choriocapillaries as seen from the retinal side. Increased tortuosity, dilation (d), outpouchings (o), narrowing (n) of the vessels, hypercellularity (open arrows), and microaneurysms (solid arrows) are shown. (From Fryczkowski et al. [20], with permission of Altier & Maynard Comm. Inc.)

Figure 22-35. Anterior view of neovascularization (open arrows), florid type, in the right eye of a 68-year-old woman with insulin-dependent diabetes mellitus (IDDM) of 25 years duration. A SEM photomicrograph of the vascular cast is shown. Dropout of the choriocapillaris is indicated by solid arrows. ×140.

Figure 22-36. 63-year-old man with proliferative diabetic retinopathy and IDDM of 28 years duration. The posterior pole, an anterior view of the retinal neovascular loops (out of focus; open arrows) originating from a vein, growing toward the vitreous is shown. V = vein. ×140.

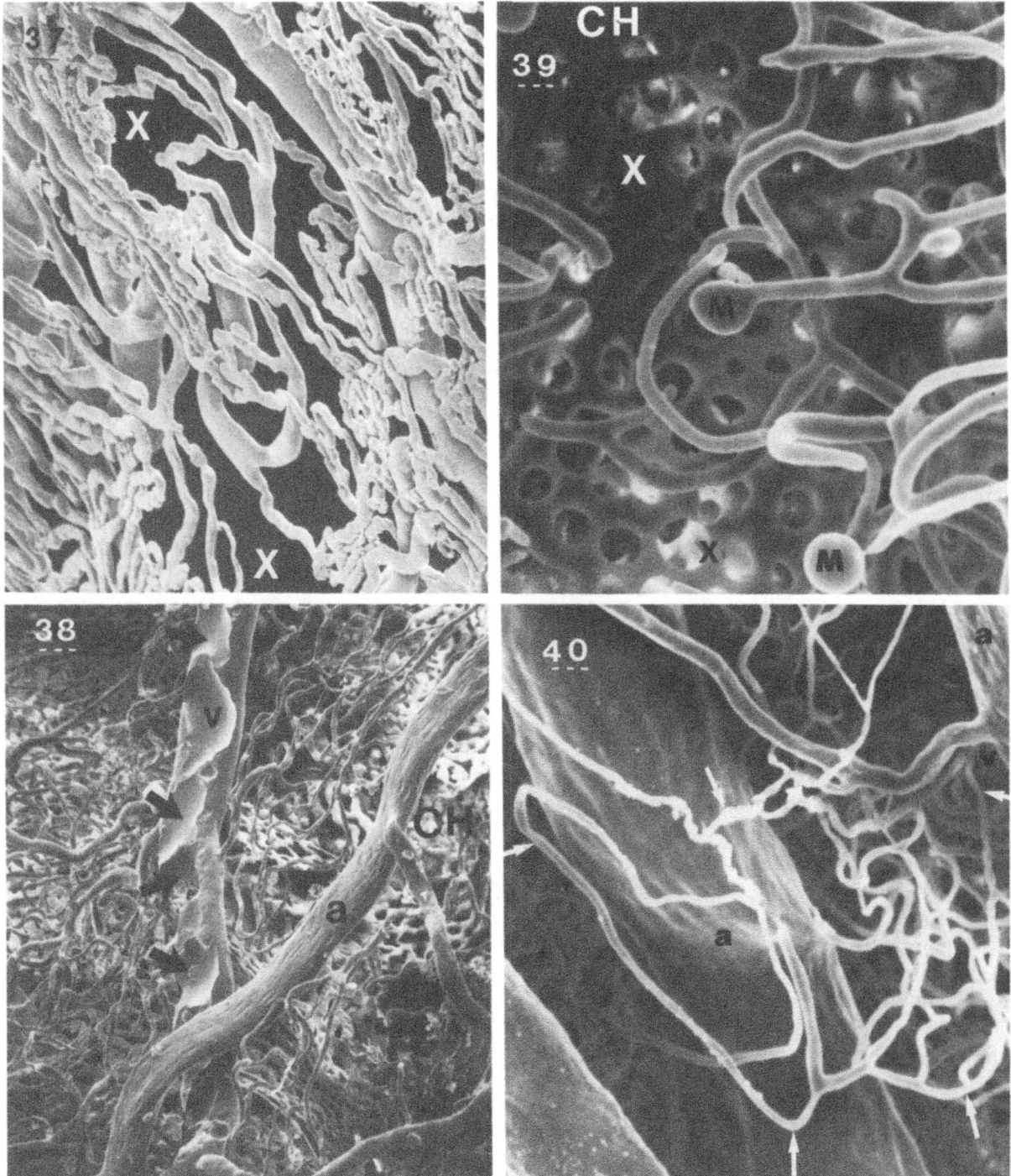

Figure 22-37. IDDM with long duration of the disease, showing a choroid, midperiphery, anterior view SEM photomicrograph of the choriocapillaries dropout in areas where 11 years earlier laser treatment was performed (X). ×312

Figure 22-38. Preproliferative diabetic retinopathy with thrombosis of the branches of the central retinal vein: posterior pole, anterior view. v = vein; a = artery; CH = background choriocapillaris. Replicas of the thrombi inside the vein are indicated arrows. ×312.

Figure 22-39. Insulin-dependent diabetic monkey with a 12-year duration of the disease: posterior pole, anterior view. The retinal microaneurysms (M) on the border of the dropout of the capillaris (X) are shown. CH = background choriocapillaris. ×312.

Figure 22-40. Insulin-dependent diabetic monkey with 15-year duration of the disease. Retinal neovascularization (arrows) close to the optic disc is shown in an anterior view. a = artery; v = vein. ×178.

from that in the nondiabetic [33], and some factors, such as endothelial cell stress, prolonged hypoxia, rheological, and biochemical changes, could be responsible for reneovascularization in the retinal and choroidal vasculature. Future studies of diabetic vascular changes in the choroid are necessary to clarify their role in the pathogenesis of diabetic retinopathy. As result of this, more effective treatment might be developed.

4. Concluding Remarks

Vascular casts from autopsy specimens from normal and diabetic human and monkey eyes were studied under scanning electron microscopy (SEM). Several SEM photomicrographs provide insight into the choroidal angioarchitecture. Regional vascular differences, such as in the lamina cribrosa, peripapillary, macular, posterior pole, equatorial, and peripheral areas, are described. The choriocapillaris, as a monolayer of continuous capillary networks, presents a different vessel configuration from the anterior (retinal) side; a dense and mostly unified net — the optic disc and macular area — through lobular areas to elongated capillaris with more regular positioning of arteries and veins on the periphery. A changing position of the feeding arteriole and collecting venules in the choroidal lobuli was documented.

The presently accepted theory [24,38] that the feeding arteriole is located in the center of the lobuli and the collecting veins are located on the periphery disagrees with the present findings. The argument that the fact that fluorescence occurs first in the center of the choroidal lobuli in fluorescein angiography supports the central location of the feeding arteriole may be misinterpreted. The fluorescein dye arrived at the choroidal arteries diluted, not in a bolus, as is implicated by supporters of this theory. Because of this, the present technical abilities to detect it are far from reality. When the concentration of the fluorescein in the lobuli in the centrally located collecting vein increases, its presence is recorded. Next, the increase of fluorescein concentration in the lobuli suggests that it spreads from the center to the periphery; however, this process is probably reversed and the evidence

of fluorescence of the lobuli depends on the increased amount of fluorescein in the lobuli. Knowing that the whole lobular circulation is limited to 1.2 seconds, it is extremely difficult at the present level of technology to decide the direction of blood flow in the lobuli. At any rate, the anatomical observations provide evidence of the location of a central collecting vein in humans and monkeys. Another explanation for the lobular circulation may arise by presence of the choroidal unit consisted of a few lobuli supplied by one or more feeding arteriole. Existence of the choroidal units may explain some "choroidal puzzles" observed in the acute ischemia and its resolution which differ from place to place.

Additional vascular channels located along the smaller, as well as larger, venous vessels were documented. These additional vascular channels can function as a safety valve for the blood outflow from the choriocapillaris, which will fit the blood dynamics demand. Direct arterio-venous anastomoses were observed. The presence of such large, direct anastomoses could explain the high blood oxygen saturation on the venous side of the choroid. Three-dimensional views of the diabetic vascular changes in the retinal and choroidal microcirculation were obtained, showing endothelial hypercellularity, increased vessel tortuosity, outpouching, microaneurysm formation, capillary dropout, microvascular abnormalities, and retinal neovascularization. Retinal neovascularization was observed originating from the vessels located in the inner layers of the retina, with florid neovascularization growing up toward the inner limiting membrane and loop neovascularization growing toward the vitreous. Choroidal microvascular changes — loop formation and choroidal microaneurysms, increased tortuosity, hypercellularity, and dropout of capillaries — were observed in insulin-dependent diabetics. We noted that smooth and regular vascular dilations between the lobuli, located primarily in the peripheral and the equatorial area, changed into sinuslike structures located primarily in the equatorial area. Such changes could contribute to more sluggish blood flow in this area, and thus decreased oxygenation and increased carboxylation of the photoreceptors/RPE complexes. In this way, chronic hypoxia in this area could be expected. This speculation could be supported by the fact that color vision and ERG in long-standing diabetics are decreased. Significant changes were seen in other larger choroidal vessels at the posterior pole, in the submacular and peripapillary areas. All these observations indicate that the choroidal vessels in diabetics are involved in the disease process, and the term *diabetic choroidopathy* is justified. The presently developed dynamic study of the choroidal blood circulation, using indocyanine green dye and high-speed digital angiography, will be extremely helpful in clarifying the remaining questions regarding the choroidal changes in diabetics.

Acknowledgments

This study was supported in part by grants from the Ohio Lions Eye Research Foundation, the Bremer Foundation, and the Juvenile Diabetes Foundation International.

References

1. Archer D, Krill AE, Newell FW. Fluorescein studies of normal choroidal circulation. *Am J Ophthalmol* 69:543–552, 1970.
2. Ashton N. Observations on the choroidal circulation. *Br J Ophthalmol* 35:465–481, 1952.
3. Ashton N. Diabetic retinopathy. A new approach. *Lancet* 2:625–630, 1959.
4. Bresnick GH, Davis MD, Myers FL, Venecia G. Clinicopathological correlations in diabetic retinopathy. II. Clinical and histological appearance of retinal capillary microaneurysm. *Arch Ophthalmol* 95:1215–1223, 1977.
5. Cogan DG, Toussaint D, Kuwabara T. Retinal vascular patterns IV. Diabetic retinopathy. *Arch Opthalmol* 66:366–378, 1961.
6. Ernest JT, Stern WH, Archer DB. Submacular choroidal circulation. *Am J Ophthalmol* 81:574–582, 1976.
7. Francois J, Fryczkowski AW. The blood supply of the optic nerve. *Adv Ophthalmol* 36:164–173, 1978.
8. Flower RW. Choroidal fluorescent dye filling patterns: A comparison of high speed indocyanine green and fluorescein angiograms. *Int Ophthalmol* 2:143–150, 1980.
9. Flower RW, Hochheimer BF. A clinical technique and apparatus for simultaneous angiography of the separate retinal and choroidal circulations. *Invest Ophthalmol Vis Sci* 12:248–255, 1973.
10. Frank RN. On the pathogenesis of diabetic retinopathy. *Ophthalmology* 91:626–634, 1984.
11. Friedman E, Smith TR, Kuwabara T. Senile choroidal vascular patterns and drusen. *Arch Opthalmol* 69:220–230, 1963.
12. Freyler H, Prskavec F, Stelzer N. Diabetische choroidopatheine retrospective fluoreszenzangiographische studie. *Klin Mbl Augenheilk* 189:144–147, 1986.
13. Fryczkowski AW. SEM of the human choroid: Normal anatomy. In: *Proceedings from Symposium on the Choroid, Ittingen, May 11–14*, pp 11–22, 1986.
14. Fryczkowski AW. Vascular casting and scanning electron microscopy in diabetes. *Scann Microsc* 1:811–816, 1987.
15. Fryczkowski AW, Chambers RB, Craig EJ, Walker J, Davidorf FH. Scanning electron microscopy study of microaneurysms in the diabetic retina. *Ann Ophthalmol*, 23:130–136, 1991.
16. Fryczkowski AW, Grimson BS, Peiffer RL. Scanning electron microscopy of vascular casts of the human scleral lamina cribrosa. *Int Opthalmol* 7:95–100, 1984.
17. Fryczkowski AW, Grimson BS, Peiffer RL. Vascular casting and scanning electron microscopy of human ocular vascular abnormalities. *Arch Ophthalmol* 103:118–120, 1985.
18. Fryczkowski AW, Hodes BL, Walker J. Diabetic choroidal and iris vasculature scanning electron microscopy findings. *Int Ophthalmol* 13:269–279, 1989.
19. Fryczkowski AW, Sato SE. Scanning electron microscopy of the ocular vasculature in diabetic retinopathy. *Ophthal Forum* 4:39–50, 1986.
20. Fryczkowski AW, Sato SE, Hodes BL. Changes in the diabetic choroidal vasculature: Scanning electron microscopy findings. *Ann Ophthalmol* 20:299–305, 1988.
21. Henkind P. New observations on radial peripapillary capillaries. *Invest Ophthalmol* 6:103–109, 1967.
22. Hayreh SS. Segmental nature of the choroidal vasculature. *Br J Ophthalmol* 59:631–648, 1975.
23. Hayreh SS. Submacular choroidal vascular pattern. *Abrecht v. Graefes Arch Klin Exp Ophthal* 192:181–196, 1974.
24. Hayreh SS. The choriocapillaris. *Albrecht v Graefes Arch Klin Exp Ophthalmol* 192:165–179, 1974.
25. Hidayat AA, Fine BS. Diabetic choroidopathy: Light and electron microscopic observations of seven cases. *Ophthalmology* 92:512–522, 1985.
26. Hyvarinen L, Flower RW. Indocyanine green fluorescence angiography. *Acta Ophthalmol* 58:528–538, 1980.
27. Krey H. Segmental vascular patterns of the choriocapillaris. *Am J Ophthalmol* 80:198–202, 1975.
28. Kuwabara T, Cogan DG. Studies of retinal vascular patterns. *Arch Ophthalmol* 64:904–910, 1960.
29. Leber T. Die Circulations und Ernahrungsverhaltnisse des Auges. In: Graef, Saemisch (eds.), *Hand. der Gesamten Augenheilk*, 2nd ed. Engelman, Leipzig, Bd. 2:1–10, 1903.
30. Little HL. The role of abnormal blood rheology in the pathogenesis of diabetic retinopathy. *Am Ophthalmol Dabetologia* 9:20–24, 1973.
31. Matusaka T. Angioarchitecture of the choroid. *Jpn J Ophthalmol* 20:330–346, 1976.
32. Novotny HR, Alvis DL. A method of photographing fluorescence in circulating blood of the human eye. *Am J Ophthalmol* 50:176–181, 1960.
33. Perry DD, Risco JM. Choroidal microvascular repair

after argon laser photocoagulation. *Am J Ophthalmol* 93:787–793, 1982.

34. Risco JM, Nopanitaya W. Ocular microcirculation: Scanning electron microscopy study. *Invest Ophthalmol Vis Sci* 19:5–12, 1980.

35. Saracco JB, Gastand P, Ridings B, Ubawd CA. La choroidopathie diabetique. *J Fr Ophtalmol* 5:231–236, 1982.

36. Scott D, Dollery CT, Hill DW, Hodge JV, Fraser TR. Fluorescein studies in diabetic retinopathy. *Br Med J* 1:811–818, 1964.

37. Shimizu K, Ujiie K. *Structure of Ocular Vessels.* Igaku-Shoin, Tokyo, pp 1–7, 50–92, 1978.

38. Torczynski E, Tso MOM. The architecture of the choriocapillaris at the posterior pole. *Am J Ophthalmol* 81:428–440, 1976.

39. Weiter JJ, Ernest JT. Anatomy of the choroidal vasculature. *Am J Ophthalmol* 78:583–590, 1974.

40. Wilson DJ, Green RW. Argon laser panretinal photocoagulation for diabetic retinopathy. *Arch Ophthalmol* 105:239–242, 1987.

41. Wise GN, Dollery CT, Henkind P. The retinal circulation. Harper and Row, New York, pp 2–18, 34–54, 1971.

42. Wolter RJ. Diabetic capillary microaneurysms of the retina. *Arch Ophthalmol* 65:847–854, 1961.

43. Woodlief NF, Eifrig DE. Initial observations on the ocular microcirculation in man: The choriocapillaris. *Ann Ophthalmol* 14:176–180, 1982.

44. Wybar KC. Anastomosis between the retinal and ciliary arterial circulations. *Br J Ophthalmol* 40:65–78, 1956.

45. Yanoff M. Ocular pathology of diabetes mellitus. *Am J Ophthalmol* 67:21–38, 1969.

46. Yoneya S, Tso MOM. Angioarchitecture of the human choroid. *Arch Ophthalmol* 105:681–687, 1987.

47. Zinn JG. *Descriptio Anatomica Oculi Humani,* Vandenhoeck, Gottingen, 1755.

Author's address:
Prof. Andrzej W. Fryczkowski
Department of Ophthalmology
The Ohio State University
456 West 10th Avenue
Columbus, OH 43210
USA

Effects of Antigen Stimulation and Irradiation on the Blood Vascular System of Murine Lymph Nodes

DOUGLAS A. STEEBER & RALPH M. ALBRECHT

1. Introduction

This chapter provides a description of the micro-vascular changes that occur in peripheral lymph nodes, especially murine, during a primary immune response. Descriptions are included for both the resting and the antigen-stimulated lymph node microvasculature, with special emphasis on the role the vasculature plays in the immune function of the lymph node. Comparisons are made between lymph nodes from normal and irradiated animals. Special reference is also made to the high-endothelial venules, which are important in lymphocyte trafficking in lymphoid tissue and which potentially play a role in certain diseases, such as myasthenia gravis and rheumatoid arthritis. Emphasis is placed on our use of scanning electron microscopy of vascular corrosion casts, although the relevant material published in this area is also reviewed to provide a comprehensive description of peripheral lymph nodes undergoing a primary immune response.

2. General Description

The lymph nodes, together with the spleen, appendix, tonsils, and Peyer's patches, make up the peripheral lymphoid tissue. The lymph nodes are distributed throughout the body, but they are larger and more concentrated in the neck, axilla, groin, mesentery, and along the spine. They are interspersed in the path of lymphatic vessels such that each lymph node serves as a lymph filter.

Lymph is composed of blood plasma and tissue fluids, which contains cells, many types of debris, and some bacteria. The lymph is collected by an extensive system of lymph capillaries that drain most of the tissue spaces of the body. The lymph capillaries join together to form the collecting lymphatics, which in turn merge to form the afferent lymphatics. Generally, several afferent lymphatics enter the lymph node through the capsule and empty into the subcapsular sinus. The lymph then drains through the trabecular sinuses and into the medullary sinuses, eventually converging into an efferent lymphatic vessel, which exits at the hilus of the node. Phagocytic cells line the sinuses and engulf and process the particulate debris and foreign matter present in the lymph. The efferent lymph is finally returned to the blood stream through either the thoracic duct or the right lymphatic duct.

The basic structure of the lymph node, with the major regions indicated, is illustrated in Fig. 23-1. The lymph nodes are generally spherical or kidney-shaped and are surrounded by a connective tissue capsule that extends numerous trabeculae into the interior of the node. The interior of the node is dominated by the presence of lymphocytes, which are dispersed throughout a reticular cell network. The lymph node interior can be separated into three major zones: the cortex, paracortex, and medulla. The cortex occupies the peripheral zone of the node, except at the hilus, and is populated primarily with B lymphocytes, many of which are organized into follicles. In a resting lymph node, most of the

Motta, P.M., Murakami, T., and Fujita, H. (eds.), Scanning Electron Microscopy of Vascular Casts: Methods and Applications.

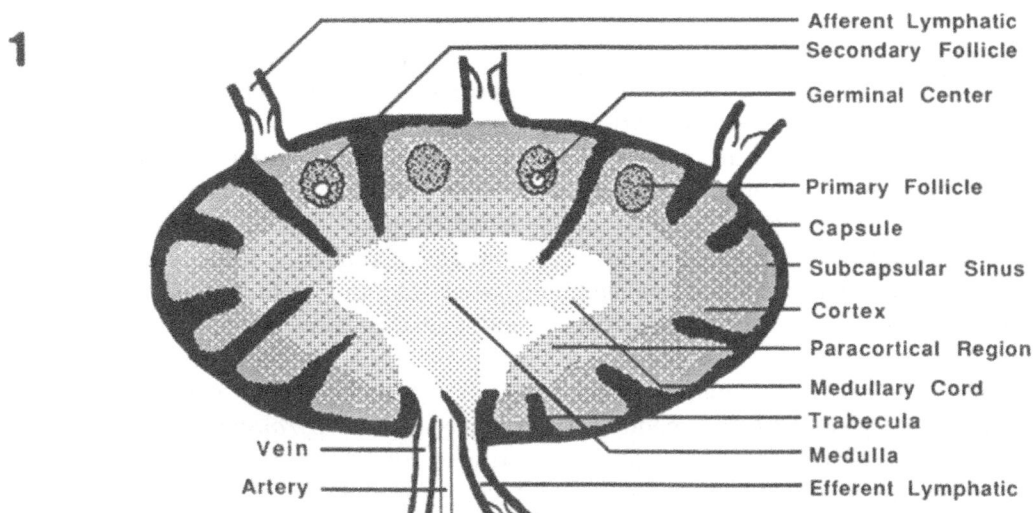

Afferent Lymphatic
Secondary Follicle
Germinal Center
Primary Follicle
Capsule
Subcapsular Sinus
Cortex
Paracortical Region
Medullary Cord
Trabecula
Vein
Artery
Medulla
Efferent Lymphatic

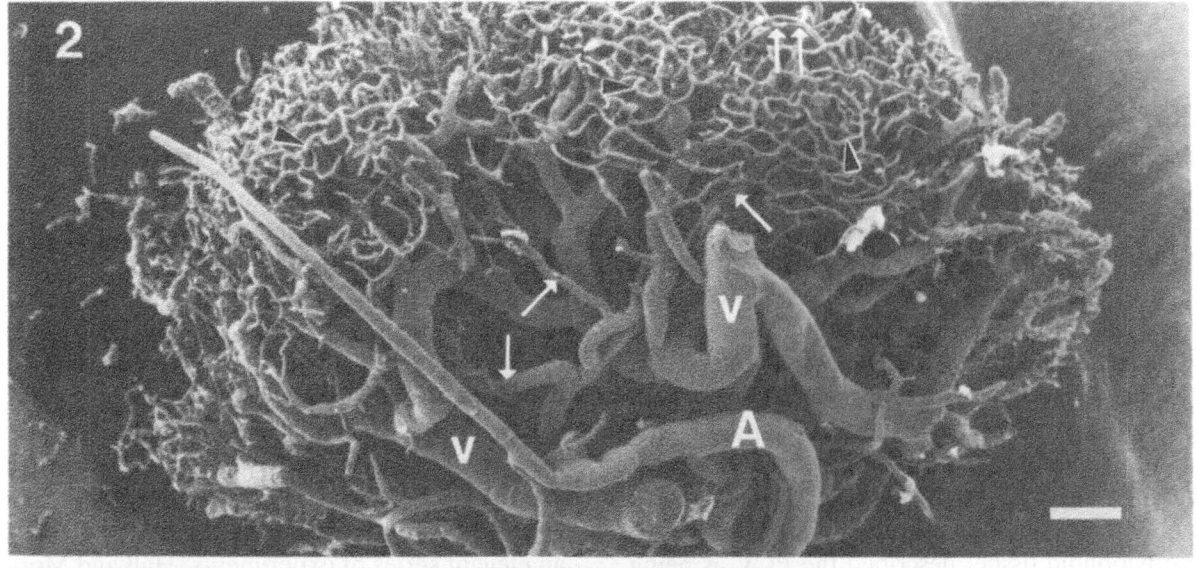

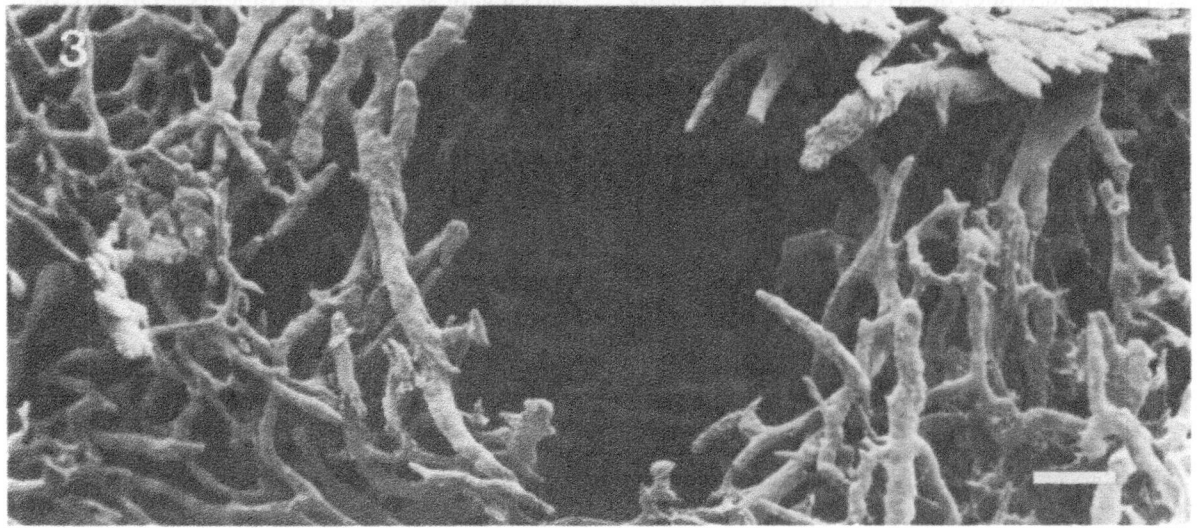

follicles have a uniform cell density and are called *primary follicles*. The paracortex, also called the *deep* or *diffuse cortex*, separates the cortex from the medulla and is filled with mostly T lymphocytes. Lymph nodes from neonatally thymectomized mice and nude mice have a severely depleted paracortex; therefore, this region has been termed the *thymus-dependent* or *T-dependent area*. The medulla occupies the interior of the node and contains both B and T lymphocytes. The majority of the plasma cells in the lymph node are located in the medulla, organized around connective tissue fibers called *medullary cords*. Lymphocytes and plasma cells leave the node by migrating toward the hilus and into the efferent lymphatic vessel.

2.1. Blood Vessels

One of the earliest detailed descriptions of the vascular system of the lymph node was reported by Calvert [1] in the late 19th century. In these studies, canine lymph nodes were perfused with contrast media, allowing the description of a relatively simple vascular system. Following this early report, Schulze [2] heightened the interest in this vascular system when he described in detail peculiar postcapillary venules (PCV). These venules are characterized by a high cuboidal endothelial lining and by the presence of a large number of lymphocytes, both in the lumen and within the vessel wall. Ehrich [3], in his studies of the lymphoid tissue, described the PCV as being present not only in the lymph node, but also commented on seeing them regularly in lymphoid tissue of the intestine, lungs, and tonsils. The presence of these venules was subsequently noted in a number of studies, and their distinction as the site of lymphocyte migration from lymphoid tissue into the blood stream was suggested.

Unique staining characteristics of the PCV were reported by Smith and Hénon [4]. They found the endothelium of the PCV, in contrast with other endothelium, to be metachromatic when stained with toluidine blue, to contain succinic dehydrogenase, and to be strongly positive for nonspecific esterase. Röpke et al. [5] demonstrated that the metachromatic staining of the endothelial cells of the PCV was due to the presence of RNA in their cytoplasm. Therefore, these endothelial cells are now more commonly referred to as *pyroninophilic* to indicate the presence of RNA. As the body of literature on these vessels increased, the term *PCV* became inadequate to identify these venules, because by definition all venules are "post" capillary in location. Since not all venules possess this unique endothelium, it was suggested by several investigators that the term *PCV* be replaced with *high-endothelial venule* (HEV) to indicate only those venules involved in lymphocyte trafficking. Therefore, HEV will be used throughout this chapter when referring to these specialized venules.

One of the most important contributions in understanding the function of the HEV came from Gowans and Knight [6] when they provided definitive evidence that, within the lymph nodes, small lymphocytes migrate continuously during their lifetime from blood to lymph and back to the blood. They demonstrated that this lymphocyte recirculation occurred specifically in the HEV. In an elegant series of experiments, Hall and Morris [7] determined the degree to which this recirculation occurred in an individual lymph node and also determined the origin of the lymphocytes leaving the node in the efferent lymph. In their studies using isolated and cannulated sheep lymph nodes, they showed that recirculation, not cell division within the lymph node, accounted for

←

Figure 23-1. Diagrammatic representation of a murine lymph node in cross section.

Figure 23-2. Mouse popliteal lymph node mounted so the hilus is visible. A single small artery (A) enters at the hilus branching (arrows) as it passes through the medulla and enters the cortex. The subcapsular capillary arcades (arrowheads) are seen to cover the periphery of the cast. Occasionally small arterioles (double arrow) are observed to loop through the subcapsular capillaries without branching into capillaries. Large efferent veins (v) are seen near the hilus. These veins would have been seen to merge into a single exiting vessel if the cast were not dissected so close to the hilus. (Reproduced from Steeber et al. [18], with permission.) Bar = 100 μm.

Figure 23-3. Vasculature surrounding a germinal center. As the germinal centers form, the blood vessels are displaced outward, resulting in a vascular "basket" surrounding the avascular center. Bar = 100 μm.

>95% of the lymphocytes present in the efferent lymph of a resting lymph node. They estimated that approximately 10% of the lymphocytes in the blood circulating through a resting lymph node migrate into the node. Thus circulating lymphocytes continuously enter the lymph node at the HEV, migrate through the node toward the hilus, exit in the efferent lymph, and return to the blood stream to repeat the process in other lymphoid tissues. In this way, the lymph node acts as a dynamic structure in which lymphocytes are continuously brought in contact with antigens that have been filtered from the lymph, allowing the initiation of both humoral and cell-mediated immune responses.

2.2. Response to Antigen

When a lymph node becomes activated by antigen that has entered either in the lymph or has been transported in by an immune cell, a series of responses occur. One of the earliest measurable responses to antigen stimulation is a rapid increase in the regional blood flow through the node, which peaks, subsides slightly, and is maintained at an increased level throughout the course of the immune response [8–10]. During this time, the number of circulating lymphocytes entering the node through the HEV is greatly increased [7,11]. While the number of lymphocytes entering the node is increased, the number exiting the node in the efferent lymph is greatly reduced [11,12]. Depending on the stimulating antigen used (e.g., viral antigens), the efferent lymph may become nearly acellular for up to 18 hours post-stimulation [11]. This early phase of activation has been termed *lymphocyte recruitment*. This recruitment process has been shown to be, at least in part, antigen specific, based upon the finding that lymphocytes reactive to the stimulating antigen disappear from the circulating lymphocyte pool by being "trapped" inside the antigen-stimulated lymphoid tissue [13]. Following the transient decrease in exiting lymphocytes, there is a large increase in the output of lymphocytes in the efferent lymph, which generally peaks between 2 and 4 days post-stimulation. Even though this increase may be as much as 10-fold, >95% of the cells have been shown to originate from the blood, as was the case in

the resting node [7,11]. During this period, the microvasculature of the stimulated lymph node undergoes a dramatic enlargement, typically a four- to six-fold volume increase, which usually peaks between 4 and 7 days following stimulation. B and T lymphocytes within the stimulated node migrate into the follicles, enlarge, undergo blast transformation, and proliferate. These primary follicles enlarge, and areas of rapidly dividing B and T cells, called *germinal centers*, are formed inside. The enlarged primary follicles containing germinal centers are now called *secondary follicles*. The germinal centers are commonly observed to be avascular, with the vessels being displaced outward and forming a circular array around their periphery. The differentiated B lymphocytes from the secondary follicles, mostly plasma cells, migrate down through the cortex and into the medulla, where they either settle into the medullary cords or exit the node in the efferent lymph.

During the course of the immune response, the activated lymph node increases in size and weight. The first detectable increase in weight occurs approximately 12 hours following antigen stimulation, with peak increases (up to 10 times that of control) usually occurring 5–7 days following stimulation [14]. This increase in both size and weight has been attributed to alterations in blood flow, edema, increased cell migration from the blood stream, and cellular proliferation within the lymph node.

As a result of these antigen-induced changes, a rapid cellular and humoral immune response to the trapped antigen can be initiated within the lymph node. The rapid increase in blood flow and lymphocyte trafficking in the stimulated lymph node helps ensure that lymphocytes responsive to the trapped antigen are present in the node, and also increases the number of reactive lymphocytes that contact the antigen. Specifically, it has been demonstrated that the number of lymphocytes perfusing through the lymphoid tissue at the time antigen is present determines the magnitude of the subsequent immune response [15]. Thus, the antigen-dependent vascular changes help provide increased numbers of antigen-reactive lymphocytes within the node that are consequently able to participate in the ensuing immune response. This is especially important, since the number of lymphocytes that are able to recognize and respond to a particular antigen has been shown to be small [16]. It has also been shown that the cells leaving the stimulated lymph node in the efferent lymph are responsible for propagating and amplifying the systemic immune response to the antigen [17].

3. Techniques Used in Microcorrosion Casting Studies

3.1. Animals

Inbred Balb/c mice between the ages of 12 and 18 weeks were used.

3.2. Irradiation

The mice that were irradiated received 800 rads total body irradiation from a [137]Ce source.

3.3. Antigen Stimulation

At predetermined time points (0, 13 hours; 2, 4, 7 days) prior to corrosion casting, the mice received footpad injections of both keyhole limpet hemocyanin (KLH) (40 mg/ml in sterile saline) and 0.9% sterile saline. KLH (0.05 ml) was injected into the right rear footpad, and an equal volume of sterile saline was injected into the left rear footpad as a control. The irradiated mice received antigen stimulation 24 hours following irradiation and were corrosion cast at the same time points described above.

3.4. Volume Measurement

Following removal from the animal, the popliteal lymph nodes were carefully dissected free of the surrounding fat and equilibrated for 1 hour in 0.9% saline. The nodes were then blotted on filter paper and suspended on a wire in a beaker of 0.9% saline resting on a tared balance. Following addition of the node, the weight of the displaced volume of saline was determined. Each node was measured three times and the average value was used. The volume of the node was then calculated by dividing the weight of displaced saline by its specific gravity (i.e., 1.0048 g/cm^3).

3.5. Cell Counts

Blood was collected from the tail veins of normal and irradiated mice daily and placed directly into 0.89% ammonium chloride. The samples were then stained with acridine orange and counted under UV illumination in a hemacytometer.

3.6. Corrosion Casting

The methods used have been described previously [18,19]. At the predetermined time points, the animals were anesthetized with 0.15 ml ketamine (100 mg/ml) that had been supplemented with rompun (0.25 mg/ml) given s.c. The thoracic cavity was opened and 0.1 ml of a mixture of 1% sodium nitrite and heparin (1000 U/ml) was immediately injected into the left ventricle and allowed to circulate. A polyethylene cannula was then inserted through the left ventricle into the aorta and held in place with a silk thread. The right atrium was opened to serve as an efferent port. The animal was then perfused manually with 20 ml of prewarmed (37°C) modified Krebs buffer (NaCl, 7.145 g/l; NaHCO$_3$, 2.15 g/l; KCl, 0.385 g/l; MgSO$_4$ 7H$_2$O, 0.301 g/l; CaCl$_2$, 0.221 g/l; dextrose, 1.7 g/l; heparin, 10,000 U/l; NaNO$_2$, 0.207 g/l; and bovine serum albumin, 10 g/l). Following perfusion, the animal was cast by manually injecting 10 ml of a Mercox® mixture (1:1 Mercox/methyl methacrylate monomer, v/v), containing 0.16 g/ml catalyst, through the cannula. We have found this low-viscosity mixture, approximately 3 CP, to produce the most complete casts of mouse lymph nodes. The use of resin with this viscosity has been suggested by others, since it is similar to that of blood. However, when Mercox is diluted with monomer, both the degree of shrinkage during polymerization and the probability of vascular leakage have been reported to increase [20]. The animal was then placed in a plastic bag and in a 45°C water bath for several hours to complete the polymerization. The popliteal lymph nodes were dissected out and corroded by alternating washes in 7.5% sodium hydroxide and 5% Triton X-100 detergent at 45°C. Following corrosion the casts were rinsed thoroughly in distilled water. The entire maceration process generally lasted from 2–3 days. The casts were dehydrated in 100%

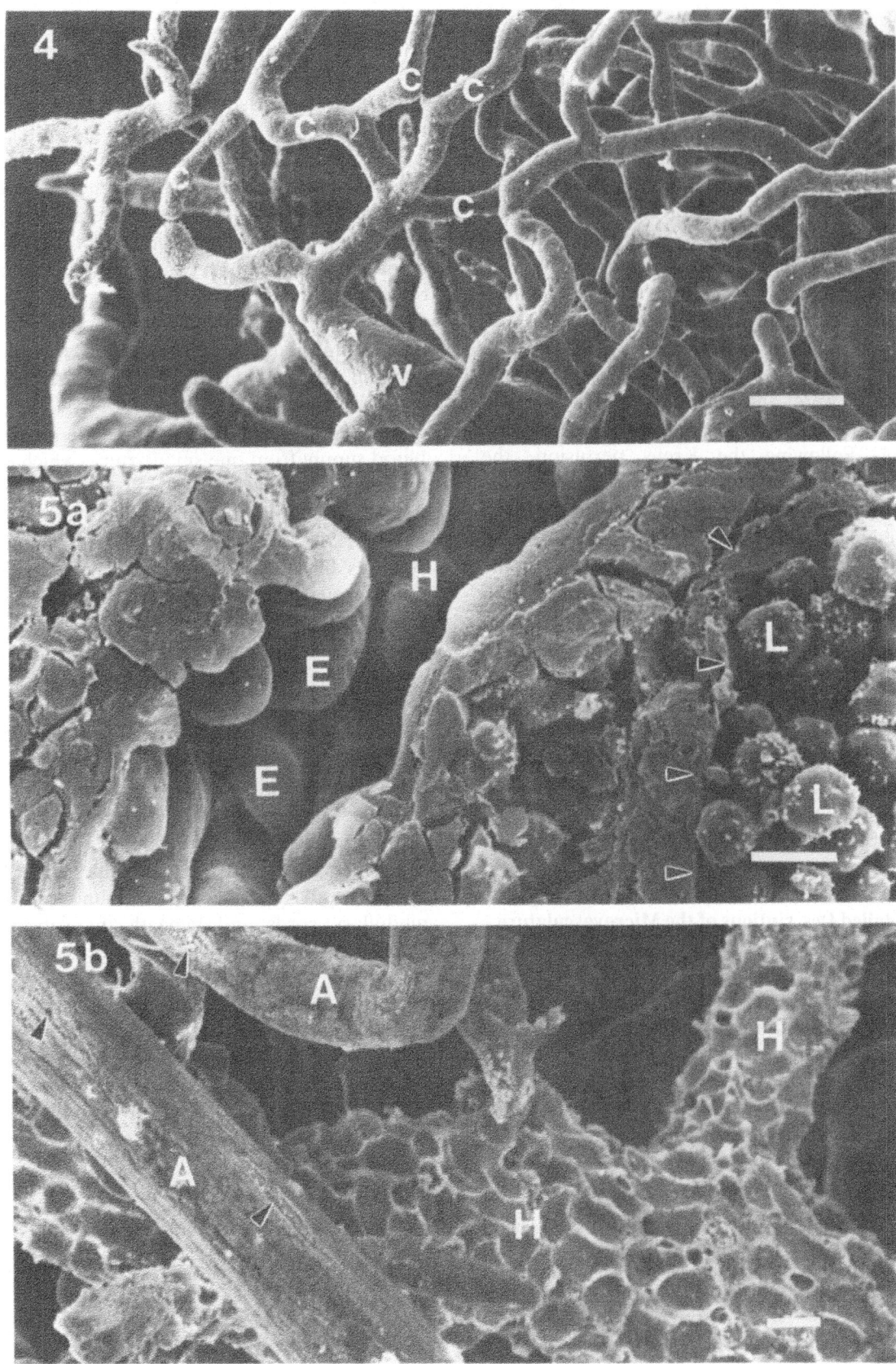

ethanol and dried by the cirtical-point method. The casts were mounted on JEOL stubs in silver conductive paint, sputter-coated with 10–20 nm gold palladium, and viewed with either a JEOL JSM-35C or Hitachi S-570 scanning electron microscope (SEM) at 10 kV accelerating voltage. Lower, 10 kV and below, accelerating voltages are desirable as the beam stability of the 1:1 resin mixture is somewhat reduced when compared to undiluted Mercox.

3.7. Preparation of Tissue for SEM

The mice were anesthetized, cannulated, and perfused as described above for corrosion casting. Following the Krebs perfusion, the animals were perfusion fixed with 20 ml of 4% paraformaldehyde/0.5% glutaraldehyde in 0.1 M phosphate buffer warmed to 37°C. The politeal lymph nodes were dissected out and placed in the above fixative for 1–2 hours at room temperature. The nodes were rinsed and dehydrated through a gradient of increasing concentrations of ethanol to 100% ethanol stored over molecular sieve. The nodes were plunged into liquid nitrogen and fractured with a precooled razor blade. The pieces were rapidly transferred back into 100% sieve-dried ethanol at 23°C and then were dried, mounted, sputter-coated, and viewed as described above for the corrosion casts.

4. Detailed Descriptions of the Microvasculature

4.1. Resting Lymph Nodes

The microvasculature of lymph nodes from many species, including human, dog, rat, sheep, rabbit, pig, guinea pig, and mouse, have been described. Several different techniques have been used to describe the microvasculature in these studies, including India ink injection, Microfil injection, microangiography, alcian blue dye staining, and microcorrosion casting. Even though different techniques were used and species differences were found, the findings of these studies are in general agreement with one another. However, only microcorrosion casting provides a complete three-dimensional replication of the microvasculature, allowing the accurate determination of the interrelationships between the different vessels. Therefore, the following description of the resting mouse lymph node microvasculature is based primarily on our studies utilizing microcorrosion casting.

In general, the murine lymph node receives its blood supply from one small artery, which enters only at the hilus (Fig. 23-2). In contrast to the mouse, up to 10–12 arteries have been observed to enter the sheep popliteal lymph node [21]. Arteries have also been observed to enter the node through the capsule rather than the hilus in sheep [21], rabbits [14], and most notably in the pig, which lacks a hilus [22]. Immediately upon entering the node, the artery divides into many smaller branches, which in turn branch as they run through the medulla and into the cortex, ultimately terminating in the subcapsular capillary arcades (Fig. 23-2). Arterioles are also observed to run up through the cortex, loop through the subcapsular capillary arcades, and run back down into the medulla (Fig. 23-2). Dense capillary networks are found within the medullary cords and beneath the subcapsular sinus. Less dense capillary networks are found throughout the undifferentiated areas of the cortex. Substantially fewer capillaries are found within the differentiated areas of the cortex (i.e., follicles). The germinal centers of secondary follicles, when present, are typically avascular (Fig. 23-3). Usually three to five capillaries join together and drain into either normal venules or

←

Figure 23-4. Typical vascular pattern observed in the outer cortex in which several capillaries (c) generally join together as they enter a venule (v). Bar = 20 μm.

Figure 23-5. a: Scanning electron micrograph of a freeze-fractured lymph node showing the lumen of a HEV (H). The bulging cuboidal endothelial cells (E) of the HEV are clearly seen. Characteristic concentric rings of lymphocytes (L) are found in perivascular sheaths (arrowheads) around the HEV. Bar = 5 μm. b: Corrosion cast of a HEV. The large cuboidal endothelial cells, as seen in Fig. 23-5a, produce characteristic deep impressions in the cast that can be used to identify the vessels as HEV (H). Two cast arterioles (A) are also present, demonstrating typical endothelial cell nuclear impressions (arrowheads) that are often observed in casts of low-endothelial vessels. Bar = 10 μm.

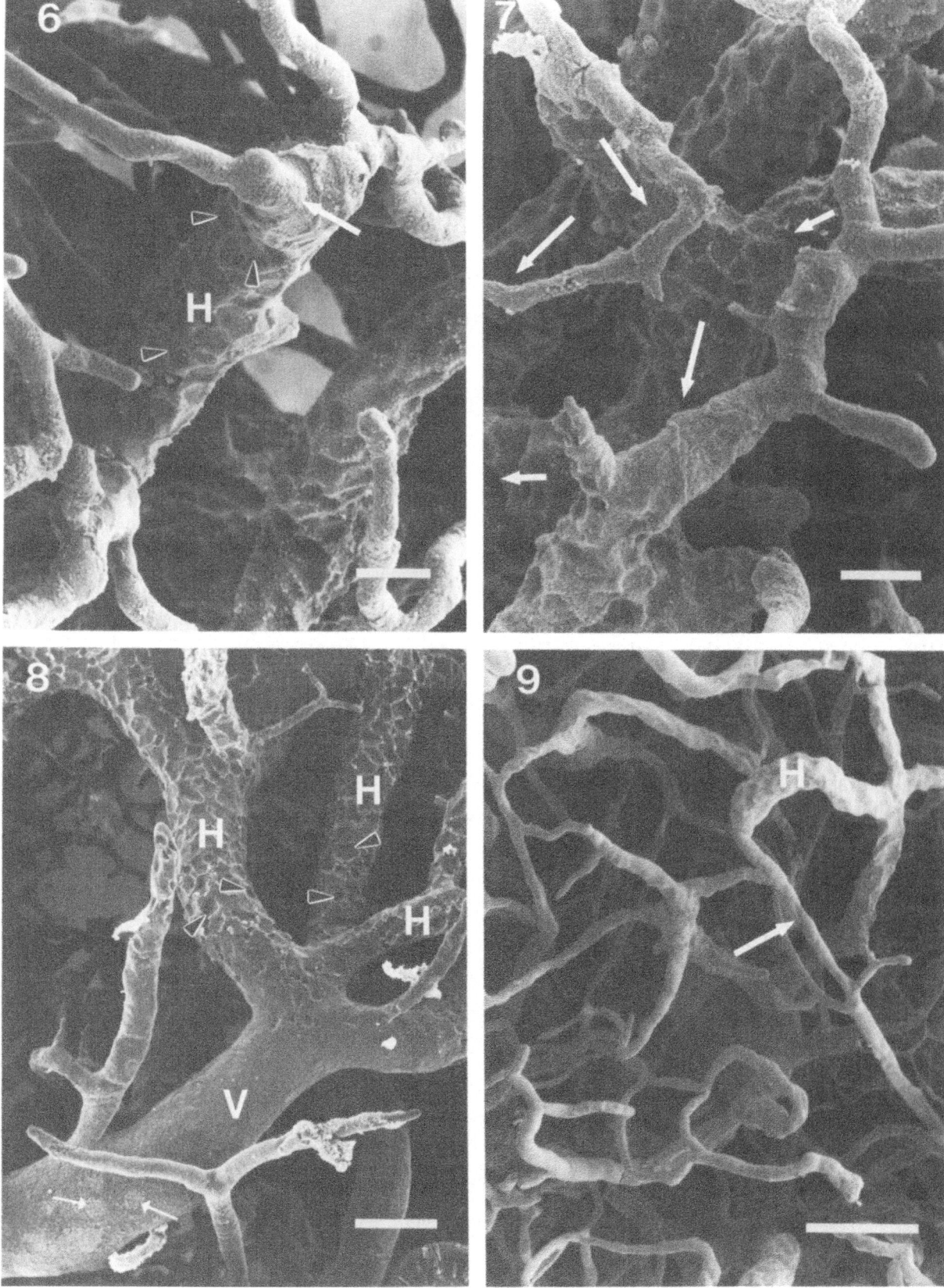

HEV (Fig. 23-4). The HEV are readily recognizable in the corrosion casts by the characteristic deep impressions produced by the high cuboidal endothelial cells (Fig. 23-5) [18,19,23]. There is some question as to whether the capillaries must first drain into normal low-endothelial venules, which then drain into HEV [24] or if the capillaries can drain directly into HEV [10,14,18,25]. We have observed both types of vascular patterns to occur in the corrosion casts and feel that the difference in findings may be due, in part, to the ability of the different techniques to distinguish between low-endothelial venules and HEV. Generally, there is an abrupt transition at the beginning of a HEV, with a large increase in lumen diameter and a change from low to high endothelium (Fig. 23-6). The HEV are primarily located in the paracortical or thymus-dependent region of the node, although they are also found in outer cortex beneath subcapsular sinus and can extend into the medulla. In the cortex, they are located randomly throughout the undifferentiated tissue and remain at the periphery of the follicles. Several small HEV join with larger diameter HEV as the vessels pass through the paracortical region toward the medulla (Fig. 23-7). Upon entering the medulla, these large HEV drain into the typical small veins of the medulla, which are lined with low endothelium.

In contrast to the abrupt transition seen at the beginning of a HEV in the cortex or paracortex, the transition from HEV into low-endothelial veins in the medulla is gradual. This is readily observed in corrosion casts as the HEV impressions become more shallow and are finally replaced by a typical venous low-endothelial cell pattern (Fig. 23-8) [18]. Interestingly, vessels having one side of their lumen lined with high endothelium and the other side with low endo-

thelium have occasionally been reported [25,26]. Contracted venous sphincters have been observed in the medullary veins at sites where they join the larger efferent veins and have been suggested to have a role in the regulation of blood flow through the HEV [24]. These venous sphincters have not been observed in corrosion casts. The medullary veins continue toward the hilus, joining larger veins that are in turn drained by a single large efferent vein that exits at the hilus (Fig. 23-2).

The lymph node also contains a secondary circulatory pathway consisting of numerous arteriovenous anastomoses (AVA), which serve to shunt the blood flow away from the cortical capillary networks directly into the HEV (Fig. 23-9) [22,24,27]. These AVA have also been observed in the thymus [28] and Peyer's patches [23]. The AVA in the lymph node are found throughout the cortex, have a luminal diameter of $6-15 \mu m$, and have been described to resemble terminal arterioles in terms of ultrastructure [24]. These shunts have been shown to play an important role in the control of regional blood flow through the lymph node following antigen stimulation [10].

4.2. Antigen-Stimulated Lymph Nodes

Alterations in the lymph node microvasculature in response to antigen stimulation have been known to occur since the early studies of Conway [29]. Since then, numerous reports describing vascular changes occurring in antigen-stimulated lymph nodes have been published [8–10,18, 25,30,31]. As with the descriptions of the resting lymph node microvasculature, several different techniques and animal models were used in these studies. We have found microcorrosion casting to

---

Figure 23-6. The beginning of a HEV (H) is clearly identified in a corrosion cast by the abrupt increase in vessel diameter (arrow), as well as by the presence of deep endothelial cell impressions (arrowheads) characteristic of HEV. Bar = 20 μm.

Figure 23-7. Several HEV located in the cortical region can be followed (arrows) as they merge to form larger diameter HEV. Bar = 20 μm.

Figure 23-8. The efferent end of a HEV is identified by the disappearance of deep endothelial cell impressions (arrowheads) along with the subsequent appearance of low endothelial cell impressions (arrows). In this micrograph three large diameter HEV (H) are seen to drain into a large diameter low endothelial vein (V). Bar = 50 μm.

Figure 23-9. An arteriovenous anastomosis (arrow) located in the outer cortex is shown to drain directly into a HEV (H). Bar = 100 μm.

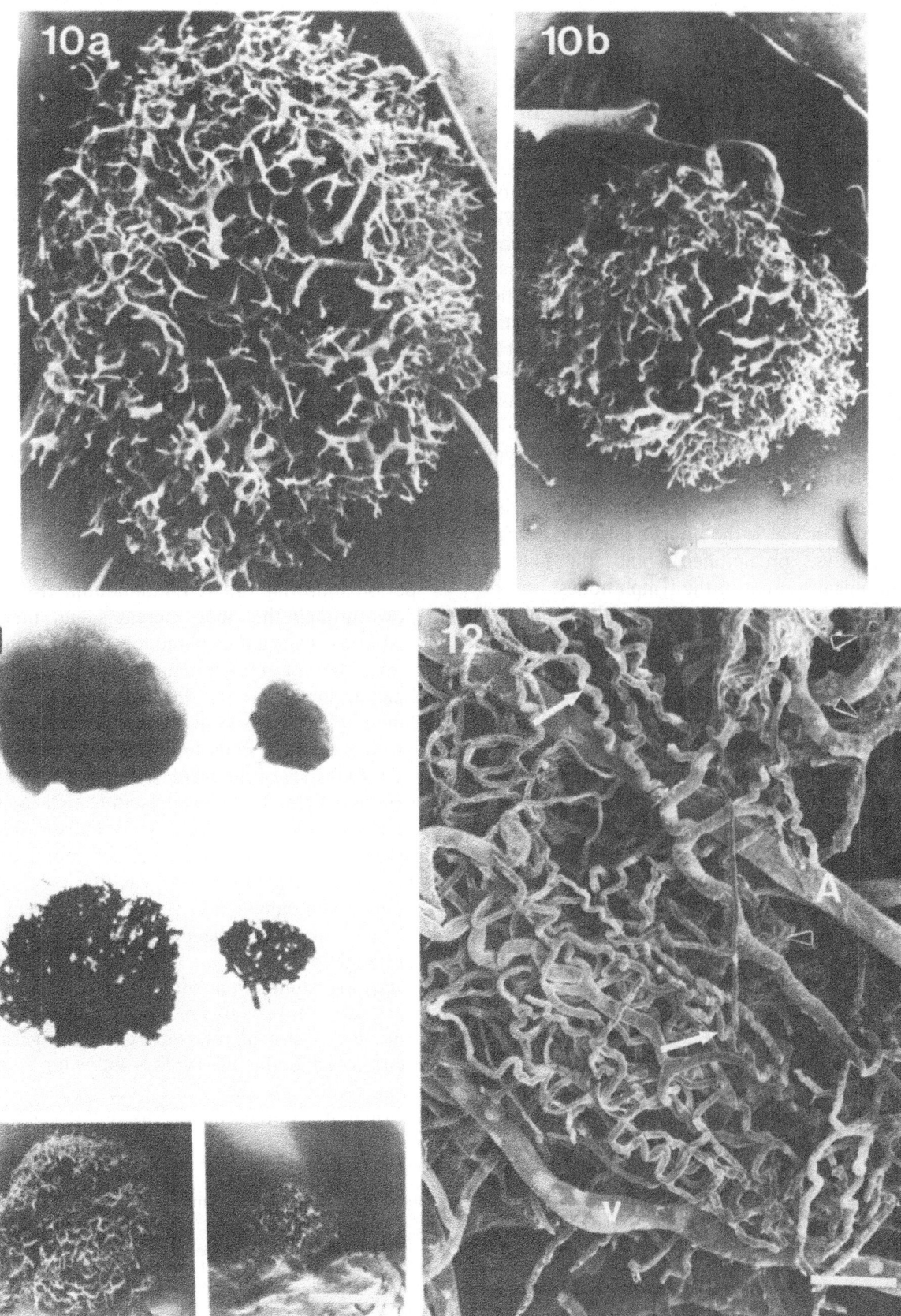

323

be an especially useful technique in comparing the vascular changes that occur following antigen stimulation, since the entire vasculature of both the stimulated and contralateral control nodes can be directly compared [18]. Other techniques, using isolated thick sections, provide only a limited three-dimensional view of the vasculature; therefore, direct comparisons are more difficult [9,14,30,31]. In the above studies, a number of consistent vascular changes have been reported to occur following antigen stimulation, and taken together, these reports represent the current understanding of this response. However, discrepancies in the findings of these reports do exist, but may be explained in part by the use of different stimulating antigens, such as thymus dependent (e.g., keyhole limpet hemocyanin) vs. thymus independent (e.g., lipopolysaccharide), the form in which the antigen is delivered (i.e., soluble vs. precipitated), and the different anatomical sources of the lymph nodes.

The initial vascular response of the lymph node following antigen stimulation involves a dilation of the subcapsular capillaries and AVA [14,30]. At this time, increased numbers of venous constrictions have also been observed [30]. As a result of these alterations, a significant redistribution and subsequent increase in regional blood flow through the node occurs [8,9,30]. Hay and Hobbs [9] demonstrated that a significant increase in the blood flow occurs as early as 1.5 hours following antigen stimulation. Blood flow peaks approximately 14 hours post-stimulation, after which it begins a transient decrease. Herman et al. [10] showed that most of the observed initial increase in regional blood flow is due to greatly increased shunt flow through the AVA, and thus through the HEV, with peak shunt activity occurring 16 hours following antigen stimulation.

Following the early response, dilation of the cortical and medullary cord capillaries occurs

and the larger vessels appear to become more uniformly spread throughout the cortex. By 2 days post-stimulation, the vasculature of the stimulated node expands to approximately twice the volume of the control node (Fig. 23-10). Four days following antigen stimulation, the size of the stimulated node is increased dramatically over that of the control node. Figure 23-11 illustrates the size increase occurring in a stimulated popliteal lymph node 4 days following antigen stimulation. Before corrosion, the control node was calculated to have a volume of 0.4 mm^3, while the enlarged stimulated node had a volume of 5.2 mm^3, a 13-fold difference. Following corrosion of the tissue, the size differential in the vasculature of the nodes is shown to duplicate that observed in the intact nodes (Fig. 23-11). Therefore, the observed volume increase in the stimulated node is paralleled by an increase in the volume of its vasculature. During the 2- to 4-day period following antigen stimulation, the blood flow through the node increases and then remains at a constant elevated level throughout the remainder of the immune response [8–10]. Specifically, it has been shown that there is a strict linear relationship between size and blood flow through a lymph node following antigen stimulation [8]. Much of the increased vasculature of the stimulated lymph node is present as rich capillary networks throughout the cortex, obliterating the avascular germinal centers, and throughout the medullary cords [14,18,30]. Vascular volume and density generally reach a peak between 4 and 7 days following antigen stimulation [14,18,30], although peak vascular density can occur later following stimulation with some antigens. The vasculature reverts to a more normal appearance, with the return of relatively avascular germinal centers, generally 10–14 days following stimulation. Although the vasculature appears more normal, it maintains an increased vascular density

Figure 23-10. Popliteal lymph node pair 2 days following footpad inoculations. a: Stimulated lymph node; b: contralateral control lymph node. Bar = 500 μm for both panels.
Figure 23-11. Popliteal lymph node pair 4 days following footpad inoculations: stimulated lymph node at left and contralateral control lymph node at right. Top: Light micrograph taken before corrosion of the tissue. Middle: Light micrograph taken following corrosion of the tissue. Bottom: Scanning electron micrographs showing the pattern of vessels in each node. Bar = 1 mm for all panels.
Figure 23-12. Cortical region of a lymph node 4 days following stimulation. Both capillaries (arrows) and some larger vessels appear to follow a torturous pathway. Arrow heads = HEV; A = arteriole; v = venule. Bar = 50 μm.

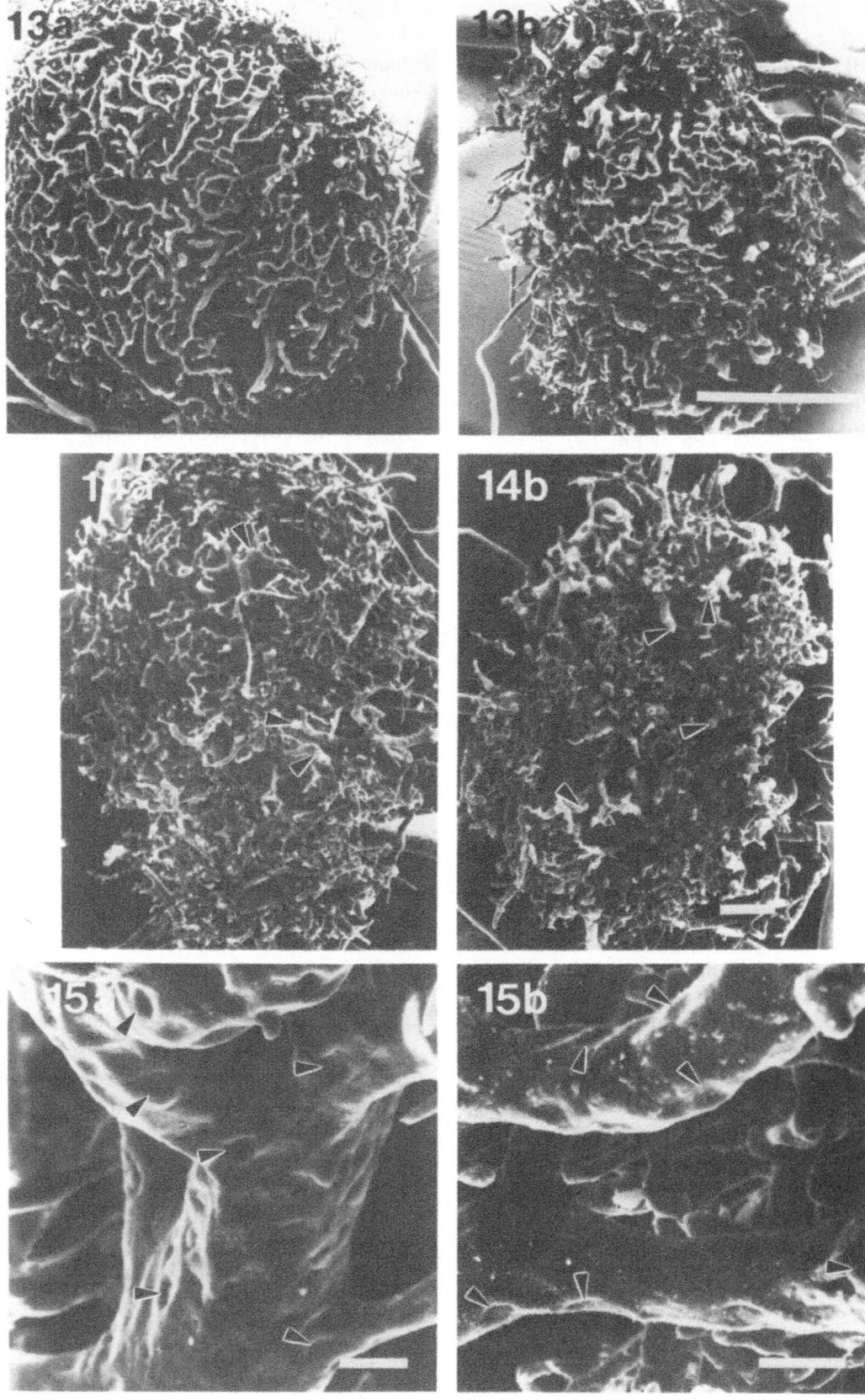

in the cortex and medullary cords for several weeks [30]. In direct comparisons of corrosion casts, the vasculature of the stimulated node has been observed to maintain an increased volume, over that of the contralateral control node, for up to 5 weeks post-stimulation. In some cases, the weight of the node may remain increased as long as 75 days following antigen stimulation [14].

4.2.1. The origin of the microvasculature: Neovascularization or redistribution?

As the lymph node responds to antigen, it undergoes a marked "hypervascularity," as described above. At present there is no general agreement as to the origin of this increased vasculature. In our vascular corrosion casting studies, in which direct comparisons of control and stimulated lymph node casts are made, it is observed that while the cast of the stimulated node is typically four to six times as large, it maintains the same relative vessel density as the control cast. Also, vessels in the stimulated node are frequently observed to follow a tortuous or "corkscrew" path, which has been described to occur during neovascularization (Fig. 23-12) [32]. These findings suggest that new vessel growth may account for much of the observed vascular increase in the stimulated node. The vasculature of the lymph node has been suggested to proliferate in response to antigen in a number of other reports [9,25,30], and endothelial cell proliferation has been documented to occur following antigen stimulation [25,30]. Other reports suggest neovascularization to be minimal in the lymph node following antigen stimulation [14,31]. In these studies, the vessels of the stimulated lymph node were observed to maintain their regular outline and branching patterns as the node increased in size, thus suggesting that a redistribution and outward expansion of the existing vessels occurred. The

above finding that vessel density remains unchanged in the stimulated node does not support this, as outward expansion of the vasculature without growth would result in a corresponding reduction in the vascular density of the stimulated node.

An increase in the relative amount of HEV in the lymph node following antigen stimulation has been reported [14,18,25,30,33], although conflicting reports also exist [34,35]. When the corrosion casts of control and stimulated lymph nodes are compared, the proportion of microvasculature occurring as HEV is observed to remain constant [18]. This suggests that the typical four- to six-fold volume increase observed in the vasculature of stimulated nodes results in a corresponding increase in the amount of HEV. In the corrosion casts, this increase in HEV is evident throughout the cortex and paracortical region, with extensive arborization observed. The apparent large increase in the amount and branching of the HEV following antigen stimulation is indicative of proliferation. Reports of increased pyroninophilia [25,30] and height [34] occurring in HEV following antigen stimulation suggest that these cells become activated. This is supported by reports of increased mitotic activity in HEV following antigen stimulation [25,30]. Specifically, Anderson et al. [30] reported finding a sixfold increase in the mitotic activity of the HEV following antigen stimulation. Furthermore, the mitotic activity was observed to be restricted to junction sites between low-endothelial venules and HEV, thereby suggesting that lengthening of the HEV occurs through endothelial cell proliferation at these sites. Proliferation at these sites would also account for the increased branching of the HEV observed in the corrosion casts. Others have attributed the observed increase in HEV to be the result of a conversion of the existing

←

Figure 23-13. Popliteal lymph node pair 3 days following irradiation (2 days following footpad inoculations). a: Stimulated lymph node. b: Contralateral lymph node. Note the small degree of enlargement in the stimulated node. Bar = 250 μm for both panels.

Figure 23-14. Popliteal lymph node pair 5 days following irradiation (4 days following footpad inoculations). a: Stimulated lymph node. b: Contralateral control lymph node. Several blunt-ended HEV (arrowheads) are present in both casts. Note the lack of size increase in the stimulated node. (Reproduced from Steeber et al. [18], with permission.) Bar = 250 μm for both panels.

Figure 23-15. a: HEV from a lymph node 3 days following irradiation. Note the shallow endothelial cell impressions (arrowheads). Bar = 20 μm. b: HEV from a lymph node 8 days following irradiation. Note the near absence of characteristic HEV impressions (arrowheads) seen whether the node is resting or antigen stimulated. (Reproduced from Steeber et al. [18], with permission.) Bar = 20 μm.

low-endothelial venules into HEV [14]. The endothelium of normal venules from nonlymphoid tissues, under various conditions such as inflamed synovia [36–38], chronic granuloma [39], and in areas surrounding malignant neoplasms [40] and delayed hypersensitivity reactions [41], has been reported to exhibit characteristics of HEV. It has also been demonstrated that cultured mouse endothelial cells derived from flat endothelium can be induced to express surface markers characteristic of HEV [39]. These findings lend support to the idea that some conversion of normal venules into HEV may occur in the lymph node following antigen stimulation.

5. Lethal Irradiation

5.1. Microvascular Response to Antigen

The effects of irradiation on lymph nodes have been described in numerous studies [18,39, 42–44]. While some of these studies have addressed the effects of irradiation on the lymph node's cellular responsiveness to antigen [42,43], very few studies concerning the effects of irradiation on the microvascular response to antigen have been performed. Therefore, we used microcorrosion casting to study the specific effects of lethal irradiation on the vascular changes observed in lymph nodes following antigen stimulation [18].

Gross examination of the antigen-stimulated and contralateral control nodes showed slight atrophy to occur in both as early as 3 days post-irradiation and to become more severe by day 8. Much of this early decrease in node size can be attributed to extensive lymphocyte depletion, both within the node and in the circulation. Circulating levels of white blood cells were found to be reduced by >80% and >98% by 2 and 8 days post-irradiation, respectively. At 2 days post-irradiation, no detectable radiation effects on the microvasculature are found in corrosion casts of either control or stimulated (1 day post-stimulation) lymph nodes. Rich subcapsular and medullary capillary arcades, as well as numerous AVA, are observed in corrosion casts from both control and stimulated nodes. At 3 days post-irradiation, the stimulated lymph nodes (2 days

post-stimulation) show little if any significant size increase in vasculature over that of the control nodes (Fig. 23-13). This lack of vascular response in the stimulated nodes is in contrast to that observed in the nonirradiated animals, in which the stimulated lymph nodes typically exhibit at least a twofold volume increase in vasculature over that of the control nodes at this time (Fig. 23-10). The cortical vessels in corrosion casts from both control and stimulated lymph nodes appear to be slightly dilated when compared to those from nonirradiated animals. This finding is consistent with a number of studies in which transient vasodilation following irradiation has been reported to occur [45,46]. This effect is most likely to be the result of released inflammatory mediators, rather than a direct irradiation effect on the endothelium.

At 5 days post-irradiation, several changes in the microvasculature are observed in corrosion casts of both stimulated (4 days post-stimulation) and control lymph nodes (Fig. 23-14). Again, in contrast to the dramatic vascular volume increase observed at this time in stimulated nodes from nonirradiated animals (Fig. 23-11), no significant vascular increase is observed in stimulated nodes from irradiated animals. At this time the capillary density appears to be decreased throughout both control and stimulated lymph nodes and AVA are less apparent. This finding may be attributed, at least in part, to a direct irradiation effect, since capillaries have been reported to be the most sensitive of the blood vessels to irradiation [45, 46]. Doses of radiation between 200 and 2000 rads have been reported to cause direct effects on capillaries, such as endothelial cell edema [46,47], decrease in length [46], decrease in number [47], and loss of patency [46,47]. Additionally, the proliferative capacity of resting capillary endothelium has been shown to be inhibited in a dose-dependent manner by irradiation in the 250–750 rad range [48]. However, it has been reported that applying the stimulation for endothelial cell proliferation after irradiation, as in our protocol, instead of before, results in a higher level of endothelial cell survival [45].

Another observation made in casts at 5 days post-irradiation (4 days post-stimulation) is that increased numbers of blunt-ended vessels, especially HEV, occur in both control and stimulated nodes. This finding is somewhat unexpected, since the larger blood vessels have been reported to be fairly resistant to irradiation, with no direct effects being observed even at doses from 1800 to >4000 rads [46,47]. However, it has been reported that the endothelial cells of the HEV, following 740 rads total body irradiation, undergo hypertrophy and possibly proliferation, resulting in apparent occlusion of the lumen by 7 days post-irradiation [44]. This occlusion would account for the increased numbers of blunt-ended HEV observed in the corrosion casts. Corrosion casts of stimulated (7 days post-stimulation) and control nodes at 8 days post-irradiation appear to be very similar to those described at 5 days post-irradiation.

5.2. Morphology of the HEV

In our lethal irradiation studies, the morphology of the HEV, as reflected by the impressions in the corrosion casts, was observed to change following irradiation [18]. The earliest discernible change in morphology occurred at 3 days post-irradiation. In corrosion casts from both control and stimulated lymph nodes, the HEV impressions were observed to be more shallow and spread out than those typically observed in corrosion casts from nonirradiated animals (Fig. 23-15). This alteration in the morphology of the impressions indicates that a "flattening" of the normally high-endothelial cells of the HEV had occurred following irradiation. This flattening of the HEV was observed to become more pronounced with time and was very evident at 8 days post-irradiation (the last time point studied) (Fig. 23-15). Other investigators have since confirmed this finding [39]. Interestingly, these investigators also found that infusion of viable lymph node cells resulted in a very rapid restoration of the HEV morphology.

One possible explanation for the observed flattening of the HEV following irradiation is that lymphocyte recirculation through these venules is necessary for the maintenance of their cuboidal morphology. Other reports of HEV flattening following various experimental regimes that also produce lymphocyte depletion, such as neonatal thymectomy [34,49–51], treatment with lympholytic agents [36], treatment with antilymphocyte sera [52], and chronic drainage of the thoracic duct [50], lend support to this idea. Restoration of HEV morphology has also been reported to occur following infusion of lymphocytes in animals subjected to neonatal thymectomy [50,51] or chronic drainage of the thoracic duct [50].

6. Concluding Remarks

It is apparent that the dynamic nature of the lymph node vasculature plays a critical role in the function of this organ. Particularly important to lymph node function are the overall three-dimensional relationships between the various vessel types that give rise to the regional specializations seen in normal and antigenically stimulated nodes. Corrosion casting has proved to be a useful technique for identifying HEV and for detecting alterations in their morphology. This technique should also prove useful in studying these venules when they occur in nonlymphoid tissues as a result of various disease states. Studies employing microvascular corrosion casting can be performed in parallel with a variety of other research techniques, such as cryofracture or embedding and sectioning of stained or labelled whole tissues. This correlative approach should provide the means to understand the complex cellular and vascular interactions occurring in lymphoid tissue.

Acknowledgments

The authors express their gratitude to Ms. Julie Oliver for her expert assistance with the scanning electron microscopy and in the preparation of the manuscript and Dr. James Walmsley for his helpful discussions and critical review of the manuscript. We also thank Mr. Jeffery Butzow for graphics work and Mr. Paul Sims for photographic work.

References

1. Calvert WJ. The blood-vessels of the lymphatic gland. *Anat Anz* 13:174–180, 1897.

2. Schulze W. Untersuchungen über die capillaren und postcapillären venen lymphatischer organe. *Z Anat Entwicklungsgesch* 76:421–462, 1925.

3. Ehrich W. Studies of the lymphatic tissue. I. The anatomy of the secondary nodules and some remarks on the lymphatic and lymphoid tissue. *Am J Anat* 43:347–383, 1929.

4. Smith C, Hénon BK. Histological and histochemical study of high endothelium of post-capillary veins of the lymph node. *Anat Rec* 135:207–213, 1959.

5. Röpke C, Jørgensen O, Claësson MH. Histochemical studies of high-endothelial venules of lymph nodes and Peyer's patches in the mouse. *Z Zellforsch* 131:287–297, 1972.

6. Gowans JL, Knight EJ. The route of re-circulation of lymphocytes in the rat. *Proc R Soc London* [Biol] 159:257–282, 1964.

7. Hall JG, Morris B. The origin of the cells in the efferent lymph from a single lymph node. *J Exp Med* 121:901–910, 1965.

8. Herman PG, Lyonnet D, Fingerhut R, Tuttle RN. Regional blood flow to the lymph node during the immune response. *Lymphology* 9:101–104, 1976.

9. Hay JB, Hobbs BB. The flow of blood to lymph nodes and its relation to lymphocyte traffic and the immune response. *J Exp Med* 145:31–44, 1977.

10. Herman PG, Utsunomiya R, Hessel SJ. Arteriovenous shunting in the lymph node before and after antigenic stimulus. *Immunology* 36:793–797, 1979.

11. Cahill RNP, Frost H, Trnka Z. The effects of antigen on the migration of recirculating lymphocytes through single lymph nodes. *J Exp Med* 143:870–888, 1976.

12. Hall JG, Morris B. The immediate effect of antigens on the cell output of a lymph node. *Br J Exp Pathol* 46:450–454, 1965.

13. Sprent J, Miller JFAP, Mitchell GF. Antigen-induced selective recruitment of circulating lymphocytes. *Cell Immunol* 2:171–181, 1971.

14. Herman PG, Yamamoto, I, Mellins HZ. Blood microcirculation in the lymph node during the primary immune response. *J Exp Med* 136:697–714, 1972.

15. Ford WL, Gowans JL. The role of lymphocytes in antibody formation. II. The influence of lymphocyte migration on the initiation of antibody formation in the isolated, perfused spleen. *Proc R Soc London* [Biol] 168:244–262, 1967.

16. Modabber F, Morikawa S, Coons AH. Antigen-binding cells in normal mouse thymus. *Science* 170:1102–1103, 1970.

17. Hall JG, Morris B, Moreno GD, Bessis MC. The ultrastructure and function of the cells in lymph following antigenic stimulation. *J Exp Med* 125:91–110, 1967.

18. Steeber DA, Erichson CM, Hodde KC, Albrecht RM. Vascular changes in popliteal lymph nodes due to antigen challenge in normal and lethally irradiated mice. *Scanning Microsc* 1:831–839, 1987.

19. Hodde KC, Steeber DA, Albrecht RM. Advances in corrosion casting methods. *Scanning Microsc* 4:693–704, 1990.

20. Weiger T, Lametschwandtner A, Stockmayer P. Technical parameters of plastics (Mercox CL-2B and various methylmethacrylates) used in scanning electron microscopy of vascular corrosion casts. *Scann Electron Microsc* 1986; I:243–252, 1986.

21. Heath T, Brando R. Lymphatic and blood vessels of the popliteal node in sheep. *Anat Rec* 207:461–472, 1983.

22. Spalding HJ, Heath TJ. Blood vessels of lymph nodes in the pig. *Res Vet Sci* 41:196–199, 1986.

23. Yamaguchi K, Schoefl GI. Blood vessels of the Peyer's patch in the mouse: III. High-endothelium venules. *Anat Rec* 206:419–438, 1983.

24. Anderson AO, Anderson ND. Studies on the structure and permeability of the microvasculature in normal rat lymph nodes. *Am J Pathol* 80:387–418, 1975.

25. Burwell RG. Studies of the primary and the secondary immune responses of lymph nodes draining homografts of fresh cancellous bone (with particular reference to mechanisms of lymph node reactivity). *Ann NY Acad Sci* 99:821–860, 1962.

26. Sainte-Marie G. The postcapillary venules in the mediastinal lymph node of ten-week-old rats. *Rev Can Biol* 25:263–284, 1966.

27. Irino S, Takasugi N, Murakami T. Vascular architecture of thymus and lymph nodes: Blood vessels, transmural passage of lymphocytes, and cell-interactions. *Scann Electron Microsc* 1981; III:89–98, 1981.

28. Blau JN. A comparative study of the microcirculation in the guinea-pig thymus, lymph nodes and Peyer's patches. *Clin Exp Immunol* 27:340–347, 1977.

29. Conway EA. Cyclic changes in lymphatic nodules. *Anat Rec* 69:487–513, 1937.

30. Anderson ND, Anderson AO, Wyllie RG. Microvascular changes in lymph nodes draining skin allografts. *Am J Pathol* 81:131–160, 1975.

31. Herman PG. Microcirculation of organized lymphoid tissues. *Monogr Allergy* 16:126–142, 1980.

32. Grunt TW, Lametschwandtner A, Karrer K, Staindl O. The angioarchitecture of the Lewis lung carcinoma in laboratory mice (a light microscopic and scanning electron microscopic study). *Scann Electron Microsc* 1986; II:557–573, 1986.

33. Hendriks HR, Eestermans IL. Disappearance and reappearance of high endothelial venules and immigrating lymphocytes in lymph nodes deprived of afferent lymphatic vessels: A possible regulatory role of macrophages in lymphocyte migration. *Eur J Immunol* 13:663–669, 1983.

34. Krüger G. Morphology of postcapillary venules under different experimental conditions. *J Natl Cancer Inst* 41:287–301, 1968.

35. Twisk AJT, Groeneveld PHP, Kraal G. The effects of bacterial lipopolysaccharide (LPS) on high endothelial venules, T lymphocytes and interdigitating cells in mouse lymph nodes. *Adv Exp Med Biol* 237:807–811, 1988.

36. Miller JJ. Studies of the phylogeny and ontogeny of the specialized lymphatic tissue venules. *Lab Invest* 21:484–490, 1969.

37. Freemont AJ, Jones CJP, Bromley M, Andrews P.

Changes in vascular endothelium related to lymphocyte collections in diseased synovia. *Arthritis Rheum* 26: 1427–1433, 1983.

38. Jalkanen S, Steere AC, Fox RI, Butcher EC. A distinct endothelial cell recognition system that controls lymphocyte traffic into inflamed synovium. *Science* 233: 556–558, 1986.

39. Duijvestijn AM, Rep M, Butcher EC, Hendriks HR, Kraal G. Regulation of functional and morphological aspects of high endothelium in mouse. *Adv Exp Med Biol* 237:491–497, 1988.

40. Freemont AJ. The small blood vessels in areas of lymphocytic infiltration around malignant neoplasms. *Br J Cancer* 46:283–288, 1982.

41. Polverini PJ, Cotran RS, Sholley MM. Endothelial proliferation in the delayed hypersensitivity reaction: An autoradiographic study. *J Immunol* 118:529–532, 1977.

42. Hall JG, Morris B. Effect of x-irradiation of the popliteal lymph-node on its output of lymphocytes and immunological responsiveness. *Lancet* 1:1077–1080, 1964.

43. Engeset A. Local irradiation of lymph nodes in rats. Morphological and functional alterations with relation to cancer therapy. *Prog Exp Tumor Res* 8:225–270, 1966.

44. Samlowski WE, Johnson HM, Hammond EH, Robertson BA, Daynes RA. Marrow ablative doses of gamma-irradiation and protracted changes in peripheral lymph node microvasculature of murine and human bone marrow transplant recipients. *Lab Invest* 56:85–95, 1987.

45. Reihold HS. Chapter II. Cell viability of the vessel wall. *Curr Top Radiat Res Q* 10:9–28, 1974.

46. Dimitrievich GS, Fischer-Dzoga K, Griem ML. Radiosensitivity of vascular tissue. I. Differential radiosensitivity of capillaries: A quantitative in vivo study. *Radiat Res* 99:511–535, 1984.

47. Fajardo LF, Stewart JR. Pathogenesis of radiation-induced myocardial fibrosis. *Lab Invest* 29:244–257, 1973.

48. Reinhold HS, Buisman GH. Radiosensitivity of capillary endothelium. *Br J Radiol* 46:54–57, 1973.

49. Parrott DMV, de Sousa MAB, East J. Thymus-dependent areas in the lymphoid organs of neonatally thymectomized mice. *J Exp Med* 123:191–204, 1966.

50. Goldschneider I, Mc Gregor DD. Migration of lymphocytes and thymocytes in the rat. I. The route of migration from blood to spleen and lymph nodes. *J Exp Med* 127:155–168, 1968.

51. Jørgensen O, Claësson MH. Studies on the post-capillary high endothelial venules of neonatally thymectomized mice. *Z Zellforsch* 132:347–355, 1972.

52. Röpke C. Effects of antithymocyte serum and antilymphocyte serum on the postnatal development of the thymo-lymphatic system in Balb/c mice. *J Reticuloendothel Soc* 13:78–89, 1973.

Author's address:
Dr. Douglas A. Steeber
University of Wisconsin
Department of Veterinary Science
1655 Linden Drive
Madison, WI 53706
USA

The Fundamental Vascular Structures in Epithelial Cancer Demonstrated in an Experimental Lung Carcinoma

THOMAS W. GRUNT

1. Introduction

Tumor biologists and clinical oncologists have long been interested in the vascularization of malignant tissues. However, this interest arose periodically, depending on the development of novel technologies, which gave fresh impetus to this area of research. A well-established method, for example, is the implantation of tumors into transparent chambers, which are then inserted into host tissue. This technique allows repeated microscopical examination of the tumor vasculature and the estimation of diverse physiological parameters in situ. Other systems, preferentially used as in vivo angiogenesis assays, comprise the application of tissues or substances to the chorioallantoic membrane of the chick embryo. All these methods were established for solving specific problems and have considerable merits in the field of tumor vascular research. They are not useful, however, for the detailed stereoscopic demonstration of the tumor vascular bed and alterations occurring during unrestricted neo-plastic growth in situ. For such questions, scanning electron microscopy (SEM) of vascular corrosion casts is the method of choice. Since the advent of this technique [1], we have gathered comprehensive knowledge about the vascular development, host vessel recruitment and displacement, and vascular degeneration occurring in malignant tissues [2–11] (see Chapters 3, 25, and 26.)

Since the pioneering work of Folkman's group [12–14], it has been well known that tumors produce substances that cause the adjacent host blood vessels to form a new microvascular bed. By inhibiting this angiogenesis, tumor growth and hematogenous metastasis can be blocked [15,16]. For therapeutic reasons it is important to realize that tumor vascularization determines the cure rates of radiotherapy and chemotherapy, as well as of treatment with hyperthermia [17,18]. Most tumors have specific vascular patterns. Therefore, angiography of the tumor vascular system provides a valuable tool for tumor diagnosis [19]. The tumor vascular bed is of significant importance for tumor metabolism, tumor growth, necrotic degeneration, and blood-borne dissemination of malignant cells. Finally, for designing the most effective therapeutic regimen for a given tumor (radiotherapy, chemotherapy, hyperthermia), the peculiarities of the vascular system of this tumor should be taken into account.

This chapter will survey the three-dimensional angioarchitecture and submicroscopic vascular structures occurring in an experimental murine lung carcinoma. Furthermore, the fundamental vascular phenomena occurring in a majority of epithelial tumors will be outlined and contrasted with those vascular features that may be specific for individual tumor types.

2. Methodological Remarks

In the present experimental work we used the Lewis lung carcinoma transplanted subcutaneously into C57Bl/6-mice. For correlative

Motta, P.M., Murakami, T., and Fujita, H. (eds.), Scanning Electron Microscopy of Vascular Casts: Methods and Applications.

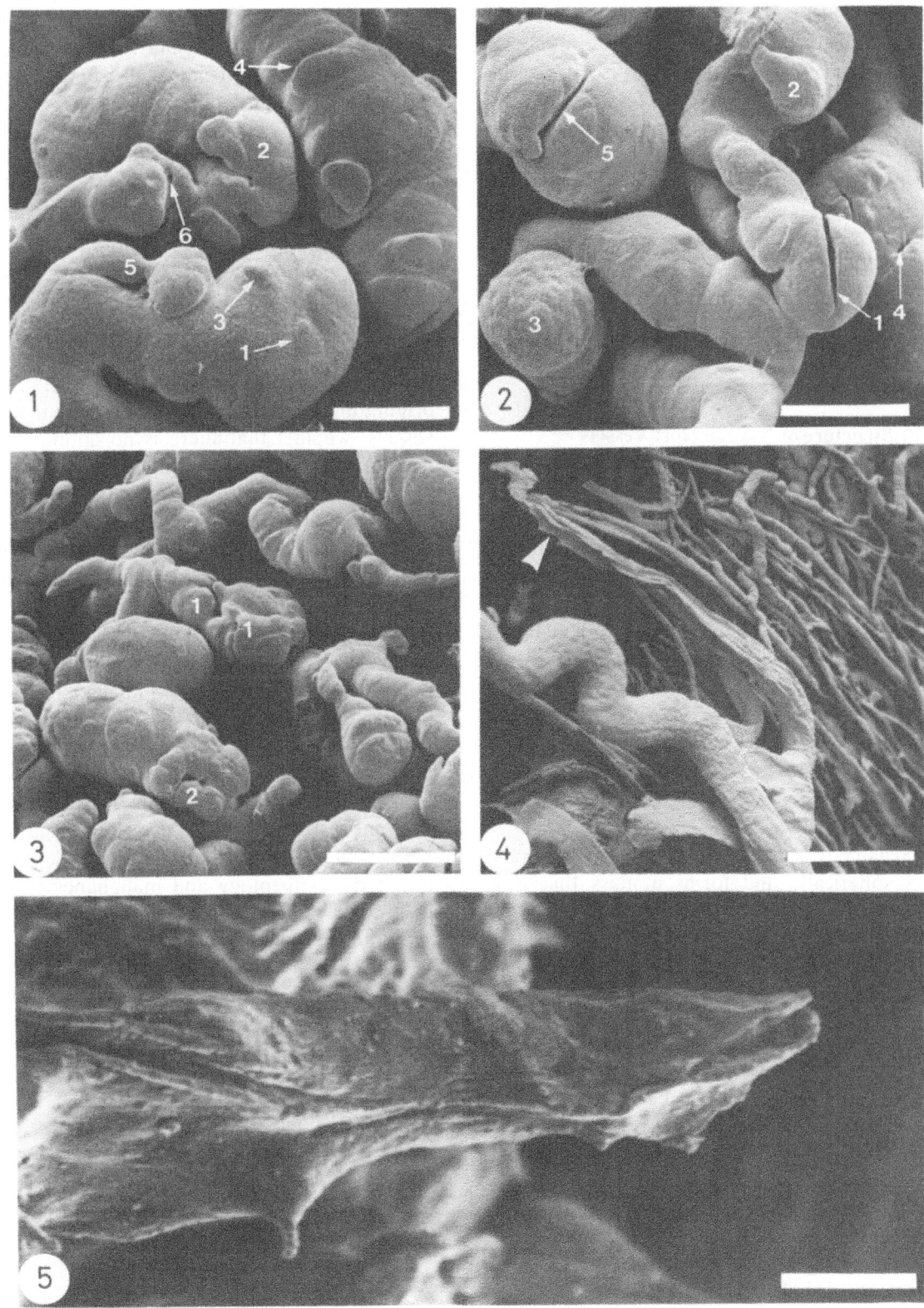

reasons, additional investigations were done in human basal cell tumors. Apart from SEM of critical-point-dried tumor tissue blocks and standard light microscopy of tissue sections, the majority of the tumors were prepared for SEM of vascular corrosion casts. Tumors of different growth stages were prepared in this manner in order to trace the chronological alterations occurring in blood vessels. The use of different perfusion pressures enabled us to investigate the influence of this parameter on the organization of the casts. The detailed protocols describing the methods used can be found in previous publications [2,4,5].

3. The Structure of Normal Host Vessels

For a correct assessment of the vascular remodelling occurring during tumor growth, it is necessary to compare the tumor blood vessels with the architecture and structure of the normal host vascular bed. Some characteristics of the subcutaneous vascular system of mice are depicted in Fig. 24-1.

4. Gross Survey of the Tumor Vascular System and its Implications in Tumor Histology and Malignancy

The axillary transplanted Lewis lung carcinoma grows spherically and forms a mass that contains only minor portions of stromal tissue components. At about the eighth day of malignant growth, the first necrotic foci develop, which then expand into a large necrotic area in the tumor center. This necrosis is caused primarily by elevations of the interstitial tissue pressure occurring during tissue expansion and by a reduction in tumor blood flow, leading to compression of the central tumor blood vessels. According to this peculiar growth pattern, the tumor vascular system soon takes the shape of a hollow sphere, comprising a central cavity and a peripheral vascular "envelope." Generally, during invasive growth the tumors make multiple contacts with the host vascular system, thereby recruiting all nearby host arteries and veins, which then serve to meet the metabolic demands of the proliferating tissue. Apart from the finding that venous vessels are more susceptible to the influences originating from the tumor tissue than arterial vessels, we feel that there is no additional preference for any particular category of host vessels to be recruited and remodelled by the malignant tissue. The only determinant is the location of the host vessels relative to the growing tumor mass.

In correlative studies, which were designed to demonstrate the interdependence between the tumor parenchyma and the tumor stroma, we compared the angioarchitecture of the normal human skin and of the "semimalignant" human basal cell tumor with the vascular patterns of the mouse skin and of the highly malignant Lewis lung carcinoma [4]. From these investigations it became evident that the architecture of the tumor vascular system strictly follows the overall growth pattern of the tumor, which in turn depends on the tumor's histology and malignancy. In basal cell tumors, the parenchyma grows in the form of separate cell aggregates, which are embedded in the stroma tissue. According to this spatial distribution, the vascular bed is arranged in mul-

---

Figures 24-1 to 24-3: Remodelling and proliferation on dilated blood vessels at the tumor base (phase 1 of tumor angiogenesis). Vascular corrosion cast. SEM.

Figure 24-1. (1) Small furrow; (2) large flap; (3) small, tapered bulge; (4) small furrow between two slight bulges; (5) bowlike collateral; (6) small hole. Bar = 20 μm.

Figure 24-2. (1) Deep, incomplete circular furrow separating a small flap; (2) slightly curved, tapered structure; (3) dome-shaped to fungiform bulge; (4) small hole; (5) deep furrow. Bar = 20 μm.

Figure 24-3. (1) complex arrangement of several bulges; (2) ringlike anastomosis. Bar = 40 μm.

Figure 24-4. Apical tumor periphery. Venous vessels of the "tumor vascular envelope" with a wrinkled, concave surface. Vascular corrosion cast. SEM. Bar = 200 μm.

Figure 24-5. A tapered tumor vessel projects into the central, avascular cavity. Note the endothelial-cell nuclear imprint at the end of the tapering vessel. A longitudinal furrow represents the borderline between two adjacent endothelial cells. Vascular corrosion cast. SEM. Bar = 5 μm.

334

tiple basket-like capillary plexuses, each of which embraces an individual tumor cell aggregate.

Sprouts were found occasionally arising from the capillary plexus and ascending into the overlying hollow space. This situation, found in the basal cell tumors, resembles the first stage of growth of the Lewis lung carcinoma (see below). The tumor-specific vascular reactions, therefore, correlate closely with the histology, growth pattern, growth rate, and malignancy of a given tumor. This opinion is supported by a recent contribution from one of our laboratories working with a rat glioma model [10]. The malignant tissue generally acts as an "organizer" of the newly formed vascular system and determines its overall architecture. During later stages of the neoplastic development, however, the microcirculation becomes insufficient and restricts further tumor expansion. For each individual tumor entity, a specific parenchyma-stroma interrelation does exist. According to Gabbert et al. [21], the tumor growth rate depends primarily on inherent properties of the cell population, while the vascularization provides the proper conditions for cell proliferation. Additionally, in a recent study we could demonstrate that even established tumor cell lines contain cell subpopulations that exhibit heterogeneous growth rates [22].

5. The Characteristic Structural Features of Tumor Blood Vessels

In this section a brief survey of basic vascular features will be given. At least four different structural categories can be distinguished: (1) dilated blood vessels with globular outpouchings, (2) blind ending blood vessels, (3) blood channels and lacunas, and (4) extravasal structures.

It has to be emphasized that by using different perfusion pressures during rinsing and by applying different flow rates for resin influx, we were able to discriminate between tumor-specific vascular structures and artifacts produced during casting of the vascular tree.

Generally, near the tumor a striking dilation of host capillaries occurs. This dilation is not confined to vessels of any particular type. Vasodilation, a decrease in intercapillary distances, vascular accumulation — sometimes padlike or

glomeruloid — and vascular twisting and kinking increase as the vessels approach the tumor region. The sinusoids are frequently dotted with globular outpouchings. Correlative light microscopy of tissue sections reveals twisted and tightly packed sinusoids, which are endowed with a continuous and intact endothelium. Located in the "reactive" zone of the tumor, they are surrounded by abundant lymphocytes and macrophages, representing important effector cells, which — in addition to the tumor cells themselves — are capable of inducing angiogenesis [23]. Originally we thought that the outpouchings found on these sinusoids were formed just by random bulging of the vessel walls. However, stimulated by the findings describing leukcocyte-triggered angiogenesis, we subsequently performed a detailed morphological analysis that allowed us to classify these structures into distinct categories, which could be combined by several developmental paths. Thus, each structure represents a particular stage of a dynamic process that leads to multiple anastomosing of two or more outgrowing pouches and/or to a special type of vasodilation, which we call *active* vasodilation, trying to thereby express that these reactions are due to endothelial-cell proliferation and vascular fusion. To produce a systematic and clear demonstration of these rather complicated processes, a simplified model comprising three closely interrelated principal courses of vascular remodelling is proposed. Transitions from one course of vascular development into another are, of course, likely to occur. Here we will give a brief survey of this system, which has already been presented in great detail [2].

1. A small furrow (Fig. 24-1 No. 1 and Fig. 24-6a) deepens to form a semicircular cleft, thereby separating the small flap from its parent vessel (Fig. 24-2 No. 1 and Fig. 24-6b). This flap subsequently enlarges (Fig. 24-1 No. 2 and Fig. 24-6c).
2. Pointed elevations on the vascular surface (Fig. 24-1 No. 3 and Fig. 24-6d) elongate (Fig. 24-2 No. 2 and Fig. 24-6e) and form sprouts with distinct endothelial-cell borderlines on their surfaces (Fig. 24-6f).
3. Slight superficial bulges (Fig. 24-1 No. 4 and Fig. 24-6g) form variously shaped protuberances (Fig. 24-2 No. 3 and Figs. 24-6h to 24-6i). With further elongation real sprouts

Table 24-1. Characteristic morphological features of vascular corrosion casts of normal murine subcutaneous blood vessels

	Arterial vessels	Venous vessels	Capillaries
Diameter	Regular, thinner than veins	Variable, thicker than arteries	Regular, 3–8 μm
Course	Straight, parallel to counterpart vessel	Curved	Arcadelike loops
Ramification	Dichotomous	High variability	Frequent, angles 20–90°
Imprints of endothelial-cell nuclei	Oval, sharply edged, shallow, parallel to vessel axis	Round, deep, troughlike, no alignment	Oval, 6–10 μm, shallow, parallel to vessel axis
Endothelial-cell border-lines	Longitudinal furrows wedged against each other		
Intercapillary distance			About 10 μm

develop that frequently give rise to secondary sprouting (Fig. 24-6j).

Mutual fusion of these elongated sprouts causes multiple anastomosing, thus establishing a dense sinusoidal plexus. Fusion of the growing vascular structures with their parent vessels, on the other hand, may give rise to collateral shunts (Figs. 24-1 No. 5, Fig. 24-3 No. 1, Figs. 24-6k, and 24-6l) or to vascular circles (Fig. 24-3 No. 2). A subsequent reduction of the tiny central cavity (Fig. 24-1 No. 6 and Fig 24-2 No. 4) transforms the circle into an excessively dilated sinusoid. Due to this careful analysis, it is now, for the first time, feasible to arrange the overwhelming variety of individual structures into a chronological system of tumor-induced vascular remodelling and to assign each structure to an individual developmental stage.

These processes are widely dispersed in the adjacent vascular system and can be found throughout all stages of tumor growth. It is worth noting that this type of vasoproliferation is responsible for the development of local hyperemic regions only and that it is not capable of vascularizing larger, nonperfused areas. This task is managed by the development of elongated, centripetally arranged, and invading sprouts, thereby establishing the initial (venous) microcirculation of the tumor implant. Subsequently, a tumor-specific angioarchitecture develops and the initial centripetal sprouting pattern is sub-stituted by peripheral centrifugally directed vasculogenesis.

Generally, these blind-ending cast structures, representing the morphological correlate of reproductive processes in the endothelium, contrast with the similar-looking vascular "dead-end" routes. These structures indicate vascular occlusions and result, in part, from elevations of interstitial tissue pressure occurring during tumor expansion. The characteristic structural features of such casts are (1) concave and wrinkled surfaces (Fig. 24-4), which correlate with the delta-shaped lumina seen in histological cross sections; (2) tapered terminations on small tumor vascular casts, sometimes exhibiting imprints of endothelial-cell nuclei and of endothelial-cell borderlines on their surfaces (Fig. 24-5); (3) large sheetlike arterial or venous casts (Fig. 24-7). Vascular compression is further supported by the occurrence of collapsed tumor blood vessels in the center of the neoplasms (Fig. 24-8). Vascular occlusion caused by either intravascular stasis or tissue pressure is just one of a number of degenerative phenomena occurring in the tumor vasculature. A very prominent feature in our lung tumor casts is the occurrence of blood channels and lacunas, which usually have direct contact with normal casts. Histological analyses reveal many erythrocytes dispersed throughout the malignant tissue, thus indicating widespread gaps and deficiencies in the endothelial lining

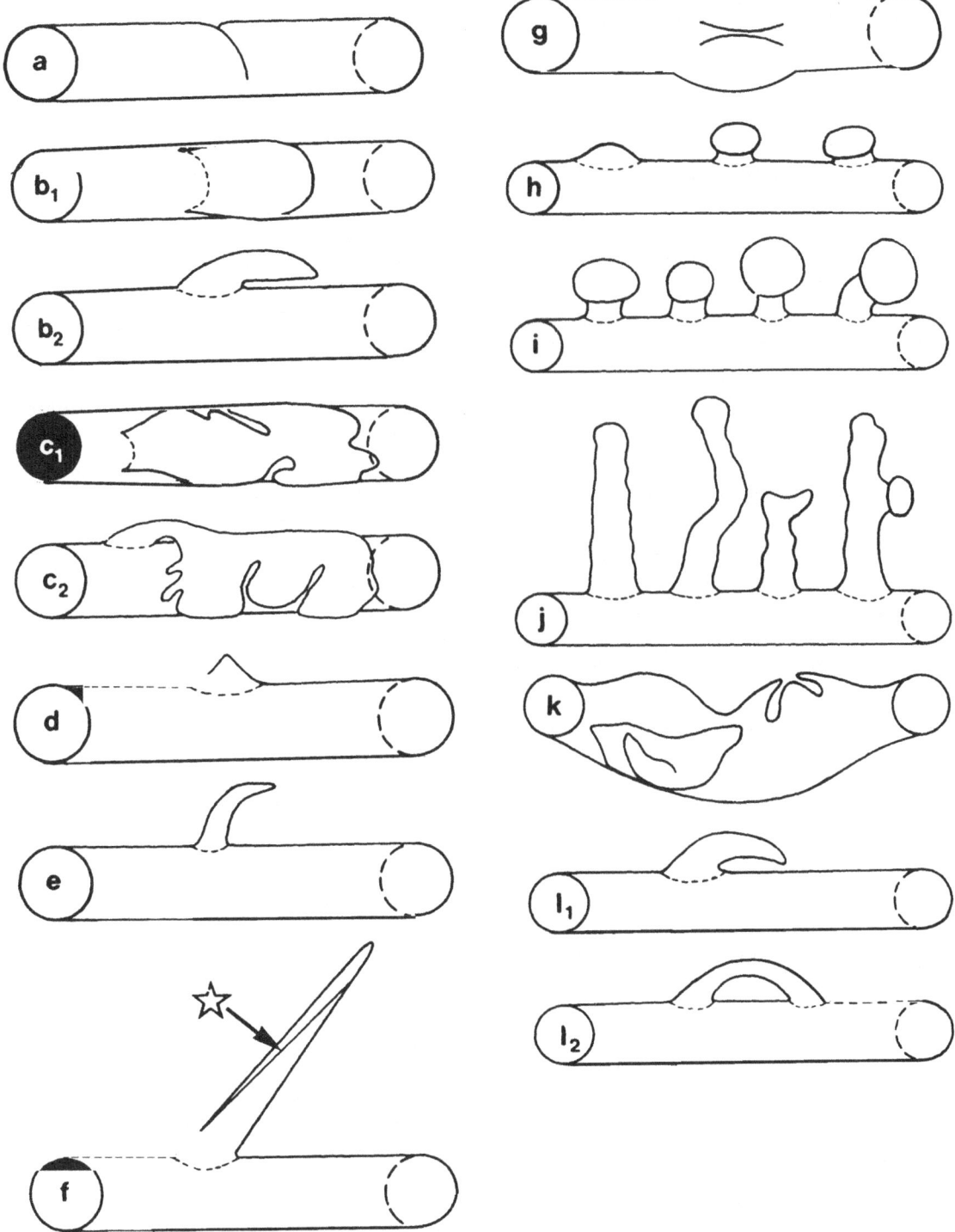

Figure 24-6. Vascular remodelling and dynamics of dilated tumor blood vessels (phase 1 of tumor angiogenesis). a: small furrow; b: small flap; b_1: plan view; b_2: side view; c: large flap; c_1: plan view; c_2: side view; d: small and pointed elevation; e: curved and pointed outgrowth; f: well-developed capillary sprout with endothelial cell border (asterisk); g: two slight bulges; h: dome-shaped to fungiform elevations; i: pedunculate dome-shaped sprouts; j: well-developed capillary sprouts, frequently with a secondary capillary sprout, as shown in h or i; k: complex arrangement of several sprouts; l: formation of collaterals; l_1: before fusion; l_2: after fusion.

(Figs. 24-9 and 24-10). Preparation of such tissues for vascular casts is, conceivably, extremely susceptible to the production of typical extravasal structures, which, therefore, represent a characteristic feature of tumor vascular casts.

Generally, arterial vessels are much more resistant to tumor-induced vascular remodelling than venous and capillary vessels. Arterial sprouting, for instance, was seen neither in the present murine lung cancer model nor in previously described human basal-cell tumors [4]. According to other investigators [25], this differential reaction may be explained, in part, by the secretion of angiogenic stimuli into the interstitial fluid, which are then drained into the microcirculatory bed, thus acting at first on capillaries and venous vessels. A second reason is found in the much more compact and tight composition of the multi-layered arterial vessel walls.

The tumor surface is covered by a plexus of straight-running arterial vessels of uniform caliber and by a system of tortuous veins with variable diameters. Both types of vessels take individual courses. These vessels are supported by layers of subcutaneous host muscle capillaries and tortuous sinusoids (Fig. 24-11). The sinusoidal vessels, which have thin and fragmentary endothelia, reveal diameters of up to 160 µm and are therefore called *giant capillaries*. In deeper lying tumor regions, the blood vessels take centripetal courses and approach the central cavity. These vessels usually are terminated by extravasal structures, tapered points, or intact capillary loops. A schematic representation of this angioarchitecture is given in Fig. 24-12.

6. The Vascular Alterations Occurring During Approach to the Tumor

For the localization of tumors by angiography, it is important to know the vascular features that may occur in the vicinity of a presumed neoplasm. Loss of the regular tissue-specific patterns of the host capillary beds, variations in vessel diameters and dilations, as well as decreases in intervascular distances and increases in vessel tortuosity constitute reliable markers for a nearby mass of malignant tissue. Furthermore, these structures also demonstrate the chronological changes

that occur if a host blood vessel falls under the influence of an expanding tumor mass. The luminal width is significant for the localization of a particular vessel relative to the tumor. The intervascular distances decrease and regions with marked venous hyperemia develop that contain oxygen-deprived, slowly flowing blood and thus are indicative of poor metabolic conditions. For the development of such regions several phenomena may be responsible: (1) vascular dilation and elongation caused by endothelial-cell multiplication; (2) blood-flow reduction and blood congestion, leading to additional passive vasodilation; (3) spatial restriction of the blood vessels caused by the growth of the tumor mass; resulting in (4) tortuous and irregularly arranged blood vessels. Longitudinal compression occurring in the blood vessels is evidenced, for instance, by the occurrence of luminal invaginations in the vessel walls, which are even found on straight-running segments.

7. The Vascular Alterations Occurring During Tumor Growth

In order to give a general view on the tremendous vascular alterations occurring during malignant growth, we defined four different stages of tumor development and related each individual vascular phenomenon to these growth stages.

Stage 1 of tumor growth covers the time of nidation of the tumor and continues from day 1 to day 5. This stage is characterized by a dilation of all nearby host vessels that embrace the still avascular implant. Vascular outgrowths develop rather quickly, and due to subsequent fusion, they constitute an initial peripheral plexus of densely packed sinusoids (*phase 1 of tumor angiogenesis*). Phase 2 of tumor angiogenesis starts between day 3 and day 4, and is characterized by the development of vascular sprouts emerging from the surrounding venous sinusoids and invading the implant in the centripetal direction. Therefore, it is just after the fifth day of growth that the tumor becomes completely vascularized. *Phase 1 of tumor angiogenesis* continues in various tumor regions until the end of tumor growth. *Phase 2 of tumor angiogenesis*, however, is rather short and is finished by the onset of **stage 2 of**

338

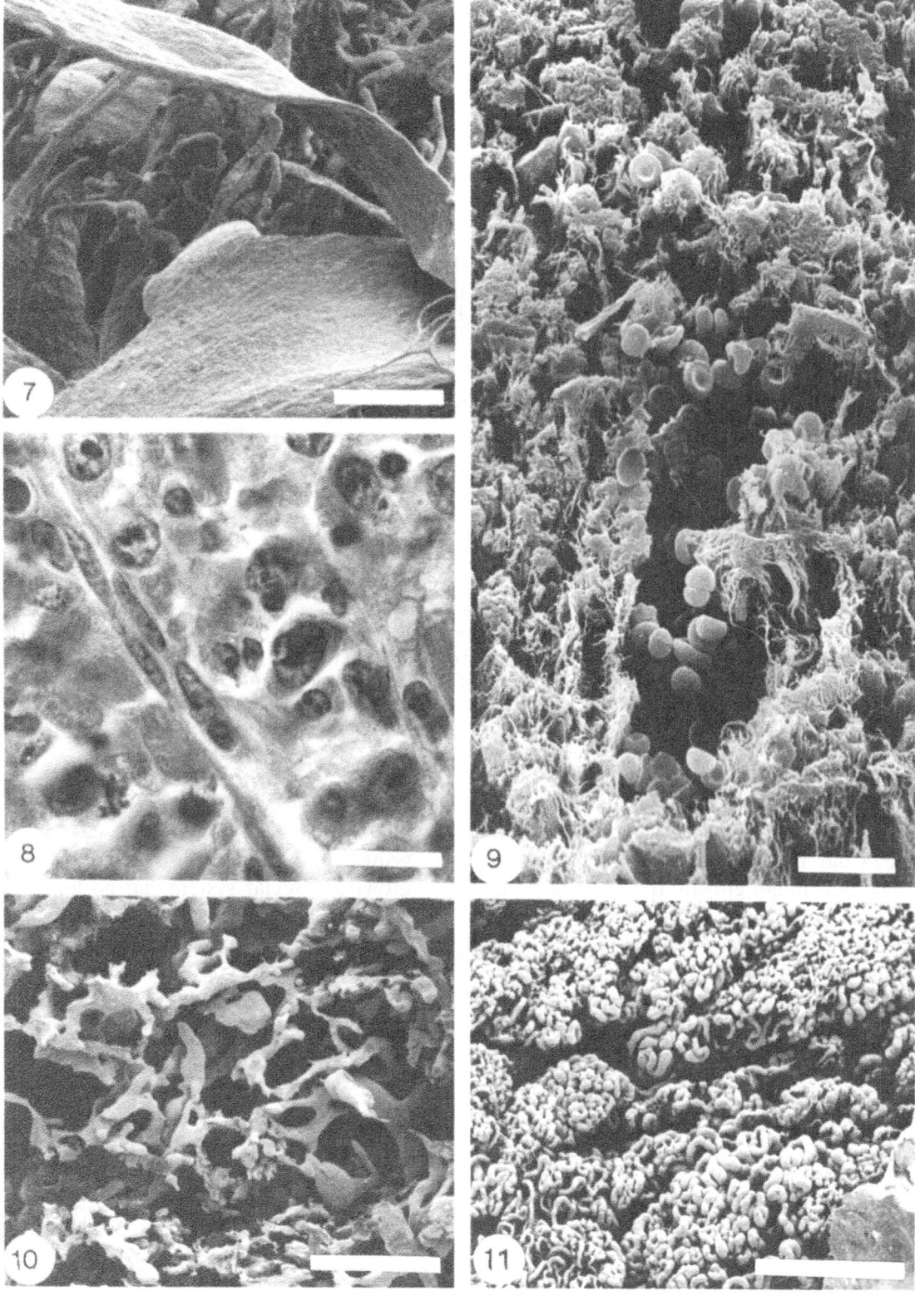

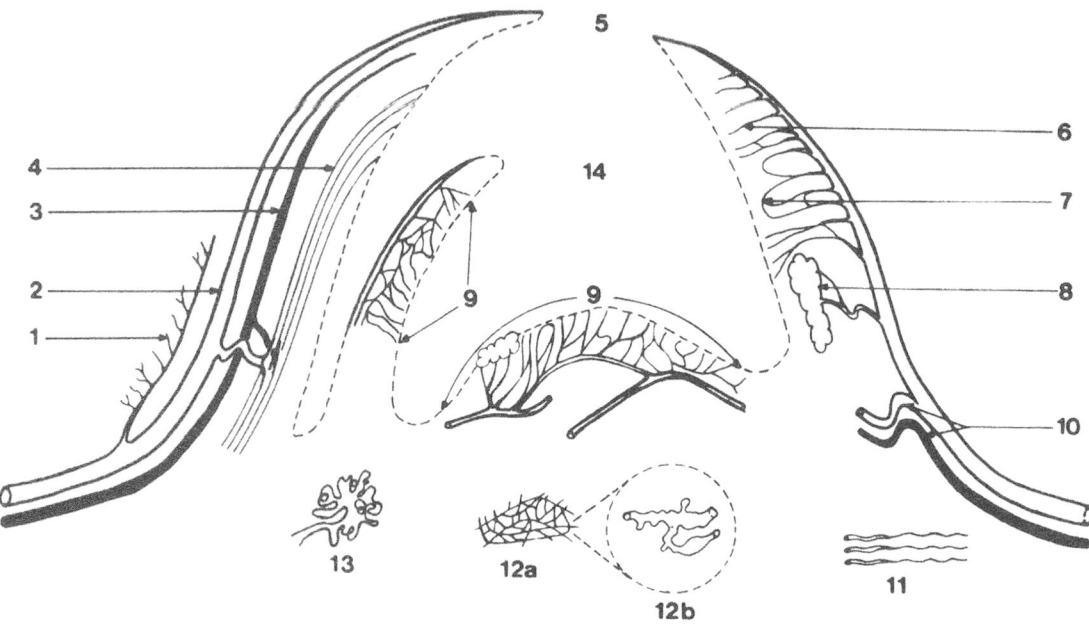

Figure 24-12. Cross section through a vascular corrosion cast of a 20-day-old, subcutaneous Lewis lung carcinoma. (1) Centrifugally growing sprouts at the tumor periphery; (2) venous host vessels (white); (3) arterial host vessels (black); (4) muscle capillaries that have been raised by the tumor; (5) apical orifice of the "tumor vascular envelope"; (6) centripetally running, blind-ending tumor vessels; (7) loop formation on centripetal tumor vessels; (8) centripetally running, blind-ending tumor vessels with terminal extravasations; (9) irregularly running and frequently blind-ending vessels of the interior tumor area; (10) deflexions of large vessels; (11) vessel distension of approaching muscle capillaries; (12a) dense plexus of dilated, proliferating capillaries at the tumor base; (12b) detail from 12a, outgrowths on dilated capillaries; (13) tortuous tumor vessels with "glomeruloidal" arrangement; (14) central, avascular cavity.

tumor growth. This stage covers the period between day 6 and day 10, and represents the stage of early tumor growth. It is characterized by necrotic degenerations in the central regions of the tumor and by the development of a central avascular cavity. Simultaneously, however, centrifugally orientated vascular sprouts occur at the margin and assist in conditioning the surrounding host tissue for further tumor invasion (*phase 3 of tumor angiogenesis*). During **stage 3 of tumor growth** (late tumor growth, days 11–21) the growing vascular system can no longer keep up with the rapid expansion of the malignant tissue. Therefore, multiple vascular degenerations develop, which continue during the prefinal stage or **stage 4 of tumor growth** (days 22–28). Finally, vascular sprouting and cell proliferation is restricted only to basal regions of the tumor.

←

Figure 24-7. Tumor base. Extremely flattened, venous vessel. Vascular corrosion cast. SEM. Bar = 50 μm.

Figure 24-8. Tumor center. Collapsed capillary. Fixation by immersion. Hematoxylin-eosin. Light microscopy. Bar = 20 μm.

Figure 24-9. "Tumor vascular envelope." Extravasal red blood cells in the interstitial space of the tumor tissue. Fixation by immersion. Critical-point-dried tumor-tissue block. SEM. Bar = 10 μm.

Figure 24-10. Interior region of the "tumor vascular envelope" near the central cavity. Plexus-like arrangement of lacunary replicae. Vascular corrosion cast. SEM. Bar = 100 μm.

Figure 24-11. Tumor base. Dilation of preexisting host capillaries to sinusoidal vessels. Vascular corrosion cast. SEM. Bar = 500 μm.

340

Folkman's group demonstrated in great detail the histological and ultrastructural alterations that occur during angiogenesis [13]. Here we depict just the most significant chronological events: (1) lysis of the endothelial basement membrane, (2) endothelial migration through the gaps in the basement membrane, (3) alignment of the endothelial cells and migration towards the angiogenic stimulus, (4) formation of an intracellular lumen, (5) mitosis in the midsection of the sprout triggered by the opening of the interendothelial-cell contacts, (6) capillary loop formation and anastomosis, (7) onset of a slow blood perfusion, (8) covering of the new tumor vessels by occasional pericytes, and finally, (9) synthesis of a new basement membrane.

From our work it becomes evident that the alterations occurring during incorporation of host vessels take an invariable chronological path. At first the vessels widen and "giant capillaries" or sinusoids develop. Then the endothelium attenuates and forms large gaps. Finally, the vessels run into interstitial lacuna. Arterial or venous differentiation of the new blood vessels could not be demonstrated in any of the tumors studied. Therefore, the circulatory sequence from arteries to capillaries and veins is not valid for tumor tissue. Furthermore, the forces occurring during tumor growth may determine the entire angioarchitecture of the tumor (Fig. 24-13).

8. Tumor Vasculature — Implications for Basic Tumor Biology and Applications to Clinical Medicine

From our correlative studies in human and murine tumors [2,4,5], as well as from work reported by other investigators [8,9,26–28], it becomes evident that dilation, globular outgrowths, blind-ending sprouts, and compressed vessels, as well as incompletely endothelialized channels, lacuna, and extravasations, represent vascular phenomena that can be found in most malignant tissues. However, the relative quantitative contribution of each of these structures to overall tumor angioarchitecture may differ between tumors of different histological origins and locations, therefore representing a tumor-specific feature.

The structures and arrangements presented here may be useful, therefore, in angiography and could help in localizing and measuring primary tumors and metastases. Furthermore, very close relations exist between the tumor vascular system and the cure rate of a particular therapeutic regimen. Hypoxic cells, for instance, are more resistant to irradiation than regularly oxygenated cells. Furthermore, most drugs are transported by the blood supply. Tissue damage by heat (hyperthermia), on the other hand, is most effective in poorly vascularized tumors.

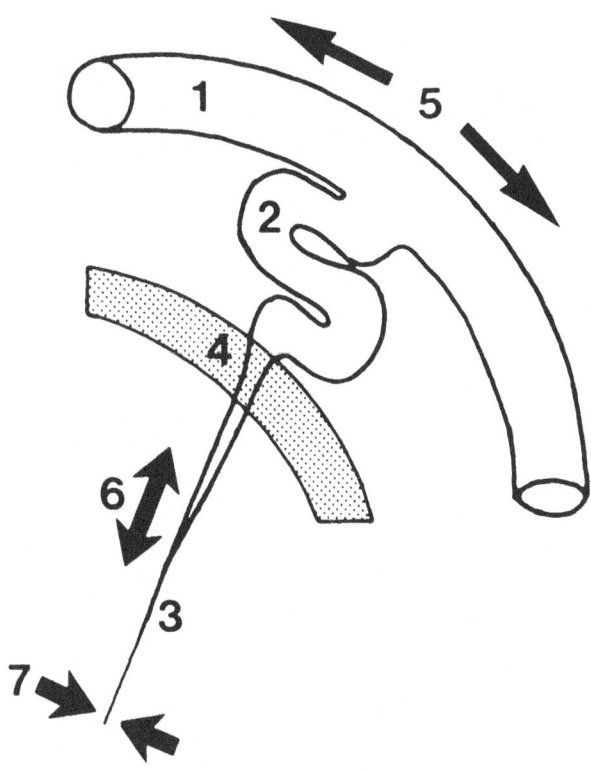

Figure 24-13. Schematic drawing demonstrating the forces that occur during tumor growth and that act on the tumor vascular bed. (1) Venous vessel of the "tumor vascular envelope"; (2) deflexion; (3) compressed, centripetal blood vessel; (4) front between vascular stretching and compression; (5) tangential stretching at the tumor periphery; (6) radial stretching in the intermediate tumor region; (7) tissue pressure at the tumor center.

The sequence of alterations occurring during angiogenesis — as reported by Folkman and coworkers [13] — apparently describes the formation of well-functioning capillaries. However, the majority of the tumor blood vessels represent irregular formed "caricatures" of normal vessels and therefore exhibit severe structural and functional deficiencies. A recent study done in melanomas [29], for instance, has demonstrated that nonendothelial cells may also be involved in angiogenic processes, especially in the melanotic form of this tumor. These results agree with data demonstrating mitosis induction in pericytes and fibroblasts during angiogenesis [13]. Some of the substances that may be responsible for such multiple stimulating effects have been well characterized. They comprise the transforming growth factors type alpha and beta, the acidic and — most importantly — the basic fibroblast growth factors, and tumor necrosis factor [30]. In malignant tissues, therefore, both paths of vasculogenesis — the angioblastic and the mesenchymal — are apparently used.

Recently great progress was made in the characterization of angiogenic effector molecules, and it was a few years ago that Folkman and coworkers purified tumor angiogenic factors with heparin-binding capacity to homogeneity and reported their amino acid sequence [31]. In 1985, a second tumor-derived angiogenic factor that does not bind heparin was purified and sequenced by Vallee's group [32–34]. Apart from these specific factors, several other effector substances, such as prostaglandins E_1 and E_2, copper ions, ceruloplasmin, heparin and heparin-derivatives, heparanase and proteases, fibronectin, and fibrin [35] were reported to modulate angiogenesis. Tontsch and Bauer [36] demonstrated conclusively the phenotypic, functional, and proliferative plasticity of cultured normal cerebral capillary endothelium. According to them, growth components, such as endothelial-cell growth supplement, were much more effective in cellular modulation than matrix components. In malignant diseases, however, the extracellular matrix and basement membranes represent important compartments for the storage of angiogenic factors. They are released from the matrix by the action of proteolytic enzymes, which are secreted by the tumor cells. Angiogenic factors can therefore be divided into "direct" and "indirect" effector substances. The former act directly on endothelial cells, while the latter need some mediator cells for exerting their action. Furthermore, other investigators [25] think that lactic acid, high pCO_2, low pH, and low pO_2 represent other types of strong angiogenic stimuli.

Apart from these advances in the characterization of angiogenic substances, similar progress was made in the identification of angiogenesis inhibitors. Among them the most promising group is represented by the "angiostatic steroids," which are most effective if they are administered together with heparin or related substances. Therefore, heparin has a dual function in that it is a modulator of both angiogenic as well as antiangiogenic processes [38].

The structural peculiarities and insufficiencies of the tumor vascular bed and the production of proteases by tumor cells exert strong, and sometimes even opposite, effects on the hemostatic steady state of the host. Experimental and human tumors, therefore, frequently show contradictory hemostatic reactions, which may range from massive bleeding to intravascular coagulation. This is explained, at least in part, by two different activities of tumor cells. According to Curatolo et al. [39], the cells of the Lewis lung carcinoma produce a procoagulant that activates the coagulation factor X. The same cells, however, may also express fibrinolytically active plasminogen activators [37]. It becomes evident, therefore, that the substances elaborated by the tumor cells may be effective, not only in altering the hemostatic steady state, but also in modulating the composition of the extracellular matrix, thereby exerting strong influences on tumor-specific vasculogenic processes and on the invasive behavior of the tumor. While blood channels are still perfused and contain loosely packed, regularly shaped erythrocytes, the blood lacuna and extravasal structures have lost contact with the circulation and are filled by tightly packed, pale red blood cells. Efficient metastasis may occur only in leaky, but still perfused, vascular areas. Pale-colored, static erythrocytes indicate the loss of the cellular hemoglobin and may explain the microangiopathic, hemolytic anemia occurring in mice bearing the Lewis lung carcinoma [40]. However, according to this investigation, the animals do not show marked signs of intravascular coagulation.

9. Concluding Remarks

This chapter summarizes recent data on the vascularization of carcinomas obtained by SEM of vascular corrosion casts and from correlative histological studies, and gives a survey of the angioarchitecture and submicroscopic vascular structures of the murine Lewis lung carcinoma. Furthermore, the basic vascular phenomena occurring in the majority of carcinomas are opposed to features that are specific for individual tumor types.

The basket-like tumor vascular system comprises a central cavity that is covered by a peripheral vascular "envelope." Multiple vascular occlusions correlate with central tissue necroses. These phenomena are caused by elevations in tissue pressure and by cessation of blood flow resulting from an increased venous blood pressure and sinusoidal dilations. Degenerations of the endothelium causes multiple leakages of blood into the interstitium. The microcirculatory bed of the tumor, therefore, acts as a sink for tumor blood flow.

Apart from these degenerations, various forms of vascular remodelling, regeneration, and proliferation are observed in cast preparations. Angiogenic processes are grouped into three phases, which could be assigned to four stages of tumor growth (tumor nidation, early growth, late growth, and prefinal stage). The first phase causes vessel widening and local hyperemia. During the second phase of angiogensis, elongated, centripetally growing vascular sprouts develop and invade the tumor implant. The third phase of angiogenesis is characterized by centrifugal sprouting at the tumor periphery, which prepares the surrounding host tissue for further tumor invasion.

In conclusion, the most peculiar and unique feature of tumor vascular systems is the simultaneous and spatially closely related occurence of development, residual differentiation, and degeneration of the endothelium and the blood vessels. This represents an atypical and abbreviated course of the normal vascular develop-ment. The SEM methods described here are characterized by unsurpassed quality in reproducing the whole angioarchitecture and submicroscopic vascular details. These features are especially useful for the demonstration of complicated tumor vascular systems. Furthermore, by correlating cast preparations with histological and ultrastructural specimens, the morphogenic effects that act on tumor vessels can be demonstrated very convincingly.

The findings presented in this chapter reflect the implications of the tumor vasculature for basic phenomena in tumor biology and medicine, and stress the significance of the tumor microcirculation in tumor diagnosis and treatment. Future experimental studies should compare microvascular effects induced by various angiogenic factors with the phenomena occurring during normal and aberrant angiogenesis. Furthermore, cooperation with angiologists should be intensified in order to establish a better transfer of knowledge between the laboratory and the clinic.

Acknowledgments

This chapter is dedicated to Dr. A. Lametschwandtner, Head of the Department of Experimental Zoology, Institute of Zoology, University of Salzburg, to whom I am very grateful for suggesting and conducting this work and for providing the apparative and laboratory facilities for performing the experiments. Valuable discussions with Dr. Ch. Dittrich (Department of Internal Medicine I, University of Vienna) and critical review of the manuscript by Dr. U. Tontsch (Institute of Molecular Biology, Austrian Academy of Sciences, Salzburg) are also gratefully acknowledged.

References

1. Murakami T. Application of the scanning electron microscope to the study of the fine distribution of the blood vessels. *Arch Hist Jpn* 32:445–454, 1971.
2. Grunt TW, Lametschwandtner A, Karrer K. The characteristic structural features of the blood vessels of the Lewis lung carcinoma. A light microscopic and scanning electron microscopic study. *Scann Electron Microsc* II:575–589, 1986.
3. Grunt TW, Lametschwandtner A, Staindl O, Adam H.

The vascular pattern of normal human skin and human basal cell tumors. A scanning electron microscopic study on microcorrosion casts. In: Csanddy A, Rohlich P, Szabo D (eds.), *Electron Microscopy*, 8th Eur. Congr. Elect. Micros., Budapest, pp 2195–2196, 1984.

4. Grunt TW, Lametschwandtner A, Staindl O. The vascular pattern of human basal cell tumors. Histology and scanning electron microscopical studies on vascular corrosion casts. *Microvasc Res* 29:371–386, 1985.

5. Grunt TW, Lametschwandtner A, Karrer K, Staindl O. The angioarchitecture of the Lewis lung carcinoma. A light microscopic and scanning electron microscopic study. *Scann Electron Microsc* II:557–573, 1986.

6. Tatematsu M, Cohen SM, Fukushima S, Shirai T, Shinohara Y, Ito N. Neovascularization in benign and malignant urinary bladder epithelial proliferative lesions of the rat observed in situ by scanning electron microscopy and autoradiography. *Cancer Res* 38:1792–1800, 1978.

7. Miodonski A, Kus J, Olszewski E, Tyrankiewicz R. Scanning electron microscopic studies on blood vessels in cancer of the larynx. *Arch Otolaryngol* 106:321–332, 1980.

8. Shah-Yukich AA, Nelson AC. Methods in laboratory investigation. Characterization of solid tumor microvasculature: A three-dimensional analysis using the polymer casting technique. *Lab Invest* 58:236–244, 1988.

9. Walmsley JG, Granter SR, Hacker MP, Moore AL, Ershler WB. Tumor vasculature in young and old hosts: Scanning electron microscopy of microcorrosion casts with microangiography, light microscopy and transmission electron microscopy. *Scann Microsc* 1:823–830, 1987.

10. Holzinger K, Lametschwandtner A. Angioarchtitecture of an experimentally induced brain tumor in the rat. In: *Proc. 1st Eur. Workshop Scann. Electr. Micros. Corr. Casting Biol. Med.*, p 10, 1989.

11. Konerding MA, Steinberg F, Budach V. The vascular system of xenotransplanted tumors — scanning electron and light microscopic studies. *Scann Microsc* 3:327–336, 1989.

12. Folkman J, Merler E, Abernathy C, Williams G. Isolation of a tumor factor responsible for angiogenesis. *J Exp Med* 133:275–288. 1971.

13. Ausprunk DH, Folkman J. Migration and proliferation of endothelial cells in preformed and newly formed blood vessels during tumor angiogenesis. *Microvasc Res* 14:53–65, 1977.

14. Brem SS, Gullino PM, Medina D. Angiogenesis: A marker for neoplastic transformation of mammary papillary hyperplasia. *Science* 195:880–881, 1977.

15. Heuser LS, Taylor SH, Folkman J. Prevention of carcinomatosis and bloody malignant ascites in the rat by an inhibitor of angiogenesis. *J Surg Res* 36:244–250, 1984.

16. Salsbury AJ, Burrage K, Hellmann K. Histological analysis of the antimetastatic effect of (±)-1,2-bis(3,5-dioxopiperazin-1-yl)propane. *Cancer Res* 34:843–849, 1974.

17. Hilmas DE, Gillettte EL. Morphometric analyses of the microvasculature of tumors during growth and after X-irradiation. *Cancer* 33:103–110, 1974.

18. Reinhold HS, Endrich B. Tumour microcirculation as a target for hyperthermia. *Int J Hyperthermia* 2:111–137, 1986.

19. Carlsson G, Ekelund L, Stigsson L, Hafström L. Vascularization and tumour volume estimations in solitary liver tumours in rats. *Ann Chir Gynaecol* 72:187–191, 1983.

20. Taylor S, Folkman J. Protamine is an inhibitor of angiogenesis. *Nature* 297:307–312, 1982.

21. Gabbert H, Wagner R, Höhn P. The relation between tumor cell proliferation and vascularization in differentiated and undifferentiated colon carcinomas in the rat. *Virchows Arch (Cell Pathol)* 41:119–131, 1982.

22. Grunt TW, Somay C, Dittrich E, Dittrich C. Determination of the clonogenic cell pool and of differentiated cell phenotypes in a human ovarian cancer cell line (HOC-7). In: Dittrich C, Aapro MS (eds.), *Drugs, Cells and Cancer*, G. Welley, Vienna, pp 34–38, 1989.

23. Polverini PJ, Cotran RS, Gimbrone MA, Jr. Unanue ER. Activated macrophages induce vascular proliferation. *Nature* 269:804–806, 1977.

24. Folkman J, Taylor S, Spillberg C. The role of heparin in angiogenesis. In: Ciba Foundation (ed.), *Development of the Vascular System*, Sympos. 100, Pitman Books, London, 1983.

25. Reinhold HS, Van Den Berg-Blok AE. Factors influencing the neovascularization of experimental tumors. *Biorheology* 21:493–501, 1984.

26. Egawa J, Ishioka K, Ogata T. Vascular structure of experimental tumours. *Acta Radiol Oncol* 18:367–375, 1979.

27. Miodonski A, Kus J, Tyrankiewicz R. SEM blood vessel casts analysis. In: DiDio LJA, Motta PM, Allen DJ (eds.), *Three-Dimensional Microanatomy of Cells and Tissue Surfaces*, Elsevier North Holland 1981.

28. Ahlstrom H, Christofferson R, Lorelius LE. Vascularization of the continuous human colonic cancer cell line LS 174 T deposited subcutaneously in nude rats. *APMIS* 96:701–710, 1988.

29. Hammersen F, Osterkamp-Baust U, Endrich B. Ein Beitrag zum Feinbau terminaler Stombahnen und ihrer Entstehung in bösartigen Tumoren. In: Messmer K, Hammersen F (eds.), *Mikrozirkulation in Forschung und Klinik*, Vol. 2, S. Karger, Munich, pp 15–51, 1983.

30. Folkman J, Klagsbrun M. Angiogenic factors. *Science* 235:442–447, 1987.

31. Shing Y, Folkman J, Sullivan R, Butterfield C, Murray J, Klagsbrun M. Heparin affinity: Purification of a tumor-derived capillary endothelial cell growth factor. *Science* 223:1296–1299, 1984.

32. Fett WJ, Strydon DJ, Lobb RR, Alderman EM, Bethune JL, Riordan JF, Vallee BL. Isolation and characterization of angiogenin, an angiogenic protein from human carcinoma cells. *Biochemistry* 24:5480–5486, 1985.

33. Kurachi K, Davie EW, Strydon DJ, Riordan JF, Vallee BL. Sequence of the cDNA and gene for angiogenin, a human angiogenesis factor. *Biochemistry* 24:5494–5499, 1985.

34. Strydon DJ, Fett JW, Lobb RR, Alderman EM, Bethune

344

JL, Riordan JF, Vallee BL. Amino acid sequence of human tumor derived angiogenin. *Biochemistry* 24: 5486–5494, 1985.

35. Ungari S, Katari RS, Alessandri G, Gullino PM. Cooperation between fibronectin and heparin in the mobilization of capillary endothelium. *Invas Metast* 5: 193–205, 1985.

36. Tontsch U, Bauer H-C. Isolation, characterization, and long-term cultivation of porcine and murine cerebral capillary endothelial cells. *Microvasc Res* 37:148–161, 1989.

37. Skriver L, Larsson LI, Kielberg V, Nielsen LS, Andersen PB, Kristensen P, Danö K. Immunocytochemical localization of urokinase-type plasminogen activator in Lewis lung carcinoma. *J Cell Biol* 99:753–758, 1984.

38. Folkman J, Weisz PB, Joullié, Li WW, Ewing WR. Control of angiogenesis with synthetic heparin substitutes. *Science* 243:1490–1493, 1989.

39. Curatolo L, Colucci M, Cambini AL, Poggi A, Morasca L, Donati MB, Semeraro M. Evidence that cells from experimental tumors can activate coagulation factor X. *Br J Cancer* 40:228–233, 1979.

40. Poggi A, Polentarutti N, Donati MB, DeGaetano G, Garattini S. Blood coagulation changes in mice bearing Lewis lung carcinoma, a metastasizing tumor. *Cancer Res* 37:272–277, 1977.

Author's address:
Dr. Thomas W. Grunt
Laboratory for Applied and Experimental Tumor Cell Biology
Division of Oncology
Department of Internal Medicine I
Waehringer Guertel 18–20
A-1090 Vienna
Austria

CHAPTER 25

Angiomorphology of the Human Larynx and Renal Carcinoma a Comparative Study

ADAM MIODOŃSKI, JAN KUŚ, MARIA NOWOGRODZKA-ZAGÓRSKA, EUGENIUSZ OLSZEWSKI, & ANDRZEJ BUGAJSKI

1. Introduction

Spontaneously developing human carcinomas, solid tumors of epithelial origin, are composed of malignant parenchyma and supporting tissues. In the majority of cases, the supporting structures are normal host tissues that continuously provide the metabolic and structural demands of the growing tumor. They enable the tumor to acquire its characteristic architecture and viability.

Carcinomas cannot however, proliferate beyond a certain size without developing an adequate network of blood vessels [1–3]. The microcirculation, which is subjected to un-controlled and continuous stimulation by diffusible angiogenic factor(s) (discharged directly or liberated from supporting tissues by cancer cells), plays a very important role in this respect [4–10]. It is of crucial importance not only for nutrient supply and waste product removal during growth of the tumor, but also for tumor spread (invasion and metastases) and sensitivity to therapy (radiotherapy, chemotherapy, hyperthermia, and even surgery).

The vascularization of carcinomas differs in many respects from the vascular system of the organ or tissue from which it arises. Unlike most regular tissues, the tumor vasculature is strongly heterogenous and does not conform to the normal vascular organization and morphology, although it comprises both preexisting, mostly pathologically altered vessels as well as new vessels resulting from tumor-induced angiogenesis [11–21].

We know at present that there exist close and mostly tumor-specific interrelations between malignant tissue, neoplastic vasculature, and extracellular matrix [22–29]. Experimental data indicate that the angioarchitecture of a neoplastic lesion is modulated rather than determined by the cancer tissue growth pattern, which in turn depends on the histology, and to some extent the malignancy, of the tumor [6,11,17,19,20,30–34]. Moreover, the vascular pattern of a spontaneously growing tumor is also influenced to a certain degree by the features characterizing both the specificity of the preexisting vascular bed and the anatomo-histological organization of the host organ [5,6,11,12,15,17–21].

The cancer cells that apparently act as an organizer not only influence the vascular density of the tumor [5,17,26,35], but also are responsible for adaptive changes of the vessels, and consequently, for continuous remodelling of the tumor vascular bed, especially at the level of the microcirculation [17,19,20,36,37].

Basically the macroscopical architecture of tumor vasculature can be divided into two patterns: peripheral and central, which according to Falk's classification, are characteristic for *tense* and *lax* tumors, respectively [38,39]. However, a tumor can be composed of many modules, each of which may exhibit a prevalence of one of the two patterns.

The vascularization and microcirculation of primary solid tumors have been the focuse of research for investigators for a long time [40]. This interest, as well as progress in knowledge of the vascularization of solid tumors, has resulted

Motta, P.M., Murakami, T., and Fujita, H. (eds.), Scanning Electron Microscopy of Vascular Casts: Methods and Applications.

from the development of new techniques that have opened new possibilities in this area of research. Apart from the widely used and very fruitful experimental technique of transplanting tumors into a transparent chamber [41-44], one of the most promising methods, allowing the investigation of basic characteristics of tumor and different therapeutic approaches, is the recently developed transplanting of human solid tumors into immunodeficient athymic nude mice [45]. This model offers some morphological similarities to the situation encountered in the clinic. Nevertheless, the results obtained [37,46,47] justify a certain extrapolation only when applied to human spontaneously growing tumors that are accompained by immunological as well as inflammatory events. Therefore, further morphological studies of human spontaneously growing tumor vasculature are required in order to gain more information necessary for understanding the pathophysiological aspects, such as angiogenesis, tumor growth and spread, and thus for improving diagnostic and therapeutic efficiency. In this respect, the recently developed method of vascular corrosion casting [48] is well suited for detailed morphological analysis of both normal and pathologically altered vascular systems. Scanning electron microscopy (SEM) of vascular corrosion casts gives a quasi-three-dimensional image in which arteries, veins, and capillaries can be discerned with high accuracy, at least in mammals [49].

The aim of our study was to visualize and compare the angiomorphology of the human primary larynx and renal carcinomas using SEM of vascular corrosion casts, conventional SEM of critical-point-dried tissues, and light microscopy of silicon-rubber-injected specimens.

2. Materials and Methods

Studies were performed on 16 human larynges obtained at operation because of extensive supraglottic and glottic cancer, and 16 surgically removed human kidneys with advanced renal-cell carcinoma (histopathologically confirmed as larynx squamous-cell carcinoma and as renal clear cell-carcinoma, respectively).

The specimens were immediately placed in prewarmed (37°C) heparinized saline and perfused via the superior laryngeal arteries or via the renal artery with the same saline containing 3% dextran, mol. wt, 70,000 [50]. Each specimen was prefixed by perfusion with 0.66% paraformaldehyde/0.08% glutaraldehyde in 0.1 M cacodylate buffer, pH 7.3, containing 0.2% lignocaine [51]. All specimens were then divided into three groups and subjected to the procedures described below. Eight larynges, as well as three kidneys, were injected with Microfil-MV 112 yellow silicon rubber (Canton Biomedical Products, Boulder, CO, USA). After solidification of the rubber, they were transferred to 70% ethanol, and cut with a microtome and a small bone saw into blocks 2–3 mm thick in the horizontal or vertical plane, depending on the external shape of the tumor and its relation to the anatomical parts of the larynx. The kidneys were cut sagittally into slices 1–1.5 mm thick. In both cases the tissue blocks were dehydrated in graded series of ethanol, cleared in a mixture of absolute ethanol/methyl salicylate, placed in pure methyl salicylate, and examined under a stereomicroscope.

Four larynges and six kidneys were fixed by perfusion and simultaneous immersion with modified Karnovsky's solution composed of 1% paraformaldehyde and 1.25% glutarladehyde in 0.1 M cacodylate buffer, pH 7.3. After fixation larynges were split in the midsagittal plane into two halves, each of which was then cut horizontally into slices ca. 2–2.5 mm thick and stored overnight at 4°C in the same fixative. The fixed kidneys were cut sagittally and larger blocks, ca. 2 × 2 × 1 cm, from different regions of tumor, were excised and placed in the same fixative for at least 20 hours at 4°C. After a thorough rinse in several baths of 0.1 M cacodyalte buffer, pH 7.3, all blocks were postfixed in 2% osmium tetroxide for 9 hours at 4°C, dehydrated in ethanol and acetone, and critical-point dried in liquid carbon dioxide. The blocks obtained from the larynges were mounted on stubs so that their adepiglottical surface was, after coating with carbon and gold, exposed for inspection. The dried kidney fragments were similarly attached to stubs and coated with carbon and gold. The surface morphology of the specimens was examined in a Jeol JSM-35-CF SEM at 15–25 kV.

Four larynges and six kidneys were injected

with a low-viscosity resin, Mercox CL-2R (Japan Vilene Comp. Ltd., Tokyo) and left in a warm water bath (ca. 60°C) for several hours to allow the resin to polymerize. They were next repeatedly macerated in 15–20% sodium hydroxide at room temperature or at 37°C, and washed in warm (ca. 60°C) tap water for a few days. Afterwards, the casts were washed for another few days in multiple changes of distilled water, cleaned with 3–5% trichloracetic acid, washed again in distilled water, and freeze dried. Some casts were examined under the stereomicroscope, frozen again in distilled water, fractured along the desired planes in order to expose the intratumor vasculature, and freeze dried. All casts were coated with gold and examined by SEM. The entire procedure has been described in detail elsewhere [52–54].

3. Results

3.1. Human Larynx Cancer

Squamous-cell carcinoma of the larynx is a solid tumor arising either from stratified squamous or ciliated pseudostratified columnar epithelium covering the mucous membrane that lines the endolaryngeal surface. Advanced carcinomas of the supraglottic or glottic regions are well-vascularized, most frequently exophytic tumors composed of smaller tumors, varying in size, that grow out of the underlying mass of the malignant lesion. On the basis of such a growth mode, they can be included in the group of "lax" tumors according to Falk's classification [38,39]. Nevertheless, on their surface they can show the presence of more or less deep ulcerations surrounded by an inflammatory infiltration.

When inspected under low magnification in the stereomicroscope, the vertically cut slices of the larynx injected with silicon rubber reveal general features of the vascular network characteristic of exophytic squamous-cell carcinoma (Figs. 25-1A and 25-1B). The free margin of the vocal cord exhibits vascular hypertrophy, particularly in areas adjacent to the tumor. Atypical and distended larger vessels of the vocal muscle stand out against a background of dispersed capillaries. The texture of cancer vessels is characterized by

chaotic distribution and proliferation, especially abundant in the peripheral parts of the neoplasm, which are composed of multiform small tumors growing out of the true mass of the malignant lesion (Fig. 25-1B). In the very center of the tumor there arises a second vascular system derived from centrally located vessels of larger diameter (Fig. 25-1B). In general the picture is dominated by twisted, shorter or longer vessels forming a dense tangle in which avascular fields corresponding to foci of cancer cells are dispersed.

SEM reveals groups of cancer cells varying in shape and diameter, which are frequently arranged in a nestlike aggregations (Figs. 25-1C). They are surrounded by stromal tissue containing numerous thin-walled capillaries and venous vessels of variable diameter (Figs. 25-1B and 25-1C). These vessels are surrounded and also compressed by closely packed cancer cells (Fig. 25-1C). The endothelium of smaller and medium-sized vessels, mainly venules, exhibit features of proliferation: irregular distribution, variable size, overlapping, and a relatively rough abluminal surface, frequently decorated by attached leukocytes (Fig. 25-1D). The leukocytes, which are mainly observed within the venules, exhibit signs of active diapedesis. The proliferating cells of the endothelial lining often aggregate around the outlet of new capillaries sprouting out of the parent vessels (Figs. 25-1D and 25-1E). In the deeper layers of the main tumor mass occur pseudo-vessels, composed solely of the strands of cancer cells remaining in direct contact with the blood (Fig. 25-1F).

Vascular corrosion casts show an irregular distribution and density of atypical blood vessels, which are the characteristic features of larynx carcinoma. These vessels, mostly capillaries or pseudocapillaries, are particularly abundant in peripheral parts of the tumor. They have a form of vascular tufts that correspond to small tumors observed in intact tissues, either protruding out of the main tumor mass into the lumen of the larynx or located at the advancing edge of the cancer (Figs. 25-2A, 25-2C, and 25-2D). These tufts, as a rule, include one or more thick vascular trunks, which emerge from the base and constitute the core of a tuft. They give off in all directions smaller, densely packed vessels, which on the surface of the tuft pass into tightly arranged,

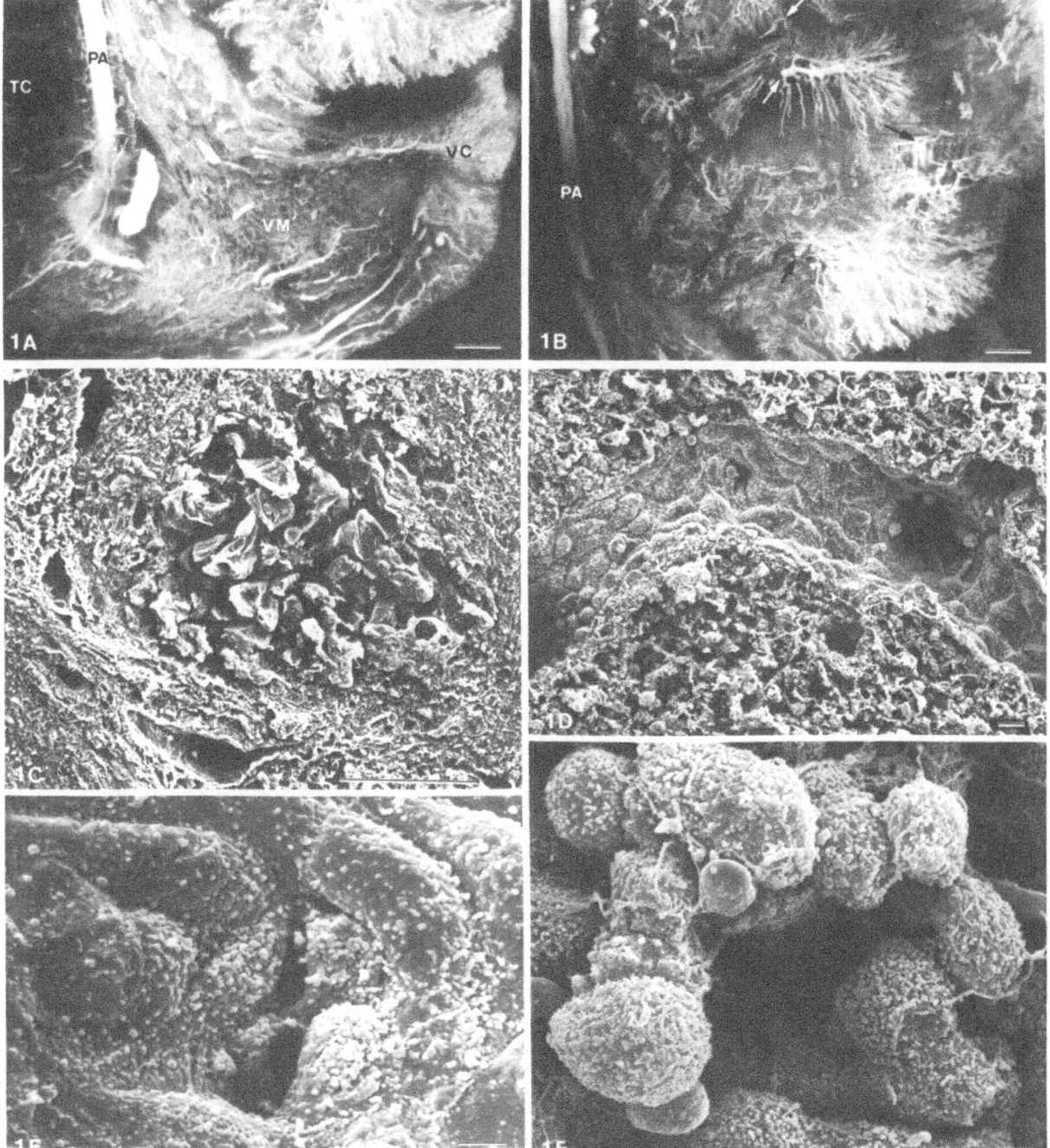

Figure 25-1A. Panoramic view of the vertical section through the center of the right half of the larynx (plane perpendicular to thyroid cartilage). Egzophytic supraglottic squamous-cell carcinoma. Note vascular organization of the invaded vocal muscle (VM) and hypertrophy of capillaries on the free edge of the vocal cord (VC); TC = thyroid cartilage; PA = posterior arch of superior laryngeal artery. Microfil injection. Bar = 1.5 mm.

Figure 25-1B. Vascularization of the tumor shown in Fig. 25-1A. The picture is shifted upwards. Note the presence of numerous capillary loops on the periphery of the tumor and a second system derived from thicker, centrally situated vessels (arrow). Microfil injection. PA = posterior arch of superior laryngeal artery. Bar = 1.5 mm.

Figure 25-1C. A group of cancer cells surrounded by stromal matrix containing numerous medium and small-sized vessels. Bar = 100 μm. SEM.

Figure 25-1D. Opened profile of a thin-walled venule exhibiting a proliferative reaction. The endothelial cells differ in size and shape, and show irregular distribution and overlapping. Cancer cells remain in direct contact with the vessel wall. Note the lymphocytes glued to the endothelial surface. Bar = 10 μm. SEM.

Figure 25-1E. Endothelium of a venule exhibiting a proliferative reaction in a close vicinity to the site where a new vessel is

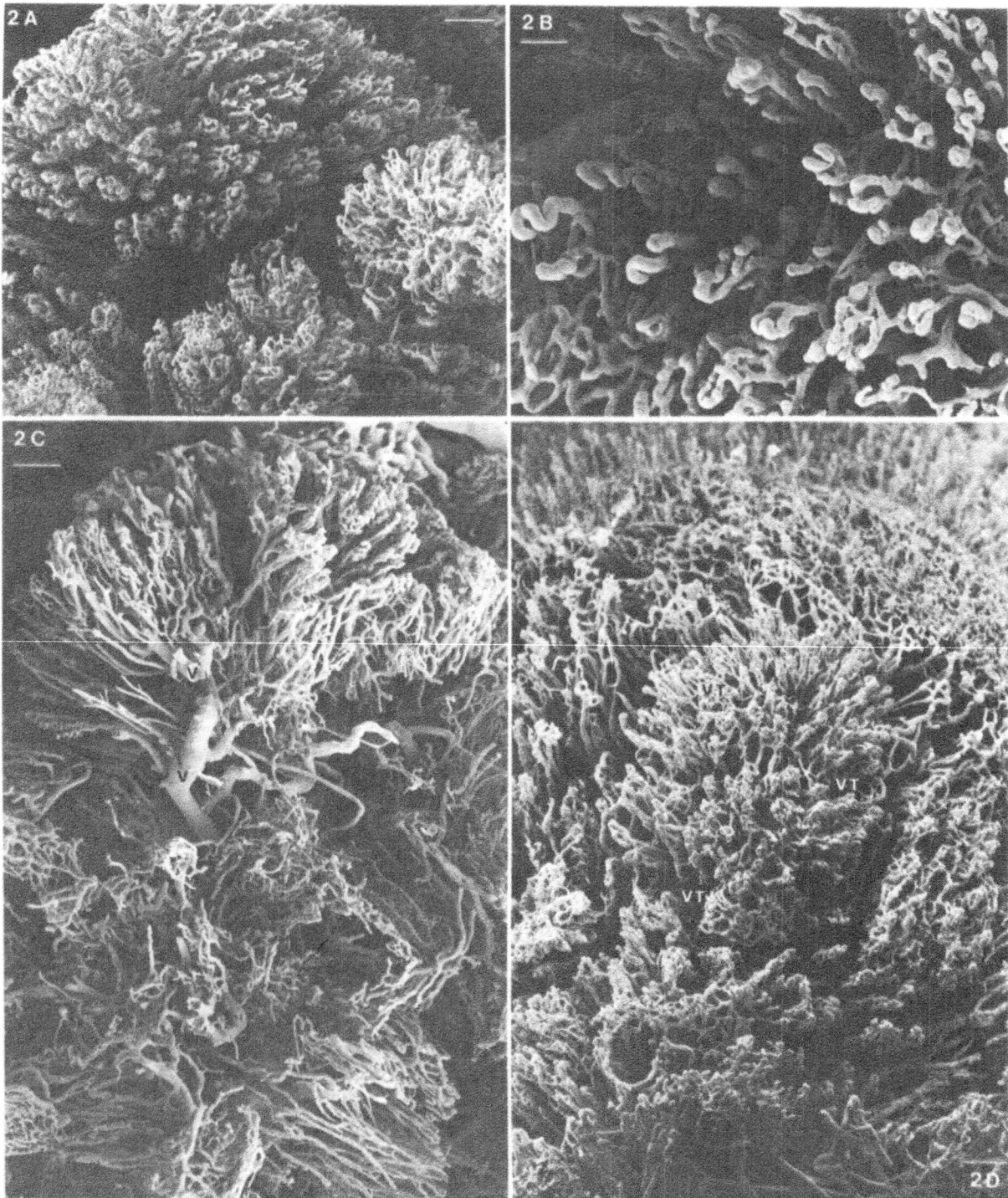

Figure 25-2A. Vasculature of small tumors growing out of the main mass of cancer in the form of vascular tufts with densely arranged capillary loops on their surface. Corrosion cast. Bar = 400 μm. SEM.

Figure 25-2B. A higher magnification of the surface of the vascular tuft shown in Fig. 25-2A. Note the small sinusoidal dilatations or glomerular loops on their tops. Corrosion cast. Bar = 150 μm. SEM.

Figure 25-2C. A specimen fractured parallelly to the long axis of the vascular tuft. Note relations between the main vessel trunk (V = vein) and secondary medium-sized branches passing into densely arranged capillary loops. Corrosion cast. Bar = 400 μm. SEM.

Figure 25-2D. A vascular tuft (VT) growing on the periphery of the larynx tumor. The vascular tuft is delimited from the adjacent tissues by a fairly deep and narrow belt of lower vascular density. Note different vascular patterns of the tuft and the surrounding tissue. Corrosion cast. Bar = 350 μm. SEM.

dilated, and spiral capillary loops (Figs. 25-2B and 25-2C). Their surface shows the presence of sinusoidal dilatations (Figs. 25-2A, 25-2B, and 25-2C).

Three zones differing in vascular organization could be distinguished on the basis of vascular texture: (1) the inner zone of the tumor, (2) the marginal zone or the advancing edge of the tumor, (3) the border zone between the tumor and surrounding tissues (Fig. 25-3A). Vascular zone 1 is characterized by vascular tufts of various shape and size, which are separated from each other by smaller or larger "empty" spaces devoid of vessels (Fig. 25-3B). The bottom regions of these spaces are occupied by irregularly arranged vessels of different caliber, mostly veins, and also by remnants of altered submucous vessels that show signs of proliferation in the form of densely packed nodular outgrowths of variable size (Fig. 25-3B).

The second zone, constituting a region of intense tumor growth and local spread, is characterized by the occurrence of an ample network of neoplastic blood vessels, also most frequently acquiring the form of vascular tufts (Fig. 25-3C). The tufts of various size are composed of rather long, radially oriented hairpin-shaped vessels that are tightly arranged side by side in bundles. As a rule, they are spiralized and arborized distally into a superficially located capillary network. The hairpin vessels are interconnected by numerous randomly directed anastomoses, especially abundant within the superficial capillary network of the tuft, where they very often form arcadelike systems (Figs. 25-2D, 25-3A, and 25-3C). Vascular branches directed towards the main tumor mass remain in contact with the vessels of the subepithelial network, as well as with deeper vessels extending from the mucosa (Figs. 25-3A and 25-3C). On the periphery, the vascular tufts are frequently surrounded and at the same time are separated by a rather narrow but fairly deep belt of lower vascular density (Fig. 25-2D and 25-3C). From the outside, this zone exhibits a preponderance of single, elongated, and spiral capillary loops interconnected by short anastomoses and together forming a wreath.

The zone of tissues directly surrounding the main tumor mass is characterized by a specific vascular reaction, most probably caused by the action of a diffusible angiogenic factor(s). The vessels located just outside the advancing edge of the tumor are densely packed, proliferating capillaries or pseudocapillaries (Figs. 25-3C and 25-3E). They have the form of spirally twisted, relatively long hairpins, running perpendicular to the plane of the submucous vascular plexus, which is their origin. Among capillary loops that constitute distinct units, single dispersed capillaries that have not yet formed the loop can frequently be observed (Figs. 25-3C, 25-3D, and 25-3E). The further away the vessels are from the advancing edge of the tumor, the smaller is the length of these vessels and the more numerous

→

Figure 25-3A. The angioarchitectonic organization of three zones distinguished in larynx carcinoma. A: inner zone of tumor; B: border zone of the advancing edge of tumor; C: zone of surrounding adjacent tissues. Corrosion cast. Bar = 240 μm. SEM.

Figure 25-3B. A fragment of the inner zone of the tumor. Note the presence of vascular tufts (rosettes) separated by sizeable empty spaces, which show irregularly distributed vascular trunks of various caliber at their bottoms and also remnants of a strongly altered submucous plexus (arrowheads). Corrosion cast. Bar = 500 μm. SEM.

Figure 25-3C. A panoramic picture of the border zone of the tumor (B) viewed towards the zone of surrounding tissues. Note a vascular tuft of the advancing edge of tumor (rosettes) and the proliferative reaction of the submucous plexus (arrowheads). Corrosion cast. Bar = 550 μm. SEM.

Figure 25-3D. A panoramic picture showing the proliferative reaction of vessels within the zone of surrounding tissues (C) viewed towards the advancing edge of the tumor (B). Corrosion cast. Bar = 550 μm. SEM.

Figure 25-3E. A higher magnification of the zone of surrounding tissues showing numerous newly formed capillary loops and single capillaries. Corrosion cast. Bar = 65 μm. SEM.

Figure 25-3F. A capillary loop showing a simple spiral formed by a vessel of a smaller diameter around a vessel of a larger diameter. Corrosion cast. Bar = 17 μm. SEM.

Figure 25-3G. A capillary loop with the smaller vessel showing several alternating U bends. Note the irregular diameter of the smaller vessel. Corrosion cast. Bar = 30 μm. SEM.

Figure 25-3H. A strongly twisted and convoluted capillary loop resembling a pseudo-glomerulus. Corrosion cast. Bar = 20 μm. SEM.

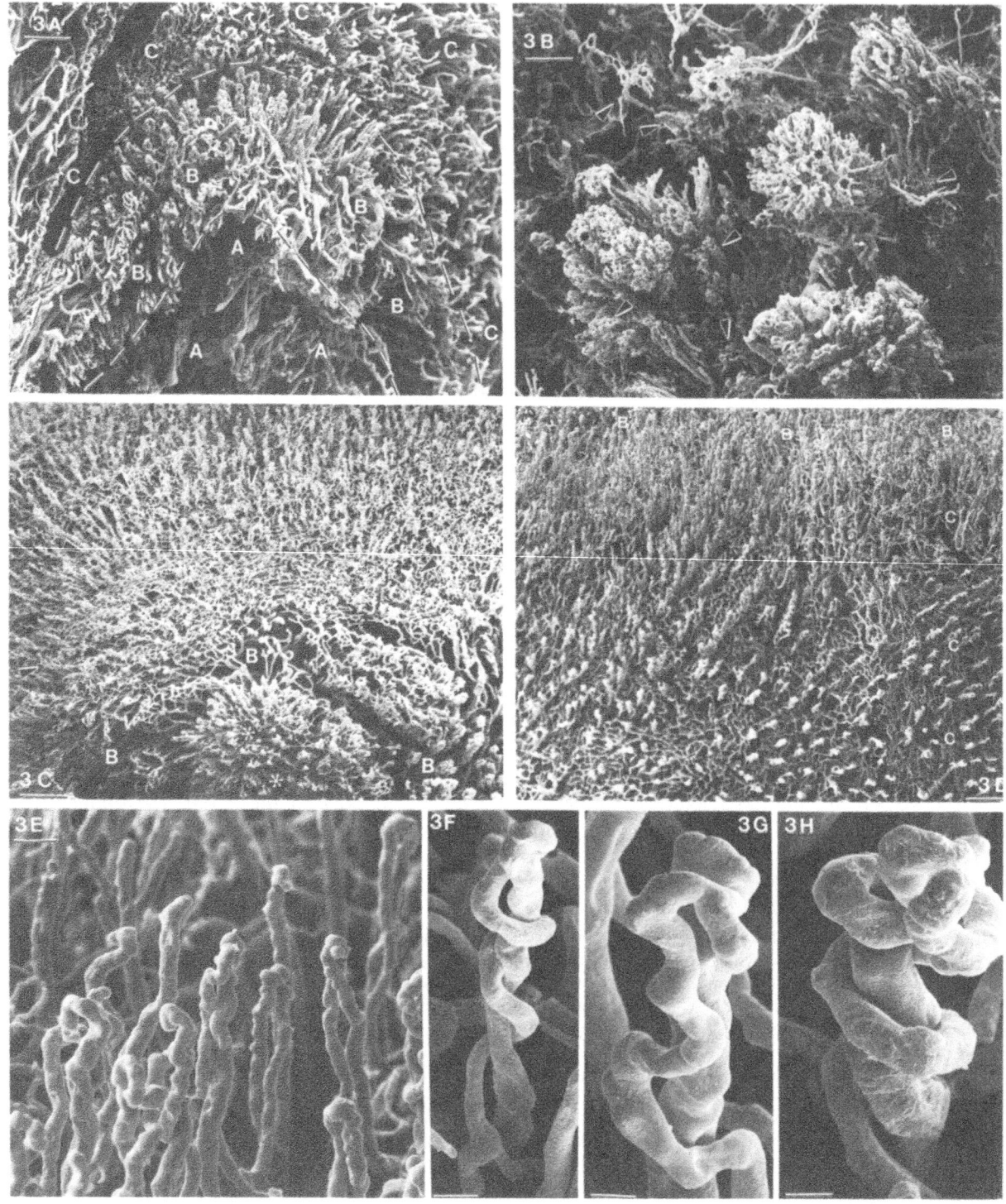

are the atypical, spirally twisted, short capillaries that appear among them (Figs. 25-3D and 25-3F–25-3H). They stand out clearly against the well-preserved and only slightly changed submucous plexus and its deeper supplying vessels (Fig. 25-3D). In the regions situated at somewhat farther away from the margin of the tumor, only short, newly formed capillaries are encountered. These capillaries or pseudocapillaries (because of their larger diameter) appear as hairpins spirally twisted along their axis. The twisting or looping, even resulting in a form of pseudoglomeruli, is most conspicuous in their distal segments (Figs. 25-3F–25-3H).

3.2. Human Renal Cancer

Human renal clear-cell carcinoma is histogenetically a malignant lesion arising from the proximal tubular cell. Its ultrastructral and antigenic pattern have been found to be consistent with that of the straight portion of the proximal tubule [55–57]. The tumor grows spontaneously as a solid ball-like mass composed of nodules that have an acinar appearance. The usually well-vascularized tumor is characterized by a high interstitial pressure and can therefore be included, according to Falk's classification, in the group of "tense" tumors [38,39]. This high pressure influences the shape and functional state of the tumor vasculature, and is thus an important factor responsible for ischemia and necrotic foci that very often occur within central areas of the tumor [14,19,21,37]. Irrespective of the mode of growth, the hematogenous spread of the tumor is the most important and frequent metastatic route, since solid tumors are usually devoid of their own lymphatics [4,5,21].

In light microscopy, kidney slices injected with silicon rubber show clear-cut angioarchitectonic differences between the normal renal parenchyma and the vascular bed of the tumor (Fig. 25-4A). The neoplastic tissue is surrounded and separated from regular (albeit compressed) renal tissue by a distinct vascular envelope. The peripheral areas of the tumor show the highest density of blood vessels and assume the form of avascular nodules encircled by thin vascular sheaths and separated by narrow, dense vascular strands. The deeper areas of the tumor that are less densely vascularized have mostly irregularly distributed atypical vessels of variable shapes, sizes, and courses. These parts of the tumor, as a rule, contain foci of thrombosis/necrosis, which also can occasionally occur near the superficial vascular coat.

SEM of conventionally prepared tissue blocks reveals tumor cells arranged tightly in the form of trabeculae or sheets. They are separated by delicate stromal tissue containing numerous capillaries and venous vessels. At the periphery of the tumor, the stromal tissue is more condensed and forms a superficial vasculo-fibrous coat (Fig. 25-4B). The vessels are surrounded and strongly compressed by densely packed cancer cells, which consequently have very irregular shapes and diameters. Alterations caused by compression are particularly pronounced in venous vessels which exhibit a sinusoidal character (Fig. 25-4C). They are tightly surrounded by neoplastic cells that remain in close contact with the vessel wall or with an inconspicuous layer of collagen fibers located on their adluminal surface (Fig. 25-4D). The endothelium of medium and small vessels displays features of proliferation: irregular distribution, variable size, overlapping, and a rather

→

Figure 25-4A. Panoramic view presenting the interface between tumor tissue and normal renal parenchyma (interrupted line). Note the superficial vascular coat (arrows) and the nodular pattern of the tumor. Microfil injection. Bar = 3 mm.

Figure 25-4B. Panoramic view showing the border between the tumor (T) and the renal parenchyma (R). The superficial vascular coat (SVC) is composed of large and medium-sized vessels that covers the tumor. Bar = 500 μm. SEM.

Figure 25-4C. The tumor parenchyma: opened profile of a compressed, irregular thin-wall venule. Bar = 100 μm. SEM.

Figure 25-4D. A proliferative reaction of the endothelium in small, thin-walled venule. The endothelial cells, different in size, show overlapping and the presence of scarse microvilli. Cancer cells remain in a direct contact with the vessel wall. Bar = 10 μm. SEM.

Figure 25-4E. The superficial vascular coat covering the anterior surface of a tumor occupying the lower pole of the right kidney. Note a dense vascular network, mainly composed of pathologically altered stellate veins. Corrosion cast. Bar = 11 mm.

Figure 25-4F. The superficial vascular coat covering the posterior surface of the tumor shown in Fig. 25-4E. Note distinct angioarchitectonical differences in the vascular pattern, as compared to Fig. 25-4E. Corrosion cast. Bar = 11 mm.

353

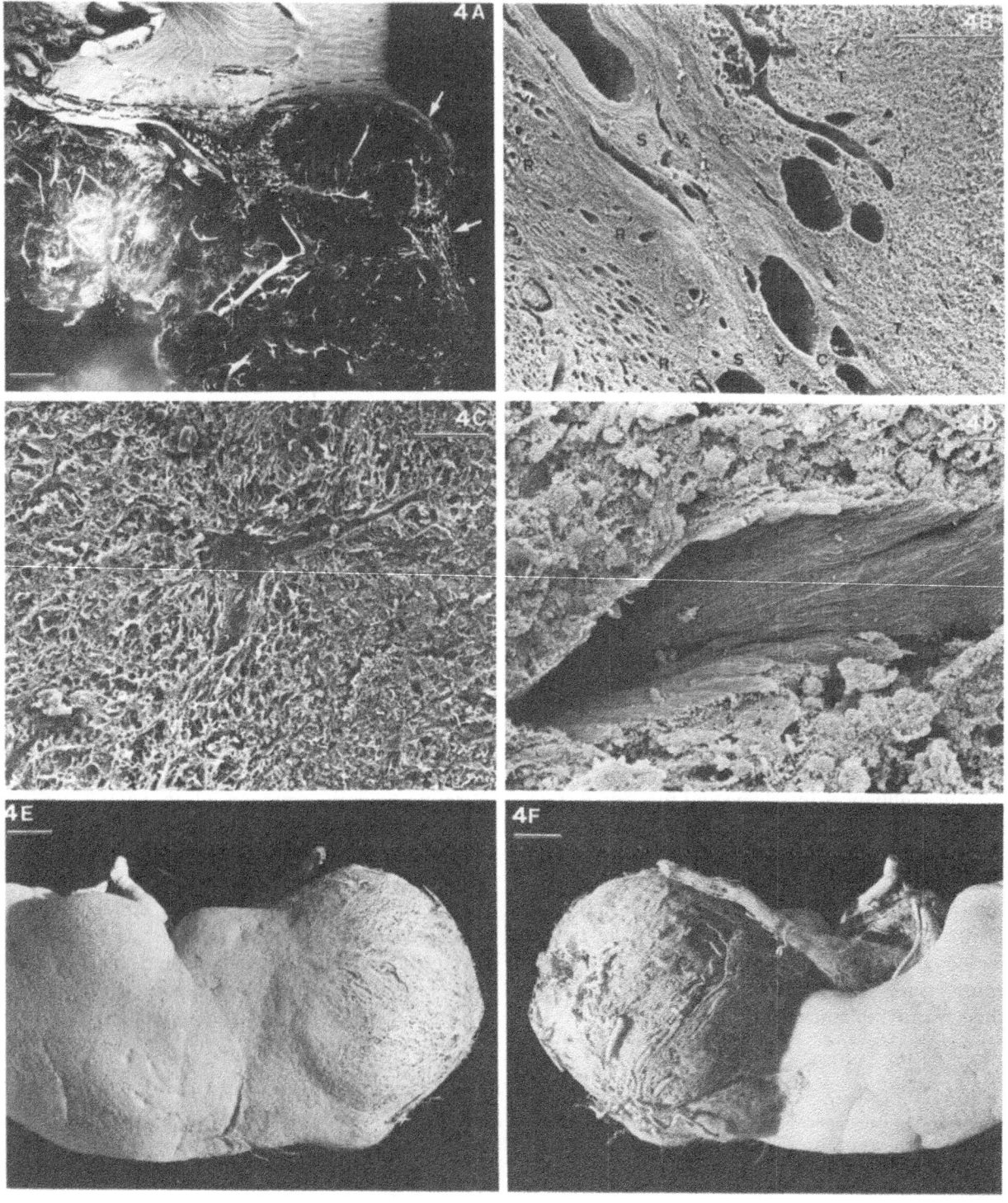

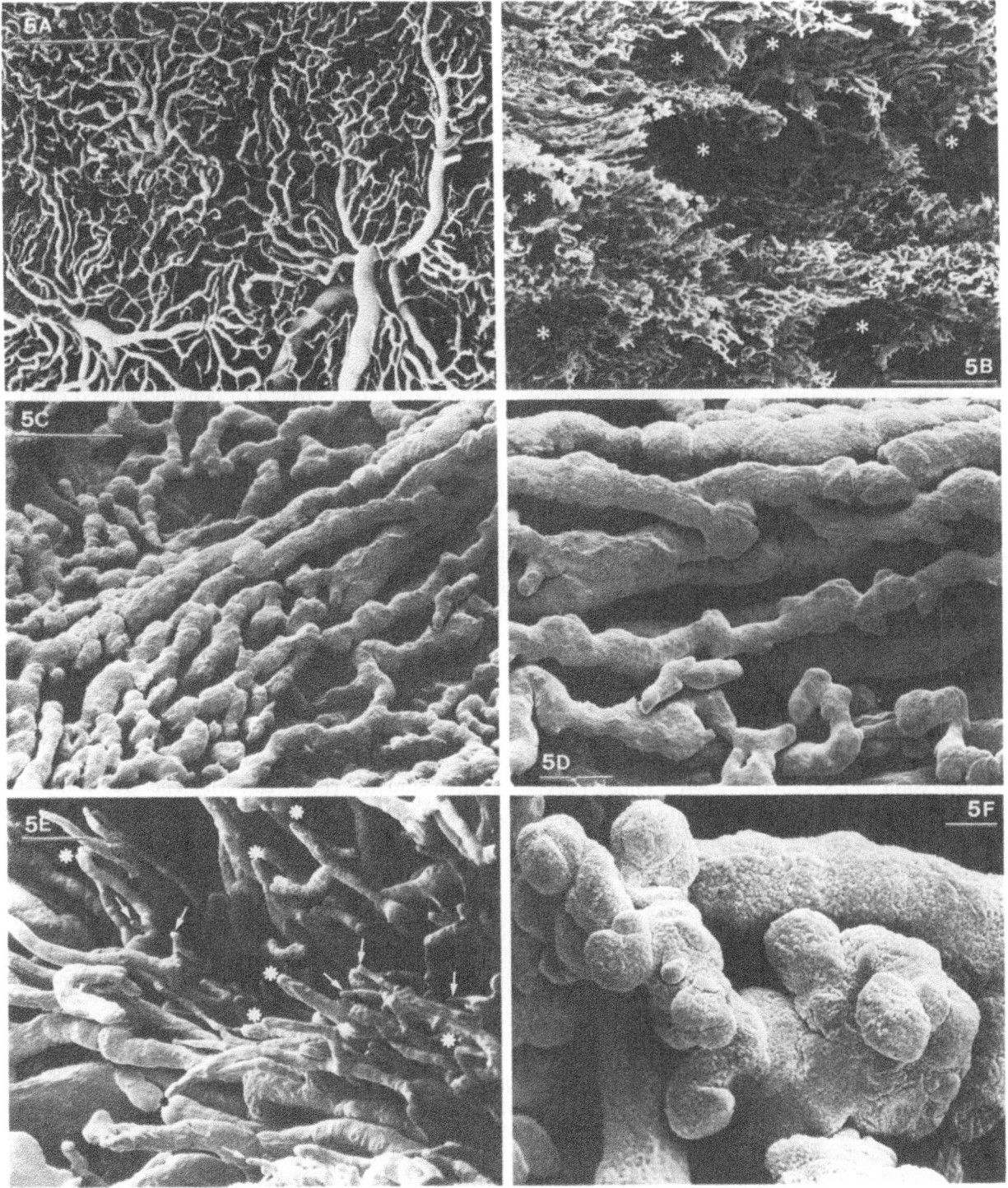

uneven adluminal surface decorated by smaller or larger platelet aggregations (Fig. 25-4D). On the contrary, larger arteries and veins have a regular pattern of the endothelial lining, which frequently shows the presence of attached leukocytes.

Vascular casts inspected under low magnification clearly reveal the organization of the superficial vesicular envelope, which is especially well developed in large tumors. A characteristic feature of this coat, which was observed in all specimens examined, is the angioarchitectonic differentiation between the anterior and posterior surface of the tumor (Figs. 25-4E and 25-4F). The anterior surface facing the peritoneum is coated by a dense and generally regular network composed of medium-sized vessels, mostly veins. These vessels, together with their small tributaries, follow a "riverine" course (Fig. 25-4E). In contrast, the vascular coat covering the posterior surface of the tumor, facing the iliopsoas muscle, is much more irregular and exhibits a peculiar horizontal arrangement of large vessels (mainly veins, but also arteries), originating from both tumor vasculature and collateral circulation.

The veins of the vascular coat comprise the main tributaries of the stellate veins, altered by angiogenic reaction and also by mechanical deformation: displacement, stretching, and compression (Figs. 25-4E, 25-4F, and 25-5A). The proliferative reaction of stellate veins, especially prominent on the anterior side of the tumor, declines towards the unaffected part of the kidney (Fig. 25-4E). The arteries of the superficial vascular envelope have undulated courses and comprise not only displaced and distorted capsular arteries, but also arteries captured from the capsular collateral circulation and from nearby structures (Fig. 25-4F). The large components of the vascular coat are located over a dense capillary network with irregular meshes spread in the background. The capillary plexus is composed of strongly dilated and contorted small twigs of stellate and capsular veins and, occasionally, of local arteries (Figs. 25-4E, 25-4F, and 25-5A). The coexistence of altered venous and arterial capillaries that are incidentally connected with larger vessels of the vascular coat is a prerequisite for the formation of arterio-venous shunts.

The intratumor vasculature has a remarkably heterogenous organization, caused not only by the progressive growth of densely packed cancer cells, but also by angiogenic processes. As a rule, this vasculature shows the presence of high-raising vascular areas interposed between adjacent avascular spaces, which acquire the form of nodules (Fig. 25-5B). This pattern observed in the casts of small fragments of the vascular bed corresponds to the nodular structure of the tumor observed in the specimens injected with Microfil. The avascular nodules are covered with their own vascular plexuses, mostly composed of small atypical lacunar vessels. Their surface has numerous nodular or fingerlike outgrowths, reflecting an angiogenic response to factor(s) present in the environment (Fig. 25-5C). The vessels, representing altered components of the peritubular plexus, are not only mutually connected, but also communicate with the interposed vessels by anastomosing channels of various diameters.

The inclined surfaces of the avascular spaces that pass into high-raising vascular areas are composed of the preexisting, atypical capillaries of

Figure 25-5A. Stellate veins altered by a proliferative reaction. Note irregularities and tortusity of larger and smaller branches. The latter branches form a dense plexus with irregular meshes. Corrosion cast. Bar = 500 μm. SEM.

Figure 25-5B. A panoramic view of the intratumor vasculature. The avascular spaces (rosettes) are interposed with adjacent high-rising vascular areas (asterisks). Corrosion cast. Bar = 1000 μm. SEM.

Figure 25-5C. A fairly regular and flat vascular plexus surrounding a ball-shaped avascular space. The surface of smaller and larger vessels are decorated with globular outgrowths. Corrosion cast. Bar = 100 μm. SEM.

Figure 25-5D. A fragment of the plexus shown in Fig. 25-5C. Note a nearly parallel arrangement of vessels and the presence of fingerlike outgrowths (a stage preceding their fusion and the formation of anastomoses). Corrosion cast. Bar = 50 μm. SEM.

Figure 25-5E. A fragment of the elevated vascular territory. Note the proliferative reaction of the displaced renal vessels in the form of fingerlike sprouts (arrows) and newly formed loops (rosettes) directed towards the avascular space and differentiated in length and diameter. Corrosion cast. Bar = 100 μm. SEM.

Figure 25-5F. A higher magnification of venules with their surface decorated by outgrowths of variable shape and diameter. Note outgrowths prior to fusion with each other or with the parent vessel. Corrosion cast. Bar = 10 μm. SEM.

356

the peritubular plexus (Fig. 25-5D), displaced and markedly malformed interlobular veins, and nearly unchanged interlobular arteries. The outer surfaces of veins and capillaries have numerous short vascular sprouts, varying in form and diameter, which is characteristic of angiogenesis (Fig. 25-5E). They frequently fuse with each other to form wider or narrower loops directed towards the avascular spaces.

The vascular territories spread from the periphery of the tumor, where they join the superficial vascular envelope, to its central areas, towards which they are mostly directed. They consist of interlobular, interlobar, as well as peripheral segments of arcuate veins and arteries. All these vessels, displaced and strongly altered by compression and stretching, are densely packed and irregularly arranged. In this region, the venous vessels also have numerous outgrowths, varying in shape and in diameter. They mainly result in lacunar dilatations decorated by numerous globular protrusions (Fig. 25-5F). These protrusions represent successive stages in their growth and fusion. The incorporation of these protrusions into the walls of the vessels produces changes in their shape and diameter. This process finally leads to the formation of irregular lacunar dilatations or to the development of anastomosing channels between closely located vessels.

4. Discussion

Many factors that affect the growth of solid tumors and their response to therapy depend on the nutrition of tumor cells and hence on their blood supply. The angiomorphological differences that can be encountered between solid tumors, such as the cancers of the larynx and kidney that were studied here, are associated with the growth pattern, as well as the biochemical characteristics, of their cells and with the morphohistological construction of the organs in which they are growing.

The vascular network of human larynx carcinoma reflects its usually egzophytic mode of growth, on one hand, and influences exerted by a fairly rigid framework of the larynx, on the other, which is divided into anatomic compart-

ments bordered by membranes and ligaments of fibroelastic tissue [13,58,59]. Nevertheless, larynx carcinoma can spread from its primary site to other parts of the larynx via the blood vessels or the lymphatics. Recent investigations indicate that larynx carcinoma frequently also spreads by direct infiltration of surrounding tissues [58, 59]. Larynx tissues are susceptible to neoplastic invasion and spread to a different degree. They either facilitate the spread of cancer in some direction or, for a certain period of time, form a barrier to its spread (e.g., nonossified parts of laryngeal cartilages) [12,13,58,59]. Direct invasion and hematogenous spread, however, seem to play a principal role, since solid tumors are usually devoid of their own lymphatics [4,5,60].

The vascular system of larynx carcinoma comprises both the preexisting, albeit pathologically altered, vascular bed as well as vessels induced by angiogenesis [12,13,15,16,61–63]. The angiogenetic reaction is mainly manifested by budding and sprouting of new blood vessels from capillaries, postcapillary venules, and veins of the host organ. Larynx cancer, as a tumor of epithelial origin, first evokes very strong proliferation of the mucosal capillary bed, and then, after infiltration beyond the basement membrane, also induces an angiogenetic response of the veins and, much more rarely the arteries of the submucosa. As cancer of the larynx usually exhibits an exophytic growth pattern, acquiring a form of smaller tumors protruding into the larynx lumen, its vascular network is dominated by multiform vascular tufts, mainly composed of densely arranged atypical capillaries interconnected by a system of arcades. These tufts, which form separate tumor units or modules, receive their blood supply from their bases and, therefore, can be regarded as small "lax" tumors [38,39]. Since the regular mucosal capillary bed, a source of the vascular tufts, shows a random orientation and spatial distribution in particular areas of the larynx, e.g., in the epiglottis, false cord, or ventricle [62], the arrangement and density of newly formed capillaries is also very irregular. They are separated by strands or foci of densely packed cancer cells. The vascular network of larynx cancer is mainly composed of altered preexisting capillaries and/or of vessels induced by angiogenesis [4,12,13,15,16]. They appear in all zones distinguished within the tumor, but as a rule are most plentiful at its periphery [4,6, 12,15,17,35]. Usually they form loops varying markedly in both length and degree of spiralization, and always exhibiting a variable diameter that changes along their length. They can assume the form of single "pipes" of variable diameter, with a bulbous dilatation at their distal end, very often arched or bifurcated into thinner, tapering tips. Usually, however, they assume the shape of long, spirally twisted hairpins, resembling a corkscrew, or strongly spiralled and convoluted loops forming pseudo-glomeruli. Within the deeper, submucous strata of the tumor, where its expansion is influenced by the fibroelastic anatomical compartments or the cartilaginous framework of the larynx, the multiplicating, tightly packed cancer cells mainly alter venous (but also arterial) vessels by compression and — to a lesser degree — by displacement. The vascular texture in those parts of the tumor has vascular tufts altered by compression and separated in corrosion casts by sizeable, "empty," irregular areas devoid of vessels. In the background of those spaces are remnants of the submucous vascular plexus, revealing a proliferative reaction of its small- and medium-sized venous vessels: densely packed nodular outgrowths that decorate their surfaces and irregularly extending, mainly venous vascular trunks of various calibers [15,18,19,21]. The compression of vessels that supply the vascular tufts at their base leads to alterations in their morphology, but also to a decrease in the blood flow, which cannot be compensated by proliferation of the endothelial cells [17–21,30]. This leads to the formation of pseudo-vessels, composed solely of cancer cells, and to necrosis in the form of ulcerations, which are irregular in size and depth [5,60,64–65].

We have found it remarkable that the extending edge of larynx carcinoma is characterized by a specific vascular response. This border area, including both the tumor periphery and adjacent tissues, has a rich capillary network, mostly in the form of vascular tufts surrounded by a narrow, but fairly deep, belt of low vessel density. Just outside this border are proliferating, densely arranged capillaries, forming long spiral loops or single twigs. The further away they are

from the tumor edge, the smaller is the length of these looped capillaries and the more numerous are the spirally twisted, short atypical vessels that appear among them. Their surfaces are frequently covered by small nodular outgrowths. This picture reflects the active influence exerted at a distance by cancer cells, which thus "prepare" the adjacent tissues for their further invasion and spread. The described features are consistent with the already known fact that angiogenic factor(s) can diffuse in tissues for a distance of a few millimeters [3,7,9,10] and that endothelial-cell divisions predominate at the tumor edge [17, 35,47]. Moreover, it is also known that tumor microvessels exhibit a strikingly increased permeability to plasma proteins, including plasminogen and fibrinogen [4,11,23,24,60]. Recently it was demonstrated that human tumor cells actively secrete vascular permeability factor (VPF). This factor, being extremely potent in enhancing the permeability of capillaries and venules [29], causes leakage of plasma proteins, extravascular coagulation, and fibrin gel deposition, creating a provisional stroma for tumor cells, regulating the infiltration of macrophages and probably other inflammatory cells, as well as — which seems to be very important — facilitating the inward migration of new blood vessels. All these data are consistent with our observation that the postcapillary portion of the larynx-cancer vascular bed is the territory within which the migration of malignant cells and cellular elements of the host immune system, between the vascular network and tissues, has been observed most frequently [64].

The arrangement of the vascular network of renal clear-cell carcinoma is influenced by forces generated during its growth within a parenchymatous, very richly vascularized organ enclosed by a fibroelastic capsule. Renal carcinomas, which grow as "tense" tumors, push and displace the preexisting vascular network by their advancing perimeter, receiving their blood supply mainly from the superficial vascular coat occupying their periphery [21,38,39]. Two different vascular patterns of the coat observed on the anterior and posterior sides of the tumor depend on different characteristics of the adjacent tissues. The nearly regular arrangement of vessels on the anterior (peritoneal) surface occurs due to the remodeling of the local renal vasculature. Large vessels observed on the posterior surface include blood vessels incorporated from the renal capsule and the underlying iliopsoas muscle.

Blood vessels of the superficial vascular envelope and their intermediate segments directed towards the center of the tumor [19–21,38] are stretched tangentially and radially so that they become contorted. Simultaneously, the central segments of tumor vessels are compressed perpendicularly to their long axis and are stretched lengthwise. These alterations, induced by elevated interstitial pressure, thus lead to damage of the endothelium [18–21,30,60,65]. This process cannot, however, be compensated by endothelial-cell proliferation [17–19,21] and induces a drop in the blood flow within those vessels, leading to a randomly distributed, focus of ischemia and anoxia, manifested by foci of thrombotic necrotic changes, mainly located centrally [5,8,17,34, 44,60]. Similar necrotic foci, however, can also be present in the most richly vascularized areas of the tumor, in the vicinity of the peripheral vascular coat, where the formation of new vessels is very intense [3,7,17–19,21,26,35,37,47]. The vasculature of renal carcinoma exhibits a prevalence of venous and sinusoidal-like smaller vessels over practically unchanged arteries [4,11, 21,36,65]. Nevertheless, the further exponential growth of the tumor depends, to some degree, on the formation of delicate, highly permeable new vessels [6,7,9,11,14,19–22,25,27–29,31,32,65].

Our observations provide morphological evidence for intense angiogenesis in different regions of the tumor, reflected by numerous vascular outgrowths present on cast surfaces of veins and capillaries. They represent progressive proliferation and fusion of endothelial cells in the course of sprouting of the new vessels [18–21]. The angiogenetic response is especially pronounced in stellate veins, which are formed by distal segments of numerous interlobular and superficial cortical veins collecting fine radicles of the peritubular plexus. All these vessels are transformed by angiogenic factor(s), as well as VPF, and are annexed into the three-dimensional framework of the vascular bed.

Interestingly, both examined tumors did not proliferate and regrow into the areas of necrosis [15,21,26,28,32]. This suggests that angiogenic factor(s) can operate most actively at the periphery of the tumor, where the vessels exhibit a strikingly increased permeability to plasma proteins [4,8,23–25,28,29,31,60], which is perpetually maintained by cancer cells secreting VPF [29]. Leaking plasma proteins, in cooperation with other biologically active substances, such as plasminogen activator, PGE_1, and PGE_2 liberated by monocytes/macrophages, and clotting factors produced by cancer cells, can regulate the generation of tumor stroma from fibrin-fibronectin deposits, thereby influencing further invasion and spread of the neoplasm [29]. In the central parts of the tumor, the steadily increasing interstitial pressure and pH (due to progressive acidification) causes compression, blood coagulation, and finally occlusion of thin-walled capillaries and veins, thus leading to marked pathological alterations of their endothelial lining and, most probably, to cessation of the development of new highly permeable vessels [21, 22,26,28,29,32], as well as to the formation of scarlike desmoplasia [26,28,29,32].

5. Conclusions

The vascular network of both of the human carcinomas studied share certain common features and also here distinct characteristics [33]. Both carcinomas show (1) intense angiogenesis, especially at the periphery (more strongly pronounced in cancer of the larynx); (2) proliferation of new highly permeable, vessels, which are most intense at the tumor periphery; (3) the prevalence of thin-walled vessels (capillaries and pseudocapillaries), mainly located on the perimeter; (4) the highest density of blood vessels occuring in peripheral areas of tumor, and poor vascularization in deeper regions; (5) adaptative changes of vessels and continuous remodelling of the tumor vascular network due to tumor-induced angiogenesis; (6) absence of tumor regrowth into areas of necrosis.

The features that characterize larynx carcinoma are summarized as follows: (1) it exhibits, as a rule, an exophytic pattern of growth in the form of size-differentiated, smaller tumors with a vasculature typical of "lax" tumors; (2) the prevalence of capillaries and pseudocapillaries, which are especially abundant at the periphery of the tumor; (3) a border zone at the advancing perimeter of the tumor characterized by a specific vascular response reflecting the "preparation" of the adjacent tissues for further local, direct expansion and spread of the tumor; (4) necrosis, if developed, in the form of ulcerations that are variable in depth and size, and are mostly accompanied by inflammation.

Renal carcinomas exhibit the following characteristic features: (1) they are usually well vascularized and grow as "tense" tumors, in the form of an irregular mass composed of avascular nodules surrounded by thin vascular sheaths and separated by narrow, dense vascular septa; (2) they show high interstitial pressure; (3) a well-developed superficial vascular coat is always present on their perimeter, and therefore they receive their blood supply from the periphery; (4) their vascular bed has many thin-walled venous vessels, deformed by compression, stretching, and displacement, which acquire a sinusoidal character; (5) a specific proliferative reaction of stellate veins.

All of these findings also suggest that further development of solid tumors following the formation of new vessels may depend not only on the oxygen and nutrient supply, but also on the presence of an outer endothelial surface covered by a basal lamina and organized in the form of a three-dimensional network, permitting malignant cells to adhere and promoting their migration. Therefore, the perivascular migration of tumor cells at the periphery of the tumor may also be a route for expansion [24,26,29,31].

References

1. Green HSN. Heterologous transplantation of mammalian tumors. *J Exp Med* 73:461–474, 1941.
2. Gimbrone MA Jr, Leapman SB, Cotran RS, Folkman J. Tumor dormancy in vivo by prevention of neovascularization. *J Exp Med* 136:261–276, 1972.
3. Folkman J. Tumor angiogenesis factor. *Cancer Res* 34:2109–2113, 1972.
4. Warren BA. Tumor angiogenesis. In: Peterson HI (ed.), *Tumor Blood Circulation*. CRC Press, Boca Raton, FL, pp 49–75, 1979.
5. Gullino PM. Consideration on blood supply and fluid exchange in tumors. In: *Biomedical Thermology*, Alan R. Liss, New York, pp 1–20, 1982.
6. Shubik P. Vascularization of tumors: A review. *J Cancer Res Clin Oncol* 103:211–226, 1982.
7. Folkman J. Angiogenesis and its inhibitors. In: DeVita V, Hellan S, Rosenberg S (eds.), *Importanat Advances in Oncology*. JB Lippincott, Philadelphia, pp 42–62, 1985.
8. Jain RK. Determinants of tumor blood flow: A review. *Cancer Res* 48:2641–2658, 1988.
9. Spławiński J. Tumor-induced angiogenesis; a rapidly maturing idea. *Drugs Today* 24:821–835, 1988.
10. Folkman J, Waston K, Ingber D, Hanahan D. Induction of angiogenesis during the transition from hyperplasia to neoplasia. *Nature* 339:58–61, 1989.
11. Warren BA. The vascular morphology of tumors. In: Peterson HI (ed.), *Tumor Blood Circulation*. CRC Press, Boca Raton, FL, pp 1–47, 1979.
12. Olszewski E. Blood vascular system of cancer of the larynx. *Arch Otolaryngol* 102:65–70, 1976.
13. Olszewski E. Vascualarization of ossified cartilage and spread of cancer in larynx. *Arch Otolaryngol* 102:200–203, 1979.
14. Osteoux M, Jeanmart L. Kidney vascularization: Morphology and angiogenesis a microangiographic experimental study. In: Lohr E (ed.), *Renal and Adrenal Tumors*. Springer, Berlin, pp 69–77, 1979.
15. Miodoński AJ, Kuś J, Olszewski E. SEM studies on blood vessels in cancer of the larynx. *Arch Otolaryngol* 106:321–332, 1980.
16. Kuś J, Miodoński AJ, Olszewski E, Tyrankiewicz R. Morphology of arteries, veins and capillaries in cancer of the larynx. *J Cancer Res Clin Oncol* 100:271–283, 1981.
17. Reinhold HS, Van der Berg-Blok A. Vascularization of experimental tumors. In: *Development of Vascular System*. Ciba Fundation (ed.), Symposium No. 100, Pitman, London, pp 100–110, 1983.
18. Grunt WT, Lametschwandtner A, Staindl O. The vascular pattern of basal cell tumors, LM and SEM study of vascular corrosion casts. *Microvasc Res* 29:371–386, 1985.
19. Grunt WT, Lametschwandtner A, Karrer K, Staindl O. The angioarchitecture of the Lewis lung carcinoma in laboratory mice. *Scann Electron Microsc* II:557–573, 1986.
20. Grunt WT, Lametschwandtner A, Karrer K. The characteristic structural features of the blood vessels in the Lewis lung carcinoma. *Scann Electron Micros* II:575–589, 1986.
21. Bugajski A, Nowogrodzka-Zagórska M, Leńko J, Miodoński AJ. Angiomorphology of the human renal clear cell carcinoma: A LM and SEM study. *Virchows Archiv A Pathol-Anat* 415:103–113, 1989.
22. Frucht LT. Critical factors controlling angiogenesis: Cell products, cell matrix and growth factors. *Lab Invest* 55:505–509, 1986.
23. Dvorak HF. Tumors: Wounds that do not heal. *N Engl J Med* 315:1650–1659, 1986.
24. Dvorak HF, Harvey VS, Estrella P, Brown LF, McDonagh J, Dvorak AM. Fibrin containing gels induced angiogenesis. Implication for tumor stroma generation and wound healing. *Lab Invest* 57:673–686, 1987.
25. Nicosia RF, Madri JA. Developmental changes during angiogenesis in the aortic ring-plasma clot model. *Am J*

Pathol 128:78–90, 1987.

26. Thompson WD, Schiach KJ, Fraser RA, McIntosh LC, Simpson JC. Tumors acquire their vascularization by vessel incorporation, not vessel ingrowth. *J Pathol* 151: 323–332, 1987.

27. Pauli BU, Knudson W. Tumor invasion: A consequence of destructive and compositional matrix alterations. *Human Pathol* 19:628–639, 1988.

28. Beranek JF. Ingrowth of hyperplastic capillary sprouts into fibrin clot: Further evidence in favour of the angiogenic hypothesis of repair and fibrosis. *Med Hypothesis* 28:271–273, 1989.

29. Nagy JA, Brown LF, Senger DR, Lanir N, Van de Water L, Dworak AM, Dvorak HF. Pathogenesis of tumor stroma generation: A critical role for leaky blood vessels and fibrin deposition. *Biochem Biophys Acta* 948: 305–326, 1989.

30. Grunt WT, Lametschwandtner A, Adam H. The development and degradation of blood vessels in tumors. *Int J Microcirc Clin Exp* 3:343, 1984.

31. Nicosia RF, Tchao R, Leighton J. Interactions between newly formed endothelial channels and carcinoma cells in plasma clot culture. *Clin Expl Metastasis* 4:91–104, 1986.

32. Knierim M, Paweletz N, Finze EM. Tumor-related reconstructive vascularization, an ultractructural study. *Anticancer Res* 6:1305–1316, 1986.

33. Vaupel P, Gabbert H. Evidence for and against a tumor type specific vascularity. *Stralentherapie und Onkologie* 162:633–638, 1986.

34. Kallinowski F, Vaupel P. pH distribution in spontaneous and isotransplanted rat tumors. *Br J Cancer* 58:314–321, 1988.

35. Denekamp J. Vascular endothelium as the vulnerable element in tumors. *Acta Radiol Oncol* 23:217–225, 1984.

36. Shah-Yukich AS, Nelson AC. Characterization of solid tumor vasculature: A three-dimensional analysis using the polymer casting technique. *Lab Invest* 58:236–244, 1988.

37. Konerding MA, Steiberg F. Bundoch V. The vascular system of xenotransplanted tumors — scanning electron and light microscopic study. *Scann Microsc* 3:327–336, 1989.

38. Falk P. Pattern of vasculature in two pairs of related fibrosarcomas in the rat and their relation to tumor response to single large dose of radiation. *Eur J Cancer* 14:237–250, 1978.

39. Falk P. Differences in vascular pattern between the spontaneous and transplanted C3H mouse mammary carcinoma. *Eur J Cancer* 18:155–165, 1982.

40. Virchow R. *Die krankhaften Geschwülste*. August Hirschwald, Berlin, 1863.

41. Algire GH, Chalkley HW. Vascular reactions of normal and malignant tissues in vivo: I. Vascular reactions of mice to wounds and to normal and neoplastic transplants. *J Natl Cancer Inst* 6:73–85, 1945.

42. Sanders AG, Shubik P. A transparent window for use in the Syrian hamster. *Isr J Exp Med* 11:118, 1964.

43. Goodall CM, Sanders AG, Shubik P. Studies of vascular patterns in living tumors with a transparent chamber inserted in the hamster cheek pouch. *J Natl Cancer Inst* 35:497–521, 1965.

44. Reinhold HS, Blachiewicz B, Blok A. Oxygenation and reoxygenation in "sandwich" tumors. *Bibl Anat* 15: 270–272, 1977.

45. Flanagan SP. "Nude" a new hairless gene with pleiotropic effects in the mouse. *Genet Res* 8:295, 1966.

46. Giovanella BC, Siehlin JS, Williams IJ, Lee SS, Shepard RC. Heterotransplantation of human cancers into mice. *Cancer* 42:2269–2281, 1978.

47. Hirst DG, Denekamp J. Tumor cell proliferation in relation to the vasculature. *Cell Tissue Kinet* 12:31–42, 1979.

48. Murakami T. Application of the scanning electron microscope to the study of the fine distribution of the blood vessels. *Arch Histol Jpn* 32:445–454, 1971.

49. Miodoński AJ, Hodde CK, Bakker C. Rasterelektronenmikroskopie von Plastik-Korrosion-Präparaten: Morphologische Unterschiede zwischen Arterien und Venen. *Beitr Elektronenmikr Direktabb Oberfl* (Münster) 9:436–442, 1976.

50. Gannon PJ. Vascular casting. In: Hayat MA (ed.), *Principles of Scanning Electron Microscopy*. Van Nostrans Reinhold, New York, pp 170–193, 1978.

51. Paine CJ, Low FN. Scanning electron microscopy of cardiac endothelium of the dog. *Am J Anat* 142:137–158, 1975.

52. Hodde CK, Nowell JA. Scanning electron microscopy of microcorrosion casts. *Scann Electron Microsc* II: 89–106, 1980.

53. Miodoński AJ, Kuś J. Tyrankiewicz R. Scanning electron microscopy blood vessel casts analysis. In: Allen DJ, Motta PM, DiDio LJA (eds.), *Three-Dimensional Microanatomy of Cells and Tissue Surfaces*. Elsevier/North-Holland, New York, pp 71–87, 1981.

54. Lametschwandtner A, Lametschwandtner U, Weiger T. Scanning electron microscopy of vascular corrosion casts: Technique and application. *Scann Electron Microsc* II: 663–695, 1984.

55. Terreros DA, Behbehani A, Cuppage FE. Evidence for proximal tubular cell origin of a sarcomatoid variant of human renal cell carcinoma. *Virchows Arch A Pathol-Anat* 408:623–636, 1986.

56. Thoenes W, Stoerkel S, Rumplet HJ. Histopathology and classification of renal cell tumors (adenomas, oncocytomas and carcinomas): The basic cytological and histopathological elements and their use for diagnosis. *Pathol Res Pract* 181:125–143, 1986.

57. Bander NH. Monoclonal antibodies in urologic oncology. *Cancer* 60 (Suppl. 3):658–667, 1987.

58. Kirchner JA, Carter D. Intralaryngeal barriers to the spread of cancer. *Acta Otolaryngol* (Stockholm) 103: 503–513, 1987.

59. Welsh LW, Welsh JJ, Rizzo TA. Internal anatomy of the larynx and the spread of cancer. *Ann Otol Rhinol Laryngol* 98:228–234, 1989.

60. Vaupel P, Müller-Kleiser W. Intterstitieller Raum und Mikro-millieu in malignen Tumoren. In: Vaupel P, Hammersen F, (eds.), *Progress in Applied Microcirculation*, Vol. 2 S. Karger, Basel, pp 78–90, 1983.

61. Oki T. The distribution of blood vessels in the larynx. *J Otorhinolaryngol Soc Jpn* 61:1827–1840, 1958.

62. Pearson BW. Laryngeal microcirculation and pathwawys of cancer spread. *Laryngoscope* 85:700–713, 1975.

63. Andrea M. Vasculature of the anterior commissure. *Ann Otol Rhinol Laryngol* 90:18–20, 1981.

64. Kuś J, Miodoński AJ, Olszewski E, Sekula J. SEM observations on the cellular transmigration through the wall of blood vessels in cancer of the human larynx. *Folia Histochem Cytobiol* 23:43–50, 1985.

65. Hammersen F, Endrich B, Messmer K. The fine structure of tumor blood vessels. *Int J Microcirc Clin Exp* 4:31–43, 1985.

Author's address:

Prof. A.J. Miodoński, M.D., Ph.D.
SEM Laboratory, Department of Otolaryngology
N. Copernicus Academy of Medicine
Kopernika 23a
PL-31-501 Kraków
Poland

Interpretation of Endothelial Structure Related to Tumor and Atherosclerotic Blood Vessels

JAMES G. WALMSLEY(†)

1. Introduction and Scope

Scanning electron microscopy (SEM) of vascular corrosion casts provides an opportunity to gain specific information about the luminal cellular lining of blood conduits. In addition to observation of the pattern, distribution, and density of this cellular layer, more subtle changes may be detectable. However, the determination of any of this information requires appropriate preparative procedures for highlighting the features being evaluated. For the purposes of this chapter, the discussion is primarily limited to the cellular lining of blood vessels, the vascular endothelium, and the alterations that are associated with experimental tumors and atherosclerosis. It is then possible to use the proper interpretation of endothelial surface structure to comment on the disease process itself.

In order to put the results obtained from vascular casts into perspective with respect to other data, it is helpful to outline (1) the proposed mechanisms and range of approaches that have been used to describe vascular remodelling associated with endothelial proliferation, tumors, and atherosclerosis; (2) the specific techniques employed to highlight the endothelial structure in vascular casts; (3) the observations made with SEM; and (4) the evaluation of these observations with emphasis on quantitative morphometry or stereology for present and future applications.

2. Vascular Remodelling

2.1. Endothelial Alteration and Proliferation

In order to understand the role of endothelial proliferation in development and changes in blood vessels, studies of capillary growth and endothelial regeneration are of paramount importance. There are early endothelial adaptations that appear as cellular alterations, some of which are observable by using SEM, e.g., cell shape.

The emphasis on capillary growth is due to the fact that the repair of dead or damaged tissue by less specialized tissue involves the invasion by capillaries. This capillary growth is seen in tumors and in extreme cases of atherosclerosis. The endothelial cells lining the surface of the capillary form a close relationship with the fibrous supporting structures and the basement membrane. In the mature capillary, the basement membrane on the luminal side is connected to the endothelial cells by hemidesmosomes, and on the periluminal side it is supported by collagen fibers. While in the developing capillary, the basement membrane and collagen fibers are altered and sprout formation is associated with endothelial cell mitosis, either directly or proximal to the sprout. There are basically two types of sprouts: the tapering pointed type and the saccular type.

As indicated by early studies using the rabbit ear chamber [1], endothelial cells of the tapered

Motta, P.M., Murakami, T., and Fujita, H. (eds.), Scanning Electron Microscopy of Vascular Casts: Methods and Applications.

and saccular type differ from those of mature vessels in numerous ways [2]. The cells are thicker and intracellularly demonstrate a more fibrillar outer region, a ribosome-decorated endoplasmic reticulum, and numerous mitochondria. The endothelial cells are not anchored to the basement membrane, which is often incomplete, and thus, may slide freely over one another. These cells have been observed to overlap with each other, particularly in tapered sprouts with a small lumen, compared to saccular sprouts. They are often found in close association with leukocytes and platelets.

The emphasis on endothelial regeneration is due to the fact that the repair of the damaged intima in established mature vessels invariably involves the proliferation of endothelial cells. There is substantial evidence that endothelial injury does not necessarily result in endothelial cell loss and denudation of the subendothelium [3]. However, the response of the endothelial cells can be desquamation [4,5]. The repair in areas of cell desquamation is rapid in order to maintain the structural integrity of the endothelial surface and involves a series of processes, which include endothelial spreading and translocation, in addition to proliferation [6]. These processes obviously involve the cell cytoskeleton elements and progress in a well-defined sequence. Cell replication may occur in order to replace any lost cells such that there is an appropriate cell density, but it is yet unclear under what conditions the replacement cell density approaches the initial density. Detailed quantitative studies are required for clarification of this problem. Although the factors that regulate these cellular events are not well known, they are most likely soluble agents released at the site of injury from activated endothelial cells, platelets, and subendothelial smooth muscle. The release of these factors may be modulated by substances in the serum in conjunction with hemodynamic parameters like shear stress. It is clear that growth factors play a major role in endothelial proliferation and angiogenesis [7] in both tumors [8] and atherosclerosis [9]; necrosis factors play an opposing role in tumors [10].

In order to understand the role of endothelial injury in arterial disease, it is necessary to recognize if the endothelium has been altered and how this relates to normal function. SEM is particularly useful in that it permits large areas to be examined and observed changes to be related to the overall anatomy. In addition to morphological investigations, studies in vivo include the evaluation of the response to a variety of stimuli by examining radiolabelled tracers [11], the uptake of dyes [12], PGI_2 production [13], and endothelial cell replication [14]. Injury to the endothelium is an important initiating factor for a broad spectrum of vascular diseases, including atherosclerosis [15,16], disseminated intravascular coagulation [17], and perhaps hypertension [18].

The role of flow patterns and shear stresses in the determination of normal orientation, realignment, and damage of the endothelium has been known for some time [19]. Studies using various alterations in flow have clearly indicated, using SEM, that the nucleus and the entire cell are reoriented. One example, an investigation of the rabbit with unilateral renal ligation or aortic constriction, indicated rounding of cells with reduced flow and randomization of cell configuration with disrupted irregular flow patterns [20]. The challenge of determining how hemodynamics affects the endothelium at bifurcations can only be dealt with by optimizing available preparative and quantitative techniques with a view towards future innovations. These approaches are addressed under the appropriate headings below.

2.2. Tumor Angiogenesis

There are two required components of the circulatory system for any mass of tissue, namely, blood vessels responsible for distribution of the blood and those responsible for diffusion from the blood to the tissue. The feeding arteries and collecting veins are mainly for distribution, and the microcirculatory vascular bed is required for diffusion. Tumor angiogenesis occurs primarily at the level of the microcirculation and can be described in terms of the formation of capillary sprouts in association with endothelial proliferation.

2.2.1. Types and classifications. Warren [2] describes two types of tumor angiogenesis in terms of the time of the events following inoculation of metastatic cells on surfaces and organs.

Table 26-1. Classification of blood vessels in neoplasms [21] with corresponding endothelial characteristics and imprints in vascular casts.

Class	Blood vessels	Endothelium	Endothelial profiles
1a	Arteries	Normal	Elongated cells, elliptical nuclei
1b	Arterioles	Normal	Less elongated cells, less elliptical nuclei
2	Capillaries with BM	Normal & contracted	Almost circular cells and nuclei
3	Capillary sprouts	Developing	Bulbous or tapering
4	Sinusoidal vessels	Interrupted lining	Flat between cells
5	Blood channels	No lining	Rough acellular texture
6	Giant capillaries	Comprises wall	Tumor cells, penetrating gaps
7	Fenestrated capillaries	Underlying particles	Cell boarder flaps
8	Venules and veins	Saccular dilations	No elongated cell or nuclei axis
9	Arteriovenous anastomoses	Normal	Protruding nuclei

The primary angiogenesis is the initial vascularization of the mass of proliferating tumor cells. This is essential for the survival and growth of the tumor cell population. The secondary angiogenesis is that which occurs in cycles at the periphery of a growing tumor. This is essential for distribution to new microcirculatory beds in the expanding tumor.

As indicated by Warren [21], the classification system given in Table 26-1 facilitates the description of all the vessels within the tumor. Of course, the network pattern may vary from tumor to tumor. Typically the arterial to venous connections from the periphery through a melanoma can be written as follows, with the appropriate class indicated in parenthesis: artery (1a), arteriole (1b), capillary with basement membrane (2), sinusoid (4), blood channel (5), giant capillary (6), fenestrated capillaries (7), venule and vein (8), or arteriovenous anastomoses (9).

Where it is possible, an effort will be made to restrict the discussion of tumor blood vessels to the effect of host factors on the malignant melanoma. A broader discussion of the tumor vasculature is included in Chapter 24. In spite of the vast number of reports on tumor vascularization and angiogenesis, the proportion of studies examining the ultrastructure of tumor vessels is small. There have been two recent studies of melanomas [22,23] that emphasize host factors in different mouse models.

2.2.2. Murine melanoma model. For tumor studies performed in this laboratory the F10 subline of B16 murine melanoma was used. The F10 strain was chosen because preliminary studies revealed that rapid and reproducible local growth occurred after implantation in young mice. This B16 murine melanoma line has been well characterized and has been used for related investigations [41,42]. Young C57BL/6 male mice (2–3 months old) were purchased from Charles River laboratories (N. Wilmington, MA). Old C57BL/6 male mice were donated from the aging colonies of the National Institute on Aging maintained at Charles River Laboratories. The SEM casting and other histological procedures employed are described below (see Section 3).

The model that offers many morphological correlations with the clinical situation is the immune-deficient, athymic nude mouse, which has been described by Konerding and colleagues [22]. These authors drew on their own investigations and the work of others to provide a critical review of some of the basic assumptions concerning the vascularization pattern. Their investigations concentrated on vascular ultrastructure and topography by examination of corrosion casts, and on vascular density and distribution by quantitation of India-ink-injected vasculature. Another investigation of melanomas grown in the athymic nude mice demonstrated a wide variation of vascular features that have been detailed using quantitative stereology [43]. Casts have demonstrated that there is heterogeneity in the vascular pattern in melanomas, with well-formed vessels apparently outside the margin of the tumor [22,23]. Many of the vascular features described by the above authors have been classified and diagramed in detail for Lewis lung carcinomas [44,45], making it obvious that tumor growth is strongly dependent on its position in relation to surrounding blood vessels.

Reduction of tumor blood flow can be therapeutically advantageous. The sensitivity of the tumor vasculature to radiotherapy and chemotherapy most likely plays a major role in their effectiveness [46]. For some time it has been suggested that the vascular endothelium is the vulnerable element in these tumor therapies [47,48]. This vulnerability may also be true in the treatment of hyperthermia [49,50]. SEM studies of tumor vasculature can provide detailed information in sites affected by tumor therapy. The uniqueness of human tumors with respect to their vascular system is reflected in the nude mouse model [22,51].

2.3. Atherosclerotic Lesions

An overview of the characteristics of atherosclerotic lesions [24] is helpful for understanding endothelial structural alterations that can be observed with vascular corrosion casts. Primarily, the site under investigation must be specified, since atherosclerosis is a patchy disease that exhibits a variety of patterns. Although there may be advanced disease in one artery, comparable arteries may not be similarly involved.

There have been hypotheses in the literature suggesting that the distribution of altered endothelial cells, and adherent serum factors and blood cells, are related to hemodynamic factors, including flow patterns and stagnation points. These views have been supported in studies of human atherosclerosis, similar to observations of carotid bifurcations obtained at autopsy [25]. The contribution of high-shear or low-shear stresses to this distribution has been a matter of some debate. The development of lesions in the hypercholesterolemic rabbit has been correlated to hemodynamic stress, with the initial distribution of injury at the lateral margins and at the inflow region of branch vessels [26,27]. Localization of damage may be consistent with early fibrin formation as an indication of fibrin-mediated endothelial injury [28].

2.3.1. Definitions and grading. The following definitions [24] were combined with one general grading system [29] to devise Table 26-2. *Severity* implies the degree to which the disease has narrowed the lumen and compromised flow. Usually severity is quantitated in terms of the percentage stenosis, but a more detailed analysis of wall dimensions has been proposed [30]. There are problems with *in vivo* techniques and with nonpressurized histological specimens. Vascular corrosion casts, when optimally prepared, can be used to obtain accurate data on changes in luminal diameter. *Extent* means the mass or bulk of atherosclerotic intimal tissue accumulated in a specified arterial segment, artery, region, or entire arterial system. Quantitative estimates

Table 26-2. Classification of atherosclerotic lesions [24,29] with corresponding endothelial characteristics and imprints in vascular casts.

Grade	Criteria of severity	Extent	Composition	Complication	Endothelial profiles
1+	Luminal surface contour defect	Intimal streaking	Fibrous plaque lipids, foam cells		Increased number of nuclei at bifurcations
2+	Lumen diameter <20% reduction	Medial lipid accumulation	Fibrous and/or fatty plaque, intimal muscle, lipids		Variable nuclei distribution
3+	Lumen diameter 20–50% reduction	Medial muscle involvement	Fatty-fibrous SM proliferation, lipids, collagen	Cholesterol deposition	Disrupted cell borders
4+	Lumen diameter >50% reduction	Intimal and medial remodeling	Atheromatous plaque calcium, collagen	Necrosis, thrombosis, calcification	Missing cells, variable spreading and protrusion, channels

by planimetry, or mapping of the proportion or pattern of the luminal surface covered by lesions, may provide a reasonable approximation of extent for some purposes [30,31]. Yet, such procedures do not usually take account of the thickness of lesions and are therefore not really quantitative determinations of the volume of diseased tissue present. Surface dimensions measured from SEM of casts can give precise data, while wall thickness estimates require tissue measurement.

Composition of a lesion includes the nature, consistency, and distribution of lesion components. Quantitative stereology is the most efficient approach for a precise description of proportional composition or volume fraction of the various components [32]. SEM would only be useful for the determination of a real fraction of components at the surface of the lesion or beyond. *Complication* is used to designate evidence that a lesion has a disrupted organization, which can involve fragmentation, hemorrhage, splitting, ulceration, thrombosis, or vascularization. Evidence of some of these complications with advanced disease can be seen with the appropriate preparation of casts for SEM [33].

2.3.2. Rabbit hypercholesterolemic model. Because of the greatly reduced metabolism of cholesterol in the common laboratory rabbit (usually the New Zealand White), the cholesterol-fed rabbit has become the most common animal model for the development and regression of atherosclerosis [34]. The alternative rabbit model is the Watanabe heritable hyperlipidemic rabbit, which has been compared experimentally to the hypercholesterolemic rabbit [35,36]. SEM studies of the arterial response to hypercholesterolemia and hyperlipidemia have been quite numerous. Early studies [37] revealed changes in cell shape and size, which were interpreted as evidence for endothelial damage and subsequent repair. Endothelial cells protruding from vessel orifices have been observed within 2 days of a 2% cholesterol supplemented diet in rabbits [38]. Many of the observation from earlier studies may have been subject to artifacts or perhaps unique conditions. The hyperlipidemic monkey model resulted in different interpretations of the initial response, as indicated by correlative transmission and scan-

ning EM [39,40]. This initial response was an accumulation of white cells on the endothelial surface, followed by infiltration into the sub-endothelial space. Again, there was no endothelial denudation in the early stages, but this loss could be detected at later stages.

For our results, reported below, male New Zealand White rabbits approximately 2.5 kg were fed a 2% cholesterol diet (Purina 5731C-9) for 3 weeks and matched to controls. After 13 weeks, the weight, serum cholesterol, mean arterial pressure (transformed ear-cuff pressure), and heart rate were determined, and the rabbits were sacrificed with an overdose of sodium pentobarbital. The common carotid arteries were often isolated and cannulated for subsequent perfusion fixation and casting of the cerebral and ear vasculature.

3. SEM Procedures for Vasculature and Endothelial Studies

3.1. Standard Perfusion and Immersion Fixation

The location of insertion of the cannula is dependent on the animal model being used. For whole-animal perfusions, the cannula tip is placed in the aorta as close to the heart as possible. In the mouse or rat, this is best achieved by insertion of the cannula through the left ventricle of the heart. In order to dilate and clear blood vessels, the vasculature is perfused at 100 mmHg with Krebs solution containing sodium nitrite (3 mM) and heparin (10 USP Us/ml). Fixation is followed immediately for 20 minutes with neutral buffered formalin at the same pressure. A solution of 1.25% paraformaldehyde and 2% glutaraldehyde in 0.1 M cacodylate buffer was used for post-fixation in the preparation of tissue for electron microscopy.

3.2. Vascular Casts and SEM

The casts for SEM were prepared and analyzed for verification of the preservation of blood vessel geometry and luminal structures. These methods have been used in this laboratory for some time with various modifications [23]. An accelerator was added to the partly prepolymerized resin

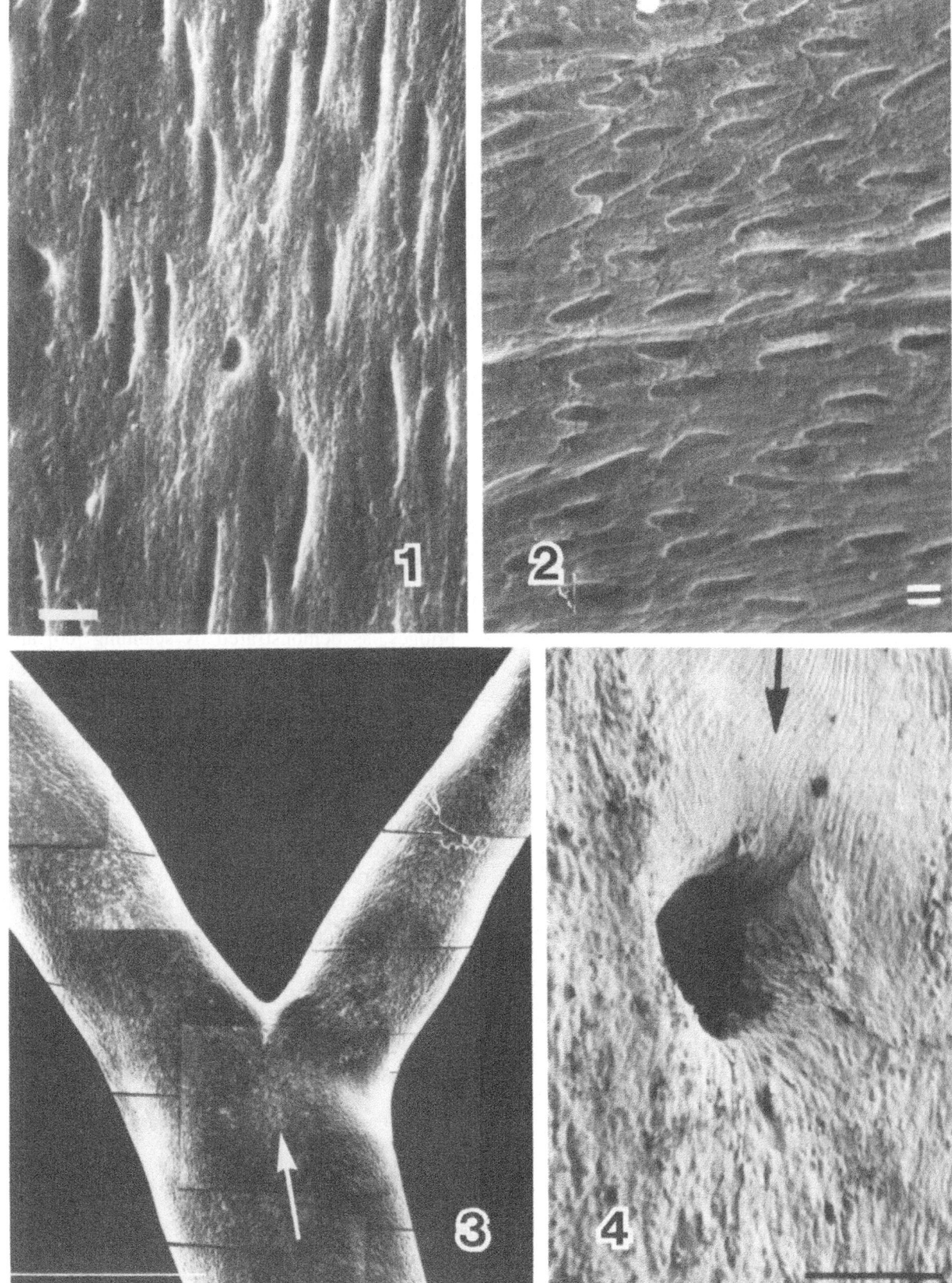

(Mercox CL-2B; Ladd Research, Burlington, VT) immediately before injection into the fixation cannula. By accounting for shrinkage, accurate *in vivo* diameters have been estimated and compared to dimensions found from sections and tissue mounts. Complete digestion of arterial tissue leaves the cast with endothelial nuclear imprints, as shown in Figs. 26-1 and 26-2.

Casts were carefully dissected and dehydrated in an increasing ethanol series. These casts were critical-point dried or air dried over a sieve for 36 hours, then mounted on stubs and grounded with silver paint. The samples were given a 10 nm coating of gold-palladium in a sputter coater (or vacuum evaporator) and viewed with a Cambridge Stereoscan 100T or ISI Super II scanning electron microscope at an accelerating voltage of 15 kV. Morphometric measurements of the cell (nuclear) density were made from SEM photographs, like those put together as a montage of a bifurcation (Fig. 26-3).

3.3. *Limitations of Tissue Preparations*

SEM of the intimal surface has and continues to be used in many laboratories, because possible artifacts caused by fixation or drying and mechanical distortion of the cells can be minimized [52]. The practice of pinning the unfixed aorta for determination of endothelial pattern or lesion distribution at bifurcations causes unacceptable distortions, except for small branches, such as the intercostals [53]. An SEM of one such intercostal branch from the hypercholesterolemic rabbit is shown in Fig. 26-4. Note the downstream lesion with surface deposition, which would be dificult to visualize using replication casting. Although the use of vascular casts has provided information about arterial geometry and endothelial cell shape, at present, standard casting alone does not help the visualization of the detailed cell surface structure [54]. The most important methodological factor altering the appearance of endothelial cells appears to be the method of prefixation and fixation perfusion parameters [55]. The electrolyte composition in conjunction with the time of preperfusion can effect crater formation, white cell deposition, and the expression of microvilli [3]. Severe folding of the lumen, which could reflect convolutions of the internal elastic lamina, is eliminated by perfusion fixation [56]. However, care must also be taken with regard to the solutions employed, in that contraction or spasm induced in smooth muscle cells can cause pronounced folding [57]. Another factor that may not be accounted for in animal models is the consequence of physiological stress, which may result in apparent damage of the endothelium as an indication of injury [58,59]. Arterial or venous spasm has been demonstrated to cause endothelial cell-to-cell adhesion, which remain after relaxation and apparently form intercellular bridges, as demonstrated by scanning [60] and transmission [61] electron microscopy. Although some of the observed alterations in the endothelial cell layer are due to monocytes entering the subendothelial layer, the accumulation of platelets is not always correlated with endothelial or subendothelial denudation, which can be observed in casts [62].

Methods have been developed for the recovery and observation of blood-vessel tissue surrounding the cast. Arterial tissue was first removed without damage using controlled techniques in 1979 [52]. Observations of wall components of tissue left on the cast after partial dissection or digestion have been reviewed by Miller et al. [63].

Figure 26-1. SEM of vascular cast of the perfusion-fixed aortic endothelial surface without any highlighting. The endothelial nuclei protrude into the cast. The bar represents 10 μm.

Figure 26-2. SEM of vascular cast of an aortic endothelial surface that has been treated with silver nitrite and perfusion fixed such that the endothelial cell borders protrude into the cast. Note the variable shape of the cell profiles compared to the almost uniform shape of the nuclei. The bar represents 10 μm.

Figure 26-3. Montage from SEM micrographs illustrating the nuclear imprints of a complete bifurcation from the rabbit ear (arrow). The bar represents 500 μm.

Figure 26-4. SEM of aortic endothelial surface from hyperlipidemic rabbit showing the opening to an intercostal artery. Flow was in the direction indicated by the arrow, such that the intimal deposits are lateral and distal to the bifurcation. The bar represents 500 μm.

3.4. Advantages of Various Vascular Cast Preparations

After the first application of SEM for the visualization of vascular casts [64], there were a series of studies in the 1970s that produced vascular casts on which the imprints of the endothelial nuclei were observable [65–69]. Furthermore, detailed vascular casting techniques for the replication of endothelial cell boundaries and detailed topography was first reported as early as 1977 [66,70]. There have been many significant modifications in the development and evaluation of various techniques [52,71–73]. The results of these findings, related to the key parameters of viscosity and shrinkage, have been summarized by Roach and colleagues [72]. One of the most important implementations to lower viscosity was the dilution of Mercox with methylmethacrylate (MMA). Steeber and Albrecht (Chapter 23) have used as much as a 1:1 dilution, which lowers the viscosity to 6 cp [74], whereas a 1:4 dilution (MMA: Mercox) lowers the viscosity to 12 cp [72]. The usefulness of lowering the viscosity seems to be limited by the increased shrinkage and decreased stability under the SEM beam. When a complex vasculature involving small blood vessels is being digested, there may be a breakage of the cast at any time when water or the basic solution is removed. One way of avoiding this potential problem is to reduce the surface tension with surface-active agents before the removal of the cast from the digesting solution.

3.5. Highlighting Endothelial Cell Borders

Hodde and colleagues [66] were able to demonstrate endothelial cell borders in vascular casts without the use of silver nitrate treatment, and recent results [22] confirm this highlighting phenomenon. This apparently is made possible by rapid perfusion fixation, followed immediately by resin (Mercox) injection [22,66]. However, a typical method (modified from [75]) for reliability causing elevated enhancement of endothelial cell borders has been achieved by perfusion staining with 0.13% $AgNO_3$ in 5.26% glucose, followed by Krebs before fixation. This was the standard technique used for the rabbit arterial preparation described below in conjunction with studies of the hypercholesterolemic model.

4. SEM Related to Endothelial Structure

4.1. General Interpretation

In order to best understand the implications of structures imprinted on vascular casts of tumor and atherosclerotic blood vessels, it is informative to evaluate the features seen on more normal vessels. Castenholz has recently reviewed the interpretation of structural patterns in microvessels [76] and lymphatics (Chapter 4), with emphasis on the replicas of endothelial nuclear protrusions and cell boundaries. As recognized by Hodde, Miodonski, and colleagues [66], casts of high replication quality that are carefully

Figure 26-5. SEM of vascular corrosion cast of tumor from a young mouse 14 days after inoculation. The well-defined vascular cast (oblate spheroid) in the center of the micrograph reflects the shape of the original tumor mass. The bar represents 1 mm.

Figure 26-6. SEM of vascular corrosion cast of tumor from an old mouse 21 days after inoculation. Before digestion, the original tumor tissue extended from the vessels at the top of the micrograph just beyond the crescent-shaped microvascular network at the bottom. The bar represent 1 mm.

Figure 26-7. Young-host penetrating vessel lumens. Bifurcating vessels gives rise to fairly straight branches, which taper gradually and give off vascular sprouts of varying length. The lumens of tapering vascular (single arrowhead) and rounded (double arrowhead) sprouts are clearly shown. The bar represents 100 μm.

Figure 26-8. Old-host penetrating vessel lumens. The three-dimensional convolution of the vessels on the right is obvious. Their origin was at the margin of the tumor space, and there is microvascular structure extending from them and terminating in sprouts that are tapered (single arrowhead) or bulbous (double arrowhead). The bar represents 100 μm.

Figure 26-9. Young-host vessel cast at the periphery of an avascular zone. Predominantly bulbous sprouts, indicating budding angiogenesis, are numerous. The bar represents 100 μm.

Figure 26-10. Old-host vessel cast at the periphery of the avascular zone. There are fewer lumen buds, which would be characteristic of angiogenesis. The bar represents 100 μm.

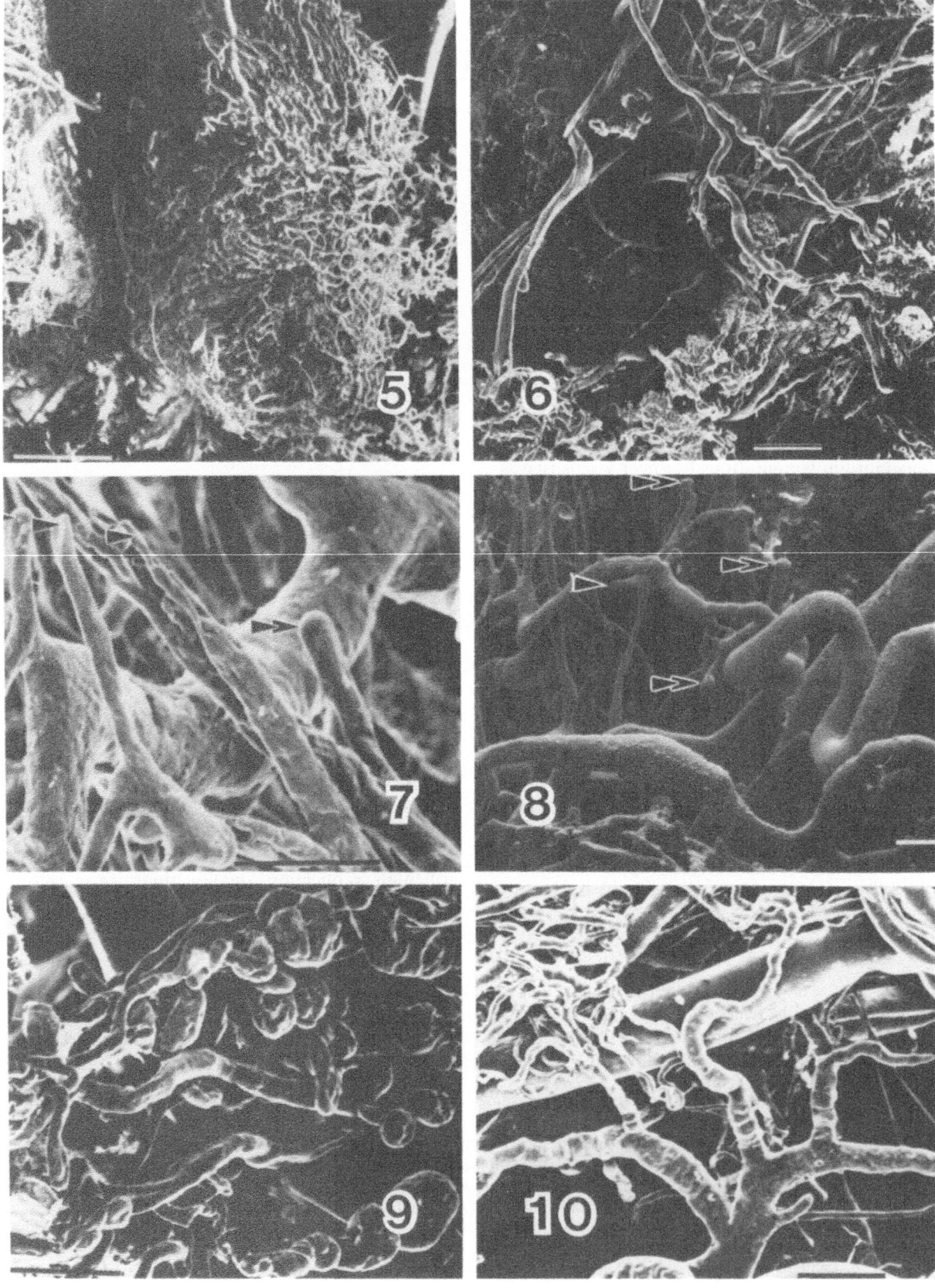

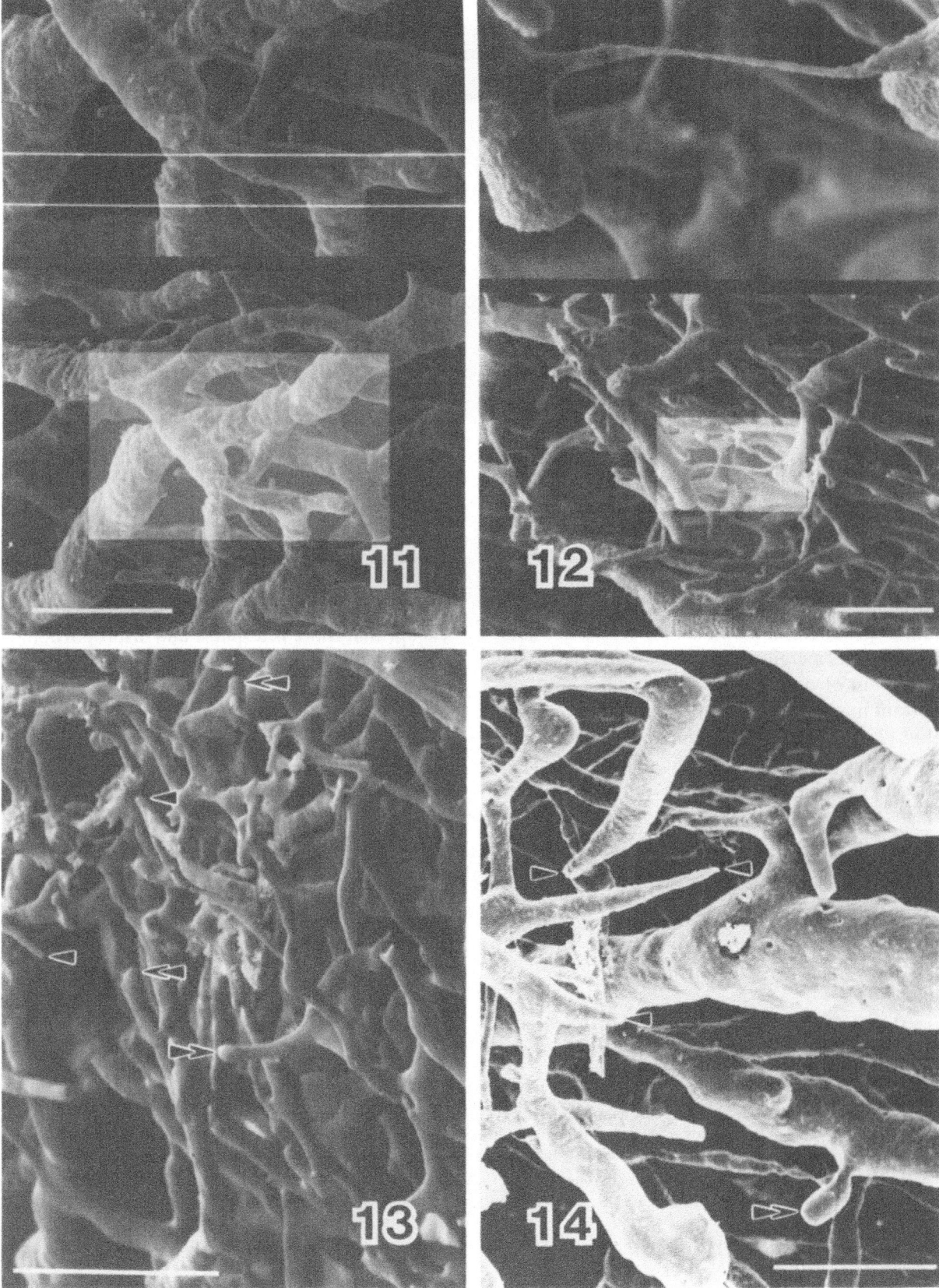

prepared, as specified above, may allow for the identification of a specific classification of vessel by analysis of the surface relief produced by the endothelium.

4.2. Tumor Angiogenesis

Tumors were grown for approximately 14 days in young hosts and 21 days in old hosts, resulting in a final tumor volume of approximately 250 mm^3. All of the results described below correspond to tumors of this size. As suggested by Lametschwandtner et al. [73], an attempt was made to determine which vessels were feeding the tumor. This was accomplished by using angiography of a radioopaque latex cast (Microfil). Except for the extreme edges of the tumor, there is uniformly less blood supply to the tumor of the old mouse. This difference between young and old is consistent with SEM observations described below. The overall SEM views of the vascular corrosion casts from young (Fig. 26-5) and old (Fig. 26-6) mice illustrate the profound difference in blood supply to the 250 mm^3 tumors. The margin of the tumor from the young mouse is obvious (Fig. 26-5), while only the bottom margin of the tumor from the old mouse (Fig. 26-6) seems to be delimited by the microvascular network. Notwithstanding the sparse overall vascularization, in the old mouse there seems to be no lack of potential vessels surrounding the open dark space in the middle, formerly occupied by the tumor tissue (Fig. 26-6). The lumen of the upper feeding vessel in the middle of the young host tumor is fairly straight and of uniform diameter, while for the old host the lumina are winding and more variable in diameter.

At the margin of the tumor, there appear to be vessels, which extend from the larger feeding vessels and either penetrate or attempt to penetrate into the tumor space (Figs. 26-7 and 26-8). For the young mice, the vessels appear to have a regularly shaped tapering lumen penetrating into the tumor (Fig. 26-7). For the old mice, the vessels are often very convoluted, abruptly taper off, and do not give rise to many sprouts (Fig. 26-8). It can be questioned whether these vessels successfully penetrate into the tumor, and if they do not, what is preventing such penetration.

As the penetrating vessels taper off, they can give rise to a dense vascular network characterized by many collaterals, rather than an ordered branching hierarchy (Figs. 26-9 and 26-10). Because of their location at the edges of the tumor, they can be called *peripheral* or *marginal vessels*. For young hosts, the bulbous and invaginated nature of the lumina is consistent with neoangiogenesis and expansion of the vascular volume (Fig. 26-9). For old hosts, the twisted string-like nature of the lumina is more characteristic of a mature microcirculatory bed, except that there is a sparse and disproportionate distribution of vessels (Fig. 26-10).

In order to detect whether there is likely to be angiogenesis near the avascular zone in the central part of the tumor, the lumenal cast of the innermost vessels was observed (Figs. 26-11 and 26-12). These views were accomplished by tilting the stage, or brushing away peripheral vessels where necessary. For young mice, the lumina of these vessels often have fungiform bulges, with holes and furrows characteristic of neoangiogenesis (Figs. 26-11–26-13). For old mice, the inner vessel casts have similar characteristics to those of the peripheral vessels, which are almost devoid of any bulges or sprouting (Fig. 26-14).

←

Figure 26-11. Young-host inner-vessel cast near avascular zone. The bar in the lower frame represents 100 μm and the upper frame is a 2X zoom of the highlighted box in the lower frame. The horizontal line separation of 40 μm would indicate that the lumen cast should be from giant capillaries.

Figure 26-12. Inner-vessel cast near avascular zone. The bar in the lower frame represents 100 μm, and the upper frame is a 4X zoom of the highlighted box in the lower frame. A tapering narrowing lumen connecting two vessels is characteristic of a sprouting connection.

Figure 26-13. The cast of a developing tumor microvasculature with lumina indicative of angiogenesis. The sprouts can be seen as rounded (double arrowhead) and tapered (single arrowhead). The bar represents 100 μm.

Figure 26-14. Cast from old mouse may be indicative of the blood-channel classification. These channels are characterized by a rough surface and variable lumen size. Sparse large sprouting features of various types are visible. The bar represents 100 μm.

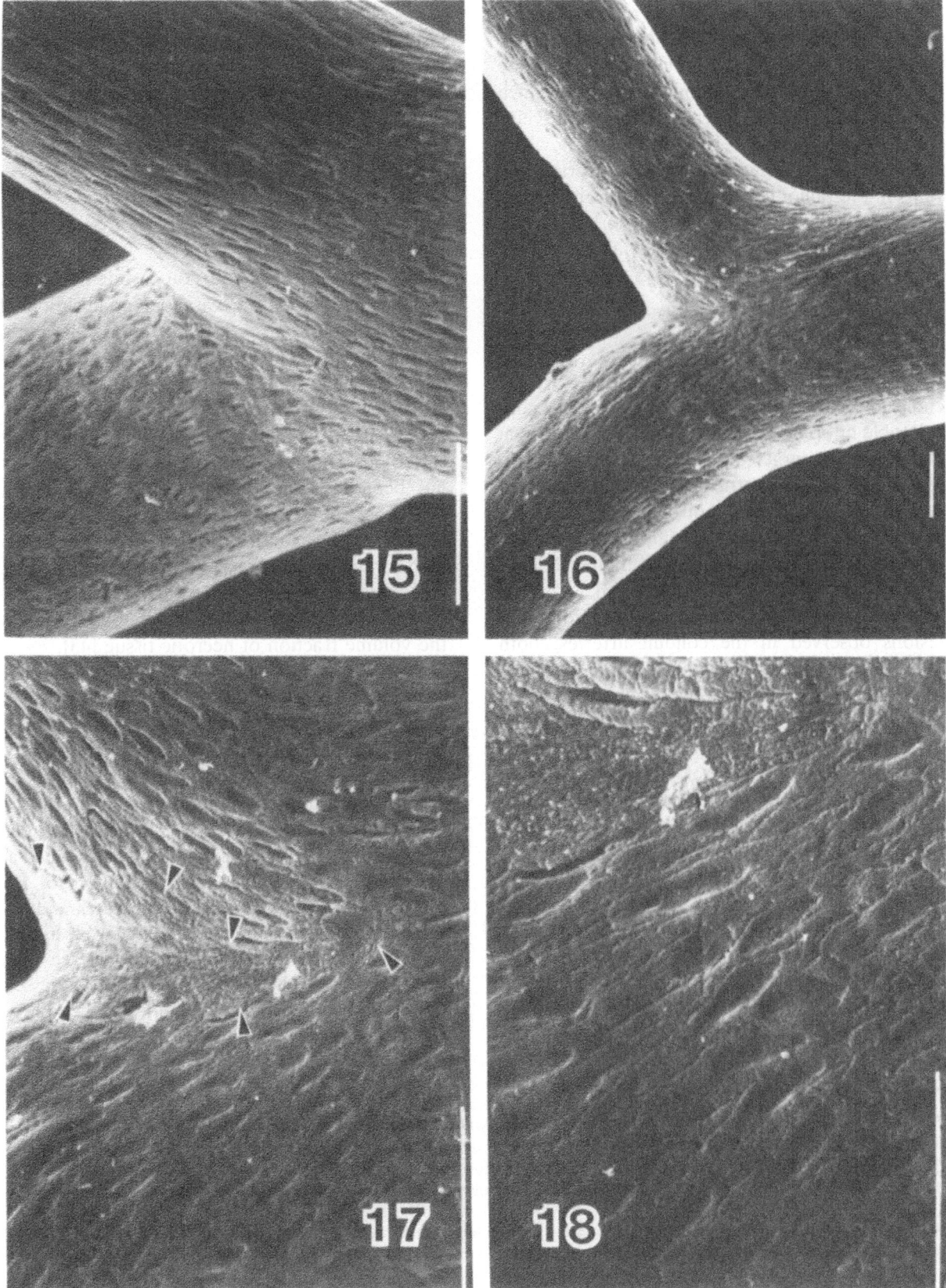

The growing vessels that penetrate or attempt to penetrate from the melanoma margin into the tumor space do not have luminal imprints, which are consistent with typical arteries and veins (Figs. 26-7 and 26-8). This observation, which implies that there are no typical arteries and veins in the tumor, is supported by recent studies [23,44,77], in contrast to earlier work [21] that included incorporated host vessels as part of the tumor vasculature.

4.3. Atherosclerotic Lesions

In atherosclerotic research the hyper-cholesterolemic (HC) rabbit has yielded considerable information on structural and functional changes in the larger distributing arteries. Histological and morphometric description of the aorta has involved the assessment of the development and regression of sudanophilic lipid deposition at branch points [76], as well as endothelial changes observed in SEM whole mounts [53] and casts [30,51,52,70,78,79]. Because of the prominent lesions observed in the conduit arteries, both pathologically in humans and experimentally in animals, atherosclerosis has been considered primarily a large-artery disease. With increased awareness of the important role of the endothelium in the variable relaxation response of small blood vessels, it has become necessary to ask whether subtle intimal changes might alter the normal function of smaller arteries that are significant for the control of vascular resistance.

In contrast to a normal bifurcation (Fig. 26-15), there is increased disruption of the endothelial patterns at HC bifurcations, but not in tubular portions of arteries in the $300\,\mu m$ range (Figs. 26-16–26-18). Although this localization at bifurcation is reminiscent of the pattern for aortic branch points, the limited extent and nature of disruption (Fig. 26-7) is not suggestive of the extensive denudation used to test experimentally for endothelial-dependent relaxation factors (EDRF).

4.4. Approaches to Quantitative Analysis

Most applications of corrosion casting have been for the analysis of the vascular organization of specific organs or disease states. Furthermore, optimal high-replication quality casting lends itself to the evaluation of the spatial relations between different classes of vessels and between structures along the vessels. Quantitative stereology applied to the melanoma vasculature has been performed using histological sections and other imaging techniques [23,24]. Standard stereological techniques [32] have been applied with certain assumptions to obtain estimates from five melanomas of vessel length, surface, and volume as a function of vessel diameter, as well as the volume fraction of necrotic tissue [43].

One distinct advantage SEM casts have over all other methods is facilitation of the quantitation of surface features. In particular, any stereological parameter that involves the counting or distribution of endothelial nuclei can be determined accurately. One stereological technique related to endothelial cell density is the mean free distance between nuclei, which can be determined as described by Underwood [32]. Of course, the estimates are limited to surface features and are subject to the interpretation of curved surface profiles and the determination of surface area from such curved surfaces. For blood vessels that approach a cylindrical geometry, the viewed surface has an area of $\pi.r.l.$, where r is the in-

---

Figure 26-15. Vascular cast of a renal resistance artery bifurcation from a rabbit on normal diet, with the the direction of blood flow upwards from the trunk to daughter branches. The bar represents $100\,\mu m$.

Figure 26-16. Vascular cast of a renal resistance artery bifurcation from a hypercholesterolemic rabbit, with the direction of blood flow upwards from the trunk to the daughter branches. The apex of the bifurcation lacks nuclear definition. The bar represents $100\,\mu m$.

Figure 26-17. The same bifurcation as in Fig. 26-16 at a higher magnification. Area around the bifurcation region lacks endothelial nuclear imprints, forming a distinct border (arrowheads). The bar represents $100\,\mu m$.

Figure 26-18. Higher magnification of the bifurcation region in Fig. 26-17 showing the abrupt ending of nuclear and cell boundary imprints. Since the endothelial border is so well defined, any features seen should not be attributed to preparation artifacts. The bar represents $50\,\mu m$.

ternal radius of the vessel and l is the length of the segment being sampled. An ingenious device for rotating rabbit aortic casts 360° in the SEM has been used to facilitate measurements of the atherosclerotic lesion area at bifurcations [80].

Cornhill and coworkers [30] developed a rigorous sampling protocol for quantitating the *en-face* size and shape of endothelial cells from rabbit aortic casts. The borders of the endothelial cell and its nucleus were outlined using a digitizing system. Measurements of area parameter, angle of orientation, length, width, length-to-area ratio, and shape index were calculated. Their main conclusion was the normal cellular orientation corresponded to flow direction; furthermore, for the interpretation of certain pathological endothelial changes, this morphometric approach has numerous possibilities.

Although the above digitization-based method is appropriate for describing an endothelial cell population of fairly uniform shape and orientation, this condition generally does not exist at arterial bifurcations, nor in tumor and atherosclerotic vessels. As indicated, endothelial orientation and shape are particularly sensitive parameters related to flow and regeneration. Unfortunately, stereological techniques used to quantify anisotropy have been limited, and resultant analysis does not always agree with the perceived orientation. Recently a new approach to estimating the area or volume orientation has been developed [81]. The distinct advantage with this method applied to endothelial cells or nuclei is that the structural profiles occur on the surface. Even though the cells may have a wide range of shapes and sizes, since the profiles are observed in two dimensions with SEM, this parametric method facilitates an estimate of alignment. In summary, each point in the endothelial feature can be assigned a local area orientation, so that if the endothelial cells have a preferred direction, the variance of the area orientation can be estimated.

As reviewed by Roach and coworkers [72], there are sophisticated techniques that can be used to account for the possible three-dimensional anisotropy of the blood vessel or tilt in the axis of viewing. Such approaches have involved the use of stereo pairs of micrographs, stereometry, or mounting of specimens on stages with three-dimensional translation and rotation. All SEM stages have some rotational and translational adjustment, so that if the micrographs are perpendicular to the normal of the surface and the region of measurement is small compared to the vessel diameter, the surface can be treated as flat [79]. Furthermore, length and area measurements made at a known tilt from normal can be corrected by the expansion in the direction of the tilt by 1/cos(tilt angle) [79]

5. Concluding Remarks and Future Challenges

If an attempt is made to go beyond the establishment of average values of complex structures, for example, sampling of individual cellular boundary length and area, it may be necessary to invoke consideration of mosaic fractals [82]. However, for most purposes two-dimensional endothelial structures that demonstrate properties of fractals [82] can still be described in terms of simplified shape, length, area, and orientation, as determined by SEM of tissue and casts. Because of the primary location of atherosclerotic lesions at bifurcations and since many of these bifurcations are nonplanar, it would be informative to incorporate the sampling of three-dimensional volume orientation [81]. This sampling may involve sectioning of the cast or arterial specimen in parallel slices, with subsequent SEM imaging and processing, but there are a number of practical problems that must be addressed.

In fact, there are a whole range of microscopic and histological techniques that can be used in combination with SEM for understanding the role of endothelial alterations. This wide range of correlative techniques has evolved precipitously over the past decade [54]. For example, the advantages of low-voltage high-resolution SEM cannot be ignored in its potential for the interpretation of the complex structures of the endothelial surface on casts. This facility can be combined with the use of a thinner coating of the specimens and the use of colloidal-gold labelling of specific features. Parallel to new ultrastructural approaches, there has been an increase in the effectiveness and number of functional techniques. Another strategy that differs from the structure-function strategy is the sequential study of identical structures by different methods. In essence, correlative techniques are used sequentially on the same specimen. For example, tissue removed from the cast and the cast itself can be observed and compared by SEM. This approach eliminates biological variation inherent in multiple specimen preparations. The most time-honored strategy is associated with rigorous investigation of preparative methods using appropriate controls. Rather than being misleading, the so-called artifacts can provide essential structural information. As summarized by Wetzel and Albrecht [54], biological structures are inherently altered by the preparative microscopic methods used to map them. As part of the investigative process, structures can be mapped by different correlative methods to visualize other cell parameters. In this manner, such structural studies make a direct comment on cellular function.

References

1. Clark ER, Clark EL. Observations on living arteriovenous anastomoses as seen in transparent chambers introduced into the rabbit's ear. *Am J Anat* 54:229–286, 1934.
2. Warren BA. Tumor angiogenesis. In: Peterson HI (ed.), *Tumor Blood Circulation: Angiogenesis, Vascular Morphology and Blood Flow of Experimental and Human Tumors*, CRC Press, Boca Raton, FC pp 49–75, 1979.
3. Richardson M, Parbtani A. Identification of nondenuding endothelial injury by scanning electron microscopy. *Scann Microsc* 1:1315–1326, 1987.
4. Reidy MA. Biology of disease. A reassessment of endothelial injury and arterial lesion formation. *Lab Invest* 53:513–520, 1985.
5. Reidy MA, Schwartz SM. Endothelial injury and regeneration. IV. Endotoxin: A nondenuding injury to aortic endothelium. *Lab Invest* 48:25–34, 1983.
6. Gotlieb AI, Wong MKK, Boden P, Fone AC. The role of the cytoskeleton in endothelial repair. *Scann Microsc* 1:1715–1726, 1987.
7. Leung DW, Cachianes G, Kuang WJ, Goeddel DV, Ferrara N. Vascular endothelial growth factor is a secreted angiogenic mitogen. *Science* 246:1306–1309, 1989.
8. Folkman J, Klagsbrun M. Angiogenic factors. *Science* 235:442–447, 1987.
9. Nilsson J. Growth factors in the pathogenesis of atherosclerosis. *Atherosclerosis* 62:185–199, 1986.
10. Old LJ. Tumor necrosis factor (TNF). *Science* 230:630–632, 1985.
11. Bell FP, Adamson IL, Schwartz CJ. Aortic endothelial permeability to albumin: Focal and regional patterns of uptake and transmural distribution of ^{131}I-albumin in the young pig. *Exper Mol Pathol* 20:57–68, 1974.
12. Bellet RE, Mastrangelo MJ. Malignant melanoma: Investigations in the nude mouse. In: Fogh J, Giovanella BC (eds.), *The Nude Mouse in Experimental and Clinical Research*, Academic Press, pp 511–5519, 1982.
13. Ingerman-Wojenski CM, Sedar AW, Nissenbaum M, Silver MJ, Klurfeld DM, Kritchevsky D. Early morphological changes in the endothelium of a peripheral artery of rabbits fed an atherogenic diet. *Exp Mol Pathol* 38:48–60, 1983.
14. Schwartz SM, Benditt EP. Cell replication in the aortic endothelium: A new method for study of the problem. *Lab Invest* 28:699–707, 1973.
15. Laschi R, Pasquinelli G, Versura P. Scanning electron microscopy application in clinical research. *Scann Microsc* 1:1771–1795, 1987.
16. Moore S. Pathogenesis of atherosclerosis. *Metabolism* 34:13–16, 1985.
17. Luscher EF. The role of blood cells and or the vessel wall in the induction of intravascular coagulation. *Klin Wochenschr* 60:710–712, 1982.
18. Haudenschild CC, Prescott MF, Chobanian AV. Effects of hypertension and its reversal on aortic intima lesions of the rat. *Hypertension* 2:33–44, 1980.
19. Langille BL, Adamson SL. Relationship between blood flow direction and endothelial cell orientation at arterial branch sites in rabbits and mice. *Circ Res* 48:481–488, 1981.
20. Reidy MA, Langille BL. The effect of local blood flow patterns on endothelial cell morphology. *Exp Mol Pathol* 32:276–289, 1980.
21. Warren BA. The vascular morphology of tumors. In: Peterson HI (ed.), *Tumor Blood Circulation: Angiogenesis, Vascular Morphology and Blood Flow of Experimental and Human Tumors*, CRC Press, Boca Raton, FL pp 1–47, 1979.
22. Konerding MA, Steinberg F, Budach V. The vascular system of xenotransplanted tumors — scanning electron and light microscopic studies. *Scann Microsc* 3:327–336, 1989.
23. Walmsley JG, Granter SR, Hacker MP, Moor AL, Ershler WB. Tumor vasculature in young and old hosts: Scanning electron microscopy of microcorrosion casts with microangiography, light microscopy and transmission electron microscopy. *Scann Microsc* 1:823–830, 1987.
24. Glagov S, Zarins CK. Quantitating atherosclerosis: Problems of definition. In: Bond MG, Insull W Jr., Glagov S, Chandler AB, Cornhill JF (eds.), *Clinical Diagnosis of Atherosclerosis: Quantitative Methods of Evaluation*, Springer-Verlag, New York, pp 11–35, 1983.
25. Zarins CK, Giddens DP, Bharadvaj BK, Sottiurai VS, Mabon RF, Glagov S. Carotid bifurcation atherosclerosis. Quantitative correlation of plaque localization with flow velocity profiles and wall shear stress. *Circ Res* 53:502–514, 1983.
26. Friedman MH, Deters OJ, Mark FF, Basgeron CB, Hutchins GM. Arterial geometry affects haemodynamics: A potential risk factor for atherosclerosis. *Atherosclerosis* 46:225–231, 1983.

378

27. McMillan DE. Blood flow and the localization of atherosclerotic plaques. *Stroke* 16:582–587, 1985.

28. Rowland FN, Donovan JJ, Picciano PT, Wilner GD, Kreutzer DL. Fibrin-mediated vascular injury. Identification of fibrin peptides that mediate endothelial cell retraction. *Am J Pathol* 117:418–428, 1984.

29. DePalma RG. Angiography in atherosclerosis: Advantages and limitations. In: Bond MG, Insull W Jr., Glagov S, Chandler AB, Cornhill JF (eds.), *Clinical Diagnosis of Atherosclerosis: Quantitative Methods of Evaluation*, Springer-Verlag, New York, pp 99–125, 1983.

30. Cornhill JF, Levesque MJ, Hendrick EF, Nerem RM, Kilman JW, Vasko JS. Quantitative study of the rabbit aortic endothelium using vascular casts. *Atherosclerosis* 35:321–337, 1980.

31. Cornhill JF, Akins D, Hutson M, Chandler AB. Localization of atherosclerotic lesions in the human basilar artery. *Atherosclerosis* 35:77–86, 1980.

32. Underwood EE. *Quantitative Stereology*. Addison-Wesley Publishing, London, 1970.

33. Laschi R. Contribution of scanning electron microscopy and associated analytical techniques to the study of atherosclerotic disease. *Scann Electron Microsc.* III: 1215–1222, 1985.

34. Cornhill JF, Bond MG. Morphology: Morphometric analysis of pathology specimens. In: Bond MG, Insull W Jr., Glagov S, Chandler AB, Cornhill JF (eds.), *Clinical Diagnosis of Atherosclerosis: Quantitative Methods of Evaluation*, Springer-Verlag, New York, pp 67–78, 1983.

35. Rosenfeld ME, Tsukada T, Chait A, Bierman EL, Gown AM, Ross. Fatty streak expansion and maturation in Watanabe heritable hyperlipemic and comparably hypercholesterolemic fat-fed rabbits. *Arteriosclerosis* 7:24–34, 1987.

36. Rosenfeld ME, Tsukada T, Gown AM, Ross R. Fatty streak initiation in Watanabe heritable hyperlipemic and comparably hypercholesterolemic fat-fed rabbits. *Arteriosclerosis* 7:9–23, 1987.

37. Goode TB, Davies PF, Reidy MA, Bowyer DE. Aortic endothelial cell morphology observed in situ by scanning electron microscopy during atherogenesis in the rabbit. *Atherosclerosis* 27:235–251, 1977.

38. Svendsen E. Focal endothelial cell injury in rabbit aorta, aggravation of injury by 2 days of cholesterol feeding. *Acta Path Microbiol Scand* 87:123–130, 1979.

39. Faggiotta A, Ross R. Studies of hypercholesterolemia in the nonhuman primate. II. Fatty streak conversion to fibrous plaque. *Arteriosclerosis* 4:341–351, 1984.

40. Faggiotta A, Ross R, Harker L. Studies of hypercholesteremia in the nonhuman primate. I. Changes that lead to fatty streak formation. *Arteriosclerosis* 4:323–340, 1984.

41. Ershler WB, Berman E, Moore AL. Slower B16 melanoma growth but geater pulmonary colonization in calorie-restricted mice. *J Nat Cancer Inst* 76:81–85, 1986.

42. Ershler WB, Gamelli RL, Moore AL, Hacker MP, Blow AJ. Experimental tumors and aging: Local factors that may account for the observed age advantage in the B16

43. Solesvik OV, Rofstad EK, Brustad T. Vascular structure of five human malignant melanomas grown in athymic nude mice. *Br J Cancer* 45:557–567, 1982.

44. Grunt TW, Lametschwandtner A, Karrer K. The characteristic structural features of the blood vessels of the Lewis lung carcinoma. (A light microscopic and scanning electron microscopic study.). *Scann Electron Microsc.* II:575–589, 1986.

45. Grunt TW, Lametschwandtner A, karrer K, Staindl O. The angioarchitecture of the Lewis lung carcinoma in laboratory mice. (A light microscopic and scanning electron microscopic study.). *Scann Electron Microsc* II: 557–573, 1986.

46. Jain RK. Determinants of tumor blood flow: A review. *Cancer Res* 48:2641–1658, 1988.

47. Rubin P, Casarett G. Microcirculation of tumors. Part II: The supervascularized state of irradiated regressing tumors. *Clin Radiol* 17:346–355, 1966.

48. Rubin P, Casarett G. Microcirculation of tumors. Part I: Anatomy, function,and necrosis. *Clin Radiol* 17:220–229, 1966.

49. Song CW. Effect of hyperthermia on vascular functions of normal tissues and experimental tumors: Brief communication. *J Natl Cancer Inst* 60:711–713, 1978.

50. Song CW, Rhee JG, Levitt SH. Blood flow in normal tissues and tumors during hyperthermia. *J Nat Cancer Inst* 64:119–124, 1980.

51. Steinberg V, Konerding MA, Korver G, Streffer C. Examination of the necrosis in xenotransplanted tumors — quantitative measurements and correlations with the vascular system, cell-proliferation and tumor growth. In: *Int. Conf. on Tumor Necrosis Factor and Related Cytotoxins* 175:102, 1987.

52. Levesque MJ, Cornhill JF, Nerem RM. Vascular casting. A new method for the study of arterial endothelium. *Atherosclerosis* 34:457–467, 1979.

53. Roach MR, Hinton P, Fletcher J. Artifacts of localization of atherosclerosis in pinned aortas. *Atherosclerosis* 31: 1–10, 1978.

54. Wetzel B, Albrecht RM. The evolution of correlative techniques for electron microscopy — An overview. *Scann Microsc* 3 (Suppl):1–6, 1989.

55. Richardson M, Hatton MWC, Buchanan MR, Moore S. Scanning electron microscopy of normal rabbit aorta: Injury or artifact. *J Ultrastruct Res* 91:159–173, 1985.

56. Tindall A, Svendsen E. Diameter changes in rabbit aorta during fixation at physiological pressure. *Atherosclerosis* 50:223–231, 1984.

57. Hirsch EZ, Chisolm GM III, Gibbons A. Quantitative assessment of changes in aortic dimensions in response to in situ perfusion fixation at the physiological pressures. *Atherosclerosis* 38:63–73, 1981.

58. Caplan BA, Schwartz CJ. Increased endothelial cell turnover in areas of in vivo Evans blue uptake in the pig aorta. *Atherosclerosis* 17:401–417, 1973.

59. Gordon D, Guyton JR, Karnovsky MJ. Intimal alterations in rat aorta induced by stressful stimuli. *Lab Invest*

45:14–19, 1981.

60. Haudenschild CC, Gould KE, Quist WC, LoGerfo FW. Protection of endothelium in vessel segments excised for grafting. *Circulation* 64 (Suppl. II):II101–II110, 1981.

61. Joris I, Majno G. Endothelial changes induced by arterial spasm. *Am J Pathol* 102:346–358, 1981.

62. Buchanan MR, Richardson M, Hass TA, Hirsch J, Madri JA. Basement membrane underlying the vascular endothelium is not thrombogenic: In vivo and in vitro studies with rabbit and human tissue. *Thromb Haemost* 58:698–704, 1987.

63. Miller BG, Evan AP, Bohlen HG. Exposure of vascular smooth muscle cells for analysis with the scanning electron microscope. *Scann Microsc.* 1:1295–1313, 1987.

64. Murakami T. Vascular arrangement of the rat renal glomerulus. A scanning electron microscope study of corrosion casts. *Arch Histol Jpn* 34:87–107, 1972.

65. Hodde KC, Miodonski A, Bakker C, Veltman WAM. Scanning electron microscopy of microcorrosion casts with special attention on arterio-venous differences and application to the rat's cochlea. *Scann Electron Microsc.* II:477–484, 1977.

66. Hodde KC, Miodonski A, Bakker C, Veltman WAM. Scanning electron microscopy of microcorrosion casts with special attention on arterio-venous differences and application to the rat's cochlea. *Scann Electron Microsc* II:477–484, 1979.

67. Kardon RH, Kessel RG. SEM studies on vascular casts of the rat ovary. *Scann Electron Microsc* III:743–750, 1979.

68. Nopanitaya W, Aghajanian JG, Gray LD. An improved plastic mixture for corrosion casting of the gastrointestinal microvascular system. *Scann Electron Microsc.* III:751–756, 1979.

69. Phillips SJ, Rosenberg A, Meir-Levi D, Pappas E. Visualization of the coronary microvascular bed by light and scanning electron microscopy and X-ray in the mammalian heart. *Scann Electron Microsc* III:735–742, 1979.

70. Reidy MA, Levesque MJ. A scanning electron microscopic study of arterial endothelial cells using vascular casts. *Atherosclerosis* 28:463–470, 1977.

71. Hodde KC, Nowell JA. SEM of micro-corrosion casts. *Scann Electron Microsc* II (Suppl.):89–106, 1980.

72. Kratky RG, Zeindler CM, Lo DKC, Roach MR. Quantitative measurements from vascular casts. *Scann Microsc* 3:937–943, 1989.

73. Lametschwandtner A, Lametschwandtner U, Weiger T. Scanning electron microscopy of vascular corrosion casts — technique and applications. *Scann Electron Microsc* II:663–695, 1984.

74. Steeber DA, Erickson CM, Hodde KC, Albrecht RM. Vascular changes in popliteal lymph nodes due to antigen challenge in normal and lethally irradiated mice. *Scann Microsc* 1:831–839, 1987.

75. Poole JCF, Sanders AG, Florey HW. The regeneration of aortic endothelium. *J Path Bact* 125:133–143, 1958.

76. Castenholz A. Interpretation of structural patterns appearing on corrosion casts of small blood and initial lymphatic vessels. *Scann Microsc* 3:315–325, 1989.

77. Konerding MA, Blank M. The vascularization of the vertebral column of rats. *Scann Microsc* 1:1727–1732, 1987.

78. Kratky RG, Roach MR. Scanning electron microscopy of early atherosclerosis in rabbits using aortic casts. *Scann Microsc* 2:465–470, 1988.

79. Kratky RG, Roach MR. Endothelial cell morphometry near branch junctions of rabbit aorta. *Can J Physiol Pharm* 65:1864–1871, 1987.

80. Zeindler CM, Kratky RG, Roach MR. Quantitative measurements of early atherosclerotic lesions on rabbit aortae from vascular casts. *Atherosclerosis* 76:245–255, 1989.

81. Odgaard A, Jensen EB, Gundersen HJG. Estimation of structural anisotropy based on volume orientation: A new concept. *J Microsc* 157:149–162, 1990.

82. Iannaccone PM. Fractal geometry in mosaic organs: A new interpretation of mosaic pattern. *FASEB J* 4:1508–1512, 1990.

Author's address:
Dr. James G. Walmsley, Ph.D.(†)
University of Illinois
College of Medicine at Rockford
Department of Biomedical Sciences
1601 Parkview Avenue
Rockford, IL 61107
USA

Index

The manufacturer's authorised representative in the EU is Springer
Nature Customer Service Centre GmbH, Europaplatz 3, 69115 Heidelberg,
Germany. If you have any concerns regarding our products, please
contact ProductSafety@springernature.com

Printed and bound by CPI Group (UK) Ltd, Croydon, CR0 4YY
23/04/2026
02095632-0002